E54.65
CPGW
LLYFRGELL
LIBRARY
ABERYSTWYTH

AF616182

MODERN PHYSICAL METHODS IN BIOCHEMISTRY, PART A

New Comprehensive Biochemistry

Volume 11A

General Editors

A. NEUBERGER
London

L.L.M. van DEENEN
Utrecht

ELSEVIER
AMSTERDAM·NEW YORK·OXFORD

Modern Physical Methods in Biochemistry
Part A

Editors

A. NEUBERGER and L.L.M. VAN DEENEN
London and Utrecht

1985
ELSEVIER
AMSTERDAM·NEW YORK·OXFORD

ISBN 0-444-80649-0 (volume)
ISBN 0-444-80303-3 (series)

Published by:
Elsevier Science Publishers B.V. (Biomedical Division)
P.O. Box 211
1000 AE Amsterdam
The Netherlands

Sole distributors for the USA and Canada:
Elsevier Science Publishing Company, Inc.
52 Vanderbilt Avenue
New York, NY 10017
USA

Library of Congress Cataloging in Publication Data
Main entry under title:

Modern physical methods in biochemistry.

(New comprehensive biochemistry; v. 11)
Bibliography: p.
Includes index.
1. Spectrum analysis. 2. Biological chemistry — Technique. I. Neuberger, Albert. II. Deenen, Laurens L. M. van. III. Series.
QD415.N48 vol. 11 574.19'2 s [574.19'283] 85-4402
[QP519.9.S6]
ISBN 0-444-80649-0

Printed in The Netherlands

Preface

The great and, one might say without exaggerating, the amazing progress which has been made in the biological sciences, particularly in biochemistry, over the last 20 years has been caused to a large extent by the development of sophisticated physical methods and their application to biological problems. Our knowledge of the structure and especially the conformation of protein and nucleic acids has been helped greatly by the use of mass spectrometry and a variety of optical methods, such as circular dichroism and the extension of optical rotary dispersion to low wavelengths. The use of electron spin resonance has been of special use in our understanding of oxidation and reduction processes, and also has been helpful in other problems affecting the structure of important organic molecules.

The use of nuclear magnetic resonance has been another very important development in biological sciences. It is even being used to an increasing extent in physiological investigations, and its application to clinical medicine is likely to be of considerable benefit. The use of X-ray crystallography goes back to the 1930s, but in recent years the techniques have been refined so that resolution has been increased to a significant extent. Therefore, it seems reasonable to describe the techniques used in a manner which is intelligible to the non-expert, and to describe at least some of the applications of these techniques to important biological problems.

The present book will be followed by a second dealing with a variety of other physical techniques. It would be quite impossible to deal with all physical methods which will be used over the next 5 or 10 years, but we hope to cover most of the major techniques which will be applied in solving important biological problems.

A. Neuberger
L.L.M. Van Deenen

Contents

Neuberger/Van Deenen (eds.) Modern Physical Methods in Biochemistry, Part A

CHAPTER 1

Nuclear magnetic resonance spectroscopy in biochemistry

JUSTIN K.M. ROBERTS and OLEG JARDETZKY

Stanford Magnetic Resonance Laboratory, Stanford University, Stanford, CA 94305, U.S.A.

1. Introduction

The absorption and re-emission of radiofrequency radiation by atomic nuclei of substances placed in a strong magnetic field is referred to as nuclear magnetic resonance (NMR). This phenomenon was first detected in bulk matter independently by the groups of Bloch and Purcell in 1946. The discovery by Knight in 1949 that the resonance frequency of a given nucleus is dependent on the chemical group in which it is located – a phenomenon known as chemical shift – led the way for NMR spectroscopy to become a powerful technique for molecular structure elucidation. Other parameters sensitive to chemical environment and molecular motions measured from NMR spectral lines (such as line splitting due to coupling of magnetic nuclei, the line width, and the related relaxation parameters, T_1, T_2, and the Nuclear Overhauser Enhancement) have also become useful probes of molecular structure and dynamics. Furthermore, kinetics of chemical reactions and exchange can be studied by a variety of NMR techniques. Because of these attributes, this form of spectroscopy occupies an important place among methods to study molecules.

The field of biological application of NMR consists of such a large body of work that it is not feasible to summarize the working knowledge of the subject in a single introductory chapter. This chapter, intended for the beginner, accordingly aims to provide no more than an orienting overview of the main directions in which the field has developed, the kinds of biochemical or biological questions which can be studied by NMR, and the major specific NMR techniques useful for this purpose. This discussion is preceded by a brief exposition of the elementary concepts of NMR and supplemented by references to the literature that treats each topic in greater depth.

Applications of NMR of interest in biochemistry can be grouped into three major categories: (1) determination of the structure of biologically active compounds – especially new natural products; (2) studies of biochemical reactions, or processes, especially in vivo; and (3) studies of macromolecular structure and dynamics. In the

first two categories of applications, NMR is used largely as an analytical tool to identify compounds, assay their concentrations and measure reaction rates. An elementary understanding of the relationship between line intensity and concentration and empirical information on chemical shifts characteristic of different molecular species suffices for most studies of this type. In the third category, NMR is used as a structural tool, and a more elaborate theoretical analysis of the experimentally measured NMR parameters is required to obtain the desired information on the details of molecular events.

2. Theory

(a) Nuclear spin

Observation of nuclear magnetic resonance relies on two properties of nuclei: charge and spin. The movement of charge in a spinning nucleus produces a magnetic field whose vector is parallel to the spin axis. In other words, the nucleus possesses a magnetic moment, μ. The fundamental property of spin is described by the nuclear spin quantum number, I (in units of $h/2$, where h is Planck's constant), its value being determined by the atomic mass number and the atomic number according to Table 1.

Thus, nuclear magnetic resonance cannot be observed in such important nuclei as ^{12}C, ^{16}O and ^{32}S. The vast majority of NMR studies in biochemistry have utilized nuclei of spin number 1/2: ^{1}H, ^{13}C, ^{15}N, ^{19}F and ^{31}P. Hence, we will consider such nuclei almost exclusively. Nuclei with $I \geqslant 1$ possess an electric quadrupole moment (from non-spherical nuclear charge distribution) leading, in general, to broad lines compared to nuclei with $I = 1/2$, due to rapid relaxation. Where the quadrupole moment is small, for example with ^{2}H and ^{11}B, broadening is not excessive, and, for certain purposes, the nuclei can be treated as if $I = 1/2$.

(b) Nuclear precession

In a stationary external magnetic field, H_0, a nucleus of spin I has $2I+1$ quantitized energy levels. This means that there is only one possible energy transition for a nucleus $I = 1/2$, a vastly simpler situation compared to energy transition of electrons in

TABLE 1
The relationship between atomic number, atomic mass and nuclear spin number

Mass number	Atomic number	Spin number, I
Odd	odd or even	half integral: 1/2, 3/2, 5/2
Even	even	0
Even	odd	integral: 1, 2, 3

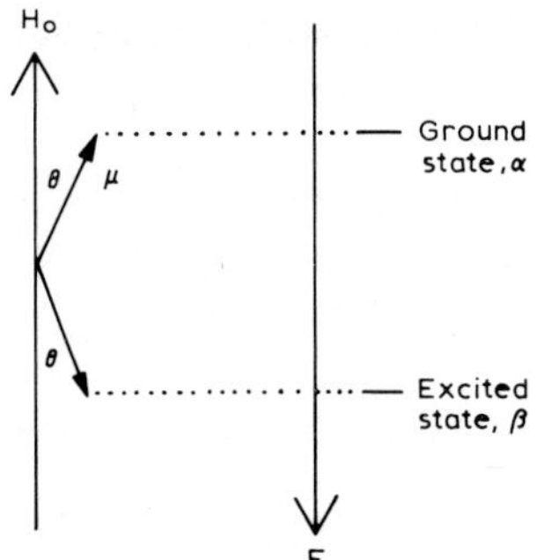

Figure 1. Quantization of the magnetic moment, μ, and the energy of interaction, E, in a magnetic field, H, for a nucleus of spin $I = 1/2$.

molecules. In the classical mechanical description of NMR, these two energy levels are considered as the alignment of μ with or against H_0 (Fig. 1).

The nucleus in Figure 1 will experience a torque, T, due to interaction of μ and H_0, expressed in vector notation as:

$$\vec{T} = \vec{\mu} \times \vec{H}_0 \tag{1}$$

Since the nucleus is spinning, the nucleus also possesses angular momentum, L, whose vector is co-linear with and linearly proportional to μ (the spinning motion being common to both nuclear charge and mass), i.e.

$$\vec{\mu} = \gamma \vec{L} \tag{2}$$

where γ is an empirically derived constant for each nucleus, the magnetogyric ratio. Newton's law of conservation of angular momentum requires that:

$$\frac{d\vec{L}}{dt} = \vec{T} \tag{3}$$

where t = time. So, from equations 1 and 2:

$$\frac{d\vec{L}}{dt} = \gamma \vec{L} \times \vec{H} \tag{4}$$

or

$$\frac{d\vec{\mu}}{dt} = \gamma \vec{\mu} \times \vec{H}_0 \tag{4a}$$

These equations indicate that at any instant, changes in μ are perpendicular to both $\vec{\mu}$

and $\vec{H}_0$, i.e., they describe the precession* of $\vec{L}$ and $\vec{\mu}$ about $\vec{H}_0$ with an angular velocity, ω_0, defined by:

$$\frac{\mathrm{d}\vec{L}}{\mathrm{d}t} = \vec{L}\omega_0 \tag{5}$$

or

$$\frac{\mathrm{d}\vec{\mu}}{\mathrm{d}t} = \vec{\mu}\omega_0 \tag{5a}$$

hence,

$$\omega_0 = \gamma H_0 \quad \text{(units of rad}\cdot\text{sec}^{-1}\text{)} \tag{6}$$

the Larmor equation. Larmor precession of a nucleus at a frequency ω_0, where:

$$\omega_0 = \frac{\gamma \vec{H}_0}{2\pi} \tag{7}$$

is shown in Figure 2.

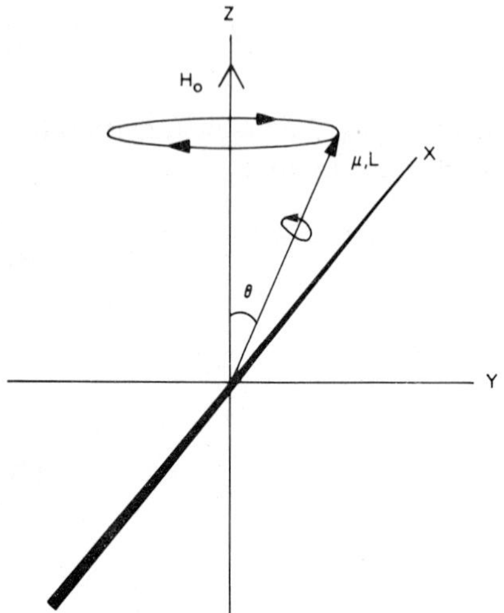

Figure 2. Nuclear precession about the magnetic field axis. The nucleus is in the ground state.

*Precession is defined as the rotation of an axis of rotation about another axis.

(c) Nuclear magnetic resonance

(i) In an isolated atomic nucleus

To each of the discrete orientations assumed by the nuclear magnetic moment vector in the external magnetic field corresponds an energy of interaction E (Fig. 1):

$$E = -\vec{\mu} \cdot \vec{H} = -\vec{\mu} \vec{H}_0 \cos\theta = -\mu_z \vec{H}_0 \tag{8}$$

where μ_z is the projection of the true nuclear magnetic moment on the z axis, the direction of the applied magnetic field, H_0. (In fact, μ is not measurable since the magnetic properties of particles can only be detected by their interaction with a magnetic field, hence magnetic moments given in tables are the maximum observable values, μ_z.) The energy ΔE associated with a transition between energy levels E_α and E_β (Fig. 1) is defined by:

$$\Delta E = E_\beta - E_\alpha = (\mu_z^\alpha - \mu_z^\beta) H_z \tag{9}$$

$(H_z = H_0)$.

If the transition is to result from the absorption of electromagnetic radiation, the frequency, ν, of this radiation must be such that the transition energy for one nucleus can be expressed as the energy of one absorbed quantum, i.e.

$$\Delta E = h\nu \tag{10}$$

Hence, equation 9 may be rearranged as:

$$\nu = \frac{\Delta E}{h} = \frac{\mu_z^\alpha - \mu_z^\beta}{h} H_z \tag{11}$$

We now want to show that the frequency of radiation necessary for a transition between nuclear energy levels is equal to the Larmor frequency, ω_0 (defined in equation 7).

The reorientation of a nuclear dipole with respect to the external field $\vec{H}_z$ is accomplished by the magnetic field component H_z of electromagnetic radiation applied to the sample, oriented in the x–y plane (Fig. 2). This field will exert a torque on the dipole according to equation 1 (H_1 substituting for H_0). In an NMR experiment, H_1 is much smaller than H_0 (by a factor of $>10^3$), so if H_1 is stationary, there will be no net torque forcing $\vec{\mu}$ into the x–y plane, because the direction of torque is reversed every 180°, as μ precesses about the external field H (in a non-quantized system, such as a gyroscope, a force equivalent to H_1 would lead to nutation: precession, together with an up and down oscillation). H_1 can only continually force toward the x–y plane if H_1 rotates about H_0 (Fig. 2) with the same angular frequency and the same sense as the precessing dipole, ω_0. This criterion is met by circularly polarized radiofrequency radiation of frequency $\omega_0/2\pi$ (although linearly polarized radiation can interact with the nuclear dipole, as it can be considered to be a superimposition of two circular polarized fields, of equal amplitude, wavelength and phase but opposite handedness – only one of these components interacting with the dipole). Thus, we may conclude that transition of a nucleus from the ground to the excited state (Fig. 1) occurs when the frequency of radiation, ν, equals the Larmor frequency ω_0 for the nucleus in a given applied magnetic field H_z. So, we can extend equation 11 as:

$$\nu = \frac{\Delta E}{h} = \frac{\mu_z^{\alpha} - \mu_z^{\beta}}{h} H_z = \frac{\gamma}{2\pi} H_z \tag{12}$$

Including a representation of precession, one may illustrate the resonance condition for a nucleus of spin 1/2, as in Figure 3.

(ii) In an assembly of identical nuclei
In practice, nuclear magnetic resonance is observed in large populations of identical nuclei ($10^{16} - 10^{18}$ per sample). The distribution of identical nuclei of spin 1/2 between the two possible energy rates shown in Figure 1 is defined, under conditions of thermal equilibrium, by the Boltzmann equation:

$$\frac{N_{\beta}}{N_{\alpha}} = e^{-\Delta E/kT} \cong 1 - \frac{\Delta E}{kT} \tag{13}$$

where N_{α} and N_{β} are the number of nuclei with their magnetic moments aligned parallel (ground state) and anti-parallel (excited state) to the external magnetic field, respectively. It should be noted that since $\Delta E \ll kT$, only a very small excess of nuclei

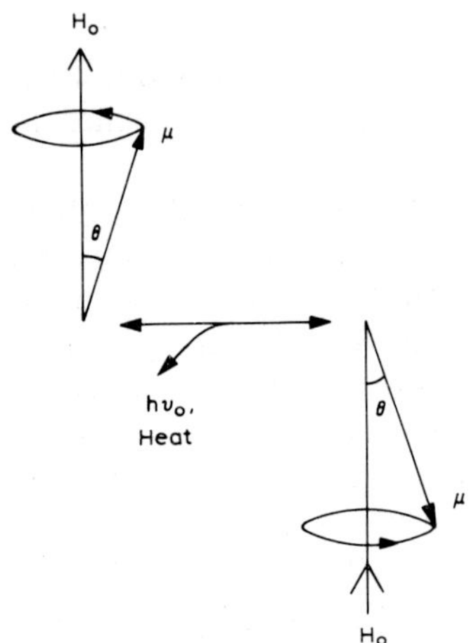

Figure 3. The resonance phenomenon.

will be in the lowest energy state at thermal equilibrium, the excess being of the order of 1 in 7×10^5 for protons in an external field of 100 kG. This excess of nuclei in the ground state gives rise to a net nuclear magnetization vector $\vec{M}$ in the direction of the external magnetic field (z axis). The absorption of radiofrequency radiation and the net excitation of a certain fraction of the population of spins results in a decrease in the z component of $\vec{M}$. According to Einstein's law of transition probabilities under the influence of a radiation field, the probabilities of excitation and emission are equal. Therefore, absorption can occur only to the extent to which there is an excess of nuclei in the lower energy state. Hence, the small excess given by the Boltzmann distribution accounts for the low sensitivity of the NMR method compared to spectroscopic methods using higher frequencies (infrared, visible) where ΔE is much larger; in a population of 10^{16} nuclei, only 10^{10} are actually 'seen' by NMR. The properties of an assembly of identical nuclei just described may be represented as in Figure 4.

The explanation of the effect that absorption of RF radiation has on this system is greatly simplified if one considers the assembly depicted in Figure 4 using a rotating coordinate system. If x and y axes of Figure 4 are rotated about the z axis with an angular velocity Ω, when Ω equals ω_0, the angular velocity of the nuclear magnetic moments in the assembly, precession of nuclear moments about z will apparently

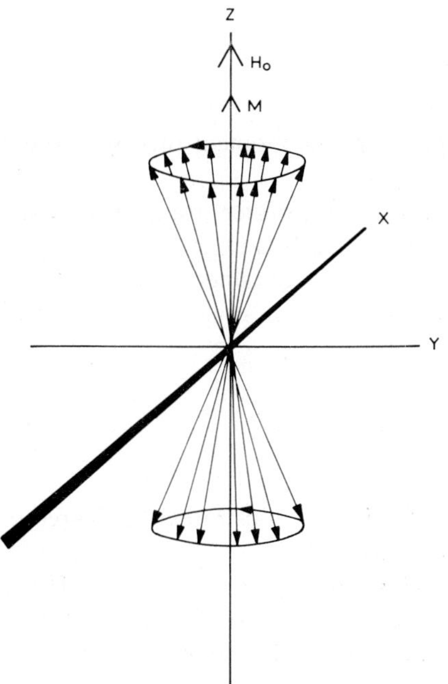

Figure 4. Precession of an ensemble of identical nuclei ($I = 1/2$) at thermal equilibrium. The net macroscopic magnetization, M, is oriented along the z axis (the direction of H), components of magnetization along x and y being zero (the dipoles are randomly oriented in the x, y plane).

cease. The external magnetic field, H_0, has therefore been effectively reduced to zero; or, in other words, the operation of rotating the x, y plane introduces a 'fictitious' magnetic field that cancels H, which, by analogy to equation 6, is equal to Ω/γ. We save space by omitting a rigorous derivation of this conclusion because it is intuitively valid (see Refs. 1 and 2). Thus, the motion of μ in the rotating frame obeys equations 4–6 (for the laboratory system) provided H_z is replaced by the effective magnetic field H_e, where:

$$H_e = H_z - \frac{\Omega}{\gamma} \tag{14}$$

Absorption of radio waves by this assembly, as discussed in the previous section and illustrated in Figure 2, occurs when the magnetic field component of the radiation, H_1, rotates in the x, y plane at the Larmor frequency $\omega_0/2\pi$. In the rotating frame just described ($\Omega = \omega_0$) H_1 will appear to be stationary; it is convenient here to arbitrarily assign H_1 along the rotating x axis, designated x'. Because, in this rotating frame, H_0 is effectively reduced to zero, individual magnetic moments μ, and the net macroscopic magnetization M, can only interact with H_1 (i.e., $H_e = H_1$). Substituting M for μ, and H_1 for H_0, equation 4a becomes:

$$\frac{d\vec{M}}{dt} = \gamma \vec{M} \times \vec{H}_1 \tag{15}$$

indicating that at resonance, the net macroscopic magnetic moment precesses about H_1.

The vast majority of NMR experiments (viz., all Fourier transform NMR techniques) are performed using short pulses of radiation. It is clear that by varying the duration of the pulse, t_p, and the field intensity H_1 contained in the pulse of radiation, one can rotate M in the zy' plane by any desired angle ψ to the z axis according to:

$$\psi = \gamma H_1 t_p \tag{16}$$

Typical values of t_p range from 1 to 50 μseconds. Figure 5 illustrates the degree of precession for two pulses of different length (H_1 constant).

Many NMR experiments are described using this model. For example, the Hahn spin-echo experiment involves measurement of the signal (or 'echo') following a 90°, τ, 180°, τ sequence, τ being the interval between two pulses. The behavior of the spin system in the spin echo experiment is shown in Figure 6.

One might now ask: how can precession of individual nuclear moments in the upper and lower quantum energy levels shown in Figure 4 permit continuous precession of the net macroscopic magnetization in the zy' plane? It is possible to obtain such

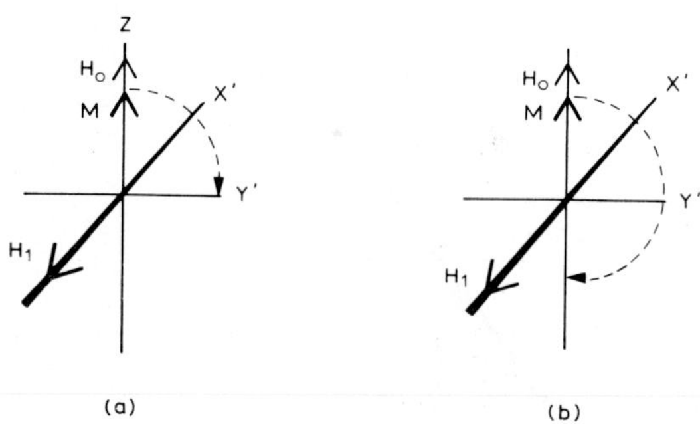

Figure 5. Precession of M about H_1 in the rotating frame following: a, 90° pulse; b, 180° pulse.

continuous precession by a combination of the excess of nuclei in the ground or excited state (Fig. 3), and the introduction of phase coherence in the precession of nuclear moments about the external magnetic field. This is illustrated in Figure 7 for different pulse angles.

Thus, the quantum mechanical and classical mechanical treatments of nuclear magnetic resonance closely correspond, as has been demonstrated mathematically [1].

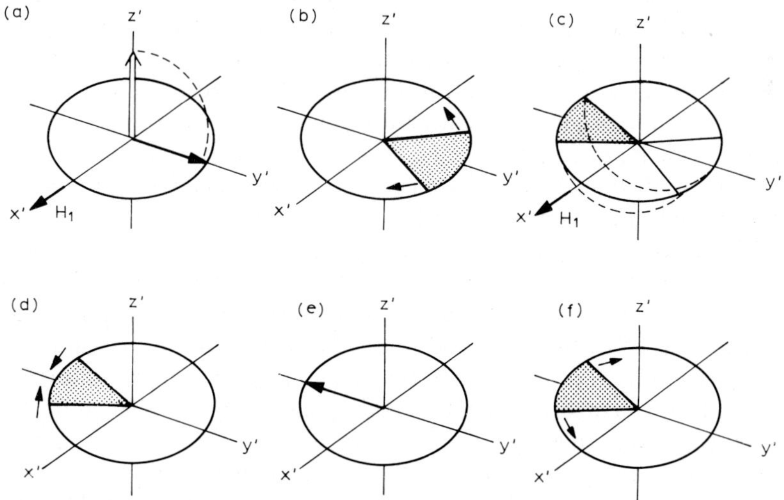

Figure 6. The Hahn spin echo experiment in the rotating frame. (a) Tipping of M into the $x'y'$ plane by 90° pulse. (b) Decrease in $M_{y'}$ as spins dephase. (c) Application of a second (180°) pulse. (d) Increase in $M_{y'}$ as spins 'refocus'. (e) Complete refocusing. (f) Decay in $M_{y'}$ as spins dephase. From [2].

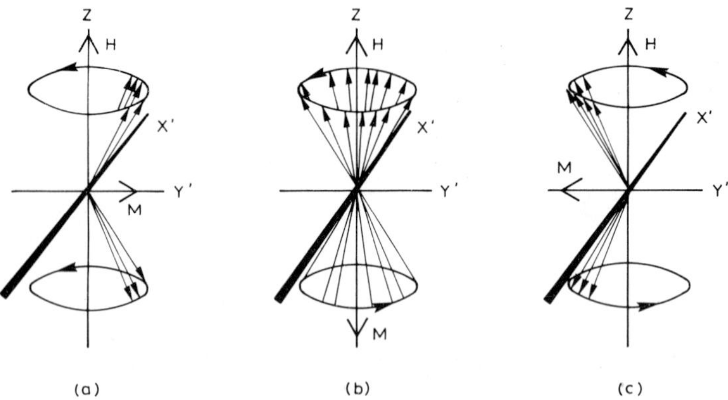

Figure 7. Positioning of individual nuclear magnetic moments to give apparent continuous precession of the net magnetic moment about x'.

(d) The free-induction decay and relaxation

In Fourier transform (FT)-NMR experiments, the signal from excited nuclei is observed following the pulse via voltage changes, induced by the net macroscopic magnetization in the $x'y'$ plane ('nuclear induction'), in a coil around the sample tuned to the resonance frequency. This signal decreases in intensity to zero with time as the nuclei return, or relax, to their original state of thermal equilibrium. Hence, the signal is termed the free-induction decay (FID). Fourier transform of the FID, or a summation of FIDs, yields a conventional absorption-type spectrum (Fig. 8). The intensity of the signal from a population of identical nuclei ('peak area') is linearly proportional to the population size, i.e., concentration (not chemical activity). In other words, Beer's law is valid over all concentrations above the detection limit of the spectrometer. Moreover, the extinction coefficient of a nuclear species is independent of its chemical environment, in contrast to the absorption of visible and ultraviolet light – hence, relative peak areas in a spectrum can be directly converted to relative concentrations (provided saturation is avoided, see Section 3(c)).

It is useful to identify two components of nuclear relaxation. One is termed spin-spin, or transverse, relaxation, by which energy is transferred from one nucleus to another (mutual spin flips or spin-spin exchange). This process leads to a decrease in the phase coherence induced by the pulse, and so to a decrease in the $x'y'$ component of the sample magnetization (i.e., the signal). Spin-spin exchange cannot affect the magnitude of the z component of the sample magnetization, for no change in the distribution of spins between the upper and lower energy levels occurs via this mechanism (i.e., no loss of energy from the sample). In homogeneous liquids, but not solids or in complex systems where there are strong interactions between different types of nuclei, this relaxation process can be described by a simple exponential decay, characterized by a time constant, T_2. Since, in the NMR experiment, the signal measured is the net magnetization in the $x'y'$ plane, $M_{y'}$, T_2 characterizes the decay of

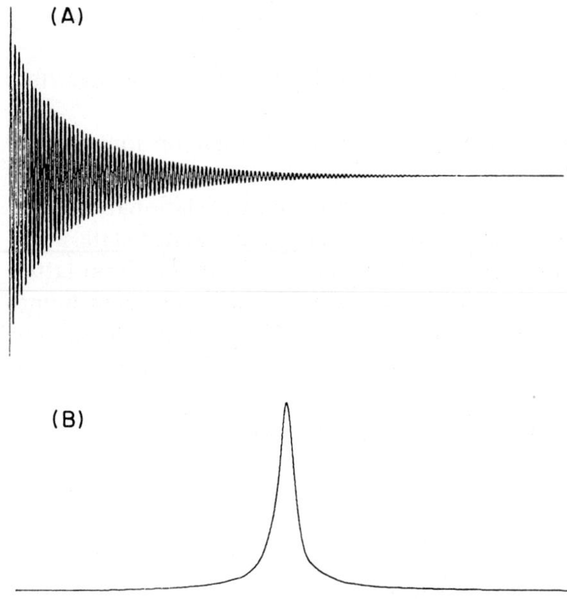

Figure 8. (A) Free induction decay. (B) Its Fourier transform, a Lorentzian line (from [61]).

the FID from a population of identical nuclei in a pulse experiment. Loss of phase coherence in the $x'y'$ plane also arises because of inhomogeneity of the stationary applied magnetic field. Such inhomogeneity results in nuclei in different portions of the sample precessing at different frequencies, since they experience different field strengths, so that the phase of one nucleus relative to others necessarily changes. Hence, if inhomogeneity effects are significant, the time constant for the decay of the FID from an assembly of identical nuclei is T_{2*}, where $T_{2*} < T_2$. It can readily be seen that as T_{2*} increases, the line-width of a resonance at half-height, $\nu_{\frac{1}{2}}$, gets narrower, in fact:

$$\nu_{\frac{1}{2}} = \frac{1}{\pi T_{2*}} \tag{17}$$

This direct effect of T_{2*} on line-widths is also evident on considering the Heisenberg uncertainty principle; when applied to the simultaneous measurement of energy and time we may write:

$$\Delta E \cdot \Delta t \geqslant h/\sqrt{2} \cdot \pi \quad \text{or} \quad \Delta \nu \cdot \Delta t \geqslant 1/\sqrt{2} \cdot \pi \tag{17a}$$

where Δ indicates the uncertainty in the measurement of parameters E, ν and t. Concerning spectroscopic lines, this relation states that the uncertainty in measurement of the frequency corresponding to a transition between two energy levels is greater than or equal to the uncertainty in the frequency of transitions between the two energy levels, characterised by $1/T_{2*}$. Hence, we can define $\nu_{1/2}$, according to equation 17.

Line-widths can also be influenced by chemical exchange processes (see Section 2(j)).

The second relaxation process is termed spin-lattice, thermal or longitudinal relaxation, in which energy contained in the nuclear spin system is lost to surrounding molecules (or 'lattice') in the form of heat (i.e., rotational and translational motion). Such energy loss leads to a decrease in the number of nuclei in the excited state, and a corresponding increase in the z component of the net magnetization, M_z. Spin-lattice relaxation, like spin-spin relaxation, is also an exponential phenomenon in homogeneous liquids, characterized by a time constant, T_1. Unlike T_2, T_1 is not influenced by magnetic field inhomogeneity. One can note that $T_1 \geqslant T_2$, for M_z cannot be at its equilibrium value before $M_{y'}$ equals zero.

Figure 9 illustrates these relaxation processes in the rotating frame.

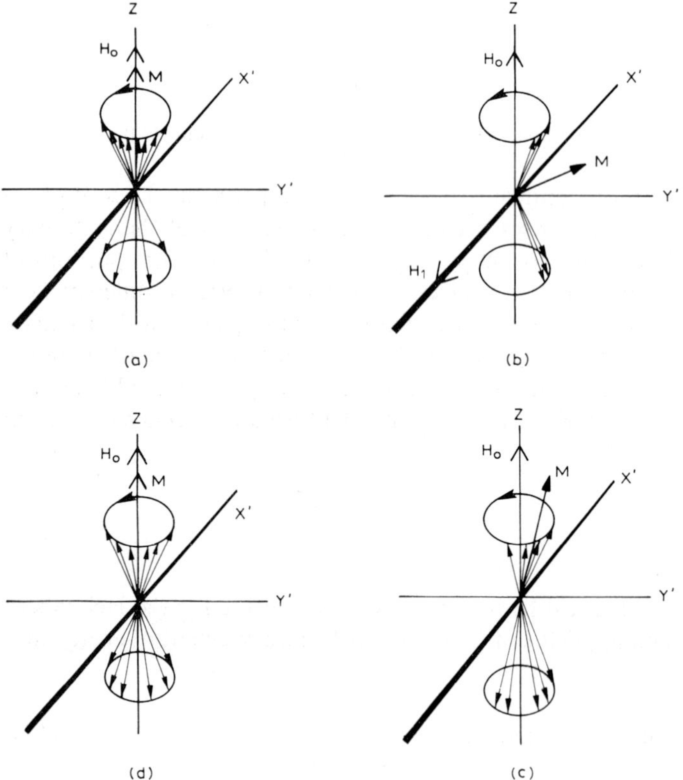

Figure 9. Excitation and relaxation in a population of spins. (a) Before pulse. (b) Induction of phase coherence along y' by H_1, and consequent tipping of macroscopic magnetization, M. (c) Dephasing of nuclear magnetic moments by spin-spin relaxation, i.e., $M_{y'} = 0$. (d) Re-establishment of the Boltzmann distribution (M_z is at its equilibrium value)(a = d).

(e) The chemical shift

Equation 7 shows that the Larmor frequency, or resonance frequency, of a nucleus depends on the magnitude of the empirical constant, γ. As Table 2 shows, γ differs greatly from isotope to isotope and so the resonance frequency of each isotope, at a given external magnetic field strength, is very different. This means that only one isotope is studied directly in an NMR experiment; there is no interference problem of one element being confused with another in NMR spectroscopy, as is possible with other analytical methods.

Equation 7 also shows that the resonance frequency of a nucleus depends on the magnetic field strength of the nucleus. In the presence of an external magnetic field the electrons around the nuclei undergo (in addition to their regular motion) a forced motion due to the field. This gives rise to an electronic magnetic moment (electromagnetic induction on an atomic scale) whose direction opposes the external magnetic field, and so the nuclei experience a field strength less than that of the applied field. The strength of this 'shielding' of nuclei from the external field will differ in different chemical groups. Hence, different chemical groups resonate at different frequencies, the so-called chemical shift. In order to compare chemical shifts determined at different magnetic field strengths, the chemical shift, δ, of a resonance is defined, in parts per million, as:

$$\delta = \frac{\nu_s - \nu_{ref}}{\nu_{ref}} \times 10^6 \qquad (18)$$

where ν_s and ν_{ref} are the absolute resonance frequencies of the sample and reference line, respectively. Figure 10 shows the correlation of chemical shift with chemical structure for ^1H, ^{13}C, ^{15}N, ^{17}O and ^{31}P resonances. Variation in δ of a particular group may result from the influence of other chemical groups in the molecule, or interactions with other molecules or ions. A precise and general theoretical explanation for the variation observed has not been formulated; this is attributable to the considerable sensitivity of chemical shifts to environmental factors.

The induction of electronic magnetic moments by an external field in materials that ordinarily have no inherent magnetic moment is termed diamagnetism, and occurs in all substances. Those substances in which only such induced moments may occur are called diamagnetic.

The presence of paramagnetic species (i.e., species containing unpaired electrons, such as certain metal ions or organic free radicals) can result in large changes in the chemical shifts of molecules, relative to their normal values. This is due to the permanent magnetic moment (large in comparison to diamagnetic moments) associated with an unpaired electron changing the magnetic field experienced by a nearby (≈ 20 Å) nucleus. Paramagnetic substances that cause such changes in the chemical shift of resonance lines of nearby nuclei are termed shift probes, examples being the lanthanides, Eu^{3+} and Dy^{3+}. Other paramagnetic species, such as Mn^{2+} and Gd^{3+}, may significantly broaden resonances of a nucleus, because large

TABLE 2
Spin resonance data for some common nuclei[a]

Isotope	Spin I in multiples of $h/2\pi$	Magnetic moment, μ, in multiples of the nuclear magneton ($eh/4\pi mc$)	Magnetogyric ratio ($\gamma/10^7$ rad·T^{-1}·s^{-1})	Electric quadrupole moment, Q, in multiples of $e \times 10^{-24}$ cm^2	NMR frequency in MHz in a field of 100 kG	Natural abundance (% by weight) of the element	Relative sensitivity[b] of nuclei at constant field
^{1}H	1/2	2.79277	26.7510	–	425.7	99.9844	1.000
^{2}H	1	0.85738	4.1064	2.77×10^{-3}	65.36	1.56×10^{-3}	9.64×10^{-3}
^{3}H[c]	1/2	2.9788	28.5335	–	454.1	–	1.21
^{7}Li	3/2	3.257	10.396	-4.2×10^{-2}	165.6	92.57	0.294
^{11}B	3/2	2.6880	8.5827	3.55×10^{-2}	136.60	81.17	0.165
^{13}C	1/2	0.7022	6.7263	–	107.1	1.108	1.59×10^{-2}
^{14}N	1	0.4036	1.9324	2×10^{-2}	30.77	99.635	1.01×10^{-3}
^{15}N	1/2	−0.2831	−2.7107	–	43.16	0.365	1.04×10^{-3}
^{17}O	5/2	−1.893	−3.6266	-4×10^{-3}	57.72	3.7×10^{-3}	2.91×10^{-2}
^{19}F	1/2	2.627	25.1665	–	400.7	100.0	0.834
^{23}Na	3/2	2.217	7.0760	0.1	112.62	100.0	9.27×10^{-2}
^{25}Mg	5/2	−0.8547	−1.6370	–	26.06	10.05	2.68×10^{-2}
^{31}P	1/2	1.131	10.829	–	172.4	100.0	6.64×10^{-2}
^{33}S	3/2	0.6429	2.0517	-6.4×10^{-2}	32.67	0.74	2.26×10^{-3}
^{35}Cl	3/2	0.8209	2.6212	-7.97×10^{-2}	41.73	75.4	4.71×10^{-3}
^{39}K	3/2	0.3910	1.2484	–	19.87	93.08	5.08×10^{-4}

^{43}Ca	7/2	−1.315	−1.7999	–	28.64	0.13	6.39×10^{-2}
^{55}Mn	5/6	3.462	6.598	0.5	105.6	100.0	0.178
^{57}Fe	–	⩽0.05	0.8644	–	–	2.245	–
^{59}Co	7/2	4.639	6.3171	0.5	101.03	100.0	0.281
^{63}Cu	3/2	2.226	7.0904	−0.15	112.85	69.09	9.38×10^{-2}
^{65}Cu	3/2	2.379	7.5958	−0.14	120.9	30.91	0.116
^{75}As	3/2	1.435	4.5816	0.3	72.93	100.0	2.51×10^{-2}
^{79}Br	3/2	2.099	6.7021	0.30	106.67	50.57	7.86×10^{-2}
^{81}Br	3/2	2.263	7.2245	0.25	114.98	49.43	9.84×10^{-2}
^{85}Rb	5/2	1.349	2.5829	–	41.11	72.8	1.05×10^{-2}
^{111}Cd	1/2	−0.5922	–	–	90.28	12.86	9.54×10^{-3}
^{113}Cd	1/2	−0.6195	−5.9330	–	94.44	12.34	1.09×10^{-2}
^{127}I	5/2	2.794	5.3521	−0.59	85.19	100.0	9.35×10^{-2}
^{133}Cs	7/2	2.564	3.5089	⩽0.3	55.85	100.0	4.74×10^{-2}
^{137}Ba	3/2	0.936	2.9729	–	47.6	11.32	6.97×10^{-3}
^{199}Hg	1/2	0.4993	4.7690	–	76.12	16.86	5.72×10^{-3}
^{201}Hg	3/2	−0.607	−1.7655	0.5	30.8	13.24	1.90×10^{-3}
^{203}Tl	1/2	1.5960	–	–	243.3	29.52	0.187
^{205}Tl	1/2	1.6114	15.438	–	245.7	70.48	0.192

[a]From the Varian Associates NMR table.
[b]For equal number of nuclei, relative to 1.00 for ^{1}H.
[c]Radioactive isotope.

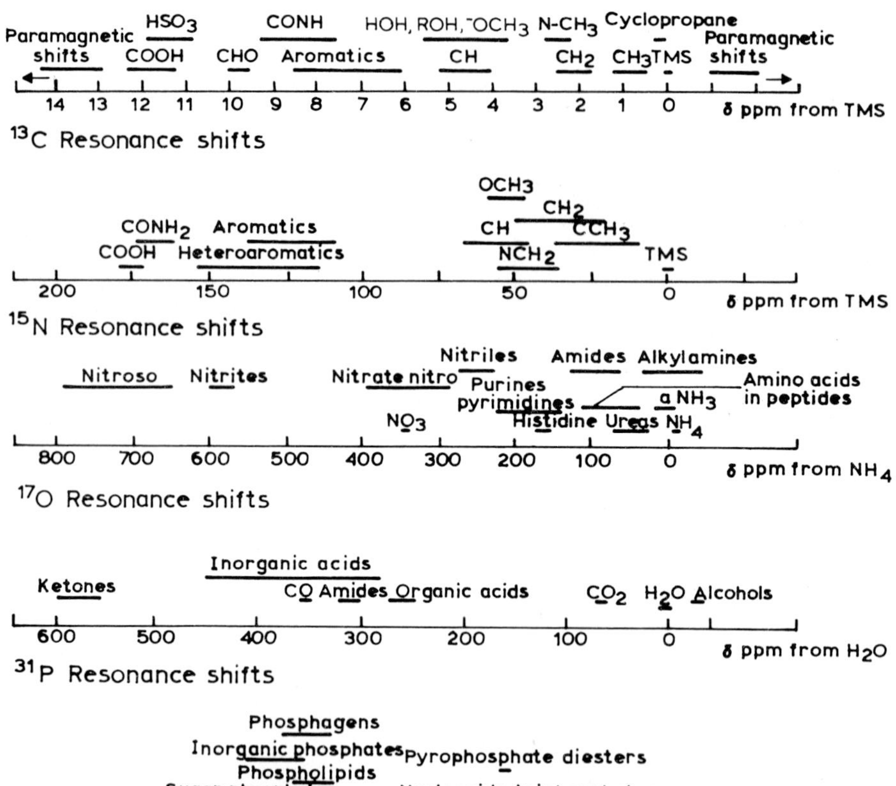

Figure 10. Ranges of chemical shifts for ^{1}H, ^{13}C, ^{15}N, ^{17}O and ^{31}P (from [61]).

oscillating local magnetic fields about the unpaired electron greatly increase the relaxation rates of nearby nuclei, so broadening lines accorded to equation 17; these species are termed relaxation probes. Shift probes are characterized by very short electronic relaxation times ($<10^{-10}$ seconds) relative to relaxation probes. Many paramagnetic species, e.g., Ni^{2+}, Fe^{3+} and Cu^{2+}, cause both line-broadening and a shift in resonance frequency of nearby nuclei. The use of paramagnetic species is considered further in Section 4(d).

Measurement of line positions in NMR spectra according to equation 18 requires use of a reference line. Experimentally, this means that a reference compound giving a sharp line(s) separate from sample resonances must be included in the NMR tube. For example, sodium 4,4-dimethyl-4-silapentane (DSS) is useful for many ^{1}H-NMR studies, while methylene diphosphonate (MDP) is a useful reference for in vivo ^{31}P-NMR work. The reference may be internal, i.e., dissolved in the same solution as the sample, or external, e.g., contained in a coaxial capillary separated from the sample. These two types of reference do not necessarily give identical chemical shift measurements, because the magnetic field experienced by a molecule depends on the bulk magnetic susceptibility, κ, of the medium around the molecule (which is determined by the chemical composition of the solution, and also the shape of the sample and its orientation with respect to the magnetic field), which may differ from an external reference solution to the sample. Hence, use of internal references is in general preferable as regards accurate chemical shift measurements directly comparable from spectrometer to spectrometer. However, internal references have disadvantages; references usually used in chemistry, such as trimethylsilane (TMS), are insoluble in water; and it must be determined whether or not the experimental conditions and variables (e.g., pH titration) affect the reference compound before chemical shift measurements can be meaningful. Furthermore, use of an internal reference obviously is not possible in spectroscopy of living systems, except where a strong naturally occurring line of chemical shift shown to be insensitive to physiological condition is present; for example, the ^{31}P line of phosphocreatine in aerobic muscle and brain has often been used as an internal reference [9]. Because of these complications, it is important to recognize that it is not always possible to compare closely chemical shifts measured in different laboratories and on different systems.

(f) Spin-spin coupling

Often NMR spectra contain multiplets of two or more clustered lines, such that the number of spectral lines exceeds the number of chemically different nuclei in the molecule under study. The frequency separation between lines in a multiplet remains constant as the applied magnetic field strength is altered, in contrast to the frequency separation between multiplets or single lines (where δ remains constant). These multiplets, then, are not attributable to the chemical shift. Rather, they result from electron-coupled interactions between the spins of magnetic nuclei connected via covalent bonds, so-called spin-spin coupling.

Figure 11 shows the two possible orientations of nuclear and valence electron spins in a covalent bond. Note that whereas nuclear spins can be parallel or antiparallel to

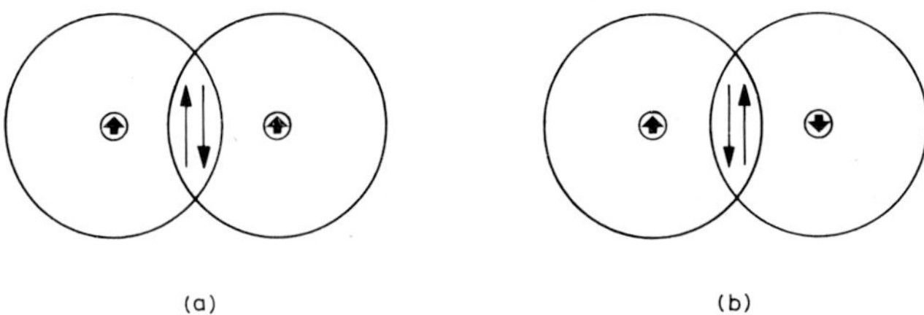

Figure 11. Electron-coupled interaction between the spins of covalently bonded nuclei. Bold-faced arrows indicate nuclear spins, light-faced arrows indicate valence electron spins.

electron spins, the electron spins in an orbital must be antiparallel, according to the Pauli exclusion principle.

It is clear that antiparallel orientation of nuclear spins, represented in Figure 11b, is of lower energy than parallel orientation (Fig. 11a), and so the energy needed to excite, i.e., reorient, one of these nuclei will depend on its orientation relative to the other nucleus. Hence, that nucleus will have two resonance frequencies. The doublet will consist of two lines of equal intensity, the probability of the situations of Figures 11a and 11b being essentially equal because of spin-spin interactions is quite weak. This explanation of line-splitting due to spin-spin coupling accounts for the fact that the separation of lines in a multiplet is independent of the applied field – Figure 11 does not involve an external magnetic field. Spin-spin coupling is characterized by the coupling constant, J, the spacing (in Hz) between the lines in a multiplet. The magnitude of J is directly proportional to the energy of the coupling between the nuclei.

The general rule for spin-spin coupling is: the maximum number of lines into which a given group of nuclei can split the absorption peak of a neighboring group is given by the number of possible orientations of their spins with respect to the external field. For a 'group' of one nucleus, this is $2I+1$. If the group consists of n identical nuclei, there will be $2nI+1$ lines, each separated by J Hz. The intensity of each line in the multiplet is determined by the number of ways the spins can be arranged to give a particular value of total spin, as shown in Figure 12 for the common case of $I=1/2$. It is apparent that the intensities are described by the coefficients of the binomial expansion (given by the Pascal triangle). The analysis obviously applies only where J is much smaller than the chemical shift difference between the coupled nuclei, a condition increasingly satisfied as higher-field spectrometers are introduced.

Spin-coupling can occur via several intervening chemical bonds, although the coupling energy, and thus J, decreases with increasing number of bonds, as can be seen in Table 3. Analysis of multiplets has been used to great advantage in structure determination in organic chemistry, and detailed treatments of the subject can be found elsewhere [3].

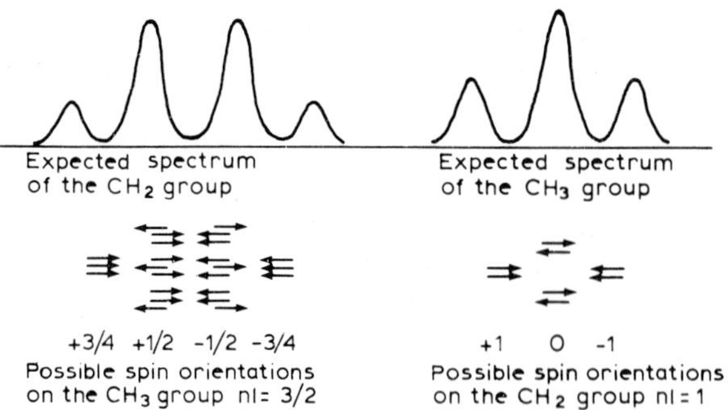

Figure 12. The origins of the multiplet structure in the ^1H spectrum of an ethyl group, and rules of binomial distribution (from [61]).

TABLE 3
Typical ranges of spin-spin coupling constants

	J (Hz)
^1H–^1H	
^1H–^{12}C–^1H	−12 to −15
^1H–^{12}C–^{17}C–^1H	2–14 (7 if free rotation)
^1H–$^{12}C=^{12}C$–^1H	10 (*cis*)
	17 (*trans*)
^1H–^{14}N–^{12}C–^1H	1–10
^1H–^{12}C–^{12}C–^{12}C–H	0–3
^1H–^{13}C	
^1H–^{13}C–(Sp^3)	110–130
^1H–^{12}C–^{13}C	−5 to +5
^1H–$^{12}C=^{12}C$	≈2
^1H–^{15}N	
^{15}N–H_3	61
^1H–^{15}N	85–95 (peptides)
^1H–^{12}C–^{15}N	15–23 (peptides)
^1H–^{31}P	
^1H–^{16}O–^{31}P	15–25
^{31}P–^{31}P	
^{31}P–^{16}O–^{31}P	10–30
^1H–^{19}F	
^1H–^{12}C–^{19}F	40–50
^1H–^{12}C–^{12}C–^{19}F	5–20
^{13}C–F	
^{13}C–F	−280 to −350

It has been possible to correlate molecular structure and stereochemistry, including parameters such as electron distributions and bond angles, to observed coupling constants. However, as is the case with chemical shifts, a firm theoretical framework for calculation of coupling constants is absent, and those semi-theoretical treatments that have been put forward must be applied with care. We will consider other aspects of spin-spin coupling in Section 4(c)(ii).

(g) Spin-decoupling

An important method used to determine which pairs of multiplets result from nuclei coupled to each other is that of spin-decoupling (a double resonance technique). Most commonly, the sample is irradiated at the frequency of one or more of the multiplets as the normal pulse FT-NMR experiment is performed. The decoupling radiation intensity is much greater than that of the pulse, and the continued precession of the nuclei about this decoupling field results in any given nucleus undergoing rapid transitions between its energy levels. Therefore, nuclei coupled to the irradiated nuclei will experience an average energy of interaction, instead of two or more interaction energies, and so these formerly coupled nuclei give a single resonance whose frequency lies at the center of the original multiplet. By examining successive spectra in which a different multiplet is irradiated, it is possible to obtain much useful information on the structure of a molecule.

One may distinguish homonuclear from heteronuclear decoupling. The former type of experiment involves decoupling spin-spin couplings between nuclei of the same isotope (most common with ^{1}H). Here, in addition to collapsing certain multiplets in the spectrum to singlets, the irradiated multiplet is lost, as the irradiation equalizes the population sizes of the two nuclear energy levels ('saturation'). Heteronuclear decoupling most commonly involves irradiation of ^{1}H resonances while the FT-NMR spectrum of another nucleus (e.g., ^{13}C, ^{31}P) is obtained. Although in principle this form of decoupling can be used for structure determination, most commonly heteronuclear decoupling is used to simplify spectra and improve the signal to noise ratio of spectra. In this case, the sample is irradiated over the whole proton frequency range (as the ^{13}C or ^{31}P, etc., spectrum is taken) using a 'noise generator' to modulate the decoupling frequency generator output. This technique is termed proton-noise or broad-band proton decoupling, and is routinely used in ^{13}C spectroscopy. Proton-noise decoupling improves the signal to noise ratio of ^{13}C spectra significantly more (in some cases, almost 3-fold) than can be accounted for by the collapse of a multiplet to a single-line. This is due to the nuclear Overhauser effect, described in Section 2(i). For a more detailed treatment of double resonance methods and theory see Ref. 3.

(h) Relaxation mechanisms

Relaxation of nuclear spins involves transfer of energy from nuclei via fluctuating magnetic or electric fields. The phenomenon of relaxation is closely analogous to resonance, for transfer of energy only occurs when these fields (near an excited nucleus) fluctuate at the Larmor frequency of the nucleus. These fluctuating fields are generated by Brownian motion, the local magnetic fields produced by nuclei and electrons in the sample moving with the molecules. Fluctuating fields with components in the x and y planes can cause longitudinal relaxation (diminution of the component of the net macroscopic magnetization, M_z); fluctuating fields with components in the x, y and z components can cause transverse relaxation (diminution of M_{xy}). The frequency distribution of these fields clearly is dependent on the frequencies of molecular motions in the sample. Therefore, relaxation rates of nuclear spins reflect the motions of the nuclei and their neighbors (nuclei and electrons). Hence, the considerable value (potentially at least) of relaxation studies to the understanding of molecular motion.

Here we enunciate the principal mechanisms of nuclear relaxation, for the mechanisms differ in the effectiveness with which they cause relaxation, such that before useful information can be obtained from a system by studying relaxation phenomena, it is essential to determine the dominant mechanism(s) responsible for relaxation. Use of relaxation data to study motions in macromolecules is considered in Section 4(e)(iii).

Five types of interaction may be distinguished:

(1) Relaxation via dipole-dipole ('dipolar') interactions between spins (nuclear-nuclear or electron-nuclear). Fluctuating magnetic fields are generated by motion of electronic or nuclear magnetic moments in tumbling molecules and, depending on their frequency distribution, can cause relaxation. This mechanism of relaxation dominates ^{1}H relaxation in most molecules, the relaxation of ^{13}C nuclei bonded to ^{1}H, and diamagnetic nuclei near ($\approx$20 Å) paramagnetic species having electronic relaxation times $>10^{-10}$ seconds (discussed above in Section 2(e)). The very large magnetic moments of the unpaired electron, compared to nuclear moments, results in these paramagnetic species (examples being dissolved oxygen and Dy^{3+}) dominating relaxation, when present. Dipole-dipole relaxation can lead to a nuclear Overhauser effect, discussed in Section 2(i).
(2) Relaxation via scalar coupling. Unlike the mechanism described under (1), scalar relaxation involves dipole-dipole interactions mediated via electrons, as occurs in spin-spin coupling (discussed previously in Section 2(f)). The scalar coupling between such nuclei can provide a mechanism for relaxation if either the coupling constant J changes over time due to chemical exchange (e.g., exchange between a chemical form permitting spin-spin coupling and one without coupling will lead to the coupled nuclei experiencing fluctuating fields, of frequency determined by the rate of exchange relative to J) or the coupled nucleus rapidly relaxes (commonly a quadrupolar nucleus, e.g., the broadening of ^{1}H coupled to ^{14}N). Scalar relaxation is relatively uncommon.
(3) Relaxation via anisotropic electronic shielding ('chemical shift anisotropy'). When molecules are placed in a magnetic field the electrons around nuclei undergo a forced motion, giving rise to an electronic magnetic moment, shielding the nuclei from the field and giving rise to the chemical shift. If the shielding is not uniform (i.e., is anisotropic) about the nuclei, they will experience rapidly changing magnetic fields as they tumble, providing a means for relaxation. This relaxation mechanism can be distinguished by its strong magnetic field dependence, T_2 being inversely proportional to the square of magnetic field strength (no other relaxation mechanism depends on the presence of an applied magnetic field). Relaxation via chemical shift anisotropy can be significant in atoms permitting distortion (nonsymmetry) of the electron clouds due to chemical bonds, e.g., ^{19}F and ^{31}P (except in symmetric molecules such as PO_4^{3-}) more than ^{13}C and ^{1}H.
(4) Relaxation via spin-rotation. A magnetic field is generated about the electrons in a molecule as the molecule moves. The magnitude of this field will increase as rotational velocity increases, hence spin rotation can be an important relaxation mechanism for very small molecules, particularly if they are symmetrical with negligible intermolecular interactions, permitting large angular velocities. A corollary of this is the increase in relaxation via spin-rotation with increasing temperature, in contrast to all the other relaxation mechanisms.
(5) Relaxation via quadrupolar coupling. Nuclei with $I \geqslant 1$ have an electric quadrupole moment (due to non-spherical nuclear charge distribution) which is capable of interacting with local electric field gradients that occur in tumbling molecules with an asymmetric electric charge distribution. Therefore this relaxation mechanism is

entirely intramolecular. It dominates relaxation of nuclei with $I \geqslant 1$, unless they are in a very symmetric environment (e.g., $^{14}NH_4^+$).

There is no straightforward and completely rigorous procedure for determining the relative combinations of the various relaxation mechanisms, except where one mechanism clearly dominates (e.g., if the maximum possible nuclear Overhauser effect (NOE) for a resonance is obtained, dipolar relaxation must dominate its relaxation; or an increase in relaxation rate in proportion to the square of the applied field must be due to chemical shift anisotropy). Hence, the study of molecular motion in proteins from relaxation data is performed most readily on ^{13}C nuclei directly bonded to ^{1}H, and so principally relaxed via dipole-dipole interactions (see Section 4(e)(iii)).

(i) Cross-relaxation and the nuclear Overhauser effect

When proton noise decoupling (Section 2(g)) is applied while ^{13}C-NMR resonances of a sample are observed, the magnitude of the ^{13}C lines frequently is found to be much larger than can be attributed to collapse of multiplets. This enhancement of signal caused by decoupling is an example of the NOE, which may be defined in general terms as the change in the NMR absorption intensity of a nuclear spin when a neighboring spin in the molecule is saturated with RF energy. The effect results from the relaxation of the saturated nucleus affecting the relaxation of the observed nucleus via dipole-dipole interactions, described in the previous section (so-called cross-relaxation), which increases the difference in population sizes between the two nuclear energy levels of the observed nucleus (and so increases M_z, hence a larger signal). As mentioned above, the NOE can only occur when relaxation of the observed nucleus occurs via intramolecular dipole-dipole relaxation, seen, for example, in the relaxation

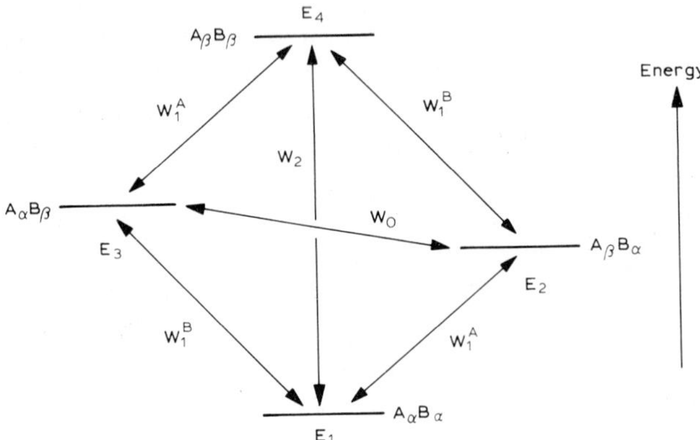

Figure 13. Energy level diagram for a system of two spins, A and B ($I = 1/2$), in which the subscripts α and β refer to their orientation with respect to the external magnetic field (ground state and excited state, respectively, as in Figure 1). W_1^A, W_1^B, etc., are the transition probabilities between energy states. (W_1^A gives rise to resonance A, W_1^B gives rise to resonance B; and W_2 and W_0 are non-photon transitions).

of ^{13}C by ^{1}H (and vice versa) in a ^{13}C–^{1}H bond, and the relaxation of ^{1}H by ^{1}H in adjacent hydrogens in a polypeptide.

The NOE was explained for a two-spin system ($I = 1/2$) by Solomon in 1955 (see Ref. 4). Figure 13 shows that such a system has four possible energy levels (= spin states) connected by six possible transitions.

Defining $N_{\alpha\alpha}$ as the number of nuclei in the energy level E_1, $N_{\beta\alpha}$ in E_2, etc., one can express the rate of change of $N_{\alpha\alpha}$ with time as:

$$\frac{dN_{\alpha\alpha}}{dt} = N_{\beta\beta}W_2 + N_{\alpha\beta}W_1^B + N_{\beta\alpha}W_1^A - N_{\alpha\alpha}(W_1^B + W_1^A + W_2) + N_{\alpha\alpha}^0 \tag{19}$$

where $N_{\alpha\alpha}^0$ is a constant, directly proportional to the Boltzmann expression, $e^{-E_1/kT}$.

Similarly, for $N_{\beta\alpha}$, $N_{\alpha\beta}$ and $N_{\beta\beta}$ we have:

$$\frac{dN_{\beta\alpha}}{dt} = N_{\beta\beta}W_1^B + N_{\alpha\alpha}W_1^A + N_{\alpha\beta}W_0 + N_{\beta\alpha}(W_0 + W_1^A + W_1^B) + N_{\beta\alpha}^0$$

$$\frac{dN_{\alpha\beta}}{dt} = N_{\beta\beta}W_1^A + N_{\alpha\alpha}W_1^B + N_{\beta\alpha}W_0 + N_{\alpha\beta}(W_0 + W_1^A + W_1^B) + N_{\alpha\beta}^0$$

$$\frac{dN_{\beta\beta}}{dt} = N_{\alpha\alpha}W_2 + N_{\alpha\beta}W_1^A + N_{\beta\alpha}W_1^B + N_{\beta\beta}(W_1^A + W_1^B + W_0) + N_{\beta\beta}^0$$

The net macroscopic magnetization along the z axis produced by the nuclei A and B in the sample, designated $\bar{A}_z$ and $\bar{B}_z$ respectively, is linearly proportional to the difference between the energy level population sizes, i.e.

$$\begin{aligned} \bar{A}_z &\propto (N_{\alpha\alpha} + N_{\alpha\beta}) - (N_{\beta\beta} + N_{\beta\alpha}) \\ \bar{B}_z &\propto (N_{\alpha\alpha} + N_{\beta\alpha}) - (N_{\beta\beta} + N_{\alpha\beta}) \end{aligned} \tag{20}$$

Identical equations apply to the equilibrium magnetization values $\bar{A}_z^0$ and $\bar{B}_z^0$, if $N_{\alpha\alpha}$ is replaced by $N_{\alpha\alpha}^0$, etc. Substitution of equations 20 into equations 19 leads finally to the expressions:

$$\begin{aligned} \frac{d\bar{A}_z}{dt} &= -(W_0 + 2W_1^A + W_2)(\bar{A}_z - \bar{A}_z^0) - (W_2 - W_0)(\bar{B}_z - \bar{B}_z^0) \\ \frac{d\bar{B}_z}{dt} &= -(W_0 + 2W_1^B + W_2)(\bar{B}_z - \bar{B}_z^0) - (W_2 + W_0)(\bar{A}_z - \bar{A}_z^0) \end{aligned} \tag{21}$$

Essentially identical equations apply to transverse relaxation $\left(\frac{d\bar{A}_x}{dt}, \frac{d\bar{B}_x}{dt}\right)$. Solutions of equations 21 show that, in general, the relaxation of $\bar{A}_z$, $\bar{B}_z$, $\bar{A}_x$ and $\bar{B}_x$ following an exciting pulse is not a simple exponential, but rather a linear combination of two exponentials. However, there are commonly situations in which the second term in equations 21 is reduced to, or near to, zero – so that relaxation can be described by a single exponential fully characterized by constants T_1 and T_2 (Section 2(d)). For example, if A and B are essentially identical ($A_z \approx B_z$) such that $W_1^A = W_1^B$; or if B relaxes very rapidly compared to A, e.g., if A is relaxed by a paramagnetic species B, $\bar{B}_z \approx B_z^0$ in equation 12; or if one of the nuclei is saturated (e.g., $\bar{B}_z = 0$), as in proton noise decoupling of ^{13}C resonances. Often the parameters T_1 and T_2 are 'measured', even though relaxation does not follow a single exponential [61].

Saturation of one of the spins in this system, in addition to rendering relaxation of the other spin exponential, also changes the equilibrium magnetization of the second spin, giving rise to the NOE. Thus, if $\bar{B}_z = 0$ by saturation of spin B, and A_z is observed at equilibrium ($\mathrm{d}\bar{A}_z/\mathrm{d}t = 0$), equation 21 becomes:

$$0 = -(W_0 + 2W_1^A + W_2)(\bar{A}_z - \bar{A}_z^0) + (W_2 - W_0)\bar{B}_z^0$$

$$\frac{\bar{A}_z}{\bar{A}_z^0} = 1 + \frac{(W_2 - W_0)}{(W_0 + 2W_1^A + W_2)} \frac{\bar{B}_z^0}{\bar{A}_z^0} \tag{22}$$

Equation 22 shows that the equilibrium magnetization can be changed relative to $\bar{A}_z^0$, the change being expressed as the ratio $\bar{A}_z/\bar{A}_z^0$ (equal to the NOE). From equations 12 and 13, equation 22 may be rewritten as:

$$\mathrm{NOE} = \frac{\bar{A}_z}{\bar{A}_z^0} = 1 + \eta = 1 + \frac{(W_2 - W_0)}{(W_0 + 2W_1^A + W_2)} \frac{\gamma_B}{\gamma_A} \tag{23}$$

η being called the nuclear Overhauser enhancement parameter, and γ_A and γ_B are the gyromagnetic ratios of nuclei A and B, respectively. Consequently, larger NOEs will be observed with larger γ_B/γ_A ratios, e.g., ^{15}N NOEs are larger than ^{13}C NOEs when ^1H bonded to these nuclei is saturated. However, just as relaxation mechanisms are sensitive to molecular motion, so is the NOE, such that the value η for a particular dipolar interaction can go to zero, or even change sign, as the relative values of W_2, W_0, etc., change. Hence, the absence of an NOE does not necessarily mean that the dominant relaxation mechanism is not dipolar. This effect of motion on the NOE is discussed further in Section 3(e)(iii).

(j) Chemical exchange

There are many instances in which nuclei in a given chemical environment are in equilibrium or 'chemical exchange' with nuclei in another environment. Examples include exchange between two (or more) conformations, exchange of ligands between 'bound' and 'free' forms and exchange in chemical reactions (e.g., CO_2

$+H_2O \rightleftharpoons HCO_3^- + H^+$). Here we describe how chemical exchange affects NMR spectra.

Figure 14 shows ^1H spectra of *N,N*-dimethylacetamide at various temperatures. It can be seen that at lower temperatures, resonances attributable to the two N–CH_3 groups are well resolved and sharp. However, as the temperature is raised, the two lines first broaden, then begin to move towards each other (while continuing to broaden), coalescing to a single broad resonance which becomes sharper as the

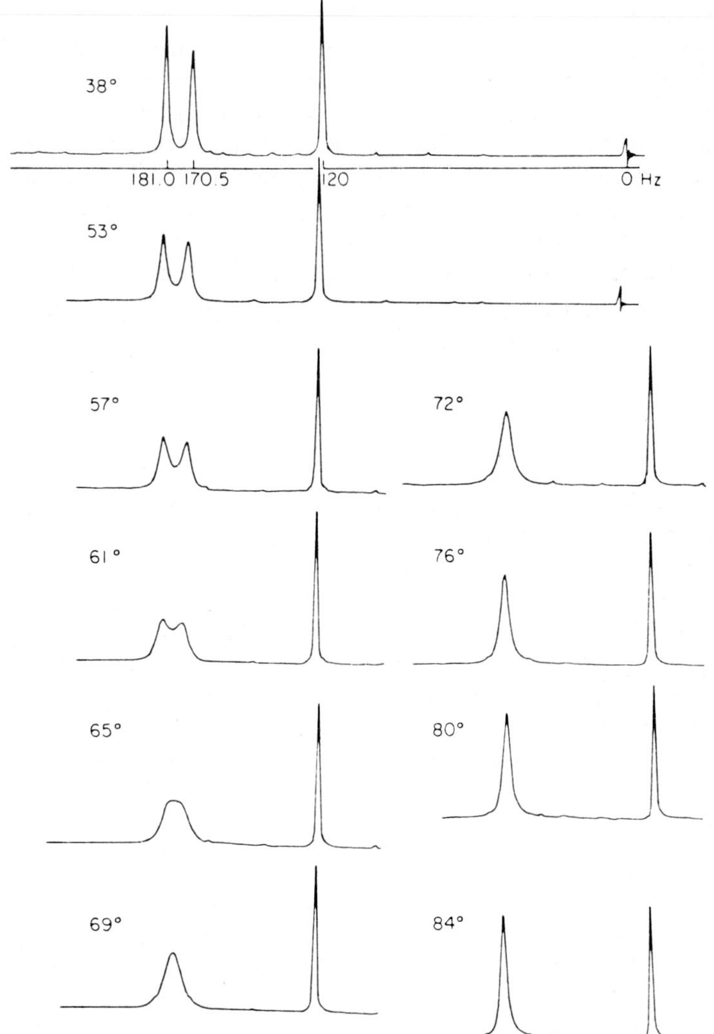

Figure 14. The 60 MHz ^1H spectrum of *N,N*-dimethylacetamide at the temperatures indicated (°C) (from [5], reprinted by permission of John Wiley and Sons Ltd.).

temperature is raised further. These changes in the spectrum result from an increase in the rate of rotation of the N–CH_3 groups about the central bond, described by the equilibrium [5]

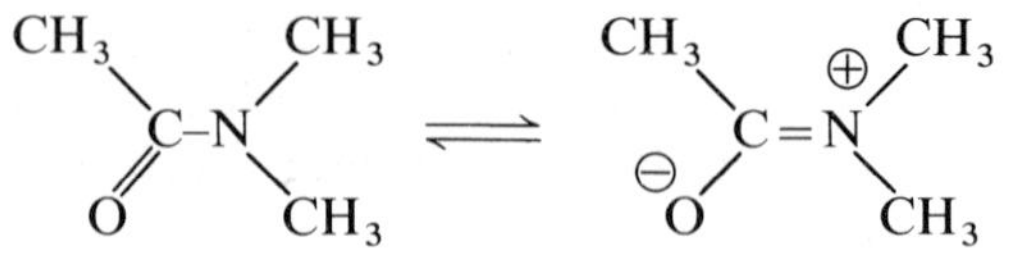

$$\begin{matrix} CH_3 & & CH_3 \\ & C{-}N & \\ O & & CH_3 \end{matrix} \rightleftharpoons \begin{matrix} CH_3 & & \oplus\, CH_3 \\ & C{=}N & \\ \ominus\, O & & CH_3 \end{matrix}$$

with increasing temperature.

The broadening of the two separate resonances during slow exchange is accounted for by realizing that chemical exchange leads to dephasing of spins in the $x'y'$ plane, i.e., decreases the apparent transverse relaxation time, so increasing line-width, as described in Section 2(d) (equation 17). Thus, because nuclei in each of the exchanging sites precess at different frequencies, when nuclei at one site are transferred to another site they will be out of phase (in the $x'y'$ plane) with the population of nuclei already there, so the effective T_{2*} is decreased. The broadening of separate signals A and B by this process clearly is linearly proportional to the rates of exchange, k_1 and k_2, respectively, according to $A \underset{k_2}{\overset{k_1}{\rightleftharpoons}} B$ $(k < |\nu_1 - \nu_2|)$, and so the line-width at half-height of A is given by

$$\nu_{1/2} = \frac{1}{\pi T_{2*}} + \frac{1}{\tau_A} \tag{24}$$

where τ_A is the average lifetime of nucleic in environment A.

The coalescence of the two broadening resonances at intermediate exchange rates $(k \sim |\nu_A - \nu_B|)$ as the temperature increases further can be considered to be a consequence of the uncertainty principle (Eqn. 17a). As τ (Δt in Eqn. 17a) decreases, the energy corresponding to the separation of the two lines measured in the absence of exchange becomes less discernible. Initially the two lines move together, but eventually under conditions of fast exchange $(k > |\nu_1 - \nu_2|)$ a single resonance forms, which narrows as τ decreases with higher temperature, until chemical exchange no longer contributes to line-width $(k > 50|\nu_1 - \nu_2|)$. From Eqn. 17a, we may deduce that coalescence of the two lines occurs when

$$\tau = \frac{1}{\sqrt{2}\pi|\nu_1 - \nu_2|} \tag{25}$$

The complete description of the observed separation between the lines, $|\nu_1^{obs} - \nu_2^{obs}|$, is given by

$$|\nu_1^{obs} - \nu_2^{obs}| = |\nu_1^0 - \nu_2^0| \left(1 - \frac{1}{2\pi^2\tau^2(\nu_1^0 - \nu_2^0)^2}\right)^{1/2} \tag{26}$$

where ν^0 denotes the frequency of the line in the absence of exchange.

At very fast rates of exchange, there is no distinction between nuclei in the exchanging system; the 'mixing' effectively permits all these nuclei to experience the same magnetic fields defining the environments of the exchange sites, over the time period of the free induction decay (seconds). This averaging of the magnetic fields experienced by nuclei is clearly inversely proportional to τ. In addition, line-widths in rapidly exchanging systems are also proportional to $(\nu_A^0 - \nu_B^0)^2$. Thus, when $1/\tau > |\nu_A^0 - \nu_B^0|$ the contribution of exchange to line-width, $\nu_{1/2}^{ex}$, is given by:

$$\nu_{1/2}^{ex} \propto |\nu_A^0 - \nu_B^0|^2 \tau \qquad (27)$$

In other words, collapse of the two lines to a single sharp line occurs at slower exchange rates if $|\nu_A^0 - \nu_B^0|$ is smaller.

Study of the effect of exchange on line-widths and frequencies constitutes a major branch of NMR spectroscopy, and is discussed in detail in a number of reviews (e.g., Refs. 6, 7). There are circumstances in which analysis of chemical exchange by NMR is relatively straightforward, as in slow exchange described by Eqn. 24 and fast exchange described by Eqn. 27. However, in many situations, notably at intermediate exchange rates, complex lineshape analysis is required for accurate quantitation, often necessitating assumptions or approximations that are difficult to verify.

(k) The spectrometer

Figure 15 shows the basic design of an NMR spectrometer, which consists of:
(a) A magnet to align the magnetic nuclei. Important attributes of the magnet are (1) low field inhomogeneity ($\leqslant 1$ in 10^9) so that the line-widths of resonances are not broadened (Eqn. 17); (2) good stability ($< 1 \times 10^{-9}$/hour) to allow experiments to be run over extended periods without frequencies of resonances changing; (3) large field strength: in general, the larger the better (provided (1) and (2) are satisfied) because of increased separation of spectral lines, measured in Hz (Eqn. 18) ('resolution' in the context of NMR), and increased sensitivity (Eqns. 12 and 13). In the last decade, advances in the application of NMR to biochemical problems have been

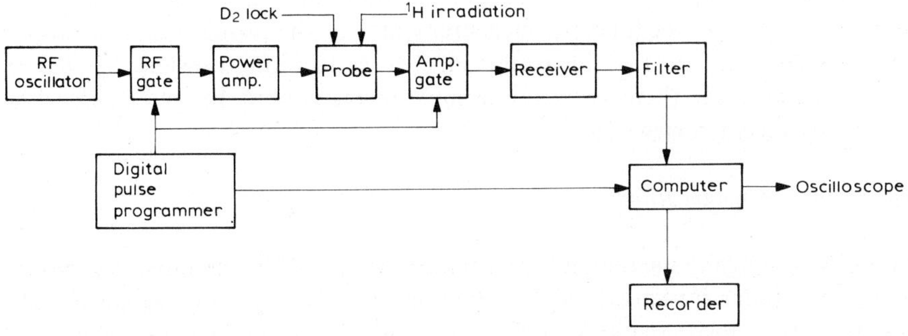

Figure 15. Block diagram of pulsed Fourier transform NMR spectrometer (from [120]).

closely coupled to successes in the construction of more powerful magnets. Most of the cost of the spectrometer is due to the magnet.
(b) A sample holder, the probe, consisting of one or more tuned RF coils that surround the sample, for exciting the nuclei in the sample and detecting their signal after excitation.
(c) An RF transmitter system coupled to a pulse programmer, that delivers, most commonly, short intense pulses of electromagnetic radiation of desired frequency range (cf. Eqn. 16).
(d) A signal receiver system, which amplifies and filters the signal detected by the probe.
(e) A computer system which, in addition to controlling experimental parameters, such as the pulse length, performs the manipulations of accumulated data such as Fourier transformation.
(f) A data display system, consisting of a cathode ray tube and a chart recorder.

3. Biochemistry in vivo

(a) Introduction

High-resolution NMR studies of living systems represent a major branch of NMR spectroscopy. Under suitable circumstances the method permits monitoring of metabolite concentrations, pH, rates of reactions, and metabolic pathways in vivo. The ability to make such measurements and observations without perturbing the system (in most cases) represents a considerable advantage over conventional, invasive, biochemical approaches. Moreover, only freely mobile molecular species, i.e., only metabolites free to participate in metabolism, are observed – in contrast to observations made using conventional analytical procedures which may confuse 'bound' and 'free' metabolites [8]. In vivo NMR suffers from two principal disadvantages. First, the inherent low sensitivity of NMR spectroscopy limits in vivo observations to metabolites present at relatively high concentrations ($>10^{-4}$ M). Hence, many metabolic phenomena cannot be examined by NMR. Second, interpretation of NMR spectral parameters often requires assumptions about the intracellular environment (e.g., for intracellular pH measurements), or the properties of metabolic reactions in vivo (e.g., for rate measurements by saturation transfer). Therefore, potential consequences of these assumptions must be recognised. In vivo NMR is the subject of many recent reviews [9–15].

(b) Experimental considerations

To obtain physiologically meaningful information using NMR one must first ensure that the sample under investigation is maintained in a defined physiological state. Satisfying this prerequisite requires careful attention. For example, dense packing of tissue or cell suspensions, desirable for maximising signal to noise ratio, increases the possibilities for hypoxia. And because commercial NMR spectrometer probes rarely permit perfusion of the sample, most in vivo NMR studies require the use of specially built probes, which may also permit specific treatment or monitoring of the sample during the experiment [16].

Two types of in vivo NMR experiment may be distinguished. ^{1}H and ^{31}P in vivo NMR simply involves observation of naturally occurring metabolites; the sample is treated as in any other physiological experiment. This contrasts with most ^{13}C and ^{15}N studies, where isotopically enriched substrates are fed to the sample, only metabolites derived from the particular substrate being observed.

(c) Observation and quantitation of metabolites

(i) Assignment of resonances

Once spectra have been obtained, it is necessary to obtain valid assignments for each resonance. As a first step, comparison of chemical shifts of spectral lines with shifts of known metabolites obtained in vitro under physiological conditions (see Fig. 10) permits tentative assignments. These tentative assignments can be checked by conventional analysis of tissue extracts, to see if the amount observed in the extract corresponds to the amount indicated by the in vivo spectrum. Another approach is to obtain spectra of extracts as a function of pH, assigning on the basis of pK_a [17]. These approaches become less reliable as spectra become more complex (as is often the case for in vivo ^{13}C- (see, e.g., Ref. 18) and ^{1}H-NMR [19] (Fig. 16)), or broad, overlapping lines are encountered (e.g., the monophosphate ester region of ^{31}P-NMR spectra).

(ii) Quantitation of metabolites

Because the intensity of a resonance due to a particular chemical group is linearly proportional to its concentration, intracellular (mobile) metabolite concentrations can be determined from in vivo NMR spectra. Two factors can make the proportionality constant differ from resonance to resonance. First, in vivo spectra are usually obtained using pulse repetition rates such that most resonances are partially saturated (i.e., pulse interval $<5 \cdot T_1$), to maximise spectral signal to noise ratio. Since T_1 differs from resonance to resonance (e.g., ATP interacts much more strongly with Mn^{2+} present in tissue than with P_i, and so will relax more rapidly [20]), the degree of saturation among spectral lines will differ. Second, when heteronuclear decoupling is used, as is usually essential for in vivo ^{13}C-NMR, line intensities will also be partly determined by nuclear Overhauser effects (see Section 2(i)). The magnitude of the NOE usually will differ from resonance to resonance. These two factors can be quantitatively accounted for either by determining T_1 and the magnitude of the NOE for each line, and inserting these parameters (and the pulse interval and pulse angle) into standard equations [21], to obtain the appropriate correction factors (proportionality constants). Or, the correction factors can be determined empirically, by comparing line intensities obtained under normal conditions (rapid pulse repetition rate and/or heteronuclear decoupling) with intensities obtained under non-saturating, unde-

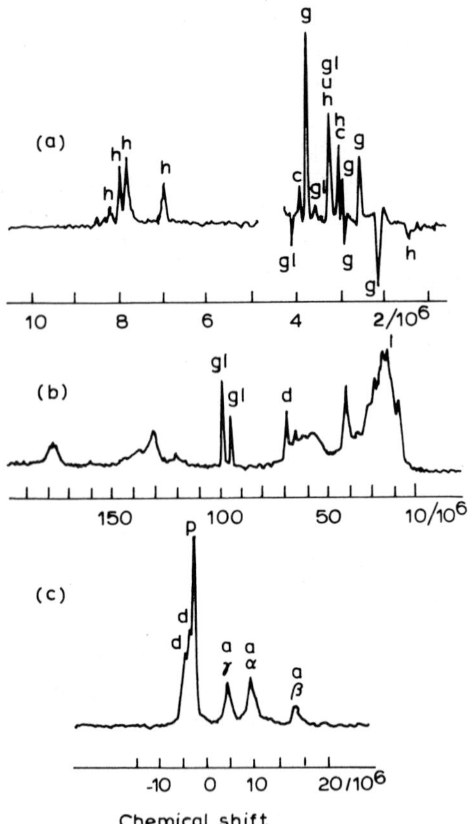

Figure 16. NMR spectra from red cells: (a) ^{1}H spin-echo spectrum shows hemoglobin (h), glucose (gl), glutathione (g), creatine (c), and unassigned (u) peaks. The magnetic field was 6.3 T, the sample volume 0.5 ml, and the accumulation time 4 minutes. The medium was $^{2}H_2O$/Krebs buffer with 10 mM glucose. (b) ^{13}C-NMR spectrum shows peaks from labelled positions: lactate C-3 (l), glucose C-1 (gl) and 2,3-diphosphoglycerate C-3 (d). The field was 4.3 T, the sample volume 3.5 ml, and the accumulation time 10 minutes. The medium was $^{1}H_2O$/Krebs buffer with 10 mM glucose enriched at the C-1 position. (c) ^{31}P-NMR spectrum shows peaks from 2,3-diphosphoglycerate (d), P_i (p) and ATP (a). Conditions were the same as for spectrum b (from [119]).

coupled conditions (see, e.g., Ref. 16). The latter approach is perhaps less prone to error, and requires less spectrometer time. The fact that these corrections are time consuming explains to a large extent why they are neglected in most studies.

The corrections discussed above allow relative concentrations of metabolites to be followed. To measure accurately absolute concentrations of freely mobile metabolites requires determination of both the spectrometer sensitivity (peak area versus concentration, using in standard solutions) and the proportion of the sample volume occupied by tissue (see, e.g., Ref. 16).

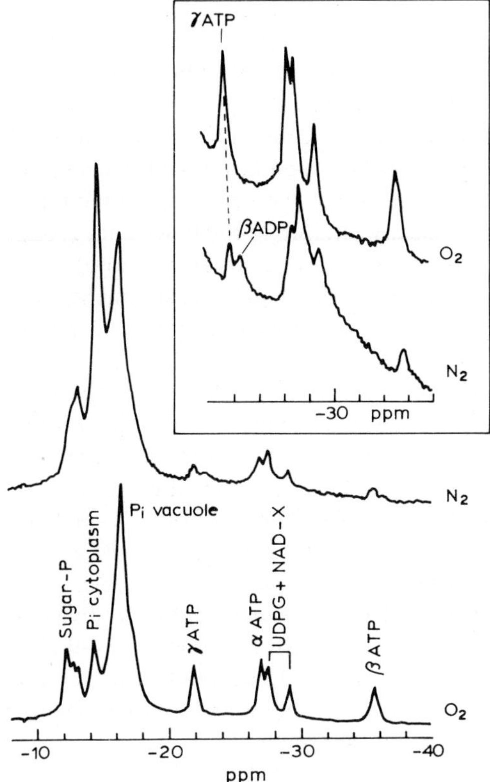

Figure 17. Effect of oxygen tension on concentrations of mobile phosphates in root tips. 145.7 MHz ^{31}P-NMR spectra of maize root tips perfused with oxygen- or nitrogen-saturated 50 mM Glc/0.1 mM $CaSO_4$. The inset is an expansion of the ATP region of the spectrum showing very high ATP/ADP ratios in aerobic tissue. The concentration of cytoplasmic P_i increases greatly in hypoxia. Reproduced, with permission, from [15], © 1984, Annual Reviews Inc.

An example of changes in metabolite concentrations observed by ^{31}P-NMR is given in Figure 17.

One important result obtained from quantitative in vivo NMR studies is the demonstration [8] that most of the ADP and, in some cases, much of the P_i in aerobic cells is bound, and therefore does not contribute to the 'phosphorylation potential' [8] in the cell. This measured phosphorylation potential corresponds to the cytosolic phosphorylation potential measured by enzymatic or chemical analysis of tissue extracts [22].

(d) Intracellular pH measurements

The chemical shift of functional groups is dependent on its ionisation state. Thus, ^{31}P chemical shifts of phosphates [23], ^{13}C chemical shifts of carboxyl groups [24], and

^1H shifts of histidine (imidazol) [25], for example, are pH-dependent. This dependence permits estimation of intracellular pH from chemical shifts of such metabolites, measured in vivo. The estimation is based on titration curve data obtained in vitro. The validity of the pH estimate rests entirely on the use of a titration curve identical to the pH titration curve in vivo. This requirement is complicated by the fact that there are many, long recognised, factors other than pH that also significantly affect chemical shifts of ionisable groups – either indirectly, by affecting the pK_a of the functional group (e.g., ionic strength), or directly, by specifically binding to the functional group (e.g., divalent cation binding), or both. Such interactions occur at physiologically significant concentrations (Fig. 18). Hence, the accuracy of pH measurements obtained by NMR is in part determined by our understanding of the extent to which factors other than pH influence chemical shifts.

Studies of these complicating interactions with respect to intracellular pH measurements by ^{31}P-NMR [26] (cf. Fig. 18) have shown that knowledge of intracellular ionic strength and free divalent cation concentration are required if the pH measurements are to be considered accurate with ± 0.05–0.1 pH unit. Uncertainties of similar magnitude can be expected to pertain to intracellular pH estimates from ^{13}C and ^1H chemical shifts. The potential for error will be greater for pH estimates based on chemical shift values at the extremes of pH titration curves, where the chemical shift becomes less sensitive to pH (but is still very dependent on ionic strength and divalent cation concentration). The fact that in vivo spectral linewidths are broad (tens of Hz), relative to the change in chemical shift per pH unit, also contributes to inaccuracies in pH measurement. It is clear from this discussion that accurate pH measurements using NMR can only be obtained if careful controls are made.

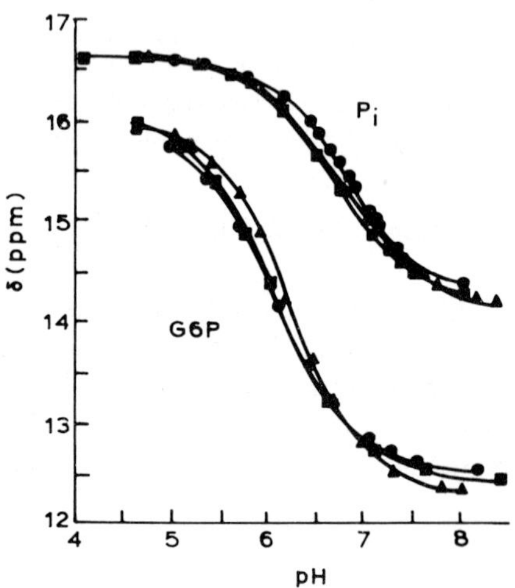

Figure 18. Titration curves of 5 mM potassium phosphate and 5 mM glucose 6-phosphate (G6P) alone (▲) or in the presence of 5 mM MgCl (●) or in the presence of 5 mM MgCl plus 0.1 M KCl (■). Reprinted, with permission, from [26], © 1981, American Chemical Society.

(e) Compartmentation of metabolites

It has long been recognised that metabolites, and metabolic activities, are distributed non-uniformly within cells. Such compartmentation is undoubtedly an important aspect of metabolic regulation. The accurate quantitation of metabolite levels in various intracellular compartments represents a major stumbling block in the study of metabolic regulation. There are two aspects to the study of metabolite compartmentation by in vivo NMR. The first, and most problematic, is the assignment of resonances to specific intracellular compartments; we focus on this aspect here. The second is the use of NMR spectroscopic parameters (e.g., intensity, chemical shift) to monitor conditions (e.g., pH, concentrations, fluxes) within specific compartments, using methods outlined in other sections of this chapter.

Assignment of resonances to specific intracellular compartments has been achieved in two ways. If a specific metabolite is present in two or more compartments, differences in environment between the compartments can lead to differences in certain spectroscopic parameters. For example, inorganic phosphate (P_i) is present in both the cytoplasm and vacuole of plant tissues [27]; because these compartments differ in pH by nearly 2 pH units, and because the chemical shift of $^{31}P_i$ is strongly dependent on pH, two P_i resonances are observed (see Fig. 17). Another example is the occurrence of ATP in both the cytoplasm and granules of blood platelets; only cytoplasmic ATP is observed by ^{31}P-NMR, the ATP present in granules being immobile, and so giving rise to extremely broad (i.e., undetectable) signals [28].

Physiological experiments, in which resonances are manipulated by various treatments, represent a second, indirect, means of assigning resonances to particular compartments. For example, comparison of normoxic with hypoxic maize root tips [29] shows that the pH estimated from the chemical shift of the glucose 6-phosphate ^{31}P resonance correlates closely with the pH estimated from the cytoplasmic P_i resonance, suggesting that these metabolites are in the same compartment.

The intracellular distribution of many enzymatic activities is well characterised. Such knowledge can be used to interpret in vivo spectra. For example, after cells take up 2-deoxyglucose, it is phosphorylated by hexokinase in the cytoplasm, and not metabolised further; therefore, the chemical shift of 2-deoxyglucose 6-phosphate can be used as a cytoplasmic pH indicator [30]. Another example comes from ^{13}C-NMR studies of organic acid [24] and amino acid [31] metabolism. In mitochondria, the C 1 and 2 of malate (and derived metabolites) can be scrambled to the C 3 and 4 due to fumarase activity in the Krebs' cycle. Thus, when certain plants are fed $^{13}CO_2$, [4-^{13}C]malate is formed in the cytoplasm via phospho*enol*pyruvate carboxylase, and then is transferred to the vacuole for storage; the significant amounts of [1-^{13}C]malate observed indicates that cytoplasmic malate exchanges with mitochondrial organic acids before reaching the vacuole [24].

(f) Measurement of unidirectional reaction rates by saturation transfer

Saturation transfer NMR experiments [32–34] enable chemical exchange rates to be measured in living tissues in the steady state. Hence, enzymatic activity can be studied under truly physiological conditions. Combination of such measurements with determinations of true intracellular concentrations of free metabolites represents a potentially powerful and important approach to the understanding of metabolic regulation.

The method can be applied only under specific circumstances. The first condition that must be satisfied is that the chemical species undergoing exchange must give rise to separate spectral lines. In other words, the rates of chemical exchange cannot be so high as to cause excessive broadening, or collapse, of the two resonances (discussed in Section 2(j)). This upper limit on exchange rates can be expressed as $R/(\nu_{A-P} - \nu_{B-P}) < 1$, ν_{A-P} and ν_{B-P} being the resonance frequencies of $A-P$ and $B-P$, respectively, and R being the exchange rate (assuming the population sizes are equal). Other limitations of the method are discussed below.

The saturation transfer experiment consists of obtaining two spectra: one spectrum in which the resonance of one of the exchanging species (e.g., $A-P$) is saturated prior to data acquisition by application of a selective, low-power RF field at its resonance frequency. The second spectrum is a control or reference spectrum, obtained under identical conditions as the first except that the selective RF at some 'control' position in the spectrum (equidistant from the resonance showing saturation transfer). Saturation transfer is observed as a decrease in the intensity of one or more line (other than $A-P$), in the first spectrum, relative to the second spectrum; it is the ratio of these intensities that is measured. An example of such an experiment is shown in Figure 19.

The method has been applied successfully to measurement of exchange reactions that can be analysed as simple equilibria, e.g., for phosphoryl exchange:

$$A-P+B \underset{k_2}{\overset{k_1}{\rightleftarrows}} A+B-P \tag{28}$$

where k_1 and k_2 are the rate constants for each unidirectional reaction, defined as:

$$k_1[\mathrm{A-P}]=k_2[\mathrm{B-P}] \tag{29}$$

If the $A-P$ line is saturated, chemical exchange will result in indirect (partial) saturation of $B-P$ (saturation transfer), this effect increasing as the rate of exchange increases. The magnitude of this effect depends on one other spectroscopic parameter, the spin-lattice relaxation time T_1, for after exchange from environment $A-P$ to $B-P$ the nuclei remain saturated only for a time on the order of T_1. Hence, the magnitude of the saturation transfer effect, for a constant exchange rate, will decrease as T_1 of $B-P$ decreases. These relationships can be described using the Bloch equations, modified to include chemical exchange [35], for the $B-P$ resonance we have:

$$\frac{\mathrm{d}}{\mathrm{d}t}M_z^B=\frac{M_2^B-M_0^B}{T_1}+k_1M_z^A-k_2M_z^B \tag{30}$$

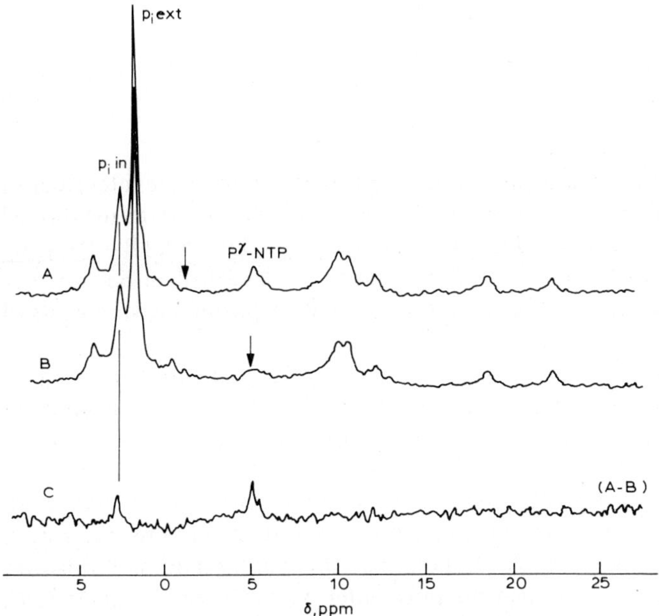

Figure 19. ^{31}P-NMR spectra of aerobic glucose-grown *Escherichia coli.* The arrows indicate the frequencies of the low-power pulses used in *B* to saturate the nucleotide triphosphate (NTP) peak. The peaks labelled P_i^{in} and P_i^{ext} correspond to intracellular and extracellular P_i, respectively. Peak NTP_γ consists of approximately 50% ATP and 50% nonadenine nucleotide triphosphates. The difference spectrum ($A - B$) shows transfer of saturation from NTP_γ to P_i (from [36]).

M_z^B and M_z^A are the instantaneous magnetizations of $B-P$ and $A-P$, respectively, M_0^B is the magnetization of $B-P$ at thermal equilibrium, and T_1^B is the spin-lattice relaxation time of $B-P$ in the absence of exchange. As described by the previous equation, chemical exchange can alter the apparent rate at which M_z^B returns to thermal equilibrium, so that T_1^B cannot be measured simply by a conventional T_1 experiment. Rather, T_1^B is measured either directly by a conventional T_1 experiment whilst inhibiting the exchange reaction in vivo [36], or indirectly [39] by determining the apparent T_1 of $B-P$ while the $A-P$ resonance is saturated. This apparent T_1 will reflect true spin-lattice relaxation (T_1) and the rate of conversion of $B-P$ to $A-P$, or, from the previous equation ($k_1 M_z^A = 0$):

$$\frac{1}{(\text{apparent } T_1)} = \frac{1}{T_1^B + k_2} \tag{31}$$

This equation can then be combined with a reduced form of Eqn. 30 applicable to the saturation transfer experiment, to solve for both k_2 and T_1. Specifically, saturation of the $A-P$ resonances makes $M_z^A = 0$, and under steady-state conditions $d/dt M_z^B = 0$, so that Eqn. 30 becomes:

$$\frac{M_z^B}{M_0^B} = \frac{1}{1 + k_2 T_z^B} \tag{32}$$

M_z^B/M_0^B is determined in the experiment, being given by the ratio of the intensities of the $B-P$ resonance in the presence or absence of saturation of $A-P$, as mentioned above. The rate of conversion of $B-P$ to $A-P$ is given by $k_2[\mathrm{B-P}]$, $[\mathrm{B-P}]$ being determined from the intensity of the $B-P$ resonance as described in Section 3(c). Analogous equations apply in the determination of $k_1[\mathrm{A-P}]$ from measurements of saturation transfer from $B-P$ to $A-P$.

Inspection of Eqn. 32 reveals that k_2 must be of comparable magnitude to T_1 for M_z^B/M_0^B to be significantly less than 1. This constraint places a lower limit on values of k_2 that can be measured by saturation transfer, approximately $0.1\ \mathrm{s}^{-1}$ if T_1^B is approximately 1 second.

The equations given above apply only under specific circumstances. First, the spectra used to measure M_z^B/M_0^B must be obtained under steady-state conditions, i.e., with a pulse interval of approximately $5 \cdot T_1$. If not, M_z^B/M_0^B will depend not only on the rate of chemical exchange, but also the pulse interval and the pulse angle [39], such that more complex equations apply. Second, this method of analysis only applies to a simple exchange reaction. There are very few such simple exchange reactions in living systems; for example, ^{31}P can be exchanged between inorganic phosphate and ATP via innumerable reactions. Despite this problem, analysis of saturation transfer data is almost invariably made on the simple framework outlined above. In some studies, notably those concerned with P_i-ATP phosphate exchange [36–38], analysis as a simple exchange reaction appears to describe the results adequately insofar as estimated forward and reverse fluxes are approximately equal. However, the uncertainties in the data are often considerable (largely due to inadequate signal to noise ratio for accurate measurement of flux in at least one direction), so this is not a rigorous criterion. The most detailed examination of forward and back reaction rates by saturation transfer have been on the phosphate exchange between ATP and phosphocreatine [39–41]. In these studies, the apparent rate of ATP formation from phosphocreatine was significantly greater than the apparent rate of phosphocreatine formation from ATP – a result at odds with the constancy of tissue phosphocreatine and ATP concentrations. Three qualitative explanations have been put forward to explain this discrepancy. One proposal, concerning creatine kinase activity in rabbit hearts [40], was that the ATP may be compartmented such that not all the ATP observed by NMR is accessible to creatine kinase; this would lead to an underestimate of the rate of phosphocreatine formation. A second proposal, coming from a study of creatine kinase in frog skeletal muscle [39], suggested that exchange of ^{31}P via the adenylate kinase reaction may account for the discrepancy; saturation transfer experiments indicate high adenylate kinase activity in many tissues [37,39]. In a more recent report [41], concerning creatine kinase activity in rat hearts, it was suggested that the existence of exchange reactions other than the creatine kinase reaction may be the cause of the discrepancy. In none of these studies was it possible to put forward quantitative arguments to support these hypotheses.

We should mention an interesting study of ATP synthesis and Na^+ transport in the rat kidney [8], which led to the conclusion that the Na^+/ATP stoichiometry of the transport ATPase is 12, rather than 3 – the value observed in erythrocytes. This study is one of the few examples in which in vivo NMR methods have yielded a novel biochemical result.

(g) Tracing metabolic pathways by ^{13}C- and ^{15}N-NMR

In order to observe carbon or nitrogen NMR signals in vivo it is first necessary to supply a substrate enriched with ^{13}C or ^{15}N (to ≈90% or greater), the only isotopes of C and N that give high-resolution NMR spectra. Which particular compounds are observed will depend on the specific labelled metabolite supplied, as well as the type of tissue under study. Therefore, ^{13}C- and ^{15}N-NMR have formal similarities to metabolic studies using radioactive tracers. Although both can yield equivalent information [42], each method has specific advantages: radioactive tracers give much greater sensitivity, allowing in principle all intermediates in a metabolic sequence to be identified and analysed, whereas the low sensitivity of NMR restricts study to only the most abundant metabolites of the major metabolic pathways. The great advantage of the NMR approach is that concentrations of labeled metabolites, and the distribution of isotopes within individual molecules and between anomers, can be determined without the laborious extraction, separation, degradation and counting procedures associated with radioactive techniques – as has been extensively demonstrated by Shulman and colleagues [43] (cf. Fig. 20). Moreover, pulse-chase experiments can be done with a single sample [43].

In vivo ^{13}C-NMR can be used to study processes such as glycolysis [43–48], gluconeogenesis [49–51], glycogen metabolism [52], CO_2 metabolism [53,54] and porphyrin biosynthesis [13]. These studies have confirmed information obtained by more conventional biochemical methods.

Very few ^{15}N-NMR studies of living cells have been reported, despite the fact that the absence of a useful radioactive nitrogen isotope has limited our understanding of nitrogen metabolism [122]. This may be attributed to the relatively low sensitivity of ^{15}N-NMR, and to the large cost (or scarcity) of ^{15}N-labeled metabolites. The most comprehensive ^{15}N-NMR study of metabolism in the literature concerned synthesis of storage proteins in soybean [55–57], performed on lyophilised material using solid-state NMR methods [58]. The availability of instruments operating at higher field, and using larger sample volumes, should permit similar studies in vivo.

An interesting approach to observe metabolism of ^{13}C-labeled substrates involves use of ^{1}H-NMR to detect protons bonded to both ^{12}C and ^{13}C [59]. Protons bonded to ^{13}C will give rise to multiplets due to spin-spin coupling, while those bonded to ^{12}C will not. Hence, not only can the concentration of ^{13}C in a metabolite be determined, but also the fractional labeling. An advantage of this method is the much greater sensitivity of ^{1}H-NMR compared to ^{13}C-NMR. A disadvantage is that the range of

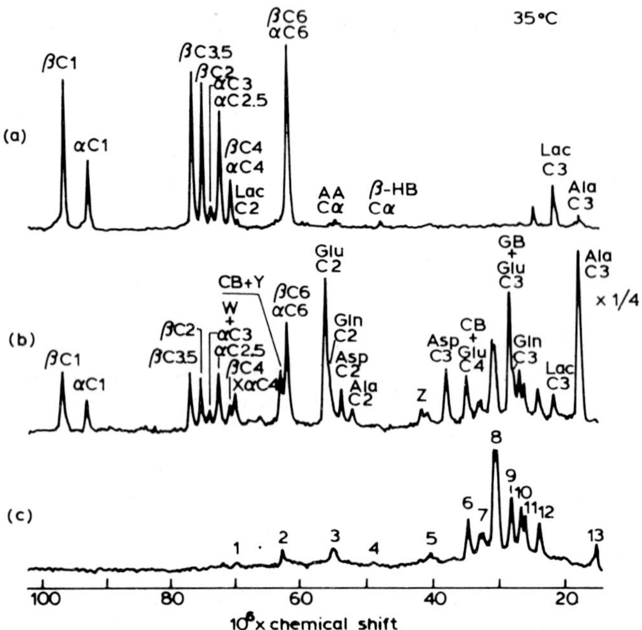

Figure 20. ^{13}C-NMR spectra from a perfused mouse liver at 35°C. (c) ^{13}C natural abundance background of this liver, accumulated before the substrate was added. The substrate, 8 mM [3-^{13}C]alanine and 20 mM unlabelled ethanol, was then added at 0 minutes and again at 120 minutes, and a series of ^{13}C-NMR spectra were taken. (b) Spectrum measured during the period 150–180 minutes (a) ^{13}C-NMR spectrum of the perfusate after the perfusion was terminated, at 240 minutes; this spectrum consisted of 5000 scans. The pulse repetition times were 0.5 seconds for b and c and 2 seconds for a. Abbreviations: βC1, αC1, βC3.5, αC4, βC6, αC6, βC2, αC2.5 and αC3, the carbons of the glucose anomers: Glu C2, glutamate C-2; Gln C2, glutamine C-2; Asp C2, aspartate C-2; Ala C2, alanine C-2; LacC3, lactate C-3; CB, cell background peak; W, X, Y and Z, unknowns; AA Cα, acetoacetate CH_2; and β-HB Cα,β-hydroxybutyrate CH_2 (from [31]).

chemical shifts in ^{1}H-NMR is much smaller, so that resolvable resonances are seen with samples that give narrow lines – such as cell suspensions.

4. *Macromolecules in vitro*

(a) Introduction

Historically, the applications of NMR to problems of macromolecular structure and dynamics have preceded extensive studies of biochemistry in vivo by more than a decade. This is a reflection of the development of magnet technology: much useful work on the molecular level could be done with the narrow-bore magnets developed first, whereas in vivo biochemical work for the most part requires magnets with a bore wide enough to accommodate at least an isolated organ, and preferably an intact organism; such magnets have become available only in the last few years. However, structural applications require a deeper understanding of the theory of the method, and therefore are discussed here last in the logical order of complexity.

The literature on applications of NMR to problems in molecular biology now encompasses several thousand entries, and space limitations preclude a complete exposition of the subject in this article. Amino acids, peptides, proteins, nucleic acids and their constituents, polysaccharides, phospholipids, membranes, and a large range of biologically active compounds have been extensively studied by NMR. The reader interested in doing research in this field will find it necessary to consult a more extensive treatise (see, e.g., Refs. 60, 61). Our aim here is to give only a brief introduction to the types of problems that have been studied and an outline of the methods and strategies developed for this purpose.

(b) Analysis of macromolecular spectra

A representative high-resolution ^{1}H-NMR spectrum of a moderate size peptide, endorphin (32 amino acid residues), is shown in Figure 21. Characteristically, a large number of individual spectral lines can be resolved, and since the molecular weight is relatively small, the lines are relatively narrow. A complete or nearly complete analysis of such a spectrum is possible, providing a wealth of information on the structure and dynamics of the peptide. In contrast, the ^{1}H-NMR spectrum of a large protein, myosin (3000 amino acid residues), is shown in Figure 22. Here, the larger number of overlapping lines and the increased width of individual lines associated with the high

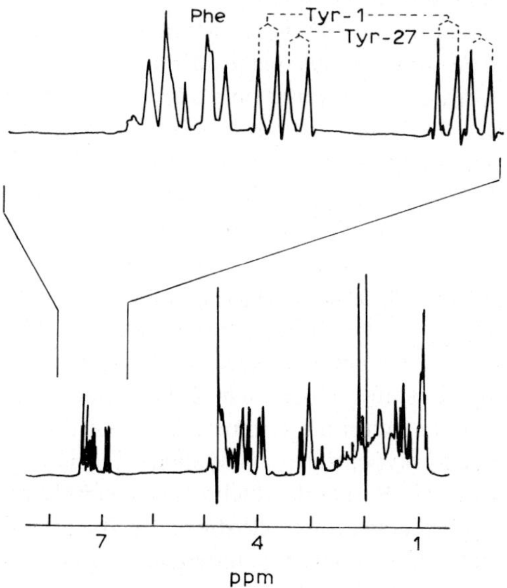

Figure 21. 270 MHz NMR spectrum of 6 mM human β-endorphin in ^2H_2O solution at pH 6.3. From [121].

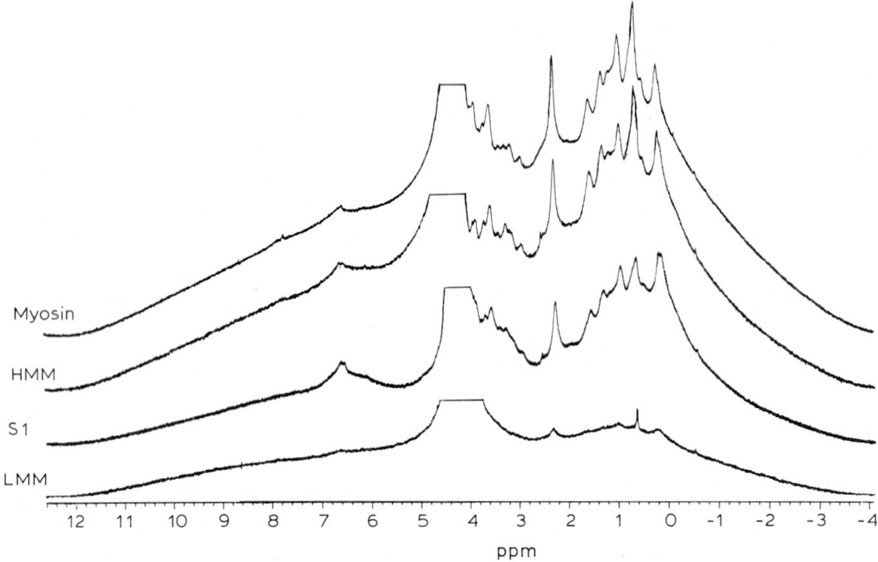

Figure 22. 360 MHz ^{1}H NMR spectra of myosin and its constituent fragments. Reprinted, with permission, from [118], © 1979, American Chemical Society.

molecular weight result in only an envelope being observable. Only very general conclusions, such as the existence of extensive mobile regions [63], can be drawn from such a spectrum, and any more detailed structural work would depend on a simplification of the spectrum, e.g., by selective isotopic labeling.

The first and essential step in the analysis of the NMR spectrum of a macromolecule is the assignment of as many of the individually resolved lines to specific residues in the sequence. This in itself is a formidable task and only a few examples of complete assignment for molecules with a molecular weight (MW) > 1000 have appeared in the literature. The largest completely assigned structure is that of the bovine pancreatic trypsin inhibitor (BPTI, MW 6000) [64]. Extensive, but far from complete, assignments are also available for lysozyme (MW 14,000) [65] and ribonuclease (MW 13,600). In the NMR literature on peptides and proteins it has become conventional to speak of two stages of assignment: (1) assignment to residue type and (2) assignment to an individual residue in the known peptide sequence. The former is relatively straightforward and can be accomplished by a systematic examination of chemical shifts and coupling constants, since each of the amino acids has a characteristic coupling pattern. No single technique is generally applicable to accomplish the latter. The spectrum of a folded polypeptide (or polynucleotide) is very much more complex than the spectrum of the random coil, which is given simply by the sum of the spectra of the constituent amino acids (or nucleotides). In the folded structure the spectral lines of chemically identical residues are shifted with respect to each other [66]. This means that the NMR spectrum of the folded structure reflects the structural features in considerable detail, but it also means that assignment of a spectral line to a specific residue in the sequence will often be predicated on using knowledge of these features in the assignment process.

The strategies for the assignment of lines in the spectra of macromolecules may be considered to fall into three categories: (1) purely spectroscopic techniques, (2) techniques dependent on the knowledge of the crystal structure and (3) combinations of chemical and spectroscopic techniques not dependent on prior knowledge of the crystal structure. Each of these approaches is subject to characteristic limitations and the choice of the strategy for a particular problem may be dictated by the properties of the molecule under study.

(i) Purely spectroscopic techniques

Assignment to residue type is accomplished readily by systematic spin decoupling, since the spectrum of each amino acid has a characteristic coupling pattern. This is illustrated in Figure 23, where the decoupled residue is identified readily as tyrosine by comparing the decoupling pattern to the spectrum of this amino acid [67]. The only requirement for the success of this procedure is extensive resolution of spectral lines and line widths sufficiently narrow to permit the identification of multiplets.

Assignment to an individual residue in the amino acid or nucleotide sequence is also possible in favorable cases by purely spectroscopic techniques. The method for such an assignment is as follows: (1) By side-chain decoupling identify all lines in the

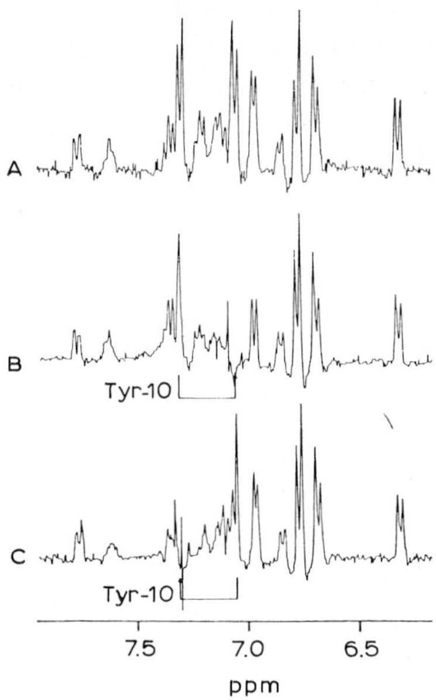

Figure 23. (A) Aromatic region of the convolution difference ^{1}H-NMR spectrum at 360 MHz of a BPTI solution in ^2H_2O, pD = 7.8 = 26°. (B) Double-resonance irradiation at 7.07 ppm causes the collapse of a doublet at 7.33 ppm. (C) Irradiation at 7.33 ppm causes the collapse of a doublet at 7.07 ppm. The assignment of this $AA'BB'$ spin system to residue 10 is from Reference 67. From [68].

spectrum of a given residue, specifically including its NH peak. (2) Elicit an NOE from this NH to the neighboring alpha CH. The NOE to the nearest neighbor can usually be identified as being the first to appear in a time-dependent NOE experiment. (3) By side-chain spin decoupling identify the neighboring residue in the sequence as to residue type. (4) Repeat to identify the next residue, etc. This method was first used by Gibbons and his colleagues [69] to assign the amino acid residues in Gramicidin S (Fig. 24). In the most favorable case of a relatively rigid structure with extensive hydrogen bonding – such as tRNA [85,86] – it is possible to 'walk up and down' the structure and obtain both the sequence and the assignments from a sequential alternation of decoupling and NOE experiments. In the ideal case, where the spectrum is resolved completely and a complete set of interresidue NOEs can be obtained, it is in principle possible to obtain the entire polymer sequence by this procedure. In most cases, however, a complete set of NOEs is not obtained, or their interpretation remains ambiguous because of overlap or inability to decouple and hence identify the NH because of broad lines, so that only short stretches (2–6 amino acids long) can be sequenced. Even in this case, however, it may be possible to obtain a complete set of assignments by matching the spectroscopically sequenced peptides to stretches in the known sequence of the polymer. This procedure, illustrated in Figure 25, is akin to the peptide mapping originally used in the chemical sequencing of proteins. It becomes subject to an increasing number of ambiguities as the molecular weight of the macromolecule under study is increased. The success of the purely spectroscopic approach to complete assignment depends strongly on the choice of the system. Its principal limitations stem from frequent failure to observe an NH–CH NOE in random coil segments, thus interrupting the identifiable sequence, and from the

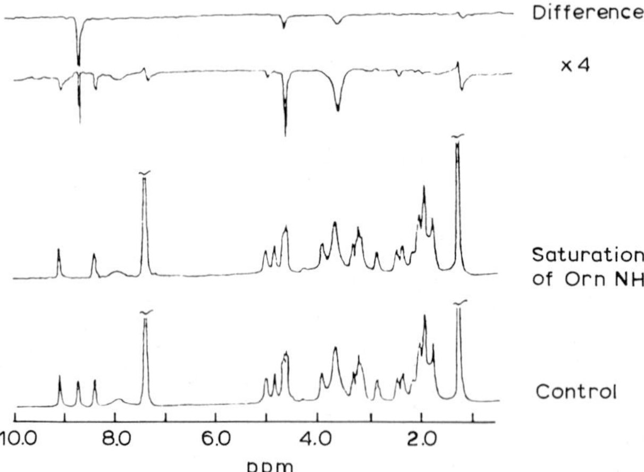

Figure 24. Bottom, the control spectrum of Gramicidin S obtained with irradiation at a frequency at which no resonances occur; middle, the spectrum obtained with irradiation at the transitions of the Orn NH; top, the difference spectrum obtained by subtracting the second from the first. The difference spectrum shows two NOEs in the $C_{\alpha}H$ region (4.5–5.5 ppm), the most prominent of which is the NOE at the Val $C_{\alpha}H$. Reprinted, with permission, from [69], ©, 1978, American Chemical Society.

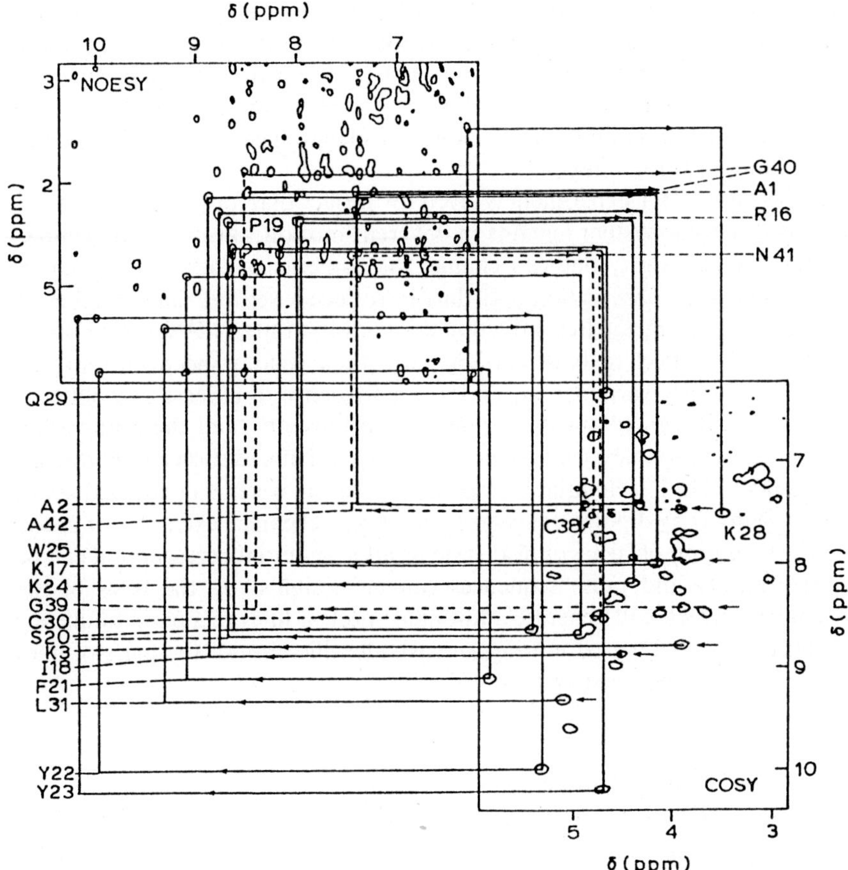

Figure 25. Combined COSY–NOESY connectivity diagram for sequential resonance assignments via NOEs between amide protons and the C_α protons of the preceding residues in inhibitor K. In the upper left region ($\omega_1 = 2.9$–5.9 ppm, $\omega_2 = 6.3$–10.3 ppm) of the ^{1}H NOESY spectrum the inhibitor K is presented. In the lower right the region ($\omega_1 = 6.3$–10.3 ppm, $\omega_2 = 2.9$–5.9 ppm) of a ^{1}H COSY spectrum recorded from the same sample under identical conditions, i.e., at 50°C and pH 3.4, is shown. The connectivities between neighboring residues in the segments 42–40 (- - -), 39–38 (- - -), 31–28 (——), 25–19 (——), 18–16 (——) and 3–1 (——) are indicated. For each segment, the start and the end of the connectivity pattern are indicated by ● ←. and by ● and identification of the terminal residue, respectively. To identify the connected cross peaks, assignments are indicated in the lower left at the amide proton chemical-shift positions. Exceptions are Asn 41 and the C-terminal residues of the assigned peptide segments, which are identified in different regions of the figure. For A1, R16, P19 and Gly40 the connectivities end in the NOESY spectrum, since the amide protons are not observed; for K28 and C38 they end in the COSY spectrum (from [117]).

overlap of NH or CH lines, which makes a unique identification of a dipeptide fragment impossible.

The development of two-dimensional Fourier transform techniques (2DFT) [70,71] has greatly increased the efficiency of collecting, presenting and analysing the data necessary for obtaining a set of assignments in a macromolecular spectrum. A detailed description of these techniques is beyond the scope of this article, and can be found elsewhere [72,73], but the three 2DFT techniques that have proved particularly useful for implementing the approach defined above – i.e., *J*-resolved 2DFT, correlated 2DFT spectroscopy (COSY) and 2DFT NOE correlated spectroscopy (NOESY) – are worth outlining in this context.

The possibility of a 2DFT experiment is given by the fact that for periods $t_1 < T_1, T_2$ a once perturbed nuclear system remains in coherent motion, effectively 'remembering' its history. By varying t_1 one can change the contents of that memory at the moment of a subsequent observation and thus introduce a second time variable, in addition to the one given by the free induction decay over a time, t_2. A double Fourier transformation with respect to both variables will then yield a spectrum that is a function of two frequency variables, ω_1 and ω_2. A sequence of at least two pulses is thus required for a 2DFT experiment, a preparation pulse, p_1, and the observation pulse, p_2, t_1 being in the simplest case the period between them, as shown in Figure 26.

The usefulness of 2DFT techniques is especially great in systems of coupled spins, whether the coupling is by dipolar or scalar interactions, chemical exchange or cross-relaxation. The evolution of such coupled systems after an initial perturbation can be predicted theoretically and pulse sequences can be designed on the basis of the theoretical analysis which will generate additional spectral lines related in frequency to both coupled lines, thereby permitting an easy identification of the pairs of lines that correspond to each coupled system.

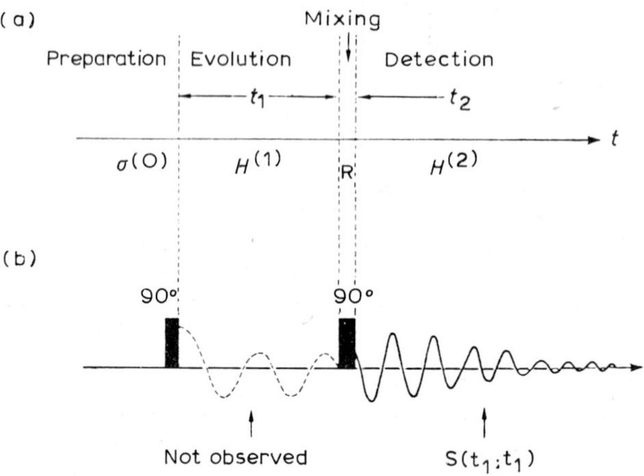

Figure 26. Experimental scheme of two-dimensional NMR spectroscopy. (a) The general scheme of two-dimensional spectroscopy. Here, t_1 and t_2 become two variables of two-dimensional response signals. $H^{(1)}$ and $H^{(2)}$ are the Hamiltonians during t_1 and t_2 periods, $\sigma(0)$ is the initial density matrix after the preparation and R represents a mixing operator. (b) One realization of two-dimensional NMR spectroscopy which elucidates the spin connectibility of coupled nuclei called 'two-dimensional correlated' NMR spectroscopy. Here, the first 90° pulse is used to prepare the initial magnetization (or initial density matrix) and the second pulse is applied to mix two transitions (precession frequencies) evolved during the two successive time periods, the evolution period and the detection period (from [73]).

It is generally useful to distinguish (1) J-spectroscopy, by which chemical shift and coupling constants are separated and appear as the two frequency variables and (2) shift correlation spectroscopy, in which the chemical shifts of coupled (e.g., H–C), exchanging (e.g., free and bound ligand) or cross-relaxing (as in the NH–CH NOE) nuclei serve as the two frequency variables. The pulse sequences needed to generate the different types of 2DFT spectra are different, but can all be described in terms of four time periods, as shown in Figure 26: (1) the preparation period, including the equilibrium and the preparation pulse p_1, (2) the evolution period t_1, (3) a mixing period, which may be absent, or may include a non-selective pulse, and (4) the detection period t_2 following the observation pulse p_2.

The pulse sequence for J-resolved spectroscopy is $(90) - t_1 - (180)$. The FID is observed as an echo, so there needs to be an additional delay t_1 before the detection period t_2, and the complete sequence is $(90) - t_1 - (180) - t_1 - t_2$, as shown in Figure 26. For COSY the basic pulse sequence is $(90) - t_1 - (90) - t_2$ and for NOESY it is $(90) - t_1 - (90) - D - (90) - t_2$, with a pulsed field gradient applied during the delay, D. A large number of other pulse sequences has been designed both to improve detection and to permit other forms of correlation spectroscopy.

All these sequences include a (90°) preparation pulse which flips the magnetization of the spin system into the xy plane. The evolution period is then the period of free precession of the spin system in the xy plane, in the Hahn spin echo experiment (Fig. 6). The almost infinite variations in the mixing strategies serve the purpose of bringing out those features of the correlated motions in the system of coupled spins which one wishes to observe.

The important feature of 2DFT spectra is the appearance of cross-peaks, i.e., absorption signals which depend on both frequencies, and must be assigned not to a single spin, but to a pair of interacting spins. In the present context these peaks are of the greatest interest, because they carry the desired information and permit the identification of a pair of spins as being part of an interacting system, e.g., an amino acid side chain or adjacent parts of a peptide backbone. Their origin can be qualitatively understood as follows. In an interacting system a pulse can affect a set of spins both directly – e.g., by flipping its magnetization through the desired (90°, 180° or any other) angle – and indirectly, by flipping the magnetization of the other set, and through the interaction, partially flipping the magnetization of the first set. With a single pulse the magnitude of this effect will always be the same, will be included in the features of the recorded spectrum and will not be separately observable. However, in a multiple pulse experiment the magnitude of the effect will vary with time, and the transitions involving the simultaneous flip of both spins become separable from the transitions involving only one set. A more precise description of the many different types of cross peaks requires the solution of the equations of motion for the specific coupled spin system under the influence of a particular pulse sequence. Knowledge of these solutions for each particular 2DFT experiment is crucial for sorting out those peaks in the spectrum which contain useful information from those that can be generated as experimental artefacts.

2DFT spectra can be presented either as stacked or as contour plots, shown in

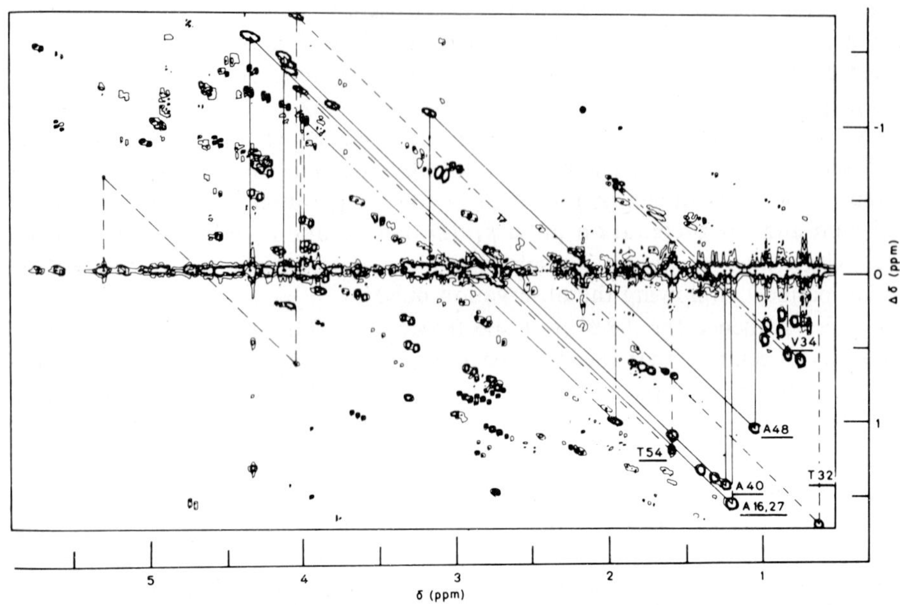

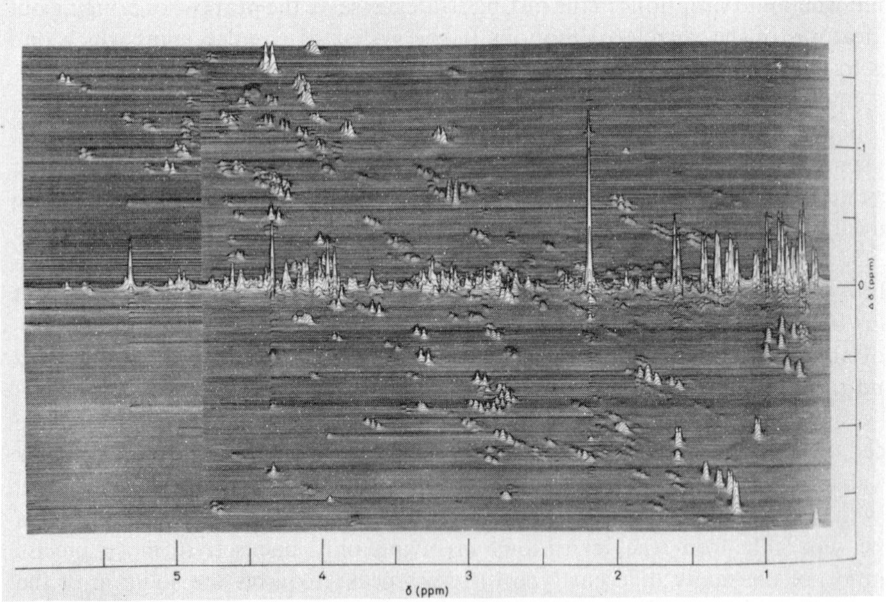

Figure 27. Lower, three-dimensional presentation of the spectral region, 0.5–6.0 ppm of a 360 MHz spin-echo-correlated ^{1}H-NMR spectrum of BPTI. The chemical shift, δ, on the horizontal axis corresponds to that in conventional, one-dimensional spectra. $\Delta\delta$ on the vertical axis stands for the difference frequencies between correlated nuclei. Cross-peaks between J-coupled nuclei are at $\pm 0.5\ \Delta\delta$. The solvent singlet resonance is at 4.35 ppm. Upper, contour plot of the same spin-echo-correlated spectrum of the inhibitor. Connectivities between the individual components of the following spin systems are indicated: ——, alanines 16, 27, 40, 48; ----, threonines 32, 54; -·-·-, Val-34 (from [116]).

Figure 27. While the information content of the two presentations is identical, the stacked plot is related more easily to a conventional spectrum, and therefore used more widely when identification of individual peaks is of primary interest, as in *J*-spectroscopy. On the other hand, a contour plot is more economical when one is attempting to establish connectivities between atoms by an observation of cross peaks, as in COSY and NOESY spectra. Here a contour projection of the normal spectrum is plotted on the diagonal, and the cross peaks can be identified readily on the projections of their coordinates, as shown in Figure 26. The cross peaks in the COSY spectrum are indicative of spin-spin coupling between the two groups on the diagonal. The cross peaks in the NOESY spectrum indicate the existence of cross-relaxation between the two groups.

The main advantage of 2DFT methods over sequential decoupling and NOE measurements described above is that all of the information on connectivities in the entire molecule is obtained and presented at once, at a considerable saving in time and effort. A significant disadvantage is that only part of the 2DFT spectrum – sometimes a rather small part – contains useful information. For example, most NOEs seen in a peptide NOESY spectrum will be between neighboring atoms in side chains. This information is redundant with – and less complete than – that obtained from decoupling or COSY experiments. Only a fraction represents new connectivities through space, as in the case of NH–CH discussed above.

2DFT methods, because of their economy and elegance, are likely to play an increasing role in studies of peptide and nucleotide structure. Their intrinsic information content and ultimate limitations are, however, the same as those of the one-dimensional methods first used to define the approach necessary to obtain assignments in polymer spectra by purely spectroscopic techniques. These limitations are given by the inherent properties of the structure under study – such as the presence of random coil segments, which interrupts the sequence of spectroscopically observable connectivities – or the lack of complete resolution in its spectrum. In the study of such structures assignment strategies that do not rely solely on spectroscopic connectivities become necessary.

(ii) Techniques dependent on the knowledge of the crystal structure

Assignments of spectral lines for proteins and nucleic acids whose crystal structures are known can be made by assuming that the structure is the same in the solution as in the crystal, and finding a feature in the spectrum that can be predicted from the structure. Among the most commonly used features are:

(1) Ring current shifts. The spectra of many proteins and most nucleic acids contain strongly shifted lines. Such large shifts can result from placing the observed group, e.g., a valine methyl, in the immediate vicinity of an aromatic ring. The resonance of a group in the plane of the ring will be shifted downfield, that of a group above or below the plane of the ring upfield. If in the crystal structure one valine lying above an aromatic ring can be found, the assignment of the upfield shifted line in the spectrum to this valine is likely to be correct. If several methyl resonances are shifted to higher fields in the spectrum and the structure contains several aliphatic residues in the

vicinity of aromatic rings, it becomes necessary to rely on ring current shift calculations to make the assignment. Because of the many approximations that have to be made in such calculations the procedure is less than wholly satisfactory and the reported assignments must be viewed with caution.

(2) Long-range NOEs. If the side chains of two residues are seen to lie in close proximity to each other in the crystal structure, an NOE between their atoms would be predicted. While in principle this is the best method for making assignments based on the crystal structure, in practice it suffers from two serious drawbacks: (a) In larger molecules there is a generalized spread of NOEs throughout the structure, known as spin diffusion, so that NOEs between non-neighboring residues will be observed in a steady-state NOE experiment. This difficulty can be overcome by making a time-dependent NOE measurement and relying solely on those NOEs that appear early [74,75]. (b) If this is done, one frequently finds that the number of long-range NOEs that can be observed before generalized spin diffusion complicates the picture is much smaller than would be expected from the number of contacts seen in the crystal structure. This probably results from extensive side-chain motion, which destroys the direct NOEs between adjacent residues and severely limits the usefulness of the procedure. Nevertheless, in the hydrophobic box of lysozyme, a relatively rigid structure, Poulsen et al. [76] have succeeded in assigning a sizeable number of spectral lines by this technique.

(3) Paramagnetic perturbations. The use of paramagnetic ions, especially lanthanides, to obtain structural information and assignments from NMR spectra of biological macromolecules has been explored very extensively. In principle, one expects to find very extensive geometric information from this type of experiment, since the observed effects are large and depend strongly on the distance of the observed residue from the ion; angular information can also sometimes be obtained. The method is described in detail in various monographs (see, e.g., Refs. 60, 61). Among the assumptions that have to be made to interpret the experimental findings unambiguously are (a) that the structure is rigid and (b) that there is only a single ion-binding site in the structure. Experience has shown that both of these assumptions are rarely satisfied in reality, and therefore the success of this approach has been rather limited.

(4) Miscellaneous perturbations. Proximity between two residues can sometimes be inferred from chemical shift changes on ligand binding, on titration or in photo-CIDNP experiments, when the process known to affect one residue is reflected in the spectrum of the other. For example, the titration of a histidine may be reflected in the spectrum of other, nearby residues. If proximity of two such residues is observed in the crystal structure, the assignment is probably reasonable. However, all procedures relying on chemical shift changes suffer from the drawback that such changes may result either from close contact or from conformational changes at a distance. The distinction between the two is extremely difficult in the absence of more direct evidence, such as an NOE. Therefore, assignments made on the basis of such observations must always be viewed with great caution.

The spectral features which permit an assignment to be made on the assumption that the structure is identical in solution and in the crystal can also be used to study the solution structure and differences between the solution and the crystal structure, provided the assignments are made by an independent method. In some published studies the assignments have been based on the crystal structure and the 'conclusion' subsequently drawn that the solution and crystal structures were identical. This clearly involves a circular argument which invalidates the conclusion, as the finding of significant differences between the solution and crystal structure would invalidate the assignments. For these reasons this entire class of assignment procedures is far less reliable than either the purely spectroscopic approach or the independent procedures described below.

(iii) Combinations of chemical and spectroscopic methods independent of the knowledge of the crystal structure

Assignments based on a comparison of chemical and spectroscopic information generally require prior knowledge of the polymer sequence, or primary structure, but not of the secondary and tertiary structure of a protein or nucleic acid. The most noteworthy of this class of procedures are:

(1) Site-specific isotopic substitution. This is the least ambiguous of all methods for achieving both complete resolution and assignment of individual resonances in a complex spectrum. Isotopic substitution of covalently bound protons – or of ^{13}C for ^{12}C and ^{15}N for ^{14}N – has never been found to lead to a significant structural change in the larger molecules and the assignment can be made simply by taking a difference spectrum of the labeled and the unlabeled molecule. The principal drawback of the technique, which has severely limited its use, is the technical difficulty and cost of carrying out the required chemical synthesis. The examples which can be cited so far [61] are limited to relatively short peptide and nucleotide sequences. Still, the technique offers the best hope for the study of larger molecules (MW $\gg$ 10,000), where the purely spectroscopic methods of assignment encounter ever increasing chances of failure because of ambiguities resulting from line overlap.

A partial assignment, to residue type, accompanied by a significant simplification of the spectrum is much more readily attainable by selective isotopic substitution using biosynthesis [77]. This procedure has been more widely used, but does not immediately lead to the assignment of a spectral line to a specific residue in the polymer sequence.

(2) Comparison of homologous proteins. This technique [78] works best when spectra of point mutants can be compared and only the lines attributable to the changed residues can be seen to differ in the spectra. Assignment by difference spectroscopy is unambiguous in such cases. However, if two or more residues are changed by mutations, more extensive spectral changes, which may obscure the assignment, may be observed [79]. Nevertheless, the technique has been successfully used to obtain extensive sets of assignments in several proteins, including hemoglobin and the cytochromes.

(3) In situ isotope exchange. The usefulness of this technique is limited to those protons that can be chemically exchanged under conditions that do not disrupt the polymer structure. It has been used successfully to assign the C(2) protons of different histidines in pancreatic ribonuclease [80]. If the exchange rates for chemically

identical residues differ in situ, assignments to specific residues can be made by correlating the decrease in area with the isotope content determined in a subsequent sequencing experiment.

(4) Selective chemical modification. The usefulness of this technique is also limited to those residues, such as lysine, methionine and tyrosine, for which modification reactions are known. In some cases assignment of spectral lines to specific residues by this technique has been possible, but more often the selectivity of the reaction in situ is inadequate to permit an unequivocal interpretation.

In summary, it needs to be said that the problem of assignment of lines in the spectra of structured polymers, such as proteins and nucleic acids, is still far from having found a general solution, although essentially complete assignments can be made by existing methods in the smaller structures (MW $<$ 6000).

(c) The information content of macromolecular spectra

The object of NMR spectroscopy on proteins, nucleic acids and other complex polymers is to obtain information on their structure and dynamics. Assignment of individual spectral lines discussed in the preceding section is a prerequisite to one's ability to decipher this information. Actually to do so it is also necessary to understand the nature of the structural and dynamic information inherent in each feature of the spectrum. And one is rightfully asked: is the nature of this information such that the result will be worth the labor required to obtain a significant set of assignments? An important part of the answer to this question is that NMR is the ONLY physical method which can provide any information at essentially atomic resolution on the structure and dynamics of macromolecules in solution, as well as in the solid state. Whatever the limits of this information, it is better than none. Knowledge of these limits, as well as of the nature of spectroscopic information, is nevertheless necessary, both to realize expectations and to avoid conclusions that go beyond the capabilities of the method. The information content and its limits can be meaningfully discussed for each measured parameter separately, and the common features summarized at the end.

(i) Chemical shift

As discussed in Section 2(e), the chemical shift of a nucleus depends on the density and geometry of the surrounding electrons and is a very sensitive indicator of local interactions and local structure. Differences in chemical shifts of chemically identical residues in a polymer reflect differences in their immediate surroundings. Therefore, the spectrum of a folded macromolecule may be taken as a fingerprint of its secondary and tertiary, as well as primary structure. Changes in the chemical shift of individual residues can be taken to indicate a change in the local structure in response to a perturbation, such as the binding of a ligand or change in the solvent composition Qualitatively, such changes may reflect either a direct interaction of the affected group with the ligand or solvent, or a conformational change. The distinction between the two usually must be made on circumstantial evidence. To demonstrate a local interaction it is sufficient to show that the observed group and the ligand or solvent are in close proximity, e.g., by observing an NOE between the two, but an additional contribution to the observed shift from a local structural change would not be ruled out by this observation. To attribute the shift to a conformational change not involving direct interaction with the perturbing ligand, it is necessary to show that the shifted group lies distant from the binding site. If the distance from the ligand can be measured – even approximately, as in the case of paramagnetic ligands – or is known from the crystal structure, and can be shown to be too large for a direct interaction shift, it is justified to attribute the shift to a conformational change. In most cases of diamagnetic ligands the distinction is difficult and, where interactions with the solvent at multiple binding sites are involved, may be impossible.

A more quantitative interpretation of chemical shift data cannot be made at present, with few exceptions, such as the calculations of ring current shifts. The theory of chemical shifts is by and large inadequate to deal with the subtleties of the observed changes. A priori calculations remain grossly inaccurate, because the large number of parameters contained in the equations cannot be derived from any experimental measurement. Even calculations of ring current shifts involve iterative fitting of these parameters to known crystal structure data. The view of molecular structure derived from a study of chemical shifts thus reveals a wealth of detail blurred in its essentials.

(ii) Coupling constants

The magnitude of the coupling constant between two nuclei depends on the number of intervening bonds and on their mutual orientation about them. The theory of coupling constants developed by Karplus [81] accurately describes the general shape of the functional dependence of the coupling constant on the angle of rotation about a single bond, the dihedral angle (Fig. 28), without being able to predict precise values. Such values can, however, be derived from extensive empirical correlations made over the past quarter of a century. They differ somewhat depending on the nuclei involved and the corrections made, but in general permit the identification of the dihedral angle in a rigid structure to within a few degrees. An example of such a 'Karplus curve' is shown in Figure 28. By measuring a coupling constant the dihedral angle about the intervening bond can be estimated, assuming the structure to be rigid. The largest problem with this approach is that most macromolecular, and especially small peptide and nucleotide structures are not rigid and the measured coupling constants are time averages. In favorable cases the time averages can be resolved and populations of individual conformations can be calculated. In many cases, however, failure to recognize the problem has led to the formulation of structures that further work has shown to have been in error. The problem of time averaging and its consequences for the interpretation of NMR data is discussed further below.

(iii) Relaxation parameters

The three relaxation parameters T_1, T_2 and NOE all depend on the same distances – e.g., between nuclei or nuclei and electrons – and on the same time constants of

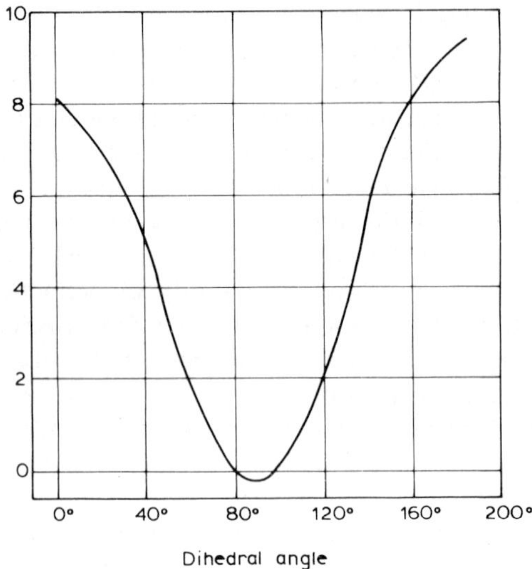

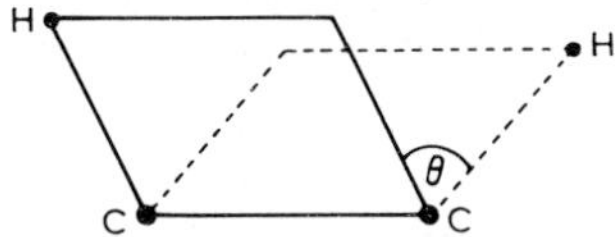

Figure 28. The dihedral angle θ formed by the H–C–C and C–C–H planes and the Karplus curve: dependence of the coupling constant J in Hz on the dihedral angle θ, defined by the planes of the H–C–C and C–C–H bonds in an H–C–CH fragment (from [61]).

molecular motion, though in somewhat different ways. The interpretation of these parameters in terms of distances and the rate of rotational diffusion is relatively straightforward for rigid spheres and this model has been widely used in biochemical work, whether applicable or not. For small molecules the model holds reasonably well, but as the number of internal degrees of motional freedom increases with the increasing number of single bonds, the problems of exact interpretation multiply rapidly. For macromolecules the problem of calculating both distances and motional frequencies and amplitudes from relaxation data is formidable and, since each interpretation depends on a set of assumptions that can rarely be verified, there is considerable disagreement in the literature about many of the published interpretations. A detailed discussion of relaxation theory for flexible molecules is beyond the scope of this article, but has been given elsewhere [61,82].

(iv) The problem of averaging

When using NMR to study structure and dynamics, it is important to recognize from the outset that the NMR phenomenon is slow compared to an optical transition or a scattering event and that all observed NMR parameters are time and space averages.

In liquid samples, where molecular motion is rapid, many states will contribute to these averages and the interpretation of the data in terms of a single state – tantamount to assuming a single rigid entity – may be very far from reality. Motional averaging imposes an inherent limitation on the interpretability of any single NMR parameter and makes it necessary to introduce information extraneous to the method or, alternatively, to rely on a correlation of a large number of different NMR observations, in order to arrive at a trustworthy conclusion.

(d) Solution structure of proteins and nucleic acids

A large amount of information on the solution structure of peptides, nucleotides, proteins and nucleic acids has been derived from NMR studies over the past 25 years. While the sum total of this information falls short of being a complete structure determination in the sense the term is used in X-ray crystallography – i.e., of uniquely defining the coordinates of each atom within the structure – many important features of molecular structures in solution, including the characteristics of their internal motions, have emerged. A structure determination by NMR is possible only to the extent to which a molecule is rigid, since all spectroscopic parameters reflect both the structure and the motions as an inseparable pair. Since there is a high degree of flexibility in most biological molecules, it is not surprising that the structural information one can obtain is only partial.

In favorable cases, NMR can provide information on all three aspects of a macromolecular structure – the sequence or primary structure, the pattern of hydrogen bonding or secondary structure and the folding pattern or tertiary structure. Primary and secondary structure are best obtained using the method described in Section 4(b), and either 1D- or 2DFT data. Using this method, Wuthrich and coworkers [83] have been able to define completely all features of the secondary structure of BPTI and its homologs, and Hare and Reid [84,85] have defined several stretches in tRNA. In principle, the tertiary structure could be obtainable by an extension of the same approach to the study of inter-chain NOEs to establish proximity relationships between neighboring pairs of atoms. In practice, however, inter-chain NOEs are not observable in sufficient numbers to give a unique definition of the tertiary structure, most likely because of extensive side-chain motions. To obtain information on the tertiary structure, one therefore has to resort to the study of other features of the high-resolution NMR spectrum.

In diamagnetic proteins the most useful spectral features that can be correlated with the characteristics of the three-dimensional structure are:

(1) Susceptibility of selected spectral lines to solvent perturbation, including photo-induced dynamic nuclear polarization [86–89]. Such selective perturbation of spectral lines can usually be interpreted in terms of the accessibility or inaccessibility of the residue in question to solvent.

(2) Ring current shifts, especially for methyl groups, which can be taken to indicate that the shifted methyl group is in the proximity of an aromatic residue [90].

(3) Titration of charged groups reflected in the chemical shifts of neighboring groups.

(4) Unusual features of the titration curves for charged groups, such as deviation from the expected Henderson-Hasselbalch pattern, indicating mutual interaction or the failure of a residue, e.g., histidine or tyrosine to titrate with the expected pK, which can be taken to reflect local interactions.
(5) Selective shifts on ligand binding. Care must be taken to distinguish between direct effects resulting from contact and indirect effects reflecting a conformational change, which generally can be accomplished only from the observation of NOEs or other knowledge of the three-dimensional structure.
(6) Slowly exchangeable NH protons either in the peptide backbone or on heterocyclic side chains, which indicate either hydrogen bonding or inaccessibility to solvent, or both, as discussed in Section 4(e)(i).
(7) Unusual chemical shifts. In the case of NH protons such shifts are often assumed to be involved in a hydrogen bond. In other cases they may reflect proximity to a charged group, especially a diamagnetic metal ion [92].

In paramagnetic proteins, and generally in macromolecules containing a single binding site for a paramagnetic ion, it is sometimes possible to estimate distances from the paramagnetic ion to specific atoms in the structure. Paramagnetic species can produce either shifts or broadening of selected lines in a macromolecular spectrum. Both effects depend on distance, with the lines of groups in the immediate vicinity of the paramagnetic species (ion or spin label) being affected most. The theory that has been used to calculate distances both from shifts (known as contact and pseudo-contact shifts) and from relaxation effects reflected in the line broadening is rather elaborate and contains many assumptions that are difficult to verify. It is beyond the scope of this article, but can be found elsewhere [60,61]. It can generally be said that the accuracy of such distance calculations is not very great and decreases with increasing distance from the paramagnetic center. Nevertheless, identification of residues in the immediate vicinity of the paramagnetic center is usually reliable. Larger distances are suspect both because the effects are smaller and because spurious effects from secondary binding sites enter into the picture, unless the ion is an integral part of the protein structure, as in the cases of hemoglobin and the cytochromes.

Paramagnetic effects on protein NMR spectra have been used to identify ligands to the paramagnetic ion, e.g., of methionine as a ligand to Fe^{2+} in cytochrome *c* [91] and histidine as a ligand to Cu^{+} in blue copper proteins [93]. The use of the paramagnetic lanthanide ions to produce shifts and relaxation effects in macromolecular spectra from which distances useful in the determination of their three-dimensional structures could be derived has been proposed as a general method for structure determination and extensively explored by R.J.P. Williams and his colleagues [94]. In many cases, useful proximity relationships have been established, but the interpretation of the findings has proved to be complicated by both the effects of motions within the structures and effects from secondary binding sites [61]. The technique cannot be recommended as a general approach to the determination of solution structures.

Limited direct comparisons of crystal and solution structures of macromolecules can be made, based on the fact that the high-resolution NMR spectrum can be regarded as a fingerprint of the total structure [95]. Since recently it has become possible to obtain high-resolution spectra of macromolecules in the solid state (for a review, see Ref. 96), the spectra of the same molecules in the two different physical states can be compared directly. The resolution achieved so far in the solid-state experiments on larger molecules is inferior to that obtained in solution and only a general statement that the structures are similar in their major features can be made. In the case of small structures, such as that of cyclo-D-Phe-Pro-Gly-D-Ala-Pro, a relatively rigid cyclic pentapeptide, a superposition of the solid state and solution spectra can be achieved, as shown in Figure 29. More specific conclusions can be drawn even on very large structures, when individual spectral features are compared, especially on isotopically (^{13}C or ^{15}N) enriched polymers. For example, the spatial orientation of labeled residues within the myoglobin molecule has been defined by measurements on magnetically ordered liquid crystals of myoglobin [97]. Dynamic information on the degree of stabilization of aromatic residues in the filamentous bacteriophage *fd* has also been obtained [98].

Similarity of substrate and inhibitor binding sites of enzymes in the crystal and in solution can often be demonstrated by a comparison of NMR and X-ray data, and

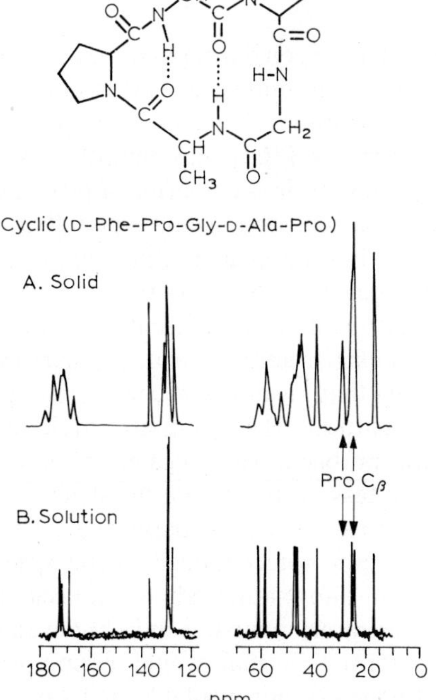

Figure 29. ^{13}C-NMR spectra of cyclo-(D-Phe-Pro-Gly-D-Ala-Pro). (A) Polycrystalline sample. (B) Peptide in C^2HCl_3 solution. The arrows point to the Pro C_β resonances. Reprinted, with permission, from [96], © 1982, Annual Reviews Inc.

provides additional evidence for the similarity of the structures in their general features [99,100].

Studies of conformational changes in response to ligand binding or changes in solvent conditions are among the most frequent applications of high-resolution NMR to biochemical problems. Among the most extensive and successful have been the studies of structural changes in hemoglobin upon oxygen binding [101]. Intermediate conformational states, not obvious from crystallographic studies of the two end-states (oxy- and deoxyhemoglobin), were shown to exist in partially oxygenated hemoglobin species. In general, the interpretation of an observed spectral change in either a protein or a nucleic acid requires, as already noted, that it be shown that the spectral change does not reflect a direct contact between the perturbing agent and the affected residue. The reverse can be shown with relative ease either by the observation of an NOE or of an effect from a suitably chosen paramagnetic ligand. However, failure to observe an NOE may result from internal motion and cannot be taken as evidence for lack of contact. Arguments from the lack of a paramagnetic perturbation are considerably stronger, provided that direct contact with other residues can be demonstrated in the same spectrum. The strongest argument that a spectral change reflects a change in macromolecular conformation can be advanced when both the three-dimensional structure and the assignments are completely known, making it obvious that the change is occurring at a site distant from the site of the primary interaction.

Perhaps the most significant general finding that has emerged from high-resolution NMR studies of macromolecular structure is that these structures have many degrees of internal motional freedom in solution. This conclusion, first based on ligand-binding studies to different parts of protein structures [102], has recently been documented in extensive studies of aromatic ring rotation in the interior of proteins [103,104], studies of ^{1}H–^{2}H exchange discussed in the next section, and especially by relaxation studies of proteins and nucleic acids, both in solution and in the solid state [82,105]. While the qualitative conclusion that extensive internal motions on time scales ranging from picoseconds to milliseconds exist in all macromolecules is readily apparent from such characteristics of macromolecular spectra as relatively narrow line widths and the appearance of simple multiplets that result from time averaging, where rigid structures would yield more complex patterns, the quantitative description of these motions presents a very difficult problem. Here again, as in all interpretations of spectroscopic measurements, the task is to separate the structural and motional contributions reflected in the measured spectroscopic parameters. This is generally possible only by introducing information not contained in the spectroscopic measurement itself. Since experimental information about molecular structure and dynamics obtained by other methods rarely compares in richness and detail to that obtainable by NMR, reliable information that would permit an unambiguous interpretation of NMR data is frequently not available and one is limited to interpretations that have to be based on unverifiable hypotheses. It is hardly surprising that numerous controversies have resulted from this state of affairs. The reader interested in this subject is referred to the more extensive discussions in recent reviews [61,82,106].

(e) Dynamics of proteins and nucleic acids

(i) Hydrogen exchange between solvent and biopolymers

In small molecules, protons that are attached to an oxygen or nitrogen often exchange with solvent water protons rapidly, on the NMR time scale, such that –OH and –NH lines merge with the large solvent resonance (see Section 2(j) concerning chemical exchange). However, in proteins and nucleic acids potentially exchangeable protons may be protected to various degrees from solvent molecules. Protection is afforded by steric hindrance of solvent access to the exchange site, or by formation of intramolecular hydrogen bonds, or by a combination of both (the two are obviously interrelated). NMR methods permit the rates of exchange of particular protons to be measured. Combination of such measurements with knowledge of the molecular structure can yield insights into the motions of the molecule in question. A prerequisite for such investigations is the ability to assign resonances to particular exchangeable protons. Imino protons of transfer RNA [107,108] and double-stranded DNA fragments [109], among nucleic acids, and amide protons of BPTI [62], among proteins, have been the most fully characterised. In this section we consider a few of the main findings of these studies. For a more detailed discussion of this subject, see [62,110].

The observed rate of exchange of a particular proton can be considered to be determined by (a) the intrinsic rate of exchange of the proton when exposed to solvent, designated EX_1, and (b) the rate at which the proton is exposed to solvent, as a result of motions within the molecule, designated EX_2. Since NH exchange can be acid- or base-catalysed, the finding that exchange of a proton is pH-independent can be taken as evidence that the latter process dominates. Such evidence has helped demonstrate that exchange rates for NH protons in helices of tRNA [107] and short DNA duplexes [109] are determined by the rate of opening of the helix ('breathing'); exchange occurs every time the helix opens. In these studies, exchange rates were measured by saturation recovery experiments, in which the apparent longitudinal relaxation times (T_1) of the exchanging protons are determined – possible because relaxation rates and hydrogen-exchange rates are of comparable magnitude. In the saturation recovery experiment, conversion of unsaturated solvent protons to NH protons due to exchange will increase the apparent rate at which the NH proton spin population recovers after selective saturation. If the true longitudinal relaxation rate is known, its subtraction from the apparent relaxation rate yields the rate of helix opening. However, there are difficulties in accurately assessing the true longitudinal relaxation rate, leading to uncertainty in measurements of absolute NH proton-exchange rates [107,109]. Relative exchange rates can be determined with less ambiguity, if the true longitudinal relaxation rates are considered to be similar among the different protons under comparison. Such comparisons have indicated higher activation energies for helix-coil opening toward the interior of a tRNA helix [107], and faster opening kinetics of a central TATA segment of a DNA dodecamer duplex, compared to an AATT segment [109], as predicted by other stability studies.

Hydrogen exchange in proteins is a more complex problem, because of the larger number of exchangeable protons, together with the larger number of types of conformation. Hydrogen exchange in proteins is usually studied by following decreases in the intensities of individual proton resonances following dissolution of the protein in ^2H_2O. Half-times for decay may vary from <1 second (not observable in this manner) to several months. The work of Wagner, Wutrich and colleagues on BPTI represents the only comprehensive study of amide exchange in proteins by NMR (see Ref. 62). They found evidence for exchange of internal amide protons by local unfolding mechanisms. Distinction between exchange due to an EX_2 mechanism from exchange due to EX_1 was possible for some protons. For adjacent protons having similar exchange rates, if exchange is correlated (i.e., when a given motion exposing the protons occurs, both enchange; EX_1) there will be no loss in the magnitude of the NOE between these protons, as they exchange with solvent deuterons. For an EX_2, the NOE will gradually disappear as remaining amide protons get more and more deuteron neighbors. Such an approach permitted observation of transitions from EX_2 to EX_1 as conditions were varied. They also found that protons in the β-sheet core exchanged more slowly than protons in α-helices. It is not clear if this is a general result.

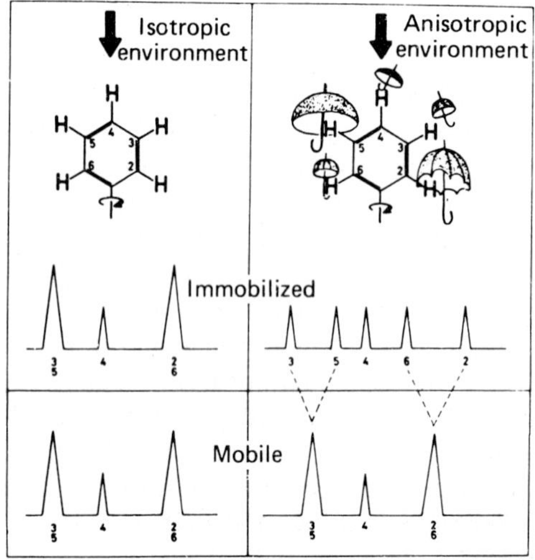

Figure 30. Schematic illustration of a phenylalanine ring in the polarizing magnetic field of a NMR spectrometer. Three limiting situations may be distinguished: (i) In an isotropic environment, the 2-fold symmetry of the covalent ring structure is manifested in the NMR spectrum by the chemical shift equivalence of the symmetry-related protons 2 and 6, and 3 and 5, respectively. (ii) In the interior of a globular protein, the individual aromatic protons have in general different nearest neighbor atoms, and hence are shielded differently against the external magnetic field. At the upper right, the magnetic field is indicated by the arrow, and differently sized umbrellas represent different shielding. As a result, the ring symmetry is masked by the asymmetric environment and different NMR chemical shifts will generally prevail for all the aromatic protons. (iii) If the phenylalanine ring in the protein interior is mobile and rotates rapidly about the C^{β}–C^{γ} bond, each of the symmetry-related 2,6- or 3,5-protons spends equal periods of time in the different environments, and hence the influence of the nearest neighbors is averaged out (from [115]).

(ii) Motion of aromatic side chains in proteins

The side chains of the amino acids tyrosine and phenylalanine contain pairs of chemically equivalent ring protons at the *ortho* and *meta* positions. When buried in the interior of proteins, these chemically equivalent protons may be shielded from the external magnetic field to different extents, and so give rise to separate lines (Fig. 30). If, however, there is motion of the ring such that both chemically equivalent protons spend equal amounts of time in the different magnetic environment, the separate lines will broaden or collapse to a single line (as described in Section 2(j)) (Fig. 30). Figure 31 shows the increase in motion of phenylalanine and tyrosine in BPTI with increasing temperature. A particularly interesting point is the fact that the aromatic side chains in BPTI appear to be tightly packed, as deduced from X-ray crystallographic studies,

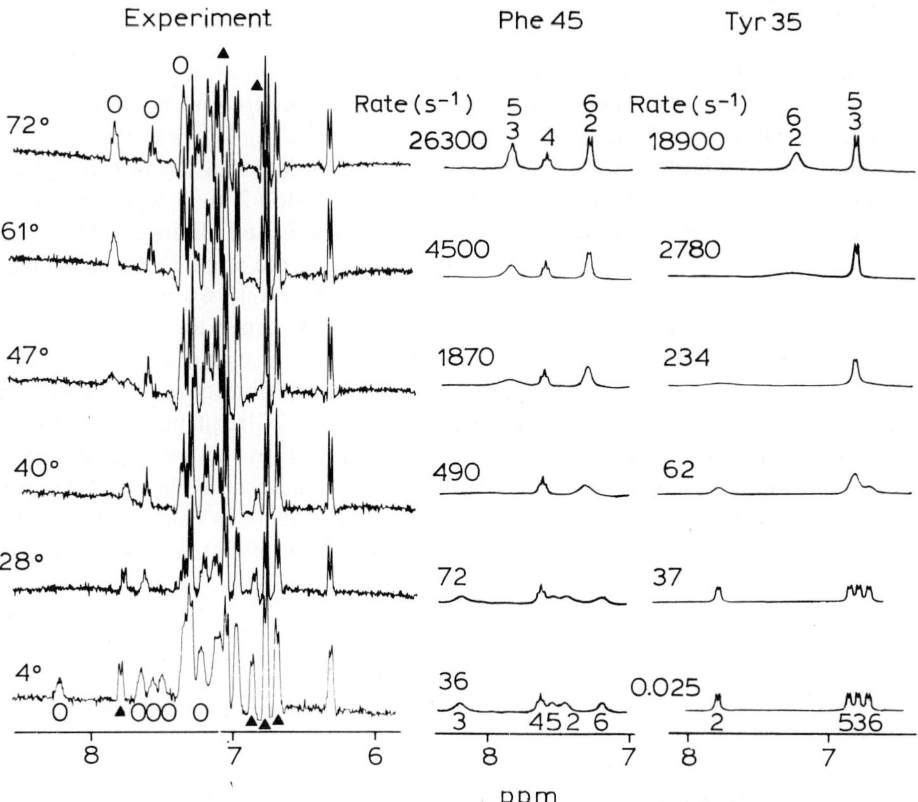

Figure 31. Temperature dependence of the aromatic resonances in the 360 MHz ^{1}H-NMR spectrum of BPTI. For Tyr 35 and Phe 45, the spectra are individually simulated and the flip rates at different temperatures obtained from the best fit with the experimental data are indicated. In the experimental spectrum at 4°C, the resonances of four protons of the Phe 45 (○) and two protons of Tyr 35 (▲) are recognized readily, whereas the other lines are masked by resonances of the other aromatics in the protein. Most of the resonance lines of Phe 45 and Tyr 35 are also resolved in the spectra at higher temperatures and the transitions from slow to rapid sign flipping is readily apparent (from [115]).

such that neighboring groups must move aside to accommodate motion of the aromatic groups [62]. That the chemical shift of the exchange averaged 2,6- and 3,5-proton resonances of the tyrosine and phenylalanine residues at higher temperatures described in Figure 31 corresponds very nearly to the average of the chemical shifts for the individual protons, together with the knowledge that 'free' rotation of these moieties would most likely be inhibited by neighboring structures, suggests that the rotational motions of these aromatic rings consist of '180° flips', in which the chemically equivalent protons exchange magnetic environment. Activation energies for these motions, determined from Arrhenius plots of rotation rates deduced by these NMR experiments, were found to be of comparable magnitude to values for the energy barriers to perturbations about the aromatic residues, obtained by semi-empirical calculation based on the crystal structure and energy minimization procedures [111,112].

(iii) Information from relaxation data

Relaxation data (magnitudes of T_1, T_2 and the NOE) contain information on motions of frequency near the Larmor frequency, i.e., $\approx 10^8\ s^{-1}$. The phenomenon of relaxation of populations of nuclear spins was considered in Section 2(d), and mechanisms of nuclear relaxation were described in Section 2(h). It was noted that dipolar interactions usually dominate the relaxation of ^{13}C and ^1H in a C–H group. For such a system, the relaxation rate will depend on (a) the strength of the dipole-dipole interaction, which will decrease with the sixth power of the dipole-dipole separation (r^6), and (b) the abundance of magnetic fields generated by the moving dipole having a frequency near the Larmor frequency. Concerning the various frequencies of motions, for a population of simple identical objects, such as a sphere, diffusing in a solvent, one might expect there to be a range of frequencies of motion – with higher frequency motions being less common than those of lower frequency – such as is shown in Figure 32. Such a curve is called a spectral density function, designated $J(\omega)$. For lighter or smaller spheres, or spheres in a less viscous solvent, the upper frequency limit for motions will be higher (Fig. 32) (note that the total energy in

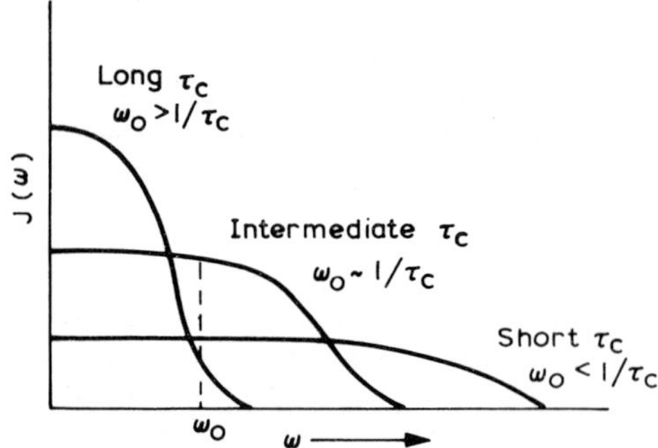

Figure 32. Spectral density functions for fast, intermediate and slow motions.

the sample, and so the area under the curves in Figure 32, is constant). The curves in Figure 32 are characterised by the parameter, τ_c, known as the correlation time, the time constant for the exponential decay process by which the positions or orientations of a molecule at times t and $t+x$ become completely uncorrelated. For water τ_c is $\approx 10^{-12}$ second, for a spherical molecule of molecule weight 2×10^4 τ_c is $\approx 10^{-8}$ second. Figure 32 also shows that, at the indicated Larmor frequency, relaxation through dipole-dipole interactions will be greatest at intermediate values of τ_c, the abundance of motions of frequency close to ω_0 being lower for both the long and short cases. This accounts for the variation in T_1 with correlation time shown in Figure 33 [113]. Note that the curves for T_1 and T_2 in Figure 33 are identical when τ_c is small. However, unlike T_1, T_2 continues to decrease as τ_c increases. This result can be considered to result from the increase in the probability of mutual spin flips as neighboring spins are 'held' together (not disturbed by motion) for longer periods of time.

The nuclear Overhauser effect, a result of dipolar interactions (Section 2(i)) is also dependent on motion. Referring to Figure 13, the NOE can be seen to be initiated during the increase in the population sizes at energy levels E_3 and E_4, $N_{\alpha\beta}$ and $N_{\beta\beta}$ respectively, that occurs when resonance B is saturated. If $W_0 \cdot N_{\alpha\beta} < W_2 \cdot N_{\beta\beta}$, then $N_{\alpha\alpha}$ will increase faster than $N_{\alpha\beta}$ decreases, so the signal from A will increase relative to the equilibrium signal (positive NOE). If $W_0 \cdot N_{\alpha\beta} > W_2 \cdot N_{\beta\beta}$, the signal from A ($N_{\alpha\alpha} + N_{\alpha\beta} - N_{\beta\beta} - N_{\beta\alpha}$) will decrease (negative NOE). The manner in which the NOE varies with motion is a result of variation of the transition probabilities W_0 and W_2 with motion. W_2 describes the probability for changes in the z components of magnetization of A and B, simultaneously, a T_1 process. Such transitions become increasingly probable when the spins are moving at frequencies close to the Larmor frequency, as described above. W_0 concerns the probability of mutual spin flips, a T_2 process. As described above, this process becomes increasingly probable as molecular motions become slower and

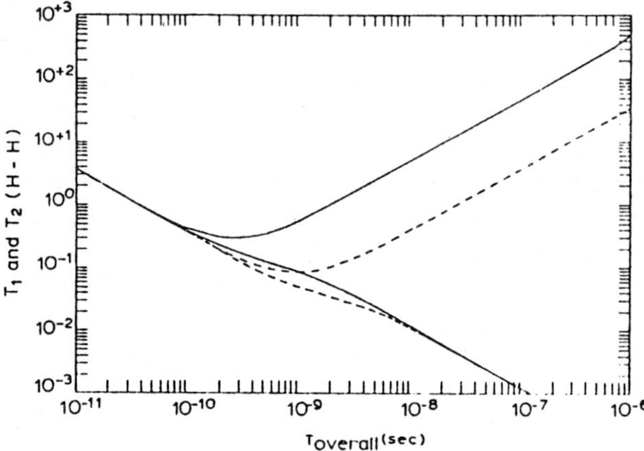

Figure 33. Dependence of the relaxation times T_1 and T_2 in seconds on τ_c at two different frequencies (---, 360 MHz; ———, 100 MHz) (from [61]).

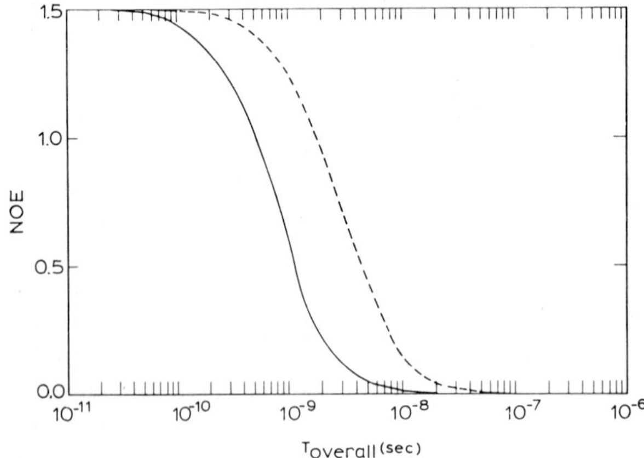

Figure 34. Dependence of NOE $= 1 + \eta$ on τ_c at two different frequencies; ——, 360 MHz; ---, 100 MHz (from [61]).

slower. Moreover, as $T_2 \rightarrow T_1$, $W_0 \rightarrow 0$ as only T_1 processes contribute to T_2. Thus, as shown in Figure 34, the NOE will decrease as molecular motion slows down, since $W_0 \gg W_2$; at high motional frequencies, the maximum NOE is observed, since $W_0 \approx 0$.

The above describes the fundamental processes that determine the relaxation behavior of spins due to dipolar interactions. Thus, magnitudes of the relaxation parameters T_1, T_2 and the NOE for a spin under consideration are determined by the number, strength and distance of neighboring dipoles, and the abundance of motions of these dipoles near the Larmor frequency of the nucleus. Quantitative determination of the contribution of molecular motions to the relaxation parameters of spins in macromolecules is fraught with complications, which we describe here in brief (see Ref. 61 for a more complete discussion).

One severe problem in ^{1}H-NMR is the dipole distance factor (r^6); another is the number of neighboring dipoles. Unless a very high resolution structure is known, neither can be determined. Moreover, even a high resolution crystal structure gives no information about how relative motions within the macromolecule, and so r^6 terms and number of neighboring dipoles change over time. A further problem is the fact that in a macromolecule nearest neighbors about a spin are surrounded by their own nearest neighbors. There is consequentially cross-relaxation (Section 2(i)) between some or all of these protons, which will also depend on the exact geometry within the molecule. Cross-relaxation allows spin energy to be dissipated by coupling throughout the macromolecule, ultimately to an efficient sink such as the solvent. This is called spin diffusion, and this process tends to equalise the observed relaxation times, so obscuring information about different motions in the molecule.

There is less of a problem in ^{13}C relaxation, at least for C–H bonds, because the geometry is fixed (bond length, ≈ 1.1 Å) and the directly bonded protons will be the only ones sufficiently close to cause substantial relaxation. In this case, relaxation data can be analysed without knowing precise geometries. It is in principle possible to resolve motions of differing frequency by measuring the relaxation parameters T_1, T_2 and the NOE at different field strengths. For example, if there is only one motion of frequency 0.1–10 τ, T_1 and T_2 measured at 100 and 360 MHz will change in the manner given in Figure 33; if there is more than one motion near this frequency range T_1 and T_2 will vary differently with magnetic field strength. The application of this principle, with reference to particular possible motions, is outside the scope of this chapter (see Ref. 82). The problems are, however, rather clear – uncertainties in the accuracy of ^{13}C relaxation data, together with difficulties in obtaining data at several field strengths, such that the best one can expect is to resolve 2–3 motions contributing to relaxation. For protein studies, one of these motions will be the overall tumbling of the macromolecule, and therefore biologically rather uninteresting. Other motions detectable may include rotation of methyl groups, again motions not expected to be

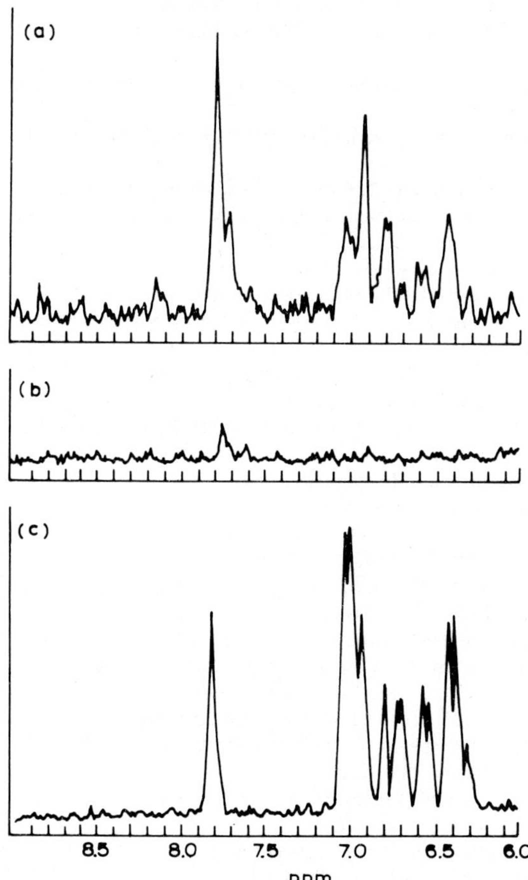

Figure 35. Resolution-enhanced ^1H-NMR spectra at 360 MHz of the 6–9 ppm region of (a) intact *lac* repressor protein; (b) T core; (c) headpiece (from [114]).

critical in the function of macromolecules. The most interesting results obtained from examination of relaxation parameters concerns identification of motions of segments within macromolecules. Such a conclusion is possible simply when observed line widths for particular resonances of a macromolecule are much narrower than expected if the molecule were a rigid object. An example of segmental flexibility in a large protein is the tetrameric *lac* repressor (Fig. 35), where the N-terminal region susceptible to tryptic cleavage is a flexible domain attached to a relatively immobile 'core' [114]. Even so, it is not clear what role these kinds of motion play in the function of macromolecules.

References

1 Slichter, C.P. (1978) Principles of Magnetic Resonance, 2nd Edn., Springer, New York.
2 Farrar, T.C. and Becker, E.D. (1971) Pulse and Fourier Transform NMR: Introduction to Theory and Methods, Academic Press, New York.
3 Becker, E.D. (1980) High Resolution NMR. Theory and Chemical Applications, 2nd Edn., Academic Press, New York.
4 Noggle, J.H. and Schirmer, R.F. (1975) The Nuclear Overhauser Effect: Chemical Applications, Academic Press, New York.
5 Abraham, R.J. and Loftus, P. (1978) Proton and Carbon-13 NMR Spectroscopy, Heyden, London.
6 Kaplan, J.I. and Fraenkal, G. (1980) NMR of Chemically Exchanging Systems, Academic Press, New York.
7 Sandstrom, J. (1982) Dynamic NMR Spectroscopy, Academic Press, London.
8 Freeman, D., Bartlett, S., Radda, G. and Ross, B. (1983) Biochim. Biophys. Acta 726, 325–336.
9 Gadian, D.G. (1982) Nuclear Magnetic Resonance and its Application to Living Systems, Oxford University Press, New York.
10 Burt, G.T., Cohen, S.M. and Barany, M. (1979) Annu. Rev. Biophys. Bioeng. 8, 1–25.
11 Gadian, D.G. and Radda, G.K. (1981) Annu. Rev. Biochem. 50, 69–83.
12 Shulman, R.G., Brown, T.R., Ugurbil, K., Ogawa, S., Cohen, S.M. and Den Hollander, J.A. (1978) Science 205, 160–166.
13 Scott, A.I. and Baxter, R.L. (1981) Annu. Rev. Biophys. Bioeng. 10, 151–174.
14 Roberts, J.K.M. and Jardetzky, O. (1980) Biochim. Biophys. Acta 639, 53–76.
15 Roberts, J.K.M. (1983) Annu. Rev. Plant Physiol. 35, 375–386.
16 Dawson, M.J., Gadian, D.G. and Wilkie, D.R. (1977) J. Physiol. 267, 703–735.
17 Ugurbil, K., Shulman, R.G. and Brown, T.R. (1979) in Biological Applications of Magnetic Resonance (Shulman, R.G., Ed.), pp. 537–589, Academic Press, New York.
18 Cohen, S.M., Shulman, R.G. and McLaughlin, A.C. (1979) Proc. Natl. Acad. Sci. U.S.A. 76, 1603–1607.
19 Brown, F.F., Cambell, I.D., Kuchel, P.W. and Rabenstein, D.L. (1977) FEBS Lett. 82, 12–16.
20 Chance, B., Eleff, S. and Leigh, J.S. (1980) Proc. Natl. Acad. Sci. U.S.A. 77, 7430–7434.
21 Martin, M.L., Martin, G.J. and Delpeuch, J.-J. (1980) Practical NMR Spectroscopy, Heyden, London.
22 Veech, R.L., Lawson, J.W.R., Cornell, N.W. and Krebs, M.A. (1979) J. Biol. Chem. 254, 6538–6547.
23 Moon, R.B. and Richards, J.H. (1973) J. Biol. Chem. 248, 7276–7278.
24 Stidham, M.A., Moreland, D.E. and Siedow, J.N. (1983) Plant Physiol. 73, 517–520.
25 Brown, F.F. and Campbell, I.D. (1976) FEBS Lett. 65, 322–326.
26 Roberts, J.K.M., Wade-Jardetzky, N.G. and Jardetzky, O. (1981) Biochemistry 20, 5389–5394.
27 Roberts, J.K.M., Ray, P.M., Wade-Jardetzky, N.G. and Jardetzky, O. (1980) Nature 283, 870–872.
28 Ugurbil, K., Hohnsen, H. and Shulman, R.G. (1979) Proc. Natl. Acad. Sci. U.S.A. 76, 2227–2231.
29 Roberts, J.K.M., Wemmer, D., Ray, P.M. and Jardetzky, O. (1982) Plant Physiol. 69, 1344–1347.
30 Bailey, I.A., Williams, S.R., Radda, G.K. and Gadian, D.G. (1981) Biochem. J. 196, 171–178.

31 Cohen, S.M., Shulman, R.G. and Mclaughlin, A.C. (1979) Proc. Natl. Acad. Sci. U.S.A. 76, 4808–4812.
32 McConnell, H.M. and Thompson, D.D. (1957) J. Chem. Phys. 26, 1189–1196.
33 Forsén, A. and Hoffman, R.A. (1963) J. Chem. Phys. 39, 2892–2901.
34 Forsén, A. and Hoffman, R.A. (1964) J. Chem. Phys. 40, 1189–1196.
35 McConnell, H.M. (1958) J. Chem. Phys. 28, 430–431.
36 Brown, T.R., Ugurbil, K. and Shulman, R.G. (1977) Proc. Natl. Acad. Sci. U.S.A. 74, 5551–5553.
37 Alger, J.R., Den Hollander, J.A. and Shulman, R.G. (1982) Biochemistry 21, 2957–2963.
38 Roberts, J.K.M., Wemmer, D. and Jardetzky, O. (1984) Plant Physiol. 74, 632–639.
39 Gadian, D.G., Radda, G.K., Brown, T.R., Chance, E.M., Davison, M.J. and Wilkie, D.R. (1981) Biochem. J. 194, 215–228.
40 Nunnally, R.D. and Hollis, D.P. (1979) Biochemistry 18, 3642–3646.
41 Matthews, P.M., Bland, J.L., Gadian, D.G. and Radda, G.K. (1982) Biochim. Biophys. Acta 721, 312–320.
42 Cohen, S.M., Rognstad, R., Shulman, R.G. and Katz, J. (1981) J. Biol. Chem. 256, 3428–3432.
43 Ugurbil, K., Brown, T.R., Den Hollander, J.A., Glynn, P. and Shulman, R.G. (1978) Proc. Natl. Acad. Sci. U.S.A. 75, 3742–3746.
44 Den Hollander, J.A., Brown, T.R., Ugurbil, K. and Shulman, R.G. (1979) Proc. Natl. Acad. Sci. U.S.A. 76, 6096–6100.
45 Styles, P., Grathwohl, G. and Brown, F.F. (1979) J. Magn. Res. 35, 329–336.
46 Eakin, R.T., Morgan, L.O., Gregg, C.T. and Matwiyoff, N.A. (1972) FEBS Lett. 28, 259–264.
47 Ezra, F.S., Lucas, D.S., Mustacich, R.V. and Russell, A.F. (1983) Biochemistry 22, 3841–3849.
48 Mackenzie, N.E., Hall, J.E., Seed, J.R. and Scott, A.I. (1982) Eur. J. Biochem. 121, 657–661.
49 Cohen, S.M., Ogawa, S. and Shulman, R.G. (1979) Proc. Natl. Acad. Sci. U.S.A. 76, 1603–1607.
50 Cohen, S.M. and Shulman, R.G. (1980) Phil. Trans. Roy. Soc. Lond. B289, 407–411.
51 Cohen, S.M., Glynn, P. and Shulman, R.G. (1981) Proc. Natl. Acad. Sci. U.S.A. 78, 60–64.
52 Neurohr, K.J., Barrett, E.J. and Shulman, R.G. (1983) Proc. Natl. Acad. Sci. U.S.A. 80, 1603–1607.
53 Matwiyoff, N.A. and Needham, T.E. (1972) Biochem. Biophys. Res. Commun. 49, 1158–1164.
54 Schaefer, J., Stejskal, E.O. and Beard, C.F. (1975) Plant Physiol. 55, 1048–1053.
55 Schaefer, J., Skokut, T.A., Stejskal, E.O., McKay, R.A. and Varner, J.E. (1981) Proc. Natl. Acad. Sci. U.S.A. 78, 5978–5982.
56 Skokut, T.A., Varner, J.E., Schaefer, J., Stejskal, E.O. and McKay, R.A. (1982) Plant Physiol. 69, 308–313.
57 Schaefer, J., Skokut, T.A., Stejskal, E.O., McKay, R.A. and Varner, J.E. (1981) J. Biol. Chem. 256, 11574–11579.
58 Schaefer, J. and Stejskal, E.O. (1979) in Topics in Carbon-13 Spectroscopy, Vol. 3 (Levy, G.C., Ed.), pp. 283–324, Academic Press, New York.
59 Brindle, K.M., Boyd, J., Campbell, I.D., Porteus, R. and Soffe, N. (1982) Biochem. Biophys. Res. Commun. 109, 864–871.
60 Dwek, R.A. (1973) Nuclear Magnetic Resonance in Biochemistry: Applications to Enzyme Systems, Oxford University Press (Clarendon), London.
61 Jardetzky, O. and Roberts, G.C.K. (1981) NMR in Molecular Biology, Academic Press, New York.
62 Wagner, G. (1983) Q. Rev. Biophys. 16, 1–57.
63 Highsmith, S. and Jardetzky, O. (1982) in Ciba Foundation Symposium 93, Mobility and Function in Proteins and Nucleic Acids, Pitman, London.
64 Wagner, G. and Wüthrich, K. (1982) J. Mol. Biol. 155, 347–366.
65 Delepierre, M., Dobson, C.M. and Poulsen, F.M. (1982) Biochemistry 21, 4756–4761.
66 Cohen, J.S. and Jardetzky, O. (1968) Proc. Natl. Acad. Sci. U.S.A. 60, 92–99.
67 Snyder, G.H., Rowan, R., Karplus, S. and Sykes, B.D. (1975) Biochemistry 14, 3765.
68 Wagner, G., De Marco, A. and Wüthrich, K. (1975) J. Magn. Res. 20, 565.
69 Jones, C.R., Sikakana, C.T., Henir, S.P., Kuo, M.C. and Gibbons, W.A. (1978) J. Am. Chem. Soc. 100, 5960.
70 Jeener, J. (1971) Ampere International Summer School, Basko Polje, Yugoslavia.
71 Ernst, R.R. (1975) Chimia 29, 179.

72 Bax, A. (1983) J. Magn. Res. 53, 149–153.
73 Nagayama, K. (1981) Adv. Biophys. 14, 139–204.
74 Wagner, G. and Wüthrich, K. (1979) J. Magn. Res. 33, 675.
75 Bothner-By, A.A. and Noggle, J.H. (1979) J. Am. Chem. Soc. 101, 5152.
76 Poulsen, F.M., Hoch, J.C. and Dobson, C.M. (1980) Biochemistry 19, 2597.
77 Markley, J.L., Putter, I. and Jardetzky, O. (1968) Science 161, 1249.
78 Roberts, G.C.K. and Jardetzky, O. (1970) Adv. Protein Chem. 24, 447.
79 Moore, G.R. and Williams, R.J.P. (1980) Eur. J. Biochem. 103, 503–512.
80 Markley, J.L. (1975) Biochemistry 14, 3546.
81 Karplus, M. (1959) J. Chem. Phys. 30, 11.
82 Jardetzky, O. (1981) Accts. Chem. Res. 14, 291–298.
83 Wüthrich, K., Wider, G., Wagner, G. and Braun, W. (1982) J. Mol. Biol. 155, 311–346.
84 Hare, D.R. and Reid, B.R. (1982) Biochemistry 21, 5129–5135.
85 Hare, D.R. and Reid, B.R. (1982) Biochemistry 21, 1835–1842.
86 McDonald, C.C. and Phillips, W.D. (1969) Biochem. Biophys. Res. Commun. 35, 43.
87 McDonald, C.C. and Phillips, W.D. (1969) J. Am. Chem. Soc. 91, 1513.
88 Kaptein, R. (1978) Nature 274, 293.
89 Kaptein, R. (1978) in Nuclear Magnetic Resonance Spectroscopy in Molecular Biology (Pullman, B., Ed.), p. 211, Reidel Publishers, Dordrecht, The Netherlands.
90 McDonald, C.C. and Phillips, W.D. (1967) in Magnetic Resonance in Biological Systems (Ehrenberg, A., Malmstrom, B.G. and Vanngard, T., Eds.), p. 3, Pergamon, Oxford.
91 McDonald, C.C. and Phillips, W.D. (1967) J. Am. Chem. Soc. 89, 6332.
92 Nelson, D.J., Opella, S.J. and Jardetzky, O. (1976) Biochemistry 15, 5552.
93 Ulrich, E.L. and Markley, J.L. (1979) Coord. Chem. Rev. 27, 109.
94 Campbell, I.D., Dobson, C.M., Williams, R.J.P. and Xavier, A.V. (1973) Ann. N.Y. Acad. Sci. 222, 163.
95 Jardetzky, O. and Wade-Jardetzky, N.G. (1980) FEBS Lett. 110, 133–135.
96 Opella, S. J., (1982) Annu. Rev. Phys. Chem. 33, 533–562.
97 Schramm, S., Kinsey, R.A., Kintanor, A., Rothgeb, T.M. and Oldfield, E. (1981) in Stereodynamics of Molecular Systems (Sarma, R.H., Ed.), Vol. 2, pp. 271–286, Adenine Press, New York.
98 Gall, C.M., Cross, T.A., DiVerdi, J.A. and Opella, S.J. (1982) Proc. Natl. Acad. Sci. U.S.A. 79, 101–105.
99 Jardetzky, O. (1965) Proc. Int. Conf. Magnetic Resonance, Tokyo, Japan, N-3-14, 1–4.
100 Meadows, D.H., Roberts, G.C.K. and Jardetzky, O. (1969) J. Mol. Biol. 45, 491–511.
101 Vigiano, G. and Ho, C. (1979) Proc. Natl. Acad. Sci. U.S.A. 76, 3673.
102 Jardetzky, O. (1964) in Advances in Chemistry and Physics, Vol. VII (Duchesne, J., Ed.), p. 499, Interscience, New York.
103 Nelson, D.J., Cozzone, P. and Jardetzky, O. (1976) in Molecular and Quantum Pharmacology (Bergman, E.D. and Pullman, B., Eds.), p. 501, Reidel Publishers, Dordrecht, The Netherlands.
104 Campbell, I.D., Lindskog, S. and White, A.I. (1977) Biochim. Biophys. Acta 484, 443.
105 Mai, M.T., Wemmer, D. and Jardetzky, O. (1983) J. Am. Chem. Soc. 105, 7149–7152.
106 London, R.E. (1980) in Magnetic Resonance in Biology (Cohen, J.S., Ed.), Vol. 1, pp. 1–69, Interscience, New York.
107 Hurd, R.E. and Reid, B.R. (1980) J. Mol. Biol. 142, 181–193.
108 Patel, D.J. (1978) Annu. Rev. Phys. Chem. 29, 337–362.
109 Patel, Ikuta, S., Kozlowski, S. and Itakura, K. (1983) Proc. Natl. Acad. Sci. U.S.A. 80, 2184–2188.
110 Woodward, C., Simon, I. and Tuchsen, E. (1982) Mol. Cell. Biochem. 48, 135–160.
111 Wagner, G., De Marco, A. and Wüthrich, K. (1976) Biophys. Struct. Mech. 2, 139–158.
112 Hetzel, R., Wüthrich, K., Deisenhofer, J. and Huber, R. (1976) Biophys. Struct. Mech. 2, 159–180.
113 Bloembergen, N., Purcell, E.M. and Pound, R.V. (1948) Phys. Rev. 73, 679.
114 Wade-Jardetzky, N.G., Bray, R.P., Conover, W.W., Jardetzky, O., Giesler, N. and Weber, K. (1979) J. Mol. Biol. 128, 259–264.
115 Wüthrich, K. and Wagner, G. (1979) Trends Biochem. Sci. 3, 227–230.
116 Nagayama, K. and Wüthrich, K. (1981) Eur. J. Biochem. 114, 365–374.

117 Keller, R.M., Bauman, R., Hunziker-Kwik, E.H., Joubert, F.J. and Wüthrich, K. (1983) J. Mol. Biol. 163, 623–646.
118 Highsmith, S., Akasaka, K., Konrad, M., Goody, R., Holmes, K., Wade-Jardetzky, N. and Jardetzky, O. (1979) Biochemistry 18, 4238–4244.
119 Brown, F.F. and Campbell, I.D. (1980) Phil. Trans. R. Soc. Lon. Ser. B 289, 395–406.
120 Willard, H.H., Merritt, L.L., Dean, J.A. and Settle, F.A. (1981) Instrumental Methods of Analysis, 6th Edn., Van Nostrand, New York.
121 Zetta, L., Hore, P.J. and Kaptein, R. (1983) Eur. J. Biochem. 134, 371–376.
122 Kanamori, K. and Robert, J.D. (1983) Acc. Chem. Res. 16, 35–41.

Neuberger/Van Deenen (eds.) Modern Physical Methods in Biochemistry, Part A

CHAPTER 2

Electron spin resonance

ROGER C. SEALY, JAMES S. HYDE
and WILLIAM E. ANTHOLINE

National Biomedical ESR Center, 8701 Watertown Plank Road, Milwaukee, WI 53226, U.S.A.

1. Introduction

From the point of view of the biochemist, there are three major classes of samples that might be investigated with ESR spectroscopy: (1) samples containing transition or lanthanide elements; (2) samples containing free radicals; (3) samples that have been extrinsically labeled by the introduction of nitroxide radical spin labels or spin probes. We provide an overview of these three subjects in this chapter. Basic references are given at the beginning of each section. The material described is by no means comprehensive; the topics reviewed have become very massive subjects. Rather, we have tried to provide information, often from our own work and experience, that we hope will be of general interest.

In the last section of the present paper we offer a selective development of some of the key aspects of ESR instrumentation. Emphasis is placed on the microwave bridge and microwave resonant structure. Most of this material has not been published previously.

There are approximately 2000 papers per year published that contain electron spin resonance data. These have been systematically reviewed since 1973 in the Specialist Periodical Reports series [1]. ESR textbooks are listed in References 2–21. The text by Carrington and McLachlan [2] provides a good theoretical background. For a less mathematical treatment intended for a more biological audience, the book by Knowles et al. [13] is useful.

While we have used a classification based on chemical structure, alternative classifications based on experimental techniques or chemical stability are frequently encountered. These afford a somewhat different and certainly useful perspective of the field, and are outlined below.

(a) Classification with respect to technique

The field of electron spin resonance comprises a range of techniques. The types of experiment that can be carried out are either linear or non-linear, continuous wave (CW) or time-domain.

Linear refers to the response of the spin system to the incident microwave power. If the latter is sufficiently low that microwave power saturation is negligible, the spin system is said to respond linearly. This implies that the ESR signal has an intensity proportional to $P_0^{1/2}$, where P_0 is the incident power, and a lineshape that is independent of power. If the spin system response shows evidence of saturation effects, it is said to respond non-linearly.

CW experiments (sometimes called 'stationary' or 'steady state') are ones in which either no modulations are used, or they are so low in frequency that no spectral complications ensue. (This is only approximately the case if 100 kHz field modulation is employed. This frequency gives rise to modulation sidebands and, under saturating conditions, rapid passage effects.) Time-domain ESR involves monitoring the spin system response as a function of time. Pulse ESR can be divided into two broad categories: the response of spin systems to sequences of microwave pulses (spin echo) and the response of spin systems to step changes in resonance conditions (saturation recovery).

There are three classes of instrumentation for time-domain or transient measurements.

(1) High-frequency field modulation, usually 100 kHz but occasionally 1 MHz, with the time response limited by the band-pass of the receiver (about 5000 Hz, typically). If the signal-to-noise needs to be improved, signal averaging can be used.

(2) Direct detection, which is useful when the repetition rate of the transient signal is greater than about 10^3 per second. A superimposed very low frequency field modulation sometimes yields improved baseline stability.

(3) Direct detection with signal gating. A problem arises when the decay is very fast and the permissible repetition rate is low. This may be the case when exciting a spin system with a high-power laser. The words 'signal-gating' imply that the receiver input is blocked or gated during the long time between pulses. Because the repetition rate is low, the system can be prone to low-frequency noise.

Most conventional ESR is linear and 'approximately' CW. Most commercial spectrometers operate at around 9 GHz, the X-band microwave region, and employ 100 kHz field modulation. However, since the resonance condition can be fulfilled for a variety of frequency-field combinations, many other frequencies are possible. The use of a range of microwave frequencies can be termed multifrequency ESR. Multifrequency ESR can (1) provide a means to increase spectral resolution by varying the interplay of Zeeman and hyperfine interaction (e.g., to separate the spectra of different species); (2) provide a critical test of theoretical simulations; (3) be used to study frequency effects on relaxation times and linewidths; (4) be used to study higher order and state mixing effects. Microwave frequency is thus an important experimental parameter. Although the cost of obtaining an array of microwave bridges has limited its use to date, experiments have been carried out at frequencies from 1 to 35 GHz. The notation L is used for the octave bandwidth 1–2 GHz, S for 2–4, X for 8.5–10 and Q for 35 GHz. Octave bandwidth spectrometers are technically feasible to about X-band. Commercial Q-band ESR bridges usually have 10% bandwidths.

Multiresonance techniques are non-linear. The two double resonance techniques in

ESR spectroscopy are electron-nuclear double resonance (ENDOR) and electron-electron double resonance (ELDOR). In ENDOR, two radio frequencies are incident on the sample, one the ESR frequency, which is at a saturating level, the other at the frequency of nuclei that are coupled to the electron. An ENDOR display is of ESR signal height versus nuclear radio frequency. When nuclear resonance is induced, the ESR signal height changes. Thus, ENDOR uses ESR to detect nuclear magnetic resonance (NMR). Its primary utility is in improving resolution. Experiments can be done on liquids, crystals and powders. ELDOR involves two microwave frequencies corresponding to two magnetic resonance features. One is at a high level (the pump frequency) and the other at a low level (the observing frequency). The effect is transfer of saturation from pump to observing. Its primary utility is in the study of relaxation processes. Various kinds of triple resonance also are possible. Multiresonance techniques can be CW, pulse or rapid passage. So-called passage effects occur from rapid sweeps of the magnetic field inducing a transient response of the spin system.

Multifrequency ESR experiments can be carried out under non-linear and pulse conditions. Similarly, multiresonance experiments can be carried out as a function of microwave frequency.

Many ESR experiments have an additional instrumental aspect. This can involve generation of paramagnetic species and/or studying the spin system response as a function of temperature, chemical environment, etc. Rapid mixing with stop-flow, studies of photoresponse, electrochemical response, etc., are all possible.

(b) Classification with respect to order, motion and stability

In biological ESR, the paramagnetic species that one encounters can have a wide range of chemical lifetimes. Reflecting this, measurements have been made over timescales ranging from around 20 ns (the limiting time-resolution of a pulse ESR spectrometer) to many hours or days. Towards one end of the lifetime scale are many free radicals, since free radicals in general are transient species. For such species a number of experimental approaches have been developed to generate them in concentrations sufficient for detection in an ESR experiment. Towards the other end of the scale are many paramagnetic metal ions: metal ion spin probes and those found in metalloenzymes.

Of course, there are exceptions to these generalizations. Nitroxide spin labels and probes are long-lived free radicals; steric hindrance around the nitroxide group greatly restricts their reactivity. Even so, in the presence of reducing agents such as ascorbate and thiols, nitroxides are easily reduced to the corresponding hydroxylamines. Because of this, nitroxides tend not to be stable indefinitely in complex systems. In cellular systems, for example, chemical or enzymatic reduction can place a restriction on the time period over which experiments can be carried out. In addition, at high hydrogen-ion concentrations the nitroxide group protonates to give a radical that decays much more rapidly than the form in which the nitroxide function is neutral. The possibility of reactions of this kind must always be borne in mind in designing experiments. Equally, not all metal-ion complexes are long-lived. Transient complexes occur as intermediates in a number of enzyme reactions; the 'fast' and 'slow' species detected from xanthine oxidase provide examples of such transient metallo complexes.

Pulse experiments afford the widest possibilities. In such experiments, following the generation of a particular paramagnetic species, its reactions to give diamagnetic products and/or secondary paramagnetic species can be followed. At short (μs) times, spin populations of slowly relaxing free radical species often have not reached Boltzmann equilibrium; this is observed in the form of chemically induced dynamic electron polarization (CIDEP) phenomena, in which resonances are found either in emission or in enhanced absorption. Analysis of CIDEP data can provide information on precursors to the detected radicals, for example, short-lived excited triplets or radical-pair states. Information of this kind is valuable in establishing the details of a reaction mechanism.

Where pulse equipment is unavailable or such experiments are impractical, other approaches can be adopted. Species with lifetimes down to about 1 ms or so can be observed in steady-state concentrations provided that a high rate of formation of the transient can be achieved. However, high formation rates may necessitate the use of flow conditions to maintain concentrations of starting materials. Species with longer lifetimes often can be detected for considerable periods of time in static systems. Whereas steady-state methods cannot provide the direct observations of chemical transformations that are possible with pulse methods, in many situations they can provide the same information in a less direct manner.

Other approaches to studying transient species involve some kind of stabilization. Freeze-quenching and chemical stabilization have been used. Freeze-quenching is quite common for macromolecular radicals and metal ions, whereas chemical stabilization (e.g., by spin trapping) is more usual for small radical species.

Many ESR experiments demand the simultaneous generation of paramagnetic species, either in a pulse or continuously. The need for accessory generating techniques puts an additional set of constraints on a system. Thus, many of the technically more difficult ESR techniques have been applied to long-lived paramagnetic species; the majority of experiments employing ENDOR, ELDOR, pulse ESR, etc., have been carried out in spin labels and metal ions. Nevertheless, pulse techniques have been used to advantage in recent studies of short-lived paramagnetic intermediates in photosynthesis.

2. *Nitroxide radical spin labels and spin probes*

Introductory spin-label material appears in Refs. 5 and 13. The two books edited by Berliner [22] contain definitive treatments of selected topics in the field of spin-labeling. Relevant chapters occasionally appear in the Plenum Press series on Biological Magnetic Resonance, also edited by Berliner (see, for example, Ref. 23). The article by Butterfield on Spin-labeling in Disease in Volume 4 of this series will be of interest to many persons. Likhtenshtein [24] provides an overview of many spin-label applications. Several general spin-label chapters that are intended for a biochemical audience have appeared in the Methods in Enzymology series (see, for example, chapters by Jost and Griffith [25] and by Berliner [26]).

(a) Labels and probes

Hindered nitroxides are a class of free radicals that are unusually long-lived. The presence of methyl groups γ to the radical center decreases the rate of self reaction to such an extent that the radical becomes kinetically stable. Hindered nitroxides can be obtained as crystalline solids or pure liquids, which in the absence of other materials are very stable. These materials have found extensive use as spin labels and spin probes. Indeed, almost all spin-label studies have employed nitroxides; spin-probe studies have used either nitroxides or paramagnetic metal ions.

The major use of nitroxides is in making a diamagnetic material paramagnetic, so that it becomes accessible to study by ESR methods. The diamagnetic material could be a polypeptide, a cofactor, a membrane, etc. A second, related, use is to introduce a paramagnetic center into a material that is already paramagnetic; here, the goal might be to obtain information on some region of the material distant from the endogenous paramagnetic center(s) or to investigate the interaction between the center and the nitroxide, for example, to determine molecular distances.

Many authors distinguish between spin-label and spin-probe studies, applying as the criterion whether the nitroxide molecule (itself often loosely called a spin label) is attached covalently or non-covalently in the system. For example, an investigation of a nitroxide covalently bound to a protein would be called a spin-label study. Such a study might provide information on, for example, conformational changes during binding of a substrate, protein aggregation, denaturation, etc. A spin-probe study involves a non-covalently bound nitroxide that provides information (reports) on a particular environment, i.e., the surrounding of the probe. For example, a nitroxide-derivative of a fatty acid intercalated into a membrane can be used to probe the fluidity and polarity of that part of the membrane with which the nitroxide interacts. Alternatively, a hydrophilic nitroxide can be used to report on the concentration of oxygen in an aqueous phase.

Approaches to introduce a spin-label into a system differ from those used for spin probes. Spin-labeling of proteins requires an amino acid side-chain –SH, $-NH_2$ or –OH group to bind the nitroxide to that site [27]. As with other protein modifications, the degree of specificity that can be achieved will depend on the number and accessibility of reactive amino acid resides, the reagent, and the conditions employed. Sometimes it is possible to label a single amino acid site with high specificity, for example, if a molecule contains a single sulfhydryl group. By far the most successful spin-label studies have been carried out on specifically labeled proteins. It becomes exceedingly difficult to analyze and interpret unambiguously data from molecules or mixtures of molecules (for example, proteins in membranes) that are not specifically labeled. The overlap of several nitroxide ESR spectra together with minor changes in response to reagents, as is frequently encountered, can be severe problems.

Introduction of a nitroxide spin probe into a membrane can be accomplished in various ways. One common procedure is to include the nitroxide in a lipid mixture prior to the preparation of liposomes by vortexing or sonication. With preformed synthetic membranes or natural membranes, incubation with either the nitroxide or the nitroxide previously intercalated into liposomes can be effective. The possibility of the nitroxide affecting the gross properties of the membrane must always be considered; high levels of spin probe should be avoided unless control studies show that the same effect is observed with lower levels, where modification of membrane properties can be discounted.

The chemical properties of hindered nitroxides make them attractive for label and probe studies. Apart from its kinetic stability, the nitroxide group is moderately polar, with the unpaired electron spin distributed between the nitrogen and oxygen atoms. To an approximation, the nitrogen hyperfine splitting and g value of the nitroxide reflect the relative amounts of neutral and ionic resonance forms present in the medium. A polar environment will favor the ionic form of the nitroxide which has spin density on nitrogen, resulting in an increased hyperfine splitting and decreased g value. The converse is true for a hydrophobic, non-polar environment. Thus, the magnetic parameters of the nitroxide provide information on the polarity of its immediate environment [28].

Nitroxides with ionizable functional groups often have spectra that are pH-dependent, so that protonation or deprotonation results in observable changes in hyperfine splitting and g value. Such molecules are of potential use as pH indicators [29].

The localization of a nitroxide in a membrane [30] depends upon the relative polarity of the nitroxide moiety and other polar groups that may be present in the molecules. If the nitroxide group is the only polar group, then the molecule will orient itself in the membrane so that the nitroxide is located in the aqueous phase. The cholestane spin label is one such example. If groups more polar than the nitroxide are present, then the nitroxide will be found within the bilayer. For example, doxyl-labeled fatty acids are oriented such that the carboxyl group is in the aqueous layer and the nitroxide is down within the membrane.

(b) Physical properties of spin labels

In this section we consider intramolecular magnetic interactions, relaxation times, and intramolecular motional modes.

Intramolecular magnetic interactions in nitroxides are (1) the nitrogen hyperfine tensor; (2) the g tensor; (3) methyl proton hyperfine tensors; (4) other proton hyperfine tensors; (5) the nitrogen quadrupole coupling; and (6) couplings to trace isotopes, mostly ^{17}O, ^{13}C. One desires to know how these properties depend on solvent and how they change among the various possible nitroxides. Since these properties are averages over various intramolecular motional modes, one can also inquire as to their temperature dependence.

Nitroxide relaxation times include (1) the electron spin lattice or longitudinal relaxation time, T_1; (2) the electron spin-spin or transverse relaxation time, T_2; (3) the nitrogen T_1; (4) methyl proton T_1 values; and (5) other proton T_1 values. In this case one desires to know particularly how these times depend on temperature or viscosity. The dependence on solvent and on molecular structure are important also.

Possible intramolecular motional modes that might in principle affect the magnetic parameters and relaxation times are (1) methyl group rotation; (2) interconversion of axial and equatorial methyls; (3) inversion of the N$\dot{O}$ bond with respect to the CNC plane; (4) interconversion of five-membered rings between twisted and planar conformations; and (5) other ring motions that may exist.

Available information on these physical parameters is summarized below. It will become apparent that major gaps in this information exist. The reader is referred to the books on Spin Labels edited by Berliner [22] for background. Forrester's 874-page tabulation of nitroxide magnetic data is an important contribution [31].

(i) Intramolecular magnetic interactions

The most significant are the Zeeman (g tensor) and nitrogen hyperfine interactions. These depend somewhat on chemical structure, solvent polarity, and temperature. An assumption has been made almost universally that magnetic parameters that are found in frozen solutions may be used to describe liquid phase spectra. This assumption should be recognized and considered carefully in precise work. It may be invalid for several reasons, including (1) altered distribution of nitroxide molecular conformations, (2) temperature-dependent solvent properties, or (3) complicated motional situations, such as fast anisotropic restricted motions superimposed on slow isotropic motions.

Typical values of magnetic parameters are given in Ref. 22, and in the extensive tables of Ref. 31. Recently, we have performed careful simulations of frozen solutions of several nitroxides (Fig. 1) at Q-band in second derivative display [32], and Tables 1–3 contain data from this paper.

Table 1 compares isotropic values of g and A obtained in the liquid phase for several typical nitroxides using a polar (glycerol) and a non-polar (*sec*-butyl benzene) solvent. It is apparent that the solvent dependence of g values would be difficult to use as a measure of polarity, but that the shift of approximately 1.6 G in values of A_{iso} is highly significant.

Table 2 makes a comparison of all components of the g and A tensors obtained in frozen solution. It is interesting that solvent effects on both g_x and A_x are very profound as measured by percent changes. This direction coincides with the NO bond. Features corresponding to the x direction are greatly overlapped and difficult to deconvolute at X-band, but can be measured with good precision at Q-band. As measured by numerical changes, the 3 G change in A_z indicates that it is the best diagnostic reporter of solvent polarity.

Table 3 compares the isotropic values of g and A obtained in two ways – from frozen solution simulation, calculating the traces of the tensors, and from the liquid phase. No significant differences in g values were observed, but the isotropic values of A change by 0.1–0.4 G.

It is sometimes possible to obtain 'rigid limit' magnetic parameters from detailed analysis of liquid phase spectra. An example is the approach of Hyde and Rao [33].

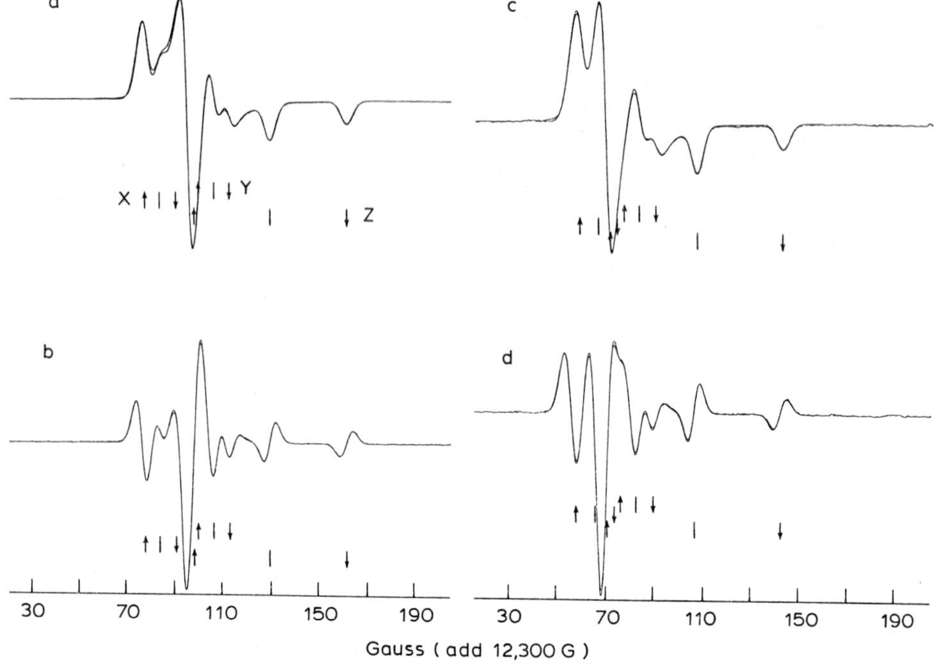

Figure 1. Experimental and simulated Q-band spectra of maleimide spin label. a,b, First and second derivatives in *sec*-butyl benzene at −99°C. c,d, First and second derivatives in glycerol at −72°C. Spectral positions for the nine 'canonical' resonance conditions are indicated. Up and down arrows correspond to $M_I = +1$ and -1 respectively; vertical bars correspond to $M_I = 0$. From [32], with permission.

These effective 'rigid limit' parameters can be expected to differ somewhat from actual rigid limit parameters.

Hyperfine couplings to the protons of spin labels, which are sometimes called 'superhyperfine couplings', have been measured by NMR directly on the free radicals. Key citations are the early papers of Briere et al. [34,35] and Kreilick [36]. The latter author reviewed this method in 1973 [37]. More recently, Volodarsky et al. [23] have provided a relevant review with numerous citations to the Soviet literature. Proton

TABLE 1
Liquid phase magnetic parameters

	sec-Butyl benzene		Glycerol	
	g_{iso}	A_{iso} (G)	g_{iso}	A_{iso} (G)
Tanol	2.00603	15.19	2.00569	16.80
ISL	2.00599	15.22	2.00582	16.83
MSL	2.00600	15.25	2.00570	16.80

For g_{iso}, error = ±1 in last figure. For A_{iso}, error = ±0.2 G.

TABLE 2
Effect of solvent on g and A tensors of maleimide spin label

Solvent	Temperature (°C)	g_x	g_y	g_z	A_x	A_y	A_z
sec-Butyl benzene	−122	2.01004	2.00623	2.00208	6.22	6.86	33.91
Glycerol	−72	2.00891	2.00606	2.00213	8.05	6.51	36.78

Values for A_x, A_y and A_z are in Gauss.

TABLE 3
Comparison of liquid and frozen solution magnetic parameters

Sample	Solvent	Temperature (°C)	g (trace)	g_{iso}	A (trace)	A_{iso}
Tanol	glycerol	−72	2.00564	2.00569	16.69	16.80
	sec-butyl benzene	−122	2.00607	2.00603	15.44	15.19
ISL	glycerol	−69	2.00580	2.00582	17.17	16.83
	sec-butyl benzene	−130	2.00607	2.00599	15.40	15.22
MSL	glycerol	−72	2.00570	2.00570	17.11	16.80
	sec-butyl benzene	−120	2.00611	2.00600	15.66	15.25
AZSL	*sec*-butyl benzene	−123	2.00591	2.00592	13.93	13.78
Doxyl	*sec*-butyl benzene	−114	2.00572	2.00572	14.20	13.99

Values for A and A_{iso} are in Gauss.

hyperfine couplings in frozen solutions can also be determined by ENDOR [38].

In favorable conditions, proton couplings can be determined by simulation of ESR spectra in solution. An example from our own work [39] is given in Figure 2. Here, both simulated and experimental spectra of the radical 3-carbamoyl-2,2,5,5-tetramethyl-3-pyrroline-1-yloxy (CTPO) are shown. The ring proton coupling is about 0.5 G and the 12 methyl proton couplings are about 0.2 G. This accidental factor of 2.5 results in an even rather than an odd number of proton hyperfine lines. However, this ring proton coupling is somewhat temperature-dependent, changing from 0.52 to 0.46 G between 10 and 80°C. At the high temperature, the departure from accidental degeneracy lowers the resolution and changes the number of lines from even to odd.

The temperature dependence of proton couplings is probably a consequence of intramolecular motional modes. It also was reported in the early NMR work of Kreilick [36].

Proton resolution of CTPO is unusually good, with best resolution near 10°C, where the accidental degeneracy is nearly perfect. This resolution is lost when oxygen is present in the system, and the effect has been used by Lai et al. [40] to measure oxygen concentration. This technique is called ‘spin-label oximetry’. Lai’s extensive calibration charts will be found useful for readers interested in this method.

In frozen solution, the proton couplings will be represented by tensors whose

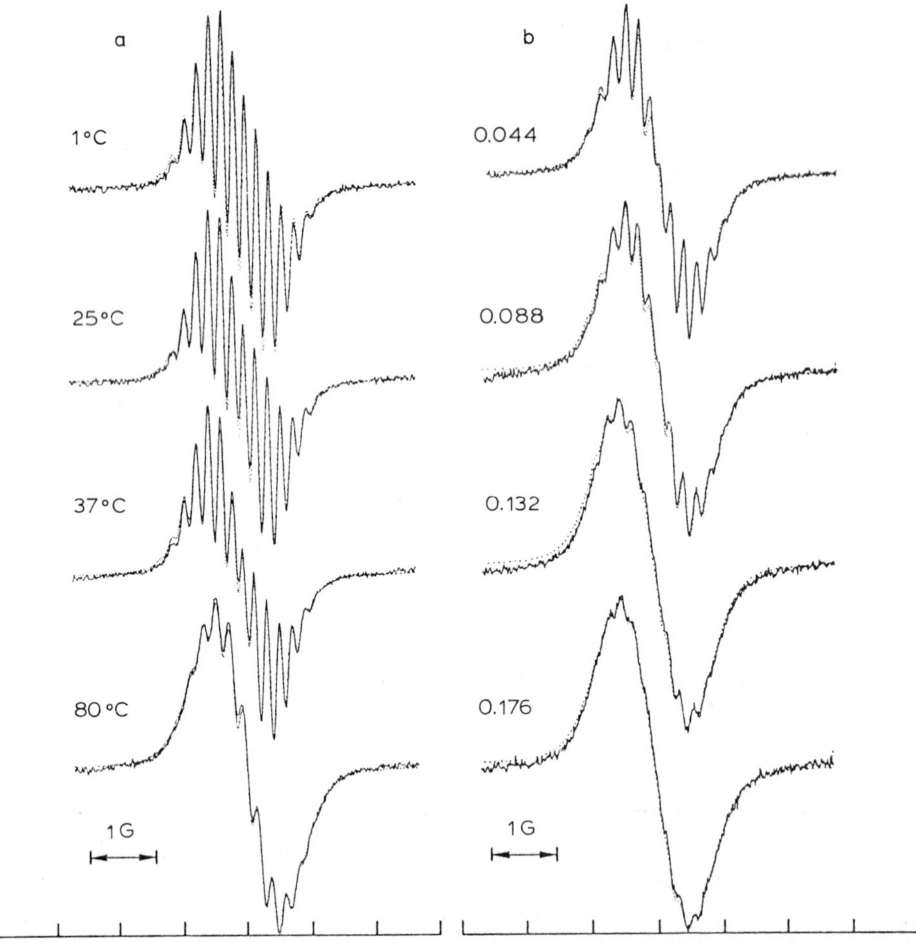

Figure 2. a, Experimental ESR spectra of CTPO in deoxygenated samples as a function of temperature (—). Simulated ESR spectra (···). b, Experimental ESR spectra of CTPO at 37°C as a function of oxygen concentration (—) ($[O_2]=0.2$ mM in air-saturated water at 37°C). Simulated ESR spectra (···) using the same coupling as for Figure 2a, at 37°C, varying only the linewidth. (All data are from the central $M_I=0$ nitrogen hyperfine line.) From [39], with permission.

principle axes will not, in general, coincide with principle axes of the g and nitrogen hyperfine tensor. The combined effect of these anisotropic weak couplings is to give a lineshape that is nearly Gaussian and that shows a variation of about 20% in width with respect to orientation. Often it is desirable to decrease the width by complete deuterium substitution. One would expect the Gaussian width to drop by a factor given by $\mu_D I_H/\mu_H I_D = 3.25$. The actual decrease may be less because of a change in character from Gaussian to Lorentzian lineshape.

Other magnetic interactions in nitroxides are of less significance from the point of view of spin labeling of biological systems. Because of the large number of methyl groups, ^{13}C coupling can be observed, and they can be of sufficient intensity to cause difficulty in careful spectral simulations. Nitrogen nuclear quadrupole coupling in nitroxides have not, to our knowledge, been reported. They are likely to give rise to nuclear state mixing at high microwave frequencies – Q-band and above – resulting in weak but observable forbidden transitions.

(ii) Relaxation times

The dominant spin-lattice relaxation mechanism for nitroxides remains unknown. Experimental saturation recovery measurements of the small nitroxides tanone and tanol in an organic solvent in the liquid phase have been published by Percival and Hyde [41]. The values fall in the 10^{-5}–10^{-6} second range and exhibit an approximately half power dependence on η/T. These authors observed that neither ^{15}N substitution, thereby removing the nuclear quadrupole interaction, nor deuterium substitution affected electron T_1 values.

Kusumi et al. [42] published T_1 values of several nitroxide membrane probes dissolved in liposomes of dimyristoylphosphatidylcholine as a function of $1/T$ (see Fig. 3). The dotted lines indicate the pre- and main phase transition temperatures. The relaxation times change at the phase transition temperature, and therefore reflect an aspect of membrane fluidity. It is notable that they change by only about a factor of 3 over a very wide range of conditions.

Figure 4 shows experimental data for maleimide-labeled hemoglobin in glycerol. A typical value of T_1 for a spin-labeled protein is 10^{-5} s.

Heisenberg exchange between spin-labels and fast relaxing paramagnetic species including dissolved molecular oxygen or transition metals provides a thermal contact with the lattice and is an effective T_1 mechanism [43]. The effect has been used to measure oxygen concentrations [44].

Relaxation of nitrogen nuclei apparently is governed by modulation of the electron-nuclear dipolar interaction, the so-called END mechanism. The nitrogen nuclear relaxation probability can be greater than the electron spin-lattice relaxation probability. See, for example, the paper by Popp and Hyde [45]. One consequence of this process is that it can alter the apparent relaxation time of the electron since it gives rise to parallel relaxation pathways. One must distinguish between apparent and actual electron spin-lattice relaxation probabilities.

Nitrogen nuclear relaxation is investigated readily using the technique of electron-electron double resonance (ELDOR). See the book by Kevan and Kispert [19] for more details. It is possible in the future that measurements of the nitrogen nuclear relaxation will become useful in obtaining motional information, since the mechanism seems well understood.

Referring to transverse relaxation processes that give rise to loss of phase coherence, three motional domains can be defined: fast tumbling, where the intrinsic linewidth is Lorentzian in character; slow tumbling, where the rotational correlation time is comparable to the inverse of the anisotropy of the magnetic interaction; very slow tumbling, where the intrinsic linewidth is once again Lorentzian, but the observed linewidth is simply a powder pattern.

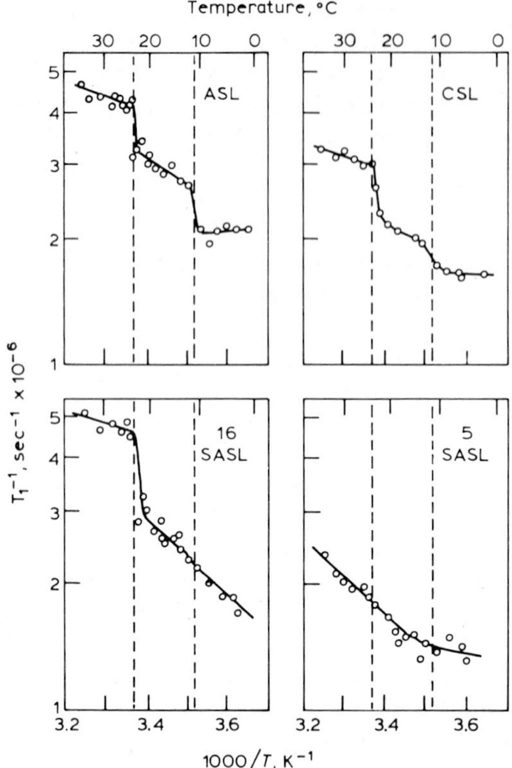

Figure 3. Rates of spin lattice relaxation (T_1^{-1}) for nitroxide spin labels in dimyristoylphosphatidylcholine liposomes as a function of temperature. ALS, androstane spin label; CSL, cholestane spin label; 5- and 16-SASL, 5- and 16-doxylstearic acid spin labels, respectively. From [42], with permission.

In the fast tumbling domain, the theory of Stone et al. [46] (and see Ref. 47) is used to obtain motional information. In the slow tumbling domain, Freed's [48] contributions are fundamental. In the very slow tumbling domain, saturation transfer methods are used [49] to obtain motional information. In theoretical analysis in this latter domain, the intrinsic linewidth is often treated as an adjustable parameter. Hyde and Hyde [50] have suggested an experimental way to determine it independently.

(iii) Intramolecular motional modes

Little information exists; it is not known whether the various motions that nitroxides can undergo affect in any significant way the information obtained from the spin-label experiment.

Lajzerowicz-Bonneteau [51] gives an overview of X-ray analysis of nitroxide conformations. NMR of spin labels can be used to obtain information on motional modes (see Briere et al. [34,35]). We have already indicated the use of ESR to observe the temperature dependence of the proton couplings of CTPO spin label.

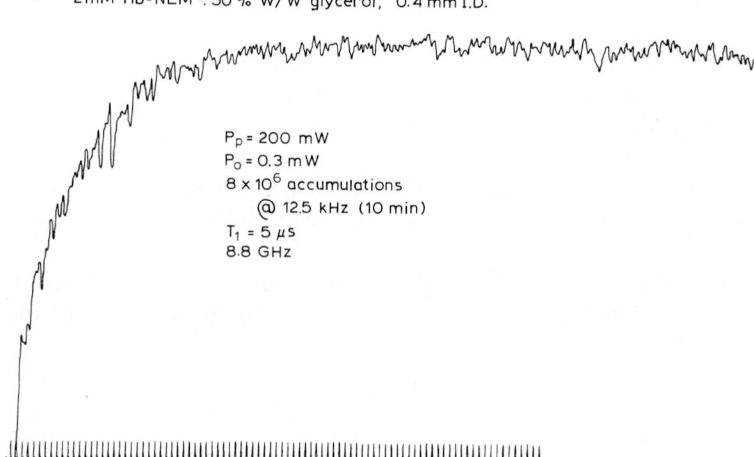

Figure 4. Typical saturation recovery signal for hemoglobin modified with *N*-ethylmaleimide spin label. The signal was obtained at 20°C in 50% aqueous glycerol with the experimental conditions indicated. P_p = pump power; P_o = observing power. (Unpublished data, by A. Kusumi, our laboratory.)

A critical matter is the planarity of the NO bond with respect to the CNC plane. Lajzerowicz-Bonneteau reviews available information. This angle varies in the X-ray studies from 0 to 30.5°. An unpublished theoretical estimate of 1.05 kcal/mole is reported for the energy difference between planar and pyramidal configurations, with shallow potential minima at $\pm 17°$. Jump diffusion between these minima might be an important intramolecular motion.

(c) Spin-label information content

In this section we provide an outline of the kinds of information that can be obtained from a nitroxide radical spin-label spectrum.

(i) Intensity

In this class of situations, all that is wanted is the amplitude of the signal. An example of this is spin-immunoassay [52], where the signal intensity is proportional to the amount of material under assay. The experimenter may be faced with the problem of obtaining the best possible signal-to-noise ratio without regard to lineshape distortions. The following is a check-list of options that can be considered.

(1) Substitution of ^{15}N for ^{14}N will increase the signal intensity by 3/2.

(2) Perdeutero nitroxides will yield lines that are about three times narrower than protonated nitroxides, which, assuming optimum field modulation amplitude is employed, will increase the signal by about three times. Because the lineshape may change from mostly Gaussian to mostly Lorentzian in character, this procedure does not always yield the expected improvement.

(3) Over modulation. The field modulation amplitude that yields the biggest possible signal should be used.
(4) Microwave power. The incident power that yields the biggest possible signal should be used, unless sample heating becomes a problem.
(5) Consider the use of the dispersion mode under rapid passage conditions using very high microwave fields [53].
(6) Consider introduction of extrinsic spin relaxers, such as molecular oxygen or elements of the lanthanide series, thereby permitting the use of higher microwave field intensities.
(7) Distinguish carefully the nature of the signal-to-noise problem: is the amount of material limited or is the concentration low? This will permit the selection of the optimum sample resonator geometry.
(8) Signal averaging of multiple spectra obtained with a short integrating time constant will permit a richer variety of smoothing and filtering procedures using a computer than can be carried out with analog filtering (i.e., use of a long time constant).

(ii) Lineshapes and rotational motions
In the range of characteristic rotational motions of 10^{-7}–10^{-11} s, the total spectral lineshape contains considerable information on motion. Deuteration increases this information content and use of ^{15}N reduces spectral overlap, thereby removing ambiguities.

The range of possible motional situations is enormous. The motional information contained in the ESR spectrum can be increased further by obtaining spectra over a range of microwave frequencies. The zeroth level of sophistication is to assume isotropic Brownian rotational diffusion, and it can safely be claimed that at this level things are well understood. A major effort is currently underway in many laboratories to learn to use spin labels to detect more complicated motional situations. To give an idea of the problem, if one has a rigid object with three moments of inertia, undergoing anisotropic rotational diffusion in an isotropic fluid, Tao [54] has shown that there are five correlation times. To this must be added the eulerian angles to relate the spin-label magnetic framework to the inertial tensor. Thus, this problem may already be too difficult to obtain unique solutions. Nevertheless, there seems little doubt but that progress is possible in certain situations. A recent example, with relevant citations, is the paper of Broido et al. [55].

The main experimental problem is that very small field modulation amplitudes must be employed in order to avoid distortion of the lineshape, which results in a poor signal intensity. It is noted that lineshape distortion is a linear process, transferring information to higher harmonics of the field modulation frequency. In principle, this information could be collected and used to correct for this distortion, but practical ways have not yet been developed.

(iii) Spectral diffusion of saturation and rotational motions
When the rotational correlation time is slower than 10^{-7} second, the spectra become insensitive to motion, resembling powder spectra. However, since the spin-lattice

relaxation time is about 10^{-5} second, motions slower than 10^{-7} second lead, in the presence of intense microwave fields, to spectral diffusion of saturation. The nature of the spectral diffusion depends on the nature of the motions. This subject is called saturation transfer spectroscopy (ST-EPR). It has been reviewed several times by one of us [49,56,57].

Spectral diffusion of saturation can be detected in various ways, including electron-electron double resonance (ELDOR), effects on adiabatic rapid passage spectra, saturation recovery, and CW saturation. Of these, ELDOR is by far the most attractive in principle, because the effect is observed directly. Unfortunately, the requisite equipment is not widely available. We look forward to increased use of ELDOR, particularly time-resolved ELDOR, to study special motions of slowly diffusing spin labels.

(iv) Translational diffusion (homospecies) and line broadening

Spin labels undergoing translational diffusion encounter each other at a frequency given by the Smoluchowski equation

$$\omega = 8\pi R D_{SL}\,[SL]$$

where D is the translational diffusion constant of the spin label, R is the interaction distance ($\approx$ 4–5 Å) and [SL] is the concentration of spin label. When this rate is comparable to the width of a spectral feature expressed in frequency units, Heisenberg exchange leads to line broadening. Quantitation can be very difficult because of the need to deconvolute the Lorentzian linewidth with the distribution from the actual inhomogeneously broadened lineshape. In order to increase the sensitivity of the method, spectral features should be as sharp as possible, which again favors use of fully deuterated spin labels.

(v) Translational diffusion (heterospecies), line broadening, and saturation

Bimolecular collisions between unlike species A and B are governed by the equation

$$\omega_A = 4\pi R(D_A + D_B)\,[B]$$

This equation gives a measure of how often an A molecule encounters B molecules. Species A might be quite dilute, so that homospecies collisions yield no observable spectral effects. There are some subtleties in distinguishing between dipolar and Heisenberg exchange mechanism for interactions during such an encounter, but the consensus seems to be that Heisenberg exchange is the dominant mechanism and that the probability for an observable event taking place during a collision is close to unity. Examples of this class of situations include collisions with dissolved molecular oxygen, which is paramagnetic, with transition metals such as Cu^{2+} or Mn^{2+}, or between ^{14}N and ^{15}N spin labels (which can be on different molecules).

The interaction between nitroxides and oxygen has been used as a basis for measurement of dissolved oxygen, a procedure that may be termed spin-label oximetry. In this approach calibration curves are constructed relating a nitroxide linewidth, or a related parameter, to the concentration of dissolved oxygen. Using such calibration curves, oxygen concentrations in a variety of systems have been determined [37,38,58].

Hyde and Sarna [43] showed that when spin labels collide with fast relaxing paramagnetic species, Heisenberg exchange provides a thermal contact between the slow relaxing species and the lattice. Thus, there is a change in effective T_{1e}. This effect has been observed directly by Kusumi et al. [42].

Heisenberg exchange between ^{14}N and ^{15}N labeled species is an ELDOR-active process, and ELDOR with ^{14}N–^{15}N can be used with great precision to investigate translational diffusion processes.

The reader is referred to the distinguished book by Molin et al. [16], on exchange interactions in ESR spectroscopy.

(vi) pH detection

S. Schreier and her colleagues have described the use of spin labels with weakly interacting protons that have pK_a values near 7 as pH indicators [29].

(vii) Polarity probes

We have already discussed the effect of polarity on the nitrogen hyperfine coupling. Labels can be used to make statements about the hydrophobicity of a binding site.

(viii) Distance determinations (fixed interaction distance)

Hyde et al. [59] discuss the roles of the various terms of the dipolar Hamiltonian to describe magnetic interactions between two paramagnets separated by a fixed distance. A notable observation is the so-called Leigh effect [60]. In certain circumstances, such as occur when a spin label and a manganese ion (Mn^{2+}) are on a protein, the interaction leads to loss of spin-label intensity. The amount of loss can be related to the distance between the two species.

In principle, very good distance measurements can be made with arguments based on the C and D terms of the dipolar Hamiltonian, which predict how spin-lattice relaxation of one species can affect spin-lattice relaxation of the other species.

Attention is called to Eaton and Eaton's comprehensive review [61] of metal–spin-label interactions. The problem of whether the dipolar or exchange interaction dominates is treated in this article.

(ix) Distance determination (distribution of fixed interaction distances)

Hyde and Rao [62], using a moment approach, consider the situation that arises if a solution of two different paramagnetic species is frozen. Their conclusion is that observable effects are always dominated by the "distance of closest approach".

(x) Concluding remarks

It is a relatively straightforward matter to go from a model to a spectrum; it is a much more difficult matter to go from a spectrum to a model and to establish the uniqueness of the model. Every possible way should be employed to increase the information content. The possible approaches are: (1) Computer filtering, making use of all known information in order to enhance the quality of the spectral content that relates to unknown information. (2) Varying microwave frequency, thereby building on the interplay of the g and A tensors. (3) Emphasizing the time domain, which can in principle distinguish between overlapping events occurring at different rates. (4) Emphasizing ELDOR – the direct observation of how one part of a spectrum speaks to another part.

We foresee a bright future for spin labels in biochemistry and biophysics, but with a much heavier instrumental and methodological content than is now prevalent.

3. Biological free radicals

This topic constitutes a major area of electron spin resonance that has been the subject of several reviews. A good account of the field with comprehensive literature coverage to 1973 is given by Borg [63] in Biological Applications of ESR. Other useful reviews can be found in chapters in the Free Radicals in Biology series. Topics that have been covered in the first six volumes include a general account of ESR [64], radicals in enzyme-substrate reactions [65], drug metabolism [66,67] and melanins [68] and the use of the spin-trapping procedure [69]. Free-radical metabolism of xenobiotics [70,71] and spin-trapping [72–74] have been subjects of several other recent accounts.

(a) Physical and chemical properties

Free radicals occur as normal metabolic intermediates in a number of enzyme reactions. These reactions may be simple one-electron transfers (either oxidation or reduction) or reactions of a more complex nature, for example, rearrangements, that are difficult to accomplish by ionic mechanisms. Radicals also may be active intermediates in the chemical transformation of drugs and xenobiotics through enzymatic and non-enzymatic (e.g., autoxidative) pathways. Due to their reactive nature, many free radicals are potentially damaging either directly or indirectly through, for example, the formation of active species of oxygen ($O_2^{\cdot -}$, $RO_2^{\cdot}$, $^{\cdot}OH$). Some other radicals, for example, those derived from antioxidants, show little tendency to cause damage; presumably, the effectiveness of antioxidants lies in their ability to remove a potentially damaging radical species from a system, replacing it with a less damaging, antioxidant-derived radical.

Because of the sensitivity and selectivity of the ESR approach, it is a powerful method for studying free radicals in low concentrations in complex systems. ESR can be helpful in two ways: to characterize and structurally identify radical intermediates and to obtain information on the kinetics and mechanisms of their reactions. These reactions can be very fast; one of the main differences between ESR of biological free radicals and ESR of spin labels and metal ions is that in the former case one generally is dealing with transient paramagnetic species.

Characterization and identification is possible from measurements of spectroscopic parameters: hyperfine splittings, *g* values, linewidths and, sometimes, microwave saturation behavior. The unpaired electron in a radical will interact with the nuclei in the environment to give, in favorable instances, a distinct hyperfine pattern and distinct *g* value. This is often the case for small radicals tumbling freely in non-viscous media, e.g., small substrate-derived radicals in fluid aqueous solution. In such a situation, the inherent anisotropy of the ESR magnetic parameters is averaged out, giving rise to narrow spectral lines and frequently a high degree of resolution. Radical identification can then be quite straightforward.

The characterization and identification of free radicals that are not tumbling rapidly may require a more sophisticated approach. If a macromolecular (e.g., enzyme-derived) radical is tumbling slowly in solution or if a small radical is immobilized, the degree of resolution is almost always very low. Under such conditions a single broad ESR line generally is observed. However, in situations such as these, sufficient information often can be obtained from conventional ESR experiments to establish a given spectrum as belonging to a radical of a particular chemical class; measurements of *g* value and linewidth, perhaps together with microwave saturation data, can provide important clues to the radical structure. Several additional chemical and physical approaches can give more information. Isotope substitution will help identify interacting nuclei that contribute to the hyperfine pattern or linewidth. ENDOR experiments may allow the measurement of hyperfine splittings that are unresolved in the ESR experiment. Finally, multi-frequency ESR experiments are extremely valuable in resolving features due to spectral anisotropy or to overlap of two or more radical species.

Although in principle all free radical species are detectable by ESR spectroscopy, in practice detection may be difficult or impossible under a given set of experimental conditions. Problems with detection of a particular species will reflect magnetic and/or kinetic factors. For example, oxygen-centered species such as $^{\cdot}OH$, $O_2^{\cdot -}$, and $RO^{\cdot}$ [75] and sulfur-centered species such as thiyl radicals ($RS^{\cdot}$) [76] cannot be detected directly in fluid solution because of extreme anisotropy in their magnetic parameters which makes their ESR signal amplitudes vanishingly small. To detect radicals such as these, it is necessary to immobilize them in frozen solutions or to resort to indirect methods of detection (see below). The same applies to radicals that have a short lifetime.

As noted, the transient nature of most free radical species is a major consideration in ESR studies of free radicals. Free radical chemistry [77] involves an initiation step in which the free radicals are formed, often followed by one or more propagation (chain) reactions before termination. Because most radical-radical termination reactions are fast, the majority of free radicals decay rapidly by self reaction, i.e., they are transient even in the absence of another species. (In non-transient, i.e., persistent, radicals the radical center is sterically hindered, thereby inhibiting self-reaction.) A comment on terminology may be appropriate at this point: many transient radicals are frequently described as stable or unreactive, which can lead to some confusion. The source of this confusion is that 'reactivity' and 'stability' are often used to denote the ability of a given radical to undergo reaction, not with itself (i.e., the usually fast self reaction) but with molecular components of the system, i.e., the tendency of the radical to engage in chain reactions. On this basis, the ascorbate radical anion, which is transient in neutral aqueous solutions [78], but shows little tendency to initiate chain reactions, is said to be 'unreactive' or 'stable'. In contrast, the hydroxyl radical, $^{\cdot}OH$, which tends to react at diffusion-controlled rates with many solutes, is transient and also 'reactive' and 'unstable'.

$$R_1{-}X \rightarrow R_1^{\cdot} + X^{\cdot} \quad \text{initiation} \qquad (1)$$

$$R_1^{\cdot} + R_2H \rightarrow R_1H + R_2^{\cdot} \quad \text{propagation} \qquad (2)$$

$$R_2^{\cdot} + R_2^{\cdot} \rightarrow R_2 - R_2 \quad \text{termination} \qquad (3)$$

Although a radical must be reactive in order to damage other components of a system, there is not necessarily a simple correlation between reactivity and the ability to cause irreversible damage to a complex structure. For example, the activity of certain enzymes is found to be inhibited more effectively by radicals of relatively low reactivity than by the reactive hydroxyl radical. This is because reactive radicals are not very selective in their reactions, having a tendency to react at many different sites in a molecule. Less reactive radicals are more selective and can be more effective at damaging a specific site. If this site happens to be essential for activity, then the less reactive radical will be more damaging.

The experimental approaches employed in free radical ESR tend to differ from those used in the other sections, again reflecting the transient nature of the species studied. Pulsed ESR methods with time resolution down to about 20 ns are now available [40,79], but for transient radicals this requires generation of radicals in very high concentrations, e.g., by a laser pulse, and this is not always applicable. More common is to generate the radicals continuously so that a steady-state situation is achieved in which the rate of radical formation is equal to the rate of decay. The steady-state radical concentration is then related to the formation rate and the radical lifetime. For a transient radical with a short lifetime, detectable steady-state levels can be achieved only if the rate of radical formation is high. A consequence of this is that starting materials will be rapidly depleted, perhaps necessitating the use of a flow system in order to maintain adequate concentrations of starting materials. This can become very costly in terms of reagents, especially for an enzyme reaction.

The ways of getting around this problem involve increasing the lifetime of the radicals by some physical or chemical means. One such approach involves stabilizing the radicals by immobilization, for example, by freeze-quenching a reaction mixture [80]. The disadvantage of this method is that an immobilized radical is generally much harder to characterize and identify than one in fluid solution. Other approaches make use of the chemical reactivity of radicals, for example, their ability to add to the double bonds in nitrones and nitroso compounds. This has led to the development of the spin-trapping procedure [81,82], in which a transient radical is reacted with the spin trap to form a spin adduct (structure 1a,b; Eqns. 4 and 5). Many biological free radicals are anions that do not react with nitrones and nitroso compounds but will react with di- and trivalent cations to form chelate complexes (structure 2; Eqn. 6). This complementary approach has been termed spin-stabilization [83]. The aim of each of these chemical methods is to convert a transient radical that is difficult to detect, either because of its short chemical lifetime or broad ESR lines, into a more long-lived radical whose ESR spectrum is characteristic of the addend. The advantage is that the reaction can be carried out in solution and that identification of the trapped radical is usually relatively straightforward.

$$R_1^{\cdot} + R_2NO \rightarrow R_1R_2NO^{\cdot} \tag{4}$$

Structure 1a

$$R_1^{\cdot} + R_2CH = \overset{+}{N}(O^-)Bu^t \rightarrow R_1R_2CH\ N(O^{\cdot})Bu^t \tag{5}$$

Structure 1b

$$R_1^{\cdot} + M^{n+} \rightarrow R_1M^{n+} \tag{6}$$

Structure 2

Even in systems where chemical stabilization is used, radicals detected in solution are usually transient. This makes quantitation more difficult in these systems than in ones where the paramagnetic species are kinetically stable. Of course, quantitation is extremely important in all radical systems. It will distinguish between situations in which a radical is an obligate intermediate in an enzyme reaction and one in which the radical is formed in a secondary reaction or side reaction of low efficiency. However, in most biological ESR to date few attempts have been made to distinguish between such possibilities.

For a steady-state situation, what is important is the rate at which radicals are being formed, R_{in}. It is this which, when compared to the rate of substrate removal or (molecular) product formation, will allow one to decide whether the overall reaction is largely free-radical or ionic in nature. The steady-state concentration, $R^{\cdot}_{ss}$, in itself has little meaning unless it can be related to the rate of radical formation. Fortunately, it is generally possible to do this, since the rate of radical formation is given by the steady-state concentration divided by the radical lifetime, τ. Radical lifetimes either can be experimentally measured in the system or, in many cases, can be obtained from literature data of termination rate constants. The relationship between the rate of radical formation and the steady-state radical concentration depends on whether the radical decays by first- or second-order kinetics.

$$R_{in} = [R^{\cdot}]_{ss}/\tau \tag{7}$$

$$\text{Second-order reaction: } R_{in} = 2k[R^{\cdot}]^2_{ss} \tag{8}$$

First-order reaction: $R_{in} = k[R^{\cdot}]_{ss}$ (9)

In the sections that follow, examples of ESR studies of biological free radicals are discussed. The first section focusses on the generation of radicals from biological materials by chemical means. Systems of this kind are important for two reasons. First, several of the chemical approaches employed to generate radicals (e.g., autoxidation, Fenton or Fenton-like systems, photooxidation and reduction) are likely to be encountered in more complex biochemical systems. Therefore, information on the mechanisms of these reactions is important. Second, they serve as model systems in which free radicals can be generated under relatively simple, controlled conditions. Such systems allow radicals to be characterized and identified and studies of their chemical reactions to be carried out. Subsequent sections cover biochemical systems involving free radicals from enzymes and their substrates, other biological free radicals such as melanins, and free radicals in drug metabolism. The behavior of radicals in multicomponent systems, for example, ones in which local concentrations of paramagnetic species are high, can differ considerably from that of the free, magnetically dilute species.

(b) Radicals from chemical oxidation/reduction

A specific free radical can be produced from a precursor molecule either in an initiation step or a propagation step in which a reagent radical reacts with the precursor. Initiation requires either removal or addition of an electron or homolysis. Chemically this can be done in a number of ways, by using one-electron oxidants or reductants or by inducing homolysis in some way: examples of these types of reactions include autoxidation [84–86], photochemical oxidation and reduction [87–90], and oxidation and reduction by metal ions and their complexes [91–93]. In propagation reactions, the reagent radical might be the hydroxyl radical, the hydrated electron, or any other suitably reactive species that will interact with the precursor molecule in the desired manner. We will consider initiation reactions first.

$$RH_2 + X \rightarrow R^{\dot{-}} + 2H^+ + X^{\dot{-}} \quad (10)$$

$(X = O_2$, $Fe(CN)_6^{3-}$, S^* (an excited state), etc.)

Autoxidation, particularly at high pH, is a simple method for generating many radicals. The initiating reaction is presumably the transfer of an electron to oxygen, leading to the formation of superoxide and a radical derived from the reagent. This procedure is effective for the generation of semiquinones and related species. At high pH, decay of semiquinone radical anions is slow and high radical concentrations can be obtained. Removal of oxygen, for example, by flushing with nitrogen, after the radicals are generated can sometimes improve hyperfine resolution if sufficient residual oxygen is present to broaden hyperfine components by Heisenberg exchange. Intense spectra of semiquinones from catechols and catecholamines [85,87] and of semiquinonimines from aminophenols such as serotonin and 5-hydroxytryptophan [86] have been obtained in this manner. There are two major disadvantages to autoxidation as a general method for radical generation: secondary radical production can be quite common (for example, addition of OH^- can occur, especially in semiquinone systems), and the ionization state of the radicals at the pH at which the autoxidation is carried out may differ from that encountered in neutral aqueous solution. These factors must be taken into account in the interpretation and comparison of spectra.

Chemical oxidation and reduction with inorganic ions also can be extremely useful. Reagents useful in aqueous solutions include $Fe(CN)_6^{3-}$, $IrCl_6^{3-}$ and MnO_4^- [91,94] as oxidants and $S_2O_4^{2-}$ [95] and Ti^{3+} [96] as reductants. $Fe(CN)_6^{3-}$, $IrCl_6^{3-}$ and MnO_4^- have been used in flow systems to generate radicals from phenolic compounds while $S_2O_4^{2-}$ is useful for the anaerobic reduction of flavins and other electron acceptors. Although $S_2O_4^-$ is in equilibrium with the radical ion $SO_2^{\cdot-}$ [95], and therefore might be expected to act solely as a one-electron reductant, it appears that both one- and two-electron transfers are possible with this reagent. Note that the presence of paramagnetic metal ions, either as reagents or reaction products, can lead to rapid relaxation of the free radicals, giving rise to broadened ESR spectra and reduced microwave saturation.

Detection of radicals in a static system, unless they are unusually long-lived, implies that radicals are being continuously generated by some mechanism or other. Probably the most common mechanism of this kind involves a redox equilibrium between oxidized and reduced molecules and the half-reduced species, the corresponding free radicals. Examples of this have been described in flavin systems [97]. Chemical reduction to give equal amounts of oxidized and reduced flavin results in the formation of significant amounts of free radical, evidently formed as a result of a redox equilibrium. A similar situation is known to exist for semiquinones [92]. If the radicals have a chelating structure, the equilibrium can be pulled to the free-radical side by the inclusion in the system of complexing metal ions [98,99].

$$R + RH_2 \rightleftharpoons 2R^{\cdot-} + 2H^+ \qquad (11)$$

(R = flavin, quinone, etc.)

Photooxidation and reduction reactions that have been carried out mostly involve reagents that absorb visible light: for example, flavins, porphyrins, melanins and various dye sensitizers. UV-photolysis, although generally less selective, can also be useful and is not uncommon. Since most photochemical reactions are strongly affected by oxygen, oxygen is normally excluded unless there are specific reasons for studying its effects.

It is probable that flavins and porphyrins act as endogenous sensitizers in cellular systems. Each can undergo either free-radical (type I) or singlet oxygen (type II) reactions depending on levels of reagents and oxygen. Photochemical reductions of flavins in the presence of mild reducing agents are well known. In anaerobic solution, triplet state flavin reacts with the reducing agent, e.g., EDTA, to form a flavin semiquinone and a radical from the reducing agent. Secondary radicals are sometimes formed [90]: photoreduction of flavin by phenylthioacetic acid yields a flavin radical together with the radical $PhS\dot{C}H_2$, formed in a decarboxylation reaction. $PhS\dot{C}H_2$ has been spin-trapped using 2-methyl-2-nitrosopropane (MNP). Anaerobic photoreduction also is an excellent means of generating semiquinones from flavoproteins. Irradiation of the flavoprotein with visible light in the presence of EDTA generally gives high yields of the corresponding semiquinone [100]. The reaction can be accelerated by addition of small amounts of deazaflavins to sensitize semiquinone production [101].

Excited-state porphyrins also are good oxidants. Photoreduction of uroporphyrin by EDTA and other electron donors [89] gives ESR spectra of the porphyrin radical anion, which has a low g value (2.002) and a broad ESR line (5 G) with no resolved hyperfine structure. The radical is transient and decays with second-order kinetics in aqueous solutions. Similar spectra have been reported from hematoporphyrin irradiated in the presence of thiols [88]. In addition, a thiol-derived radical was spin-trapped by MNP. The hematoporphyrin radical anion itself is not spin-trapped, but it will transfer an electron to MNP to form $Bu^tNHO^{\cdot}$. With reducing agents other than thiols, radicals from the reducing agent (ascorbate, catechols, *p*-phenylenediamine) were detected directly [88].

ESR has been used [102] to detect $^1O_2$ formed in photochemical reactions. The method depends on the reaction of $^1O_2$ with hindered amines to form persistent nitroxides. While mechanistic details of this reaction are not clear, it seems that the method is useful for demonstrating the production of $^1O_2$. However, there is a disadvantage: if free-radical reactions are occurring simultaneously, these free radicals may react to destroy the nitroxide [103].

Electrochemical oxidation and reduction [104] can be a simple, clean method for the generation of a wide variety of radicals, for example, radical anions from quinones and nitrocompounds.

Free radical propagation reactions have been studied extensively by ESR. The primary reagent radical is generated in an initiation step, which may be a thermal chemical reaction, photolysis, radiolysis, etc. Reaction to give the secondary radical then follows. ESR data for a very wide variety of radical species have been obtained using this approach. Typical primary radicals are $^{\cdot}OH$, alkoxyl radicals (e.g., $Bu^tO^{\cdot}$) and hydrated electrons formed in the reactions shown:

$$M^{n+} + H_2O_2 \rightarrow M^{(n+1)+} + {}^{\cdot}OH + OH^- \quad (12)$$

$$H_2O_2 \xrightarrow{h\nu} 2\,{}^{\cdot}OH \quad (13)$$

$$R_2O_2 \xrightarrow{h\nu} 2\,RO^{\cdot} \quad (14)$$

$$H_2O \rightsquigarrow {}^{\cdot}OH,\ H^{\cdot},\ e^-_{aq} \quad (15)$$

In closing this section, it should be pointed out that compilations of ESR spectral data for all classes of free radicals may be found in the Landolt-Bornstein series [105]. In addition, kinetic data for radical reactions are available from a variety of approaches, including ESR and optical methods and product analysis. Thus, it is likely that rate constants for a specific free radical reaction either are known, or can be estimated from data for radicals of related structure.

(c) Radicals from enzymes, their substrates, and other macromolecular radicals

In enzyme-substrate reactions, radicals may be formed either on the enzyme (on a prosthetic group or the polypeptide), the substrate, or both. Enzyme-derived radicals tend to be immobilized and, depending on the experimental conditions, can be long lived. In contrast, substrate-derived radicals generally are tumbling rapidly in solution and are transient. Therefore, the methods used to characterize and quantitate the two kinds of radical tend to be quite different: enzyme radicals are harder to identify with certainty but quantitation may simply require measuring a radical concentration rather than a rate of radical formation.

(i) One-electron oxidation

The above point can be illustrated taking as an example the peroxidases, a class of heme-containing enzymes that catalyze the one-electron oxidation of a wide variety of substrates [106]. With hydrogen peroxide the enzyme forms a highly oxidizing species (for most peroxidases designated compound I) that is two equivalents above the ferriheme resting state. One-electron oxidation of 2 moles of substrate (SH_2) follows, with the enzyme returning to the resting state via an intermediate oxidation state (compound II). Several other proteins, including catalase and metmyoglobin, can exhibit peroxidase-like behavior.

$$\text{peroxidase} + H_2O_2 \rightarrow \text{compound I} \tag{16}$$

$$\text{compound I} + SH_2 \rightarrow \text{compound II} + SH^{\cdot} \tag{17}$$

$$\text{compound II} + SH_2 \rightarrow \text{peroxidase} + SH^{\cdot} \tag{18}$$

In compound I the two oxidizing equivalents are separated: the iron is present in an oxyferryl ($Fe^{IV} = O$) form [107,108] while the other equivalent is a free radical which, depending on the specific peroxidase, can be either a porphyrin radical cation or a radical localized on an amino acid in the polypeptide.

For horseradish peroxidase (HRP), it was long thought that only a small fraction of the free-radical content was ESR detectable: conventional ESR spectra show a narrow component equivalent to about 0.01 spins/heme. However, recent work has demonstrated that much of the free-radical spectrum is extremely broad due to interaction of the radical with ferryl ion, probably through an anisotropic exchange interaction [109]. When these broad components are taken into account, the number of free radicals formed per heme iron is close to 1. The free radical has been identified as a porphyrin radical cation (3) by ENDOR spectroscopy. On the basis of measured proton and nitrogen hyperfine couplings it has been argued [110] that the radical is probably formed by removal of an electron from the $^2A_{2u}$ molecular orbital of the porphyrin. A porphyrin radical cation also has been shown [111] to be formed by HRP in which the iron was substituted by zinc.

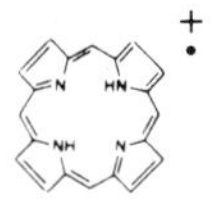

Structure 3

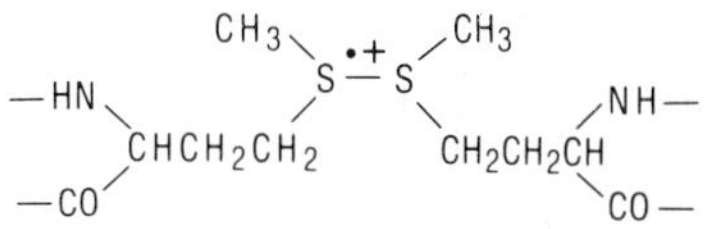

Structure 4

Other peroxidase-H_2O_2 compounds differ in both their optical and ESR spectra. Chloroperoxidase with hydrogen peroxide gives ESR spectra that have been attributed [112] to two radical species, one of which is suggested to be an A_{1u} type porphyrin cation strongly coupled to ferryl ion, while the nature of the other species is unclear. Possibly ENDOR spectroscopy can be used to clarify the situation in this system also. The hydrogen peroxide compound of cytochrome *c* peroxidase has received a great deal of attention. Because of differences in optical spectra between this and hydrogen peroxide compounds of other peroxidases, it has been designated compound ES rather than compound I [113]. The free radical detected by ESR in this case apparently has only a relatively minor interaction with the ferryl iron center. From earlier work it was suggested that the radical either was localized on an aromatic amino acid [114] or was a peroxyl radical [115]. It has since been argued [116] that neither of these suggestions can be correct – the ESR spectrum (Fig. 5) has $g_\perp < g_\parallel$, inconsistent with an aromatic π radical, and the calculated isotropic g value is extremely high (2.02). From ENDOR experiments proton hyperfine splittings have been measured, from which it was concluded [116] that the data are consistent with a radical that is sulfur-centered, perhaps a methionyl dimer-cation (4). The latter has been shown to be formed in the chemical oxidation of methionine [117]. Reaction of metmyoglobin with H_2O_2 also has been reported [118] to yield an organic free radical, in this case possibly an oxidation product of tyrosine.

Thus, no one peroxidase system is exactly like another, at least on the basis of the evidence accumulated to date. One may suppose that in each case the primary radical formed is located on the porphyrin and that the presence of oxidizable groups in the vicinity of the heme leads to transfer of the oxidizing equivalent out onto an amino acid in the peptide. Transfers of this kind may be quite common in proteins; pulse radiolysis studies using optical detection of transients have demonstrated the migration of oxidation from one amino acid site to another [119].

Turning to substrate-derived radicals in peroxidase systems, detailed measurements were first made by Yamazaki et al. [120] and later extended in a series of studies that has been recently reviewed by Yamazaki [65]. Many of the key concepts that are

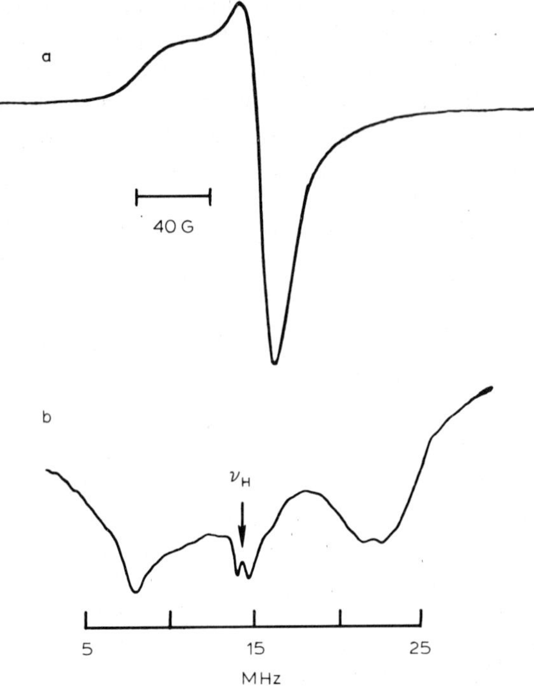

Figure 5. a, ESR spectrum of compound ES from cytochrome *c* peroxidase at 2 K; b, ENDOR spectrum at 2 K, scanned from 2 to 30 MHz. The free proton frequency is indicated by the arrow. From [116], with permission.

important in the study of substrate radicals by ESR can be found in this body of work. Anyone beginning such a study should study these papers in detail.

Initial experiments involved the use of hydroquinones, ascorbate and dihydroxy fumarate as substrates for the enzyme [120]. Radicals (e.g., 5 and 6) were detected under steady-state conditions using a flow system to minimize substrate depletion. The narrow line spectra (Fig. 6a,b) of the radical anions are identical to those generated in chemical systems, indicating that the radicals are free in solution rather than associated with the enzyme. If it is assumed that the reaction proceeds by one-electron oxidation of the substrate and that the product radicals decay by self-reaction (e.g., disproportionation), kinetic analysis predicts that the steady-state radical

Structure 5 Structure 6 Structure 7

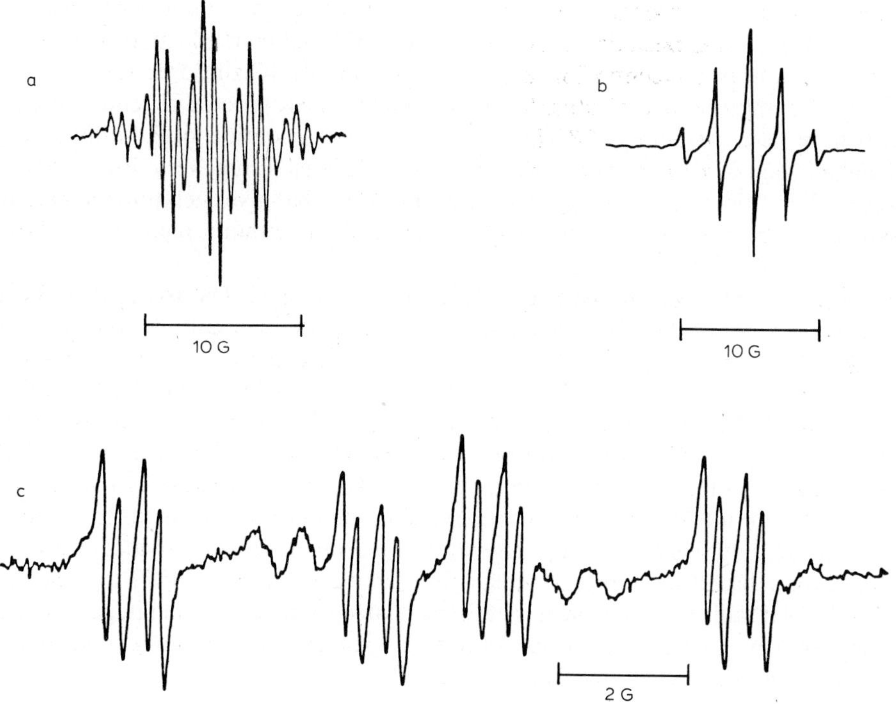

Figure 6. ESR spectra (25°C) of radicals from the horseradish peroxidase/H_2O_2 oxidation of peroxidase substrates. a, 2 mM menadione (hydroquinone form), 1 mM H_2O_2, 1.6 μM peroxidase, pH 7.5; b, 1 mM hydroquinone, 0.5 mM H_2O_2, 0.1 μM peroxidase, pH 6.5; c, 8 mM 3,4-dihydroxyphenylalanine (dopa), 0.175 mM H_2O_2 0.04 μM peroxidase, 0.225 M Zn^{2+}, pH 4.5. From [83,120], with permission.

concentration should be proportional to the square root of the enzyme and substrate concentrations. These predictions were confirmed, and a comparison of the rate of radical formation with the rate of substrate disappearance showed that close to 2 moles of free radical are produced per mole of substrate removed. Thus, all available evidence is consistent with the mechanism shown above, where the rate-determining step is reduction of compound II by substrate.

In the presence of other molecular species, a chain reaction involving conversion of a primary radical to a secondary species has sometimes been observed: for example, the *p*-methylphenoxyl radical from HRP/H_2O_2 oxidation of *p*-cresol has been shown [121] to oxidize ascorbate to the corresponding radical anion. Reactions of this kind can be extremely effective.

Because of the difficulty in carrying out the above experiments, which required a flow system to maintain adequate substrate levels, subsequent workers have rarely attempted to obtain the same degree of detailed information from ESR studies on peroxidase systems. Investigations that have been carried out have largely been restricted to observations of ESR spectra from other peroxidase substrates. Several of the spectra obtained were too weak for complete characterization and identification [122]. However, it has been demonstrated recently [83] that marked enhancements in steady-state radical concentration are possible by spin-stabilizing free radical products with complexing metal ions. In this way intense ESR spectra of semiquinones (e.g., 7) have been obtained [83] from the peroxidase oxidation of 3,4-dihydroxyphenylalanine and related molecules in static solutions using very low enzyme concentrations (Fig. 6c). For these substrates and others that give chelating radicals, it should easily be possible to obtain quantitative data without recourse to flow methods.

Several other one-electron oxidations have been investigated. The ascorbate radical anion has been detected during the oxidation of ascorbate by ascorbate oxidase and dopamine β-hydroxylase [123]. Whereas the former system has been quantitatively studied by ESR, much of the data for the latter system has come from optical experiments [124]. Lipid free radicals have been spin-trapped in prostaglandin synthetase [125] and lipoxygenase [126] systems. The radical trapped with MNP during oxidation of arachidonic acid with prostaglandin synthetase is considered [125] to be an intermediate in the conversion of arachidonic acid to prostaglandin G_2. During this reaction the prostaglandin synthetase is deactivated, with the formation of a broad immobilized signal that has been ascribed [127] to a heme-protein-derived radical. This radical is perhaps related to those detected in peroxidase systems (above). Laccase has been shown to oxidize hydroquinone and other substrates by a one-electron mechanism [128].

(ii) Rearrangement and related reactions

Many B_{12}-dependent reactions that involve substrate rearrangement evidently also proceed by free-radical mechanisms. ESR spectra [129–131] detected in several systems can be accounted for on the basis of an organic free radical that is magnetically coupled to a nearby paramagnetic Co(II) center. Spectra of these cobalamin-free radical pairs have been simulated using a mixture of isotropic exchange and dipolar coupling mechanisms [129,130]. An estimated lower limit for the cobalt (II)-radical separation is approximately 10 Å. The catalytic mechanism [132] involves homolytic cleavage of the labile Co–C bond in adenosyl cobalamin to produce an adenosyl radical ($\dot{C}H_2Ad$) which abstracts a hydrogen atom from the substrate. The key step in the mechanism is subsequent proton-catalyzed rearrangement of the substrate radical ($S^{\cdot} \rightarrow S'^{\cdot}$) [133] followed by termination with formation of product. There is strong evidence from chemical systems [134] for the proton-catalyzed rearrangement postulated to occur in the enzyme system.

$$\text{Co-CH}_2\text{Ad} \rightarrow \text{Co}^{\text{II}} + \dot{\text{C}}\text{H}_2\text{Ad} \tag{19}$$

$$\text{SH} + \dot{\text{C}}\text{H}_2\text{Ad} \rightarrow \text{S}^{\cdot} + \text{CH}_3\text{Ad} \tag{20}$$

$$\text{S}^{\cdot} \xrightarrow{\text{H}^+} \text{S}'^{\cdot} \tag{21}$$

$$\text{S}'^{\cdot} + \text{CH}_3\text{Ad} \rightarrow \text{S}'\text{H} + \dot{\text{C}}\text{H}_2\text{Ad} \tag{22}$$

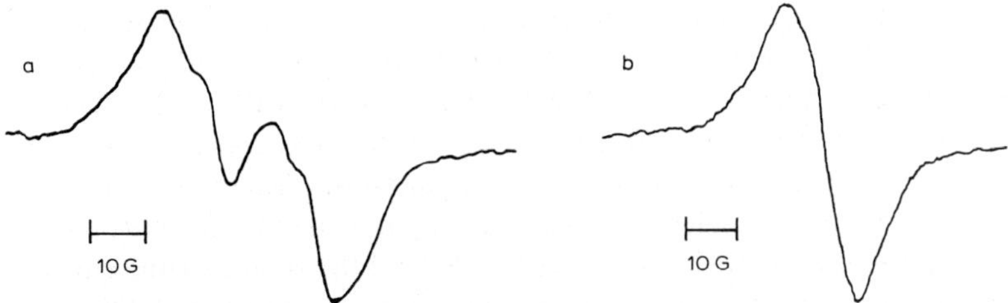

Figure 7. ESR spectra (−196°C) of tyrosyl radicals in ribonucleotide reductase from *E. coli*. a, Spectrum from enzyme grown in the presence of tyrosine; b, spectrum from enzyme grown in the presence of deuterated [β,β-$^2H_2$]tyrosine. From [137], with permission.

Cobalamin is also the cofactor in some ribonucleotide reductases. Others do not require cobalamin, but contain an organic free radical that is thought to be the functional counterpart of cobalamin. For example, ribonucleotide reductase from *Escherichia coli* has two subunits, one of which contains the organic radical [135]. This radical, which is quenched by radical scavengers such as hydroxyurea and hydroxylamine [136], is required for activity. Enzyme activities that will restore and remove the radical function have been described. The ESR spectrum of the protein is broad (as expected for an immobilized species), but shows a pronounced doublet splitting with additional minor structure (Fig. 7a). The structure of the radical was elucidated by isotope substitution experiments [137]. Deuterium substitution (Fig. 7b) revealed that the major structure is associated with methylene protons in a tyrosyl side chain, while ^{13}C substitution showed that the spin density is significantly delocalized over the tyrosyl aromatic ring. Therefore, the spectrum has been attributed either to a tyrosyl radical (8) or some other one-electron oxidation product of tyrosine. The radical is considered to be stabilized by two μ-oxo bridged antiferromagnetically coupled iron (III) ions in the protein [138]. The function of the tyrosyl radical in the enzyme is not clear at the present time, although it has been suggested that it may accept an electron from the ribose moiety during the catalytic cycle.

$^{\bullet}O-C_6H_4-CH_2CH(NH-)(CO-)$

Structure 8

(iii) One-electron reductions

Flavoprotein radicals have been extensively studied by ESR [139] and ENDOR [140] spectroscopies. Their ESR spectra, with few exceptions, are broad and featureless because of the slow rate of tumbling of the radicals, even in fluid solution. Nevertheless, some structural information is available based on the width of the observed ESR line. The spectra observed fall into three classes, having widths of about 19, 15 and 12 G. Non-covalently bound semiquinones have linewidths of either 19 or

15 G [139]. These widths correlate with the optical spectra of the radicals: semiquinones with a 19 G linewidth have pronounced absorption between 550 and 650 nm and appear blue; those with a 15 G linewidth (Fig. 8a) have spectra with peaks between 485 and 370 nm and appear red. The additional width of the 19 G spectrum is due to an exchangeable proton; it has been shown that the linewidth decreases to 15 G in D_2O solutions. Experiments on model compounds [141] indicate that the blue type of radical is a neutral semiquinone with the proton on N(5) of the isoalloxazine ring (9), and that the red species is either the semiquinone anion or the neutral O(4)-*enol* tautomer. Covalently bound semiquinones have ESR spectra that are distinctly narrowed [142–144], having a width of around 12 G (Fig. 4b). The reason for this is that the covalently bound flavins lack a methyl group at C-8, which when present makes a significant contribution to the total linewidth.

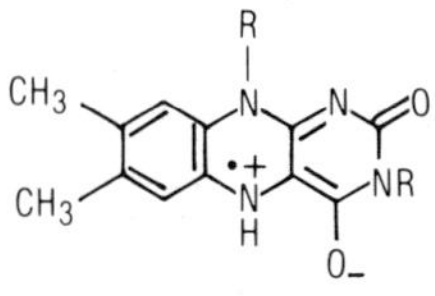

Structure 9

Flavin semiquinones in solution, being typical π radicals, ordinarily saturate very easily with microwave power. The center of the spectrum saturates more easily than do the outer wings [139,140,145]. Exceptions have been found where other paramagnetic species are in the vicinity of the semiquinone. Addition of paramagnetic metal ions to solutions containing flavin semiquinones can greatly decrease the saturability of the radical because of dipolar interactions with the metal ions [146]. Similar behavior has been encountered in enzyme systems (below).

Additional information on flavin semiquinones has been possible through the higher resolution available with the ENDOR approach. Methyl protons in flavins are easily detected by ENDOR [140,144,145] since they rotate, even at low temperatures, giving isotropic hyperfine couplings. The hyperfine splitting to $CH_3(8)$ protons in blue semiquinones is much smaller than in red semiquinones (4 G vs. 2.9 G) and is absent if the flavin is covalently bound (Fig. 8a,b). From the matrix signal, which originates from protons in the vicinity of the paramagnetic center, information on the immediate environment of the semiquinone can be obtained. The sensitivity of the matrix signal to D_2O provides a measure of the fraction of exchangeable protons that contribute to this signal. Thus, a radical in a hydrophobic environment is expected to be much less sensitive to a change from H_2O to D_2O than one in a hydrophilic environment with ready access to solvents. The range of effects of D_2O that have been found are consistent with a wide spectrum of environments for the half-reduced flavins.

Many enzymatic one-electron reduction reactions are mediated by flavoproteins. In these reactions flavoprotein semiquinones are presumably involved although this has

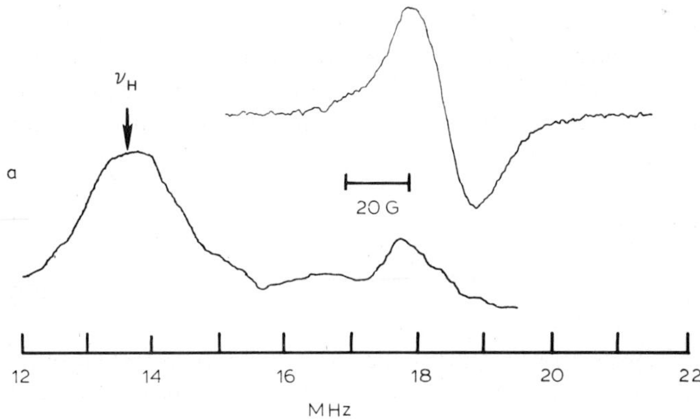

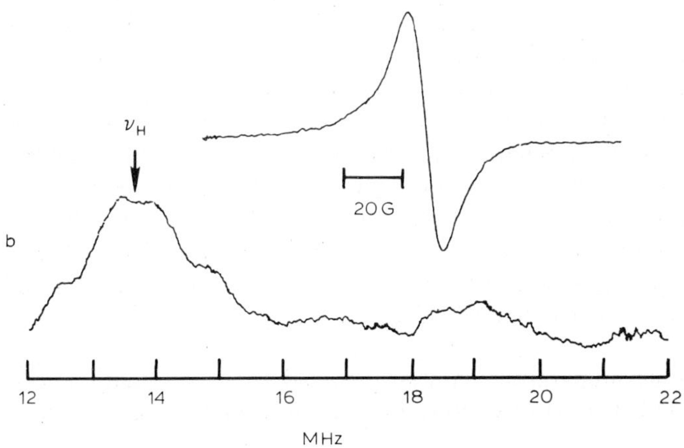

Figure 8. ESR (upper) and ENDOR (lower) spectra from flavoproteins at −160°C. a, Semiquinone radicals from the one-electron reduction of electron-transfer flavoprotein (ETF); b, semiquinone radicals from the one-electron reduction of sarcosine dehydrogenase. From [144], with permission.

not often clearly been demonstrated. Cytochrome *P*-450 reductase is an unusual flavoprotein that contains 1 mole of FAD and 1 mole of FMN. NADPH is the usual electron donor. Optical studies [147] have shown that the initial product from reaction of NADPH with the reductase is the two-electron reduced enzyme, which then can transfer one electron to oxygen to form superoxide and a long-lived blue flavosemiquinone. The semiquinone was shown by ESR [147] to be formed quantitatively in this reaction, and superoxide formation has been demonstrated by spin-trapping with DMPO. Presumably a similar one-electron transfer occurs between the two-electron reduced reductase and electron acceptors other than oxygen. Quantitative ESR data on formation of radicals from reductase substrates have been obtained for benzoquinone, 1,4-naphthoquinone and menadione [148].

The results are consistent [148] with a mechanism involving one-electron transfer from reduced flavoprotein to quinone. The resulting semiquinones from naphthoquinone and menadione subsequently transfer an electron to oxygen; benzosemiquinone does not. A similar one-electron reduction mechanism apparently operates with cytochrome-b_5 reductase.

$$Q \xrightarrow{\text{NADPH, reductase}} Q^{\dot{-}} \tag{23}$$

$$Q^{\dot{-}} + O_2 \rightleftharpoons Q + O_2^{\dot{-}} \tag{24}$$

Flavin semiquinones have been observed from succinate dehydrogenase in mitochondria [143] and in the reactions of general acyl CoA-dehydrogenase with substrate and with electron-transfer flavoprotein (ETF) [144]. For succinate dehydrogenase the semiquinone spectrum is narrow (12 G), reflecting the covalent attachment of the flavin to the protein; no ENDOR resonance attributable to methyl groups is present. Spin relaxation in the radical is faster than in related flavoprotein radicals, indicating that the flavin is in moderately close proximity to one of the iron-sulfur paramagnetic centers in the protein. It is thought [143] that this center is probably the binuclear cluster S-1, separated from the flavin by a distance of at least 12 Å.

Another important flavoprotein reaction occurs during the stimulation of neutrophils, which leads to a flux of superoxide radicals via the reduction of oxygen by a membrane-bound flavoprotein. Spin-trapping experiments with DMPO [149,150] have confirmed that superoxide radicals are released into the medium. Both $O_2^{\dot{-}}$ and $^{\cdot}OH$ (formed by secondary reactions of $O_2^{\dot{-}}$) have been spin-trapped in neutrophil systems. Spin-trapped radicals are not observed in the presence of superoxide dismutase.

Mitochondrial membrane preparations contain free radical ESR signals (Fig. 9) that have been attributed [151–153] to various forms of ubisemiquinones located in

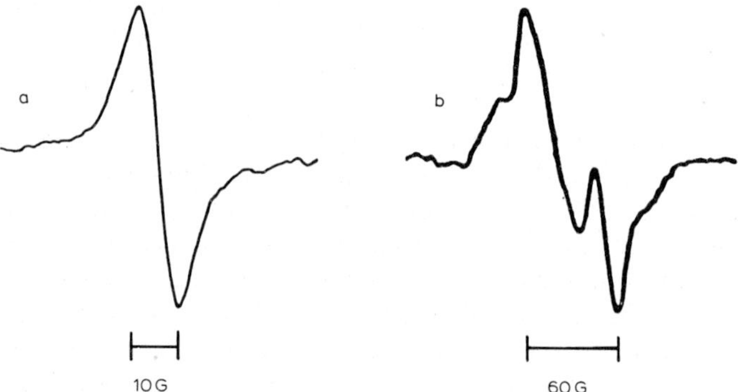

Figure 9. ESR spectra attributed to different species of ubisemiquinone in submitochondrial particles. a, Ubisemiquinone radicals at −70°C; b, dipolar-coupled ubisemiquinone radical pair at 12 K. From [153], with permission.

the membrane. The complexity of the spectra apparently reflects the high concentration and proximity of paramagnetic particles in these organized systems. Overlapping spectra almost always occur, although redox and potentiometric titrations and the use of specific complexing reagents can simplify the spectra to some extent.

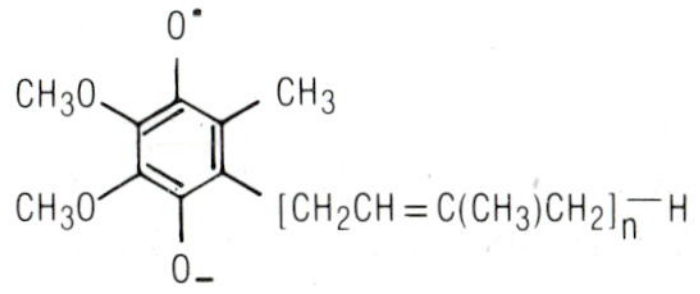

$O^{\cdot}$

CH_3O CH_3

CH_3O $[CH_2CH=C(CH_3)CH_2]_n$—H

O_-

Structure 10

It is clear from microwave saturation studies [152,153] that the ubisemiquinone population is heterogeneous: one fraction of the free-radical ESR spectrum saturates easily, whereas the other shows no saturation at the highest microwave powers attainable. The fraction that does not saturate is absent in the presence of low concentrations of the chelating agent, thenoyl trifluoroacetone. The simplest explanation of this behavior [153] is in terms of a semiquinone that is rapidly relaxed by an adjacent electron carrier in the membrane, for example, one of the iron-sulfur centers. It has been suggested that the non-saturating pool of ubisemiquinone is located in the vicinity of the succinate dehydrogenase, the other pool in the ubiquinol-cytochrome *c* reductase segment. In addition to the above, a strong spin-spin interaction between ubisemiquinone and another paramagnetic species is apparent under certain conditions. Additional features observed in the $g=2$ region, which are microwave frequency-independent and therefore cannot be ascribed to overlapping radical species, can be modeled by calculations involving a dipolar-coupled ubisemiquinone and a second radical species, either a second ubisemiquinone or a flavin semiquinone [90]. From considerations of midpoint redox potentials, it has been argued [92] that the signal is probably associated with a ubisemiquinone pair.

Immobilized semiquinones often are detected in membrane systems. In most instances analysis of their ESR spectra has been limited to measurement of a g value and a linewidth. More detailed analysis [154,155] could provide additional structural information to help identify the radical, its ionization state, and its local environment. The g value usually measured is that for the cross-over in the first derivative spectrum. While certainly useful in providing an approximate indication of the radical structure, it is generally frequency-dependent because of the anisotropy in the radical spectrum. It cannot give the more precise information on structure and environment that is available from the isotropic g value. The latter can readily be obtained from spectra of immobilized species by the method of Hyde and Pilbrow [156]. Related information is available from linewidth considerations [154]: in general, spectra for neutral semiquinones are approximately twice as broad as those for semiquinone anions and are sensitive to deuteration of solvent. Information of this kind can be of great help in assigning a spectrum of an unknown to a given radical species. The point has been made [154] that for semiquinones, as for flavin and other radicals, anisotropic saturation of the ESR spectrum of the immobilized radical does not necessarily imply that more than one radical species is present. It can simply reflect non-uniform interaction of the radical with its environment, so that its saturation behavior is different along the principal axes of the radical.

(iv) Mixed reaction mechanisms, redox equilibria

Whereas enzymatic one-electron oxidation or reduction to produce free radicals is common, one should be aware that some enzyme reactions can proceed by a mixture of one- and two-electron mechanisms, and also that free radicals can be formed indirectly subsequent to a two-electron enzymatic step.

For example, reactions of xanthine oxidase have been shown to occur by both one- and two-electron mechanisms with oxygen [157,158] and benzoquinone [159] as electron acceptors. Production of superoxide from one-electron donation to oxygen has been demonstrated by rapid-freeze ESR studies [160]. The immobilized free radical species that was detected was identified as $O_2^{\cdot-}$ by comparison with spectra obtained in chemical systems. However, the fraction of one-electron transfer that occurs depends [157] on a number of factors, including oxygen concentration, pH, and the concentration of the electron donor. The situation with benzoquinone is similarly complex: quantitative ESR studies [159] have shown that the extent of one-electron reduction depends upon the concentration of benzoquinone if xanthine is used as donor, but not if NADH is used. In addition, with NADH the reaction is very pH dependent. The apparent Michaelis constant for benzoquinone is much smaller with xanthine than with NADH. Because of the complexities of the xanthine oxidase system, it would appear that data from studies involving acceptors other than oxygen or benzoquinone must be analyzed carefully if reliable conclusions are to be drawn regarding the reaction mechanism.

In some situations substrate-derived free radicals occur by non-enzymatic reactions following product formation. This phenomenon seems to be common in enzyme reactions involving either two-electron oxidation of a hydroquinone or two-electron reduction of a quinone. In each case a mixture of quinone and hydroquinone is produced which is in chemical equilibrium with semiquinone radicals.

$$\text{quinone} + \text{hydroquinone} \rightleftharpoons 2 \text{ semiquinone radicals} \tag{25}$$

Reactions of this kind have been postulated [161] to account for semiquinone production from the oxidation of catechol by tyrosinase and the reduction of *p*-benzoquinone and menadione by DT-diaphorase [159]. The rate of radical production is very low compared with the rate of substrate depletion, indicating that the free radical is not the major primary product of the enzyme reaction. As in the peroxidase system, substrate-derived radicals having an appropriate structure can be spin-stabilized with metal ions.

Similar redox equilibria are felt [68] to be present in melanins, the final polymeric products of tyrosinase oxidation of catechols and catecholamines. Whereas natural melanins seem to be mostly derived from the oxidation of tyrosine, dopa, or the dopa metabolite cysteinyldopa, many other aromatic materials form related polymers upon enzyme or chemical oxidation. Oxidation of 5-hydroxyindoles [162] and benzidine [163] provide two such examples. Microsomal oxidations give ESR spectra that resemble those of melanins from dopa (below). Whether detailed studies can reveal differences between melanins from different precursors remains to be seen. Neuromelanin has been suggested to be derived from dopamine rather than dopa. Conceivably it might be possible to test this hypothesis with ESR.

ESR spectra of melanins are fairly characteristic. Eumelanins (derived mostly from dopa), which are the most common, show a single line with a *g* value close to 2.004 and a linewidth of approximately 4 G (Fig. 10a) [68]. Pheomelanins (from cysteinyldopa) have a quite different spectrum (Fig. 10b) which shows evidence of hyperfine interaction with nitrogen [164]. Many natural melanins are evidently derived from mixtures of dopa and cysteinyldopa and show intermediate spectra (Fig. 10c).

The free radicals found in dopa polymers are felt to be *o*-semiquinones (11), whereas those in cysteinyldopa polymers seem to be *o*-semiquinonimines (12) [164]. Character-

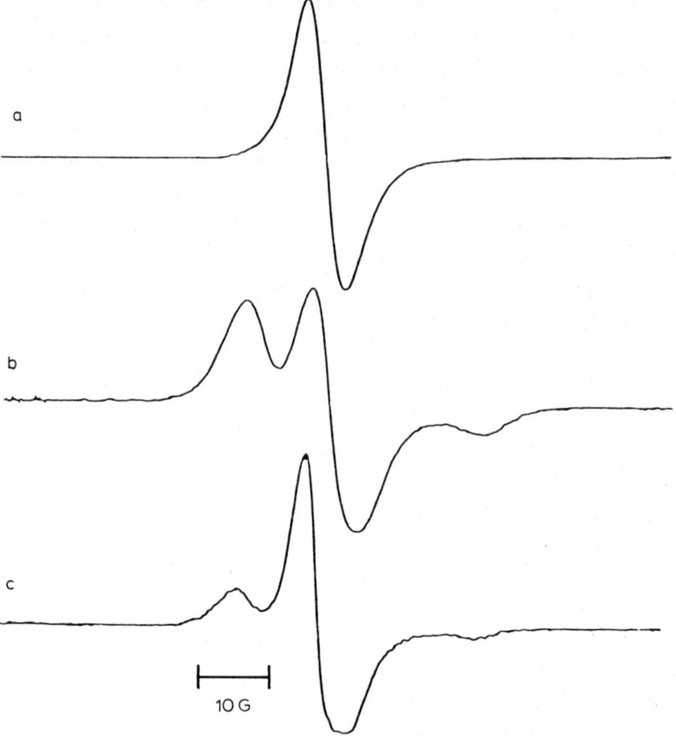

Figure 10. ESR spectra (−196°C) from different forms of melanin. a, Melanin from oxidation of dopa (a eumelanin); b, melanin from oxidation of cysteinyldopa (a pheomelanin); c, melanin from cooxidation of dopa and cysteinyldopa. From [164], with permission.

istic changes in spectra observed with diamagnetic complexing metal ions are consistent with the radicals having a chelating structure. The presence of redox equilibria in these polymers is based on evidence of changing radical concentrations in response to metal ions, changes in temperature [165] and hydrogen ion concentrations [166]. Because changes in radical concentration of this kind are uncommon with other biological free radicals, it has been proposed that such a series of chemical tests can be used to establish whether an unidentified radical or pigment has melanin-like properties.

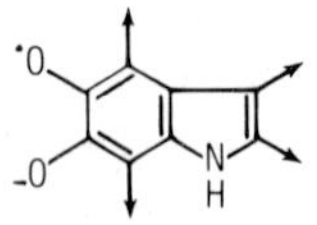

Structure 11

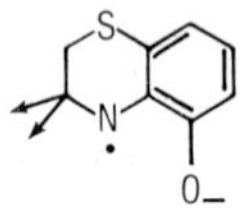

Structure 12

Melanins have photoprotective properties. Paramagnetic changes occur upon ultraviolet or visible irradiation and have been studied by ESR. Two types of reactions have been identified [161]: production of melanin free radicals and reduction of oxygen to $O_2^{\cdot-}$. Oxygen consumption in the polymer solution is strongly wavelength dependent, as shown by spin-probe measurements of oxygen concentration [167].

Almost all biological tissues contain some organic free radicals that are detectable by ESR. These radicals ('tissue radicals') are of low reactivity in the sense that they do not react readily with molecules in the system, in particular oxygen if this is present. Thus, they tend to be radicals at the end of a radical chain. Examples are the ascorbate radical anion, melanin free radicals, and some other oxygen-insensitive species, such as some flavin semiquinones. The magnetic properties of these various radicals are sufficiently distinct that their ESR spectra can be differentiated on the basis of g value and linewidth.

The spectrum of the ascorbate radical anion, which has been studied in detail in chemical systems [168], is particularly distinctive. This spectrum as normally observed in solution (Fig. 11a) is an approximate doublet of triplets (additional structure can be resolved under conditions of high resolution) with a g value of 2.0052, reflecting extensive delocalization onto oxygen. Since the radical is moderately transient at neutral pH (second-order rate constant for termination $(2k) = 5 \times 10^6\ M^{-1}\ s^{-1}$) [78], detection of the free radical in solution implies that it is being continuously generated in some way. It seems that in many systems sufficient oxidation of ascorbate is occurring to give detectable steady-state levels of the radical, especially where tissue damage has occurred (the radical is often detected in samples of blood and damaged tissue) [169]. Because ascorbate is a good antioxidant, its radical level might reflect the rate of other radical processes occurring in the tissue, but this remains to be demonstrated.

In immobilized (frozen or dried) systems, a very different spectrum (Fig. 11b) has been reported [170] from samples containing ascorbate. The spectrum is broad and

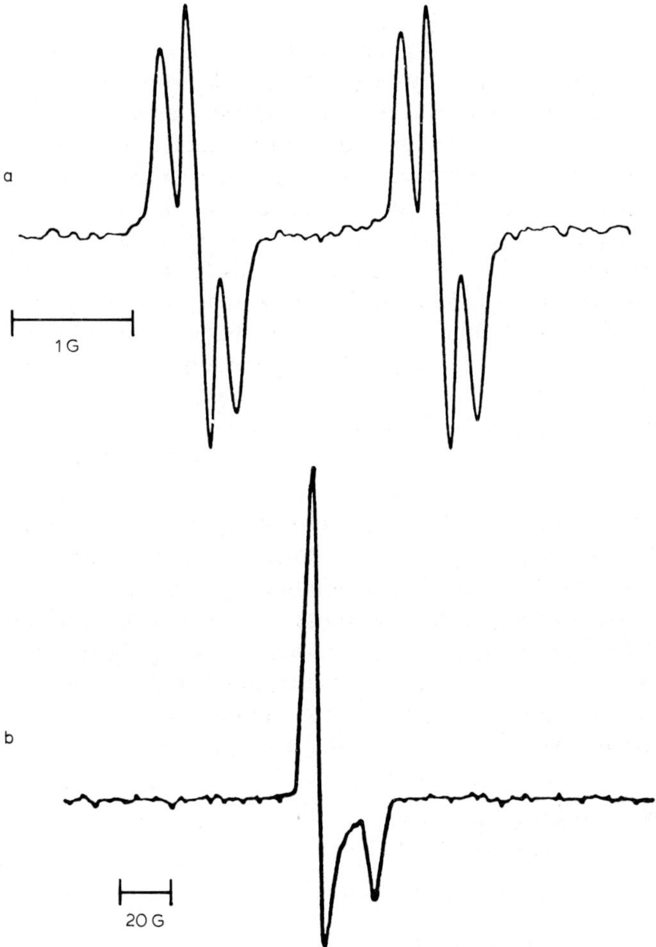

Figure 11. ESR spectra of radicals from the oxidation of ascorbate. a, Spectrum (25°C) of radicals from the photooxidation of ascorbate (20 mM) by hematoporphyrin (30 μM) at pH 7.6; b, spectrum (−150°C) from rat liver microsomes lyophilized in the presence of 0.1 mM ascorbate. From [88,170], with permission.

asymmetric and at 35 GHz appears axial, with $g_{\parallel} \simeq 2.006$ and $g_{\perp} \simeq 2.002$. It is probable that this is the immobilized spectrum of the ascorbate radical anion in which the anisotropy of the magnetic parameters no longer is averaged. This spectrum often is prominent in freeze-dried tissue samples, especially if oxygen is not rigorously excluded during the freeze-drying. Several-fold increases in ESR signal are possible during the freeze-drying procedure, particularly if protein is present. These effects must be taken into account if valid conclusions are to be drawn relating levels of ascorbate radical (or any other tissue radical) to different tissue states (e.g., normal vs. malignant).

(d) Radicals in drug metabolism

Activation of drugs to give toxic products is common. Apart from non-enzymatic activation (e.g., via autoxidation), activation by enzymatic one-electron oxidation or reduction frequently occurs. Several non-specific oxidases and reductases are encountered in mammalian tissues. Enzyme systems that have been studied in detail are peroxidases and microsomal oxidases and reductases. Xanthine oxidase also has received some attention. In many instances the end products of the reaction are critically dependent upon the presence of oxygen in the system. This is because oxygen is an excellent electron acceptor, i.e., it can oxidize donor radicals, forming superoxide in the process. In this way a redox cycle is set up in which the xenobiotic substrate is recovered. The toxic effects of the xenobiotic often can be attributed to the oxidative stress arising from such a cycle. However, it seems that for some substrates, oxidative stress of this kind can be less damaging than anaerobic reduction. Anaerobic reduction can lead to formation of further reduced products with additional toxicity.

(i) Oxidation reactions

One of the earliest demonstrations by ESR of the peroxidase oxidation of drugs involved the tranquilizer, chlorpromazine [171]. The stoichiometry of the reaction forming the chlorpromazine radical cation (13) was established, and kinetics of the transient radical were shown to be identical to those of an optically detected intermediate absorbing at 530 nm. Measurements were made of the pH-dependent radical dismutation rate and the rate of reaction of compound II with chlorpromazine. At high enzyme concentrations and low dismutation rates, the radical cation undergoes further oxidation by the enzyme, to form the thionium ion.

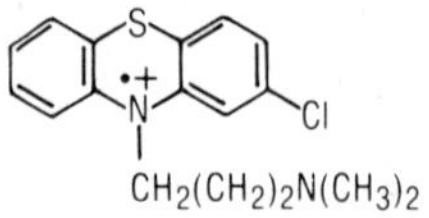

S
•+
N
Cl
$CH_2(CH_2)_2N(CH_3)_2$

Structure 13

There have been many subsequent demonstrations of radical production from peroxidase oxidation of xenobiotics, including benzidines, hydrazines, and sulfite [163,172,173]. Quantitation of the rate of radical production has not usually been carried out, even though in several cases the product radicals are sufficiently stable that this would have been quite straightforward.

Radical production from peroxidase-like systems, for example from the peroxide-supported oxidation of amino compounds catalyzed by protohemin and metmyoglobin, has received attention [174]. Both hydrogen peroxide and hydroperoxides (e.g., *t*-butyl hydroperoxide and cumene hydroperoxide) can be effective sources of oxidizing equivalents for these reactions. Since the enzyme prostaglandin synthase contains both cyclooxygenase and hydroperoxidase activities, either a substrate for the cyclooxygenase or an added hydroperoxide will support the catalyzed oxidation of substrates [175].

Phenylhydrazine is oxidized by oxyhemoglobin, initially forming a hydrazyl radical. Decomposition of this species ultimately leads to formation of a phenyl radical, which has been spin-trapped in hemoglobin-phenylhydrazine mixtures [176].

$$Hb(Fe^{2+})O_2 + PhNHNH_2 \xrightarrow{H^+} Ph\dot{N}HNH_2 + Hb(Fe^{3+}) + H_2O_2 \qquad (26)$$

ESR investigations of microsomal drug oxidation have been reported. For example, the microsomal enzyme mixed function amine oxidase converts hindered hydroxylamines to the corresponding ESR-detectable nitroxides. From a comparison of rates of nitroxide and superoxide production, it has been concluded [177] that oxidation of the hydroxylamine is mediated solely by enzyme-generated superoxide radical. In addition, it appears that some oxidations mediated by cytochrome *P*-450 may occur, at least in part, by a one-electron mechanism. Oxidation of several dihydropyridine derivatives and some substituted hydrazines in the presence of spin traps has given spectra of spin-adducts consistent with radical production from the *P*-450 substrates [178]. Cumene hydroperoxide has been shown to support the *P*-450-catalyzed oxidation of aminopyrine to its radical cation [179].

(ii) Reduction reactions

Microsomal reduction of many compounds is mediated by NADPH-cytochrome *P*-450 reductase. Quinones, nitro compounds, and azo compounds are just a few examples of materials that will accept an electron to form the corresponding radical anion. Many such radical anions have been fully characterized and identified on the basis of their hyperfine structure and *g* values. The primary radicals in each case are tumbling freely in solution, although prolonged reaction sometimes leads to immobilized species.

$$X \xrightarrow{\text{NADPH, reductase}} X^{\dot{-}} \qquad (27)$$

$$(X = Q, ArNO_2, ArN = NAr, \text{etc.})$$

Nitro-radical anions have proved to be especially accessible to ESR study [180–183]. The large nitrogen hyperfine splitting in the radical anion often leads to partly or completely resolved hyperfine structure (Fig. 12) that can be analyzed directly or by computer simulation. In the case of the nitro compound furylfuramide (AF-2), the *cis-trans* isomerization catalyzed by nitro reductases appears to proceed via the radical anion; ESR spectra of both *cis* and *trans* radical anions have been resolved during reduction of the *cis* parent compound.

Anthracycline anti-cancer drugs, such as adriamycin, also give intense ESR spectra in anaerobic microsomal incubations [184–186]. However, in these cases overlap of hyperfine lines is considerable, so that a single broad symmetric ESR line is detected, typically 4–9 G in width, with $g = 2.0035$. While it is extremely likely that this radical is indeed the primary semiquinone (i.e., the one-electron reduced drug), evidence to confirm the radical structure (as might perhaps be obtained from ENDOR measure-

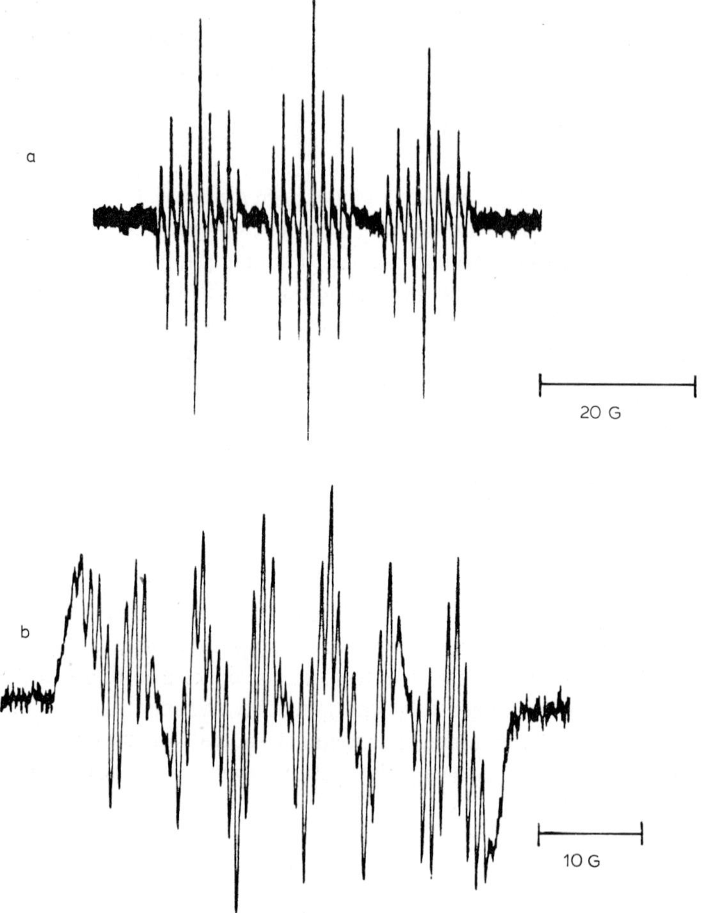

Figure 12. ESR spectra of nitro-anion free radicals from the microsomal reduction of nitro compounds. a, Radicals from the anaerobic incubation of 2 mM nitrobenzene with an NADPH-generating system and 1.5 mg/ml hepatic microsomes; b, radicals from the anaerobic incubation of 1 mM nitrofurantoin with an NADPH-generating system and 1 mg/ml hepatic microsomes. From [88,169], with permission.

ments of hyperfine splittings) would be valuable. An additional complication in these systems is that polymeric products are rapidly formed so that the spectrum of what is presumably the primary radical is soon replaced by the axial spectrum of an immobilized species, with $g_{\parallel} = 2.0046$, $g_{\perp} = 2.0021$.

Detection of primary radical anions in microsomal incubations generally is possible only under anaerobic conditions. In the presence of air a redox cycle is set up, as referred to above. Thus, in initially aerobic systems, the primary radical can be detected only after a lag time that corresponds to the time required to deplete oxygen in the medium. During this lag time the superoxide radical is generated, which can be spin trapped using the nitrone 5,5′-dimethylpyrroline-1-oxide (DMPO) [187].

Microsomal reduction of carbon tetrachloride is also a reductase-mediated reaction, giving the trichloromethyl radical $\dot{C}Cl_3$ as the primary reduction product. The $\dot{C}Cl_3$ formed has not been detected directly, but its spin-adduct with phenyl-*N*-*t*-butylnitrone (PBN) has been characterized and identified [188]. Since secondary radicals from reaction of $\dot{C}Cl_3$ (e.g. $Cl_3CO_2^{\cdot}$ and lipid radicals) also can be formed in the system and subsequently trapped [189], analysis of the ESR spectra obtained must be conducted with some care. In all spin-trapping studies it is valuable to investigate the effect of varying the concentration of the trap, since this can differentiate between spectra of trapped primary and secondary radicals: the former should be favored by high concentrations of spin trap, the latter by lower concentrations.

Not all microsomal reductions necessarily require NADPH-cytochrome *P*-450 reductase. One apparent exception is the reduction of Gentian violet and related compounds to triarylmethyl radicals, which is in part mediated by cytochrome *P*-450, as judged [190] by the inhibitory effect of CO and metyrapone.

Xanthine oxidase is another enzyme that will catalyze the transfer of electrons to anthracyclines. Both NADPH and xanthine support the reduction of adriamycin and daunorubicin. Transfer of electrons to the drug competes with transfer to oxygen. In addition, the anthracycline semiquinone will reduce oxygen to superoxide, so that the single-line ESR spectrum of the semiquinone again is detected [191] only after a lag period, during which oxygen is removed. Whether reduction of anthracyclines occurs solely by a one-electron mechanism or a mixture of one- and two-electron mechanisms (see discussion above concerning reduction of oxygen and benzoquinone) is not clear.

4. Metal ions

(a) General remarks

The third principal application of the electron spin resonance technique is to the study of paramagnetic transition metal ions in biochemical systems. Most examples are complexes of copper, iron, manganese, chromium, cobalt and molybdenum. Other metals such as titanium, vanadium and nickel are sometimes employed as structural probes. Only four of these ions, Cu^{2+}, Mn^{2+}, Gd^{3+} and VO^{2+}, are seen in ESR spectroscopy at room temperature under virtually all conditions. Therefore, they are of special importance.

Both the basic theory and techniques for studying metal complexes by conventional ESR are well established and have been extensively reviewed [2,11,13,192]. The treatise of Abragam and Bleaney [18] is the definitive reference. An earlier book by Low [193] is useful. The book edited by Yen [194], although now somewhat dated, provides a more chemically based source of information. Abragam and Bleaney use crystal field theory, breaking the fields down according to weak, intermediate and strong – discussing the effect of the fields on the 3d, 4f and 5f transition series. From a biochemical point of view, the metals of the 3d group (Ti, V, Cr, Mn, Fe, Co, Ni, Cu) are by far the most important. Yen's book is organized more from a point of view of covalently bound ligands – i.e., metal complexes – than from a crystal field perspective. Historically, crystal field theory and molecular orbital approaches to describing metal ion centers are complementary and of similar importance.

In this section, we briefly introduce the ESR parameters that are obtained for Mn^{2+}, Co^{2+}, Fe^{3+}, Ni^{2+} and Cu^{2+} complexes, using as example one liganding molecule, the antitumor agent, bleomycin, which binds to a multitude of metal ions. The section terminates with an example from the ESR spectra of cupric complexes. Although the discussion of this example is lengthy, the techniques are applicable to other paramagnetic metals. Some limitations of the standard ESR technique are discussed prior to a survey of some of the more sophisticated methods that are available.

Identification of paramagnetic metal ions is straightforward because each metal ion has sufficiently distinct ESR parameters [5–7,18,195,196]. The principal magnetic axes, defined by dimensionless g values, the interaction of the electron spin with the nuclear spin, defined by the hyperfine coupling constants, and spin-lattice relaxation times characterize each paramagnetic ion.

Transitional metal ions are tightly bound by the multidentate ligand, bleomycin. Bleomycin (Blm) is an antitumor antibiotic isolated from *Streptomyces verticillis* as a copper complex [197]. Its mechanism of action has been hypothesized to involve a strand-scission reaction of DNA whereby bleomycin intercalates DNA, scavenges ferrous ion, and, upon oxidation to ferric ion, generates damaging oxygen radicals [198–204]. This hypothesis has led to many ESR studies including investigations of the paramagnetic active antitumor agents, FeBlm and CuBlm, the inactive CoBlm, and analogs of both active and inactive antitumor agents [201,205–207]. Since preformed CuBlm, FeBlm, ZnBlm and metal-free Blm are cytotoxic to cells but only FeBlm is active in the strand-scission reaction, it is hypothesized that other metallo forms may be converted to FeBlm. Competitive binding studies have attempted to account for these phenomena [205]. Blm contains a number of moieties, including primary amine, pyrimidine, peptide and imidazole nitrogens that provide strong donor atoms for metal binding together with a number of weaker donor atoms (Fig. 13).

Whereas the binding of the different donor atoms probably depends on the properties of the metal ion, differences in binding are difficult to determine from ESR data. When Mn^{2+} is bound to ligand (Fig. 14) the asymmetry in the coordination sphere results in a zero field splitting that reduces the overlap of the five fine structure transitions. Lines are broadened, the patterns become more complex and the intensity is reduced by an order of magnitude from the free (hydrated) Mn^{2+} signal. A hyperfine coupling constant of about 93–94 Gauss is indicative of a complex that is six-coordinate octahedral or distorted octahedral. The splitting of about 80 G found for MnBlm (Fig. 14) suggests that the coordination is less than octahedral.

Like Mn^{2+}, high-spin Fe^{3+} is usually in a strongly distorted octahedral environment in biological materials and only the transition between the 1/2 magnetic levels is

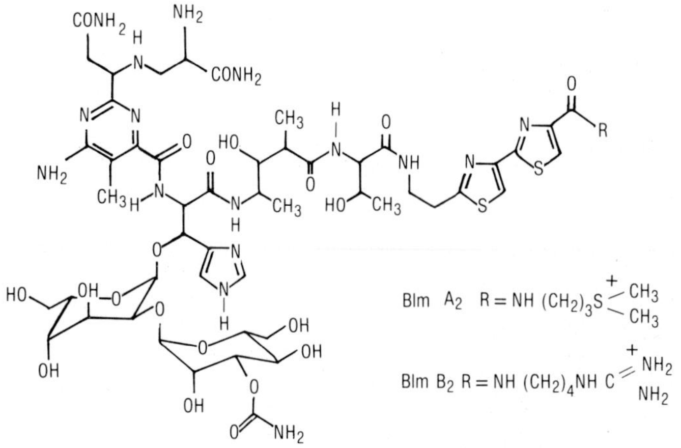

Figure 13. Structure of bleomycin (Blm) with variable R group.

observed. The g values for rhombic symmetry span $g = 2$ to about $g = 9.7$. Low spin Fe^{3+} with one unpaired electron often has rhombic symmetry and three g values which range from 1 to 3.

Fe^{3+}Blm undergoes a change from high spin ($S = 5/2$) with $g_{\perp} = 4$ and $g_{\parallel} = 2$ (not shown) to low spin ($S = 1/2$) with $g_1 = 2.45$, $g_2 = 2.18$ and $g_3 = 1.89$ as the pH is increased from about 3 to greater than 7 (Fig. 14) [201,208,209]. These g values can be used to identify the donor atoms through comparison with compounds for which the donor atoms are known [210]. The g values for Fe^{3+}Blm are similar to values for heme porphyrins and suggest four in-plane nitrogen donor atoms. The time sequence of spectra shown in Figure 14 was obtained from a system in which oxygen was added to Fe(II)Blm and the subsequent oxidation to Fe(III)Blm followed. The intermediate or 'activated form' has g values of 2.28, 2.17 and 1.94, which appear to result from quenching of the orbital angular momentum. In other Fe^{3+} complexes replacement of oxygen or nitrogen donor atoms by sulfur donor atoms is known to quench the orbital angular momentum, but the only sulfur atom in Blm is a bithiazole moiety which does not appear to be bound. This ESR spectrum has been attributed to a hydrogen peroxide-adduct, HO_2^-Fe(III)Blm [201]. The two-electron reduction is not obvious from the apparent reactant, which is either O_2-Fe(II)Blm or $^{\cdot}O_2$-FE(III)Blm. Peisach and co-workers [201] have substituted ^{57}Fe ($I = 1/2$) for ^{56}Fe(III)Blm and found a splitting of 22 G in the $g = 1.94$ region, indicating that the unpaired electron is not primarily localized on a ligand atom [201]. In addition, $^{17}O_2$ ($I = 5/2$) broadening shows that the HO_2^-Fe(III)Blm activated form is derived from O_2.

The ESR spectrum for unoxygenated Co(II)Blm is characteristic of a five-coordinate configuration with an unpaired electron in the d_z^2 orbital. The spectrum exhibits two major sets of features with $g_{\perp} \simeq 2.3$ and $g_{\parallel} \simeq 2.03$. Hyperfine coupling to Co ($I = 7/2$) splits the $g_{\parallel}$ and $g_{\perp}$ features into eight lines, but only the $g_{\parallel}$ region is resolved. Typical values for $A_{\perp}$ are 10 G and for $A_{\parallel}$ 70–90 G. If the axial ligand is a

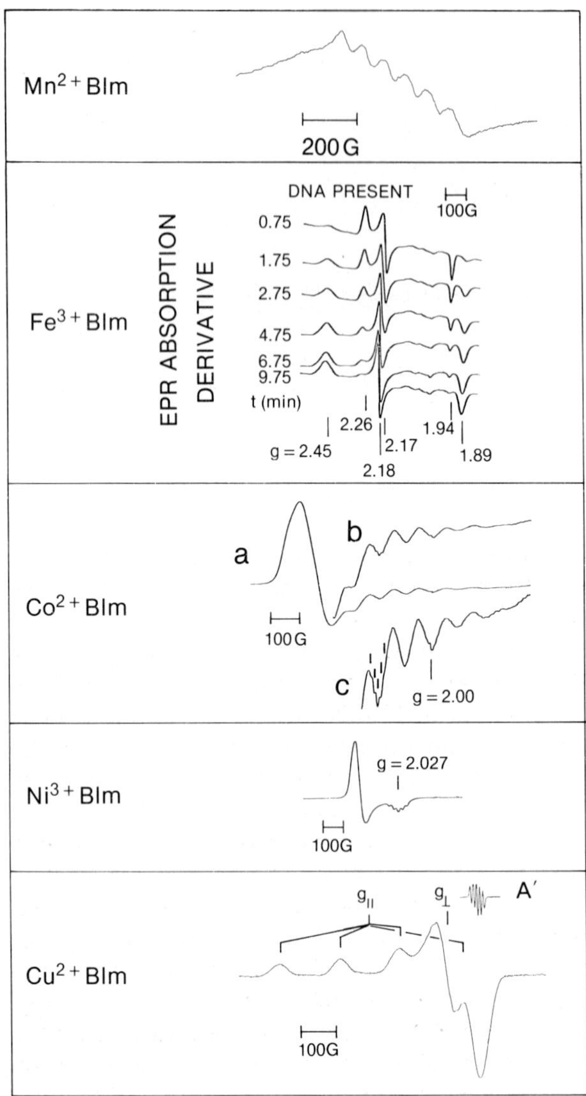

Figure 14. ESR spectra of MnBlm, FeBlm, CoBlm and CuBlm. Panels on FeBlm and NiBlm from [201,214], with permission.

strong nitrogenous base, the eight hyperfine lines in the $g_{\parallel}$ region can be split further due to superhyperfine coupling from axial ^{14}N ($I = 1$). For example, Sugiura [211] has published spectra for Co(II)Blm in the presence of DNA which show evidence of nitrogen hyperfine structure on one of the eight lines in the $g_{\parallel}$ region. A five-coordinate configuration with a nitrogenous axial ligand was inferred. The spectrum of Co(II)Blm in the absence of DNA shown in Figure 14 shows better resolution than that reported by Sugiura, having an apparent 1–2–3–2–1 pattern suggestive of a six-coordinate configuration. Addition of a second ligand to form a six-coordinate complex is known to occur only with a few nitrogenous bases, e.g., pyridines; addition of a second ligand usually alters the g values and improves resolution in the $g_{\parallel}$ region [212]. However, Co(II)Blm (Fig. 14) under the conditions given appears to have $g_{\perp}$ and $g_{\parallel}$ consistent with a five-coordinate complex, but axial hyperfine structure consistent with a six-coordinate complex. Addition of oxygen forms an adduct and alters the ESR parameters such that $g_{\parallel} \simeq 2.098 > g_{\perp} \simeq 2.0$ and $A_{\parallel}$ is reduced to about 15–25 G ($A_{\perp}$ remains $\simeq 10$ G) [210].

Ni(II) complexes of oligopeptides containing histidine are easily oxidized to Ni(III) complexes (d^7) with $Ir(IV)Cl_6^{2-}$ [213]. If the geometry for the Ni(III) complex is tetragonal, then g_x and g_y are greater than g_z, as illustrated in Figure 14. If the geometry is square planar, than g_z is greater than g_x and g_y. The Ni(III) complexes with in-plane sulfhydryl donor atoms have larger g_1 values (also referred to as g_{xx}) and more rhombic symmetry. Sulfhydryls have greater donor strength than in-plane amino nitrogens [214]. The ESR spectrum Ni(III)Blm has a five-line 1–2–3–2–1 pattern in the axial direction which also indicates the presence of two axial donor atoms (Fig. 14) [214]. Hyperfine structure from ^{61}Ni ($I = 3/2$) in ^{61}Ni-enriched hydrogenase from *Desulfovibrio desulfuricano* was used to show that the rhombic ESR signal ($g_1 = 2.32$, $g_2 = 2.16$, and $g_3 = 2.01$) observed in the oxidized state of hydrogenase is derived from a nickel complex [215].

The application of spectroscopy to copper proteins has been outlined in several texts [216]. Because the relaxation times of copper complexes are relatively long, ESR spectra are readily obtained at room temperature. Complexes with a molecular weight of less than 2000 tumble fast enough that the magnetic parameters associated with the principal magnetic axes are averaged to give isotropic values. In copper complexes, the unpaired electron occupies the $d_{x^2-y^2}$ orbital. Copper has a nuclear spin of 3/2 which interacts with this unpaired electron and splits the copper resonance into $2I + 1$ (=4) equally spaced lines (ignoring second-order shifts). The center of this four-line pattern is the isotropic g value, g_{iso}, which is less than the g value for the free electron, $g = 2.00$, because spin-orbit coupling introduces orbital motion. (Again second and higher order corrections are neglected.) If the donor atoms in the copper complex are nitrogens, one or more of the four lines may be further split due to $2I + 1$ lines where $I = 1$ for each nitrogen donor atom. The spectra in Figure 16 show such features.

If the molecular weight is greater than 2000 or if the sample is frozen without aggregation (for example, the copper complex can be immobilized in a diamagnetic matrix, preferably with glass-like properties), an isotropic spectrum is no longer observed. ESR parameters associated with the two principal magnetic axes, in-plane and perpendicular to the square plane, are marked in Figure 14 (bottom spectrum) for CuBlm. These parameters are related to the isotropic values by

$$g_{iso} = 1/3\,(g_{\parallel} + 2g_{\perp}) \quad \text{and} \quad A_{iso} = 1/3\,(A_{\parallel} + 2A_{\perp})$$

where $g_\perp$ and $g_\parallel$ refer to directions in-plane and parallel to the perpendicular to the in-plane direction, respectively [192]. If the parameters for the in-plane direction are not equivalent, then

$$g_{iso} = 1/3\,(g_1 + g_2 + g_3) \quad \text{and} \quad A_{iso} = 1/3\,(A_1 + A_2 + A_3)$$

Usually three of the four lines for the axis perpendicular to the square plane are resolved so that values for $g_\parallel$ and $A_\parallel^{Cu}$ can be determined. Structure for A^N is not always resolved. Lines from the $g_\perp$ region comprise the most intense feature and can be resolved further if nitrogen donor atoms are present. However, the interpretation of these lines can be ambiguous because the magnitudes of A^{Cu} and A^N are typically of the same order. Assuming $A_\perp^{Cu} \simeq A_\perp^N$, each of the copper hyperfine lines in the $g_\perp$ region would be split into a 1–3–6–7–6–3–1 pattern for three equivalent nitrogen donor atoms or a 1–4–10–16–19–16–10–4–1 pattern for four equivalent nitrogens. Many of the lines in the $g_\perp$ region are overlapped, resulting in a complicated pattern which is difficult to interpret. If the nitrogen couplings are inequivalent, the pattern in the $g_\perp$ region is comprised of more lines and the couplings are not as well resolved. Caution is recommended with respect to over-interpretation of such data. The presence of additional lines in the $g_\perp$ region that result from the angular dependence of the copper hyperfine lines, i.e., 'overshoot' lines, further complicates a strategy based simply on counting the number of lines in the perpendicular region [192].

Adequate computer simulation usually requires accurate values for $g_\parallel$, $A_\parallel^{Cu}$, and g_{iso}, together with good estimates for $g_\perp$, $A_\perp^{Cu}$ and A^N. Replication of the experimental spectrum should require only small changes in the input parameters. A table of $g_\parallel$ and $A_\parallel$ values for several cupric complexes is useful for comparison with the experimentally determined parameters [192]. If $A_\parallel$ is larger than 120 G, then the compound is square planar or distorted square planar; if $A_\parallel$ is less than 120 G the complex shows distortion toward a tetrahedral configuration [192]. The latter type of complex is usually characterized by a strong blue color.

Plots of $g_\parallel$ versus $A_\parallel$, generally referred to as Peisach-Blumberg plots, are indicative of the donor atoms bound to cupric ion [217]. (Plots for iron complexes are also useful indicators of the donor atoms [218].) In general, $g_\parallel$ decreases when nitrogen replaces oxygen or sulfur replaces either nitrogen or oxygen in the square planar configuration. For complexes with the same donor atoms a decrease in charge results in a decrease in $g_\parallel$ and an increase in $A_\parallel$. Although these ESR parameters provide indirect evidence for the cupric ion coordination, direct techniques are now available which eliminate some of the ambiguity which arises from these kinds of comparisons.

(b) ESR of metalloproteins and metalloenzymes

The intent of this section is to focus on various techniques employed to study ESR of metal ions. The literature cited is neither complete nor in many ways representative of the enormous volume of work utilizing ESR to study proteins and enzymes. Much of the early pioneering work done in applying ESR techniques to biochemical systems has been reviewed by Bienert [80]. His review describes the use of isotopic substitution to unravel hyperfine structure, of temperature variation and power saturation studies to separate the spectra of free radicals from those of metal centers, and of rapid-freeze techniques to determine stoichiometry and kinetics of oxidation-reduction reactions. The value of substituting ^{57}Fe ($I = 1/2$) for ^{56}Fe and obtaining iron spectra at very low temperatures to detect additional iron was discussed for non-heme iron groups. The molybdenum ion in, for example, xanthine oxidase shows very rapid, rapid, and slow ESR signals using the rapid-freeze technique. This probe gives an intense central feature (arising from 75% Mo with $I = 0$) surrounded by six less intense hyperfine lines (from the 25% Mo with $I = 5/2$) for the Mo(V) state, which is the oxidation state found in most biological materials. Recent studies with non-heme iron and molybdenum include the determination of proximate distances between Mo, Fe–S(I), Fe–S(II) and FAD [219]; the demonstration of non-random orientation of nitrate reductase in the membrane [220]; and a study of oxidation-reduction properties of the cofactor of nitrogenase [221]. Other recent studies of non-heme iron which utilize ESR concern the interconversion of Fe–S clusters [222]; the detection of a number of high-spin Fe(III) species in soybean lipoxygenase-1 [223]; the titration of porcine uteroferrin with two reducing equivalents of ferrous ion [224]; and the kinetics for the reduction of the Fe(III) ESR signal from ferrichrome A which was correlated with the rates of iron transport [225].

ESR of heme proteins has characterized both the high-spin ($S = 5/2$) and the low-spin ($S = 1/2$) forms. The review by Palmer [226] on ESR of heme proteins is particularly useful, in which the model for low-spin iron complexes (for which the d_{xy}, d_{xz} and d_{yz} are split from the d_{z^2} and $d_{x^2-y^2}$ orbitals) is clearly explained. The five electrons for low spin ferric complexes fill the d_{xy} and d_{xz} orbital and half fill the d_{yz} orbital. For a positive axial distortion, the d_{xz} and d_{yz} orbitals are raised in energy and the d_{xy} orbital is lowered in energy. A smaller rhombic component, V, splits d_{xz} and d_{yz}, raising d_{yz} and lowering d_{xz}. If there exist three different g values, the unpaired electron is in an orbital comprised of a linear combination of d_{yz} with d_{xz} and d_{xy}. The symmetry parameters Δ and V can be calculated from the three g values. These g values or symmetry terms (Δ and V) are compared to five groups obtained from Peisach-Blumberg plots of V/Δ vs. Δ/λ where λ is the spin-orbit coupling constant [210,218]. Thus, similar parameters show a correlation with structure. EPR of high-spin heme iron is usually observed only from the Zeeman splitting of the $\pm 1/2$ states and gives a line at $g_{\parallel} \simeq 2$ (coinciding with H_0 parallel to the heme normal) and $g_{\perp} \simeq 4$ or $g_{\perp} \simeq 6$ (coinciding with H_0 in the heme plane). The g values may be used to calculate the rhombicity of both high- and low-spin ferrihemoproteins [210,218].

Stable structures such as the naturally occurring ferric porphyrin complexes or porphyrins substituted with copper, cobalt, silver or vanadyl probe the active site of heme-containing enzyme [227]. Complexes of copper not associated with heme are also common. They are frequently formed at an amino terminus because the amino group provides a good primary amine donor atom. Two or three amino acid residues beginning at the N-terminus are often flexible until a more rigid portion of the polypeptide, such as the α helix, is encountered. A peptide nitrogen is available to provide stronger chelation. If the amino acid histidine is the second or third amino acid from the N-terminus, the imidazole ring provides an additional nitrogen donor atom and a stable multidentate complex is formed. ESR spectra for cupric complexes bound to the N-terminus have been described for hemoglobin [228] and serum albumin [229]. It is thought that the binding site on serum albumin functions as a carrier for cupric ion. Little is known about the function of the site on hemoglobin [230,231]. The binding site in hemoglobin is similar to, and can be modeled by, that in complexes of cupric ion bound to low-molecular-weight peptides (which also may be cupric ion carriers [232]).

The ESR spectrum for cupric ion bound to four equivalent imidazoles from carnosine (β-alanyl-L-histidine dipeptide) in frozen solution shows the quality of information that can typically be obtained [233] (Fig. 15, top spectrum). The intense line on the right split by at least nine hyperfine lines is the $g_{\perp}$ feature resolved due to hyperfine splitting to copper and to nitrogen donor atoms. Three of the four copper hyperfine lines in the $g_{\parallel}$ (2.25) region (left side) are observable with $A_{\parallel}^{\mathrm{Cu}} = 175$ G. The remaining ESR parameters can be estimated and confirmed by computer simulation. These values for carnosine are $g_{\parallel} = 2.06$, $A_{\perp}^{\mathrm{Cu}} = 15$ G and $A^{\mathrm{N}} = 15$ G. For comparison, data for cupric ion bound to hemoglobin at the N-terminus are $g_{\parallel} = 2.210$, $g_{\perp} = 2.050$, $A_{\parallel}^{\mathrm{Cu}} = 194$, $A_{\perp}^{\mathrm{Cu}} = 18$ G and $A^{\mathrm{N}} = 14.7$ G [228].

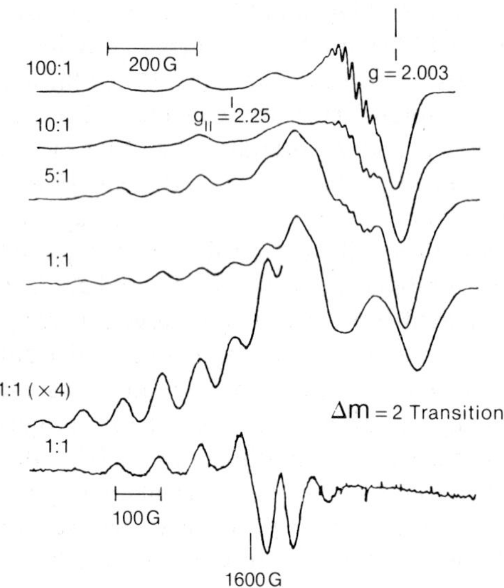

Figure 15. Transition from Cu-Cu dimer (bottom spectra) to Cu monomer (top spectra) as the ratio of ligand to metal increases. Effect of the ratio of molar concentrations of carnosine to Cu (from 100:1 to 1:1) on the ESR spectra (X-band) of frozen (77 K) aqueous solutions of 9.10×10^{-4} M copper(II) ion. All solutions contain 0.27 M sucrose and are adjusted to pH 7.2 ± 0.1. The half-field $\Delta M = 2$ transition of the copper(II) dimer is shown at the bottom of the figure. Similar half-field transitions of lower intensity also were observed at concentration ratios of 5:1 and 10:1. Substitution of 0.82 M sodium perchlorate for the sucrose has no effect on the spectra. From [233], with permission.

As the concentration of carnosine is reduced relative to that of copper, a dimer is formed (Fig. 15). Resolution of the ESR lines can be enhanced by the choice of medium (in this case one containing excess sucrose) and is less good in fluid solution [234]. The seven-line pattern (1–2–3–4–3–2–1) from two equivalent cupric ions ($I = 3/2$) is apparent in the $g_{\|}$ region around 3200 Gauss and an analogous pattern due to the half field transitions is found around 1600 Gauss. Cupric dimers have been reviewed recently by Boas et al. [235]. If one obtains a spectrum which does not appear to be indicative of a single cupric complex, one should consider the possibility of dimer formation. Alteration of the ratio of ligand to cupric ion may allow spectral separation.

(c) Complementary probes

(i) Isolated metal centers

Complementary probes are paramagnetic metal complexes for which a characteristic spectrum can be observed in biological systems at room temperature [236]. VO^{2+}, Mn^{2+}, Fe(II)–$(NO)_2$ and Cu^{2+} are examples of complementary probes. Most Fe^{3+} (both high- and low-spin) Co^{2+} and Ni^{2+} complexes have short relaxation times at room temperature that lead to extreme broadening of the spectra. ESR signals for these complexes are observed only in frozen solutions. Three situations may occur when a complementary probe such as a cupric complex is introduced into a biological system: (1) reduction (e.g., of Cu^{2+} to Cu^{1+}), (2) no reaction, and (3) adduct formation. Work with cupric complexes of known antitumor agents indicates that Cu^{2+} is reduced by excess glutathione in Ehrlich ascites tumor cells and red cells [237,238]. Isolation and non-reactivity of the cupric complex of 3-ethoxy-2-oxobutyraldehyde-bis(*N*-dimethylthiosemicarbazone) in membranes [239] has been reported. Cu^{2+} bound to the tridentate ligand, 2-formyl pyridine monothiosemicarbazone, interacts with Lewis bases (histidine, cysteine) in proteins, to form square planar adducts [236,240,241].

The case in which no reaction occurs has been somewhat neglected. In this case complementary complexes may be utilized as metallo spin probes. Cupric complexes provide information on the environment through the ESR parameters, $A_{\|}^{Cu}$, $A_{\perp}^{Cu}$, $g_{\|}$, $g_{\perp}$, A^{N}, and the rotational correlation time τ_r. Hyde and Froncisz [242] have reviewed the role of changing microwave frequency with respect to changes in liquid phase Cu^{2+} ESR spectra. They discuss the theory for the changes in linewidths and for second-order shifts of the lines in terms of A_{iso}, $\Delta A = A_{\|} - A_{\perp}$, g_{iso}, $\Delta g = g_{\|} - g_{\perp}$, τ_r, and the microwave frequency, and suggest that it is possible to determine both magnetic parameters and local viscosity from analysis of linewidth and shift data. Spectra for the copper complex of 3-ethoxy-2-oxobutyraldehyde-bis(*N*-dimethyl-thiosemicarbazone) at four different frequencies lead to sufficient independent equations to determine the ESR parameters and the rotational correlation time (Fig. 16). Because $CuKTSM_2$ partitions into non-polar solvents it is anticipated that analysis of $CuKTSM_2$ spectra will be useful for probing the membrane viscosity of cells.

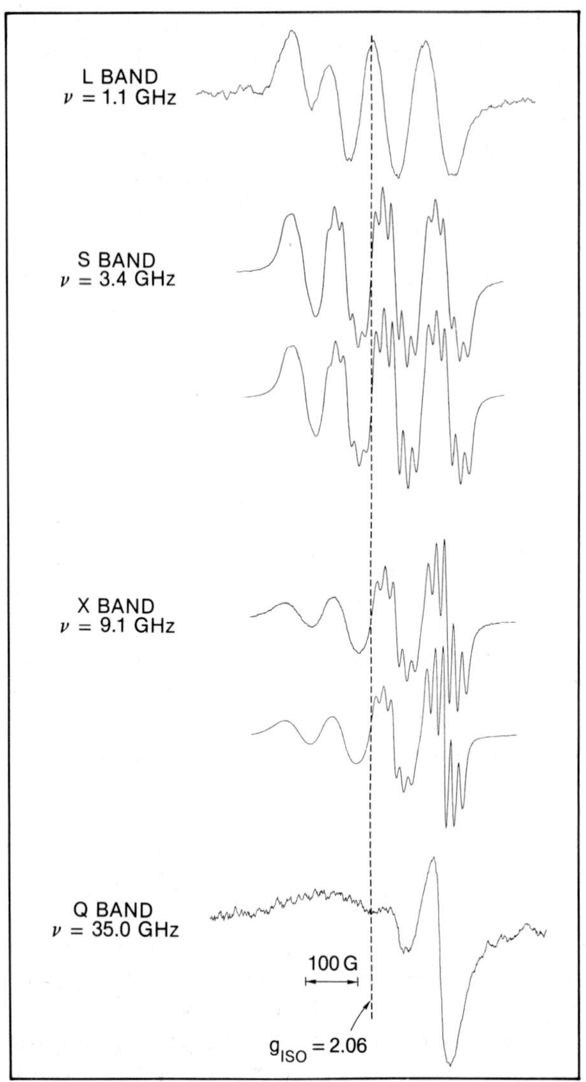

Figure 16. ESR spectra for $CuKTSM_2$ at room temperature at five different microwave frequencies. Note differences in lineshape and shifts in line positions. From [243], with permission.

Basosi et al. [244] have utilized adducts of the paramagnetic probe $Fe(I)(NO)_2$ to investigate the structural features and dynamic properties of biological ligands in solution. $Fe(I)(NO)_2$ forms adducts with two donor atoms from antithyroid drugs and two solvent molecules, to give either an elongated octahedral structure with a single unpaired electron in the d_{z^2} orbital or a flattened octahedral structure with the unpaired electron in the $d_{x^2-y^2}$ orbital. Low-spin iron complexes (d^7) give a five-line spectrum with intensities 1–2–3–2–1 due to two equivalent ^{14}N ($I = 1$) donor atoms if the drug donates oxygen or sulfur atoms. If the isotope ^{15}N is substituted for ^{14}N in the two NO groups, a three-line spectrum with intensities 1:2:1 is obtained. If the donor atoms from the drug are nitrogens, as in the structure obtained using sulfanilamide ligands (Fig. 17), a nine-line pattern is observed with ^{14}NO and a seven-line pattern with ^{15}NO. Spectra can be simulated with $A_{^{14}NO} = 2.2$ G, $A_{^{15}NO} = 3.0$ G and A_N sulfanilamide = 2.8 G. The strength of drug binding to $Fe(I)(NO)_2$ correlates with antithyroid activity. It was suggested [244] that the study of $Fe(I)(NO)_2$ complexes can determine which bases from the ligands provide the best donor atoms for metal complexation. If the binding is particularly tight, metal complexation may be expected to be involved in the pharmacological action of the drug.

Manganese (Mn^{2+}) has often been utilized to help define the magnesium binding site in proteins, enzymes, and nucleotides. The reduction in symmetry when going from aquo manganese to enzyme-bound manganese is correlated with a change from six intense, well-resolved hyperfine lines to six or more less intense poorly resolved lines. Solvent anions and/or allosteric effectors often shift the ESR lines for enzyme-bound Mn^{2+}. The ESR spectrum for Mn^{2+} is not usually sensitive to the donor atoms from the enzyme; for example, generally one cannot distinguish whether enzyme donor atoms are nitrogens or oxygen without recourse to additional strategies.

One such strategy involves isotopic substitution. A method to determine the

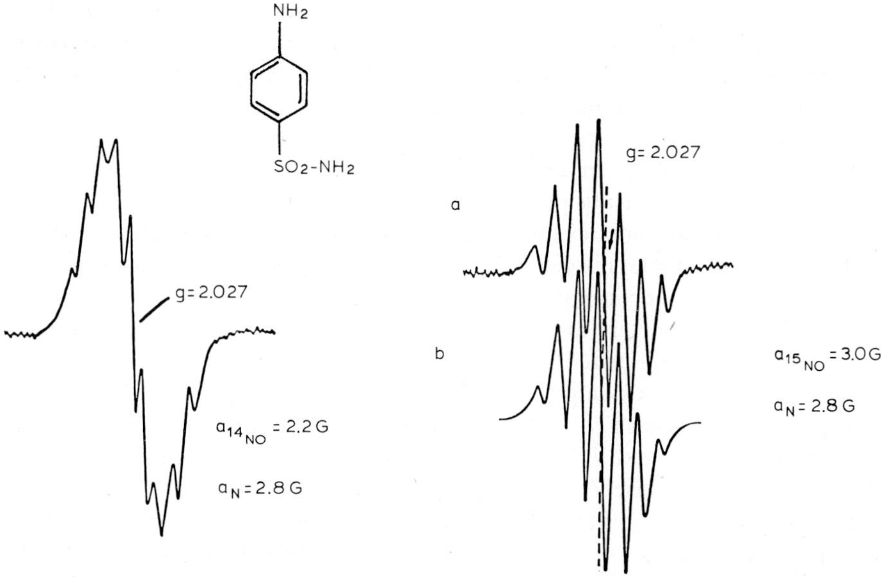

Figure 17. $Fe(I)(NO)_2$ adduct for which nitrogen hyperfine structure suggests the types of donor atoms which bind the iron center. From [244], with permission.

number of donor oxygen atoms derived from the enzyme as well as from the phosphate oxygens from the nucleotides and the waters bound to the complex has been described by Reed and his co-workers [245–247] and others [248,249]. The Mn^{2+} ESR spectra with α- or β-^{17}O-enriched ADP have been compared with that of Mn^{2+} bound to unlabeled ADP. For example, ESR spectra of myosin subfragment 1-MnADP complexes are broadened with β-^{17}O-labeled ADP but not with α-^{17}O-labeled ADP, indicating that Mn^{2+} forms a monodentate bond with the β-O of ADP [245]. The ESR signal is inhomogeneously broadened due to unresolved superhyperfine splitting from the bound ^{17}O nucleus ($I = 5/2$). The number of water ligands to Mn^{2+} is determined by comparing ESR spectra in ^{17}O-labeled water with spectra in unlabeled water, since the inhomogeneous broadening produced by $H_2{}^{17}O$ is indicative of the number of bound waters. Enriched $H_2{}^{17}O$ water is used in this experiment and the contribution from unenriched H_2O is subtracted out. The correct number of bound water ligands is related to the maximum fraction of the spectrum of Mn^{2+} in unenriched H_2O that can be subtracted from the experimental spectrum for the sample in ^{17}O-enriched water. Abnormal line shapes such as troughs in the difference spectra eliminate false combinations for the number of water ligands. For example, ESR difference spectra of myosin subfragment 1-MnADP fragment indicate that a one-water model does not fit whereas a two-water model does [245]. The combined results indicate that a monodentate β-O from ADP and two water ligands are bound in the six-coordinate Mn^{2+} complex. The remaining three sites are presumably provided by the protein.

Use of vanadyl, VO^{2+}, as a spin probe has been reviewed by Chasteen et al. [250]. The characteristics of ESR spectra for d^1 VO^{2+} complexes are similar to the ESR spectra for d^9 Cu^{2+} (where the single-hole formalism is equated with a single unpaired electron) except that each principal axis is split into eight hyperfine lines by the vanadyl nucleus ($I = 7/2$). Comparison of g values and hyperfine coupling constants with values for complexes of known structure provides information on the donor atoms coordinated to VO^{2+} in an experimental system. The intensity of aquated VO^{2+} decreases with increasing pH because hydrated species such as $(VO\text{–}OH)_2^{2+}$ form. The spectra of these species are broadened so that only the complex bound to protein (not the free complex) is detected. Binding of oxygen donor atoms is usually tighter than nitrogen donor atoms.

Analysis of ESR parameters for VO^{2+} bound to transferrin suggests that the in-plane donor atoms are comprised of a water ligand, two (or three) phenolate groups from tyrosine and a carboxylate [250]. A VO^{2+} spin probe study of apoferritin supports the hypothesis that Fe^{2+} binds to the protein as it is oxidized to Fe^{3+} [251]. Other studies have employed VO^{2+} as a structural probe of bovine insulin, carbonic anhydrase, carboxypeptidase A, serum albumin, and nucleases and phosphatases [250]. Complexes of VO^{2+} and ATP have been characterized by ESR [252] in connection with studies of the mechanism by which vanadate inhibits (Na^+, K^+)ATPase [250,253,254]. The time course of cellular uptake of vanadate can be monitored by ESR until either all the oxygen or all the glutathione is depleted (glutathione in the cytoplasm reduces vanadium (V) to vanadium (IV)) [255].

(ii) Coupled metal centers

Boas et al. [235] have reviewed the ESR studies of copper in biological systems and specifically considered coupled copper pairs. The resolution shown for the copper-carnosine dimer in Figure 15 is excellent. Usually poorly resolved lines for the allowed and forbidden transitions (half-field transitions) occur when exchange is weak. Recent studies in which two metal ions share a common ligand include cupric-cupric exchange following migration of cupric ion to the vacant zinc-binding site of bovine erythrocyte superoxide dismutase [256] and the binuclear-copper active site of mollusc and anthropod hemocyanin or *Neurospora* tyrosinase [257,258] Spin-exchange interaction between two Mn^{2+} ions in complexes of *S*-adenosylmethionine, K^+, and pyrophosphate suggests that the metal ions share a common ligand [259].

Hyde et al. [236] have reviewed the spin-probe–spin-label method, which was first described by Leigh [60] to determine the distance between a spin probe (a paramagnetic metal ion) and a spin label (usually a nitroxide). This theory is also applicable to two interacting metal ions: one a fast relaxing ‘probe’ and the other a slower relaxing ‘label’. The experimental observable is usually the decreased intensity of the slower relaxing label, *C*. If *C* and the relaxation time, T_{1k}, of the spin probe are known, then the distance, *r*, between probes may be obtained from the relationship

$$C = \frac{g_{Cu}\beta\mu^2 T_{1k}}{\hbar r^6} \tag{28}$$

where g, β, μ and $\hbar$ are fundamental constants.

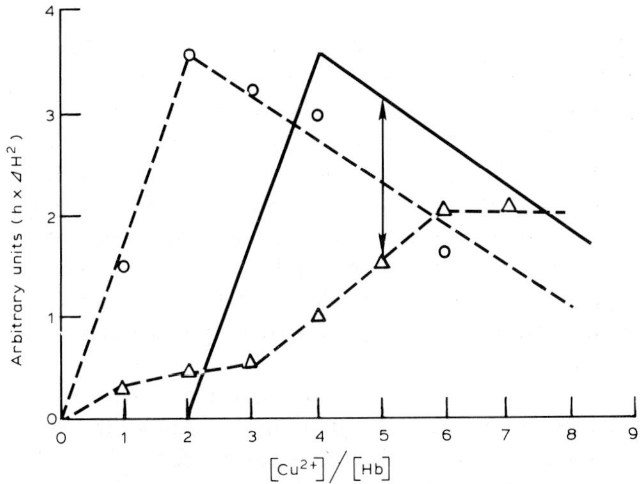

Figure 18. Arrow ↕ at a ratio of 5(Cu^{2+})/(Hb) indicates the magnitude of the decrease in the ferric heme signal due to interaction between paramagnetic Cu^{2+}. Relative intensities, I ($h\Delta H^2_{ms}$), where h is the height of a signal and ΔH_{ms} is the linewidth at the maximum slope for met human hemoglobin (△) and met cat hemoglobin (○). The solid line is expected for human hemoglobin in the absence of specific dipolar or exchange interaction after binding of two equivalents of cupric ions to the high-affinity sites. Data for the Leigh calculation are taken at a 5-fold excess of cupric ion per hemoglobin. From [260], with permission.

We have used this method to estimate that the distance between ferric ion in methemoglobin ($T_{1k} \approx 5 \times 10^{-11}$ s) and the weak binding site for Cu^{2+} ($T_1 \approx 10^{-8}$–10^{-9} s) is less than 10 Å [260] (Fig. 18). The ESR signal for cupric ion bound close to the heme iron is not observed, due to either a dipole-dipole or exchange interaction. The effect of the label on the probe is observed as a reduction of the $g = 6$ high-spin met Hb signal. The dashed line indicates the intensity expected for the met heme iron signal if cupric ion is not specifically bound but randomly interacts with the heme. Data for the met heme iron signal intensity for cat hemoglobin is consistent with this. The solid line indicates the behavior predicted for the specific binding of two equivalents of cupric ion followed by random binding of an additional two equivalents. From data for the met heme iron signal intensity for human hemoglobin the 'apparent' reduction in the intensity of the heme iron signal can be estimated. This reduction of intensity allows an estimate of the distance between the two probes to be calculated.

(d) Extensions of the standard ESR methods

(i) S-band

To obtain ESR spectra for copper complexes with optimum resolution measurements should be made at frequencies lower than those provided by the standard 9 GHz, X-band bridge. Hyde and Froncisz [242] have concluded that the width of the $M_I = -1/2$ line in the $g_{||}$ region is a minimum at a microwave frequency of 1–3 GHz whereas the $M_I = 3/2$ linewidth is a minimum for a microwave frequency of 6–8 GHz [242]. This increase in resolution is the result of a narrowing of the linewidth due to opposite signs of M_I and g dependent terms in the linewidth expression [242,261]. At S-band the observed effect is better resolution of the nitrogen fine structure on the $M_I = -1/2$ line in the $g_{||}$ region as well as resolution in the $g_{\perp}$ region. The resolved hyperfine structure for the $M_I = -1/2$ line in the $g_{||}$ region at lower frequencies can allow determination of the number of nitrogen donor atoms as well as the electron density on the nitrogen ligands. Froncisz and Aisen [262] have utilized S-band spectra (Fig. 19) to determine that a single nitrogen ligand, probably imidazole, is bound to metal in the copper-substituted transferrin-bicarbonate complex. Cupric ion bound to the low-molecular-weight dipeptide, carnosine, to hemoglobin, and to serum albumin give S-band spectra that allow the binding of four approximately equivalent nitrogen donor atoms to be confirmed (Fig. 19).

It is recommended that a single Cu^{2+} isotope be used for spectra at all frequencies in order to minimize the time necessary for spectral simulations. In addition to Cu^{2+}, a narrowing of the hyperfine lines at low frequency should be possible for Co^{2+} (^{59}Co, $I = 7/2$, 100% abundant); Cr^{2+} (^{53}Cr, $I = 3/2$, 95% abundant); Fe^{3+} (^{57}Fe, $I = 1/2$, 2.2% abundant); and Mo^{5+} (^{95}Mo or ^{97}Mo, $I = 5/2$, 15.78 or 9.6% abundant).

Optimal sample volume depends upon the cavity used. As the frequency is lowered, the dimensions of the standard rectangular cavity increase, until at 3.4 GHz the optimum sample volume is 5.0 ml. Fortunately, the loop-gap resonator structure (discussed in the next section of this chapter) reduces the optimum sample volume at

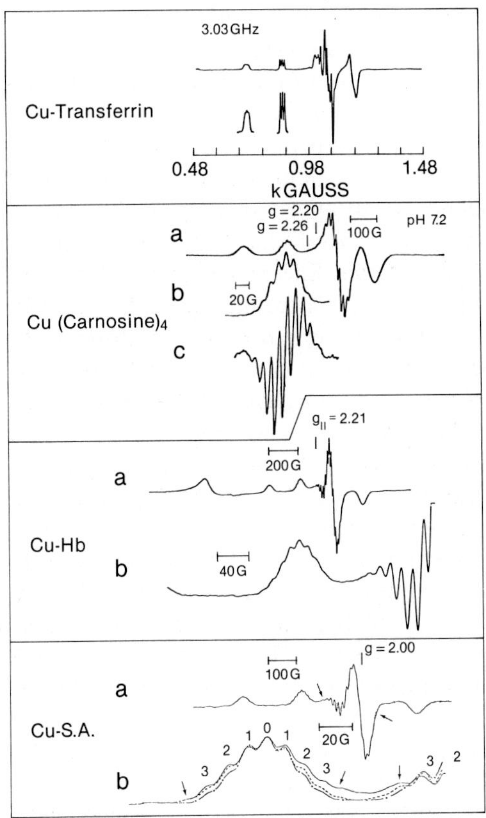

Figure 19. S-band data for Cu^{2+} bound to transferrin, carnosine, hemoglobin and serum albumin. Top panel: three-line spectrum for $M_I = -1/2$ line in $g_{\|}$ region indicates one donor nitrogen atom bound to cupric ion. Second panel: expanded spectra (b) and second harmonic (c) for $M_I = -1/2$ line in $g_{\|}$ region indicates four equivalent nitrogen donor atoms. Third and bottom panel: nine-line spectra attributed to four approximately equivalent nitrogen donor atoms (see text for further analysis). From [233,262,266], with permission.

this frequency to less than 0.5 ml. The sensitivity using the low frequency S-band bridge is comparable to that obtained using a commercial X-band bridge.

The second set of spectra in Figure 19 shows cupric ion in the presence of excess carnosine and further illustrates the advantages of S-band measurements [233]. These spectra should be compared to the top spectrum in Figure 15, also from cupric ion in the presence of excess carnosine. Both spectra are well resolved in the $g_{\perp}$ region (the most intense feature) but, as previously discussed, this region is complicated and difficult to interpret. The $M_I = -1/2$ line in the $g_{\|}$ region is resolved (due to the presence of nitrogen donor atoms) in the S-band but not the X-band spectrum. The $M_I = -1/2$ line in the $g_{\|}$ region is expanded below the full spectrum for cupric ion plus excess carnosine. The second harmonic of the expanded spectrum also is shown.

The pattern of 1–4–10–16–19–16–10–4–1 is that expected for four equivalent nitrogen donor atoms presumably from nitrogens in the imidazole ring.

The third pair of spectra are for cupric ion bound to a tight binding site near the N-terminus of the β chain of hemoglobin [213]. They demonstrate that resolution can be obtained for cupric ions bound to proteins as well as to peptides. Louro and Bemski [228] have analyzed X-band spectra for Cu^{2+} bound to hemoglobin and determined that Cu^{2+} was bound to four nitrogen ligands in an approximately square planar coordination. The expanded $M_I = -1/2$ line of the S-band spectrum is well resolved, as is the perpendicular region; the left side of this line is consistent with the 1–4–10–16–19–16–10–4–1 pattern expected for four equivalent nitrogen donor atoms. Thus, the S-band data unequivocally show that four nitrogen donor atoms form a site with a nearly square planar configuration. Although the computer simulations of X-band spectra are certainly convincing, lines in the $g_{\parallel}$ region (where the spectrum is less complicated because of fewer overlapping lines) are usually not resolved.

Determination of the number of nitrogen donor atoms from S-band data depends on the resolution of the $M_I = -1/2$ line in the $g_{\parallel}$ region. If the nitrogen hyperfine line with the lowest intensity is observed, an assignment can be made. However, determination can be difficult if the outermost line is lost in the noise. The intense central three lines for both three and four equivalent nitrogens have essentially indistinguishable patterns. One conclusion from our S-band data for cupric ion bound to serum albumin [260] is that the assignment of either three or four equivalent nitrogens should be based on the relative intensities of the second and third lines from the center of the $M_I = -1/2$ line (bottom pair of spectra in Figure 19).

The procedure that we have proposed [263] is to adjust the input parameters of the simulation program until the best possible fit of the central three lines is obtained using both three and four nitrogens, and then to select the best agreement with experimental data for the intensities of the second and third lines. This has been done in Figure 19 for the $M_I = -1/2$ line in the $g_{\parallel}$ region. The agreement on the low-field side of the line between a four-nitrogen assignment and the experiment is excellent. The agreement on the high-field side is less good but the four-nitrogen assignment remains convincing.

The examples in Figure 19 illustrate how the low-frequency technique using a loop-gap resonator provides better-resolved ESR spectra and can confirm the expected binding of nitrogen donor atoms. It is anticipated that more detailed structural information would be available if the nitrogen donor atoms were inequivalent or if proton couplings were of the same magnitude as nitrogen couplings. The only available evidence for such a situation occurs for cupric ion bound to two antitumor agents. ESR analysis indicates that 2-formylpyridine monothiosemicarbazonato copper II forms adducts with nitrogen donor atoms provided by the protein hemoglobin [241]. S-band data not only confirm the number of nitrogen donor atoms in the square planar configuration, in both the absence and presence of the additional nitrogen donor atom from the protein, but the even-line pattern observed suggests interaction with a proton having a coupling of the same magnitude as the nitrogen coupling. The antitumor agent, bleomycin, also is tightly bound to cupric ion. The pattern for the $M_I = -1/2$ line in the $g_{\|}$ region consists of an even number of lines [264]. In this case, computer simulations indicate that the nitrogen donor atoms are inequivalent. Thus, S-band data also are sensitive to inequivalent nitrogen atoms.

Only one cupric signal has been observed for 'blue' copper complexes using low frequency ESR [265]. In the ESR spectrum of cytochrome *c* oxidase, seven lines attributed to hyperfine structure in the g_x and g_y regions were evident. The g_z region was not resolved.

(ii) Spin echo spectroscopy

Poole's treatise [11] on experimental techniques and references therein provide a brief but useful introduction to the spin echo technique. After the sample is placed in the magnetic field (H in *Z* direction) to align the electron spins, a microwave pulse at the Larmor precession frequency is applied at right angles to the magnetic field. The spins are allowed to precess about the microwave-induced magnetic field (H_1 in *X* direction), which is at right angles to the main magnetic field, until the spins are tipped 90° and aligned at right angles to both the main magnetic and the microwave field (i.e., the *Y* direction). After the pulse is turned off, the spins diffuse from the positive *Y* direction into the *X*–*Y* plane. Following a time, τ, a 180° pulse is applied and the spins traverse toward the positive *Y* axis. At time 2τ the spins simultaneously arrive at the positive *Y* axis. This event at time 2τ then emits a signal or an echo.

Data from a spin echo experiment often are difficult to interpret by a person outside the field because the display is foreign to spectroscopists not involved in pulse techniques. The echo is detected as a signal which is progressively attenuated over time due to relaxation of the spins from the *X*–*Y* plane (Fig. 20). The modulation of the echo is the result of a change in amplitude due to interference between forbidden and allowed transitions [267]. Repeating spikes with a period of less than 100 nseconds often result from modulation of the envelope by nearby coupled protons from the complex and the solvent [267]. The usefulness of the data is enhanced if the weakly coupled proton(s) can be identified by deuteration. This problem arises frequently and will be discussed more fully in the ENDOR section. The modulation due to the weakly coupled nitrogen in an imidazole ring has periods of 700 and 260 nseconds for the copper diethylenetriamine-imidazole complex (Cu-det-imid) but not for the control (Cu-det-pyrimidine) [259,267,268]. The vertical lines above the figure have been added to help visualize the periods from the weakly coupled imidazole.

Spin echo measurements on metal complexes have been reviewed extensively by Mims and Peisach [267]. One use of this technique has been to study anion binding at paramagnetic metal centers. When cupric ion is substituted for ferric ion in transferrin, a pattern corresponding to the weak coupling for ^{13}C in ^{13}C-doped bicarbonate indicates that bicarbonate is bound directly to the metal ion [269]. Thus, a modulation pattern detected by spin echo can confirm that adducts with a weakly coupled nuclear spin are bound directly to the metal site.

Spin echo spectroscopy is a definitive technique for determining the binding of imidazole to copper complexes. Detection of spin echo signals is based on the weak coupling of the electron to the unbound imidazole nitrogen, rather than the bound

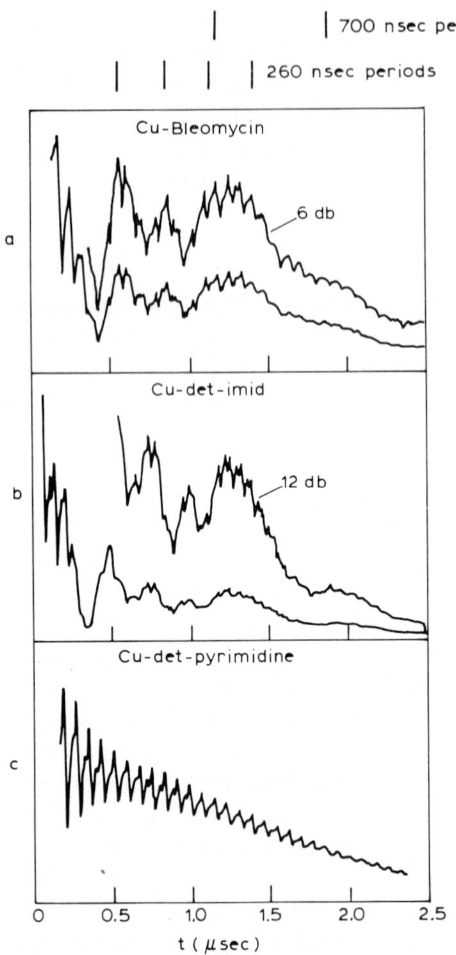

Figure 20. Two-pulse echo decay envelope of Cu(II) complexes of (a) bleomycin, (b) diethylenetriamine and imidazole, and (c) diethylenetriamine and pyrimidine. In traces a and b, one observes a modulation pattern arising from the interaction of Cu(II) with the remote ^{14}N of ligated imidazole. The respective magnetic fields and spectrometer frequencies are the following: a, 3080 G, 9247 MHz; b, 3195 G, 9251 MHz; c, 2970 G, 9225 MHz. Lines indicating the periods were added by the authors. From [268], with permission.

imidazole nitrogen. Spin echo studies have shown that cupric ion is bound to imidazole in the Cu-det-imid complex [267], to an imidazole site in azurin [267], stellacyanin [270], superoxide dismutase [271], galactose oxidase [267], transferrin [269], and the antitumor antibiotic agent, bleomycin [268]. Data for cobaltous bleomycin in the latter study confirm that imidazole is a ligand. Analysis for ferric bleomycin has confirmed the binding of nitrogen donor atoms, but here the analysis for imidazole is more complex and not yet definitive.

Newer techniques for spin echo spectroscopy have been developed recently. More sophisticated pulse sequences are described in many of the references for this section. Fourier transformation of the envelope gives a display of peaks corresponding to the hyperfine couplings.

(iii) ENDOR

In an early review, Hyde [272] discussed ENDOR in proteins, including flavoproteins, copper proteins, hemeproteins, two-iron ferrodoxin and bacteriochlorophyll. Kevan and Kispert's book [19] is an introductory text on ENDOR (electron nuclear double resonance) and ELDOR (electron electron double resonance) techniques. Poole [11] includes a chapter on double resonance techniques in his text. Schweiger [21] has covered ENDOR of transition metal complexes, including a section on biological applications. Recent reviews of ENDOR spectroscopy of chlorophylls [273], heme and heme proteins [274] and iron sulfur proteins [275] demonstrate how additional detail can be obtained from ENDOR data.

ENDOR is a magnetic resonance technique by which unresolved hyperfine couplings can be obtained, primarily for nitrogen donor atoms and protons within 10 Å of the paramagnetic metal ion. It involves both a microwave and a radiowave frequency. An ESR signal from the paramagnetic metal ion is partially saturated, i.e., the microwave power is increased, tending to equilibrate the number of unpaired electrons aligned with and against the external magnetic field. One portion of the ESR signal is then selected (through selection of a particular magnetic field) before application of the radiofrequency. The radiofrequency induces transitions between nuclear states and changes the population of the states and the intensity of the ESR signal. Therefore, if one 'sits' on a single ESR transition while sweeping the radiofrequency (i.e., carrying out an NMR experiment), the observed changes in the intensity of the ESR signal are sensitive to the energy differences between the nuclear states.

A typical ENDOR spectrum contains lines from protons as well as nitrogen ligands (Fig. 21). In this example [264] a magnetic field for a low-field line in the $g_{\|}$ region of the ESR spectrum for the copper complex was selected. The ENDOR spectra for this setting are as well resolved as single-crystal ENDOR spectra [276]. The four-line ENDOR pattern expected for nitrogen donor atoms is given by

$$v = \frac{A^{\mathrm{N}}}{2} \pm Z \pm C \tag{29}$$

where A^{N} is the hyperfine coupling constant; Z, the Zeeman splitting; and C, the quadrupole splitting. The intense peak at about 16 MHz in Figure 21 is attributed to nitrogen nuclei for which two of the four expected lines are readily apparent. This line does not move if the microwave frequency is changed and it is more intense at lower temperature (16 K) than at higher temperature (23 K). The intensity of the proton lines is less sensitive to temperature variation in this range. The field and temperature dependence help confirm the nitrogen assignment.

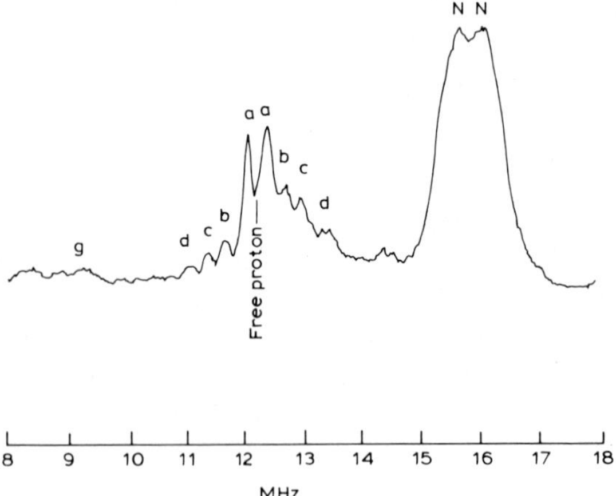

Figure 21. X-band ENDOR spectrum for CuBlm in ^2H_2O. The lines a–d and g are proton resonances. The high-field line labeled N is a nitrogen line where only two of the four expected lines are clearly resolved. From [264], with permission.

Proton ENDOR signals give pairs of resonances at

$$\nu_i = \nu_{\text{free proton}} \pm 1/2\, A_i \tag{30}$$

which are centered around the free proton frequency and separated by the proton coupling constant, A_i. Four of the five proton transitions in Figure 21 centered around the free proton frequency are clearly evident. A second criteria for proton transitions is that the proton signals move with a change in microwave frequency. A disadvantage is that one cannot determine from ENDOR data the number of equivalent nitrogen donor atoms. In addition, one has difficulty in assigning proton lines without selectively deuterated analogs.

Copper bound to proteins also has been reported using the ENDOR technique. Rist et al. [277] concluded that at least one (and probably more than one) nitrogen is bound to cupric ion in stellacyanin and the cupric ion is inaccessible to the solvent. 10 years later, Roberts et al. [278] argued that two nitrogens with $A^N/2 \approx 22$ MHz and $A^N/2 \approx 16$ MHz together with the copper hyperfine coupling (the first to be measured by ENDOR) are consistent with a flattened tetrahedral geometry. Similar arguments have been used to define a flattened tetrahedral geometry for the copper center in cytochrome *c* oxidase [279] after two strongly coupled protons and at least one nitrogen were detected from ENDOR data for the cupric center in this enzyme [280].

The major application of ENDOR has been to the study of ferric complexes in heme and iron-sulfur proteins [274,275]. Additional detail has been obtained utilizing biochemical techniques for obtaining mutant proteins and introducing isotopes, i.e., Fe^{57} for ^{56}Fe, N^{15} for N^{14} or ^{13}C for C^{12} [281–283]. For example, a shift of the ENDOR line for the nitrogen donor atom of proximal histidine bound to ferric heme in the β chain upon oxidation of the α ferrous heme monitors heme-heme interaction in Hb-Milwaukee ($\alpha_2^{II}\beta_2^{III}$ 67 Val → Glu) not detectable by ESR [281].

5. Instrumentation and methodology

The reader is referred to Poole's comprehensive treatise [11] on instrumentation and methodology. The book by Alger [4], although now somewhat out of date, contains many practical details. Wilmshurst's book [12] looks at ESR spectroscopy from the point of view of an engineer. Feher's paper [284] on sensitivity in ESR spectroscopy is a classic.

The major sub-systems of an ESR spectrometer system are: (1) the microwave bridge, sample resonator, and resonator accessories; (2) the magnet, magnet power supplies, field regulation circuits and field sweep system; (3) data acquisition, basically amplification and phase sensitive detection in the case of continuous wave spectroscopy or fast A/D converters and signal averaging in the case of pulse ESR; (4) data-processing systems, which are fundamentally filtering processes, making use of available information to enhance the quality of unknown information; (5) spectral display; (6) spectral interpretation systems, which involve comparison of the acquired spectrum with spectra obtained theoretically or from experimental models.

To this list can be added various major specialized sub-systems that are required for some of the more sophisticated experiments. Some of the basic sub-systems also require modification for these experiments. (7) Pulse programmer for time domain ESR; (8) programmable radio frequency source for electron-nuclear double resonance (ENDOR); (9) pump microwave source for electron-electron double resonance (ELDOR).

Some of the minor accessories used in ESR spectroscopy, which can be viewed logically as separate instrumental sub-systems, are: (10) temperature control systems; (11) photoexcitation systems; (12) electrochemical systems; (13) continuous and stopped flow mixing systems.

Thus, one can view an ESR spectrometer as consisting of a basic unit comprised of six subsystems, major accessories and minor accessories. A complete treatment would deal with each of these in turn, and then discuss the inter-relationships. Our approach here is more selective.

We focus to a considerable extent on category 1, because the scientist must always have a fundamental concern about the origin of his signal and also because he often has little background in microwave engineering. Restricted commentary is included about category 3, data acquisition; category 4, data processing; and categories 8 and 9.

(a) The reference arm microwave bridge

The reference arm bridge coupled to a microwave resonant structure operating in reflection has become the standard microwave geometry for ESR spectroscopy

(Fig. 22). The microwave power in the reference arm provides a phase reference for the power in the resonator arm that is reflected from the resonator and carries ESR information. Most operator difficulties with the bridge stem from an imperfect understanding of microwave phases; therefore, this aspect is emphasized below.

It is desirable for the total microwave pathlength in the reference arm to be equal to the total microwave pathlength in the resonator arm, including the path from the circulator down to the resonator iris and back to the circulator. Only the pathlengths between the first directional coupler and the second directional coupler are important in determining that the pathlengths are equal. If this condition is satisfied, the relative phases in the reference and resonator arms are independent of incident microwave frequency. Attention is drawn to the arm length equalizer of Figure 22, which is approximately twice the length from circulator to resonator. If an unusual resonator pathlength is employed by the scientist, as when using a non-standard table or magnet, an appropriate change in the arm length equalizer is desirable.

As a rule, microwave attenuators introduce phase shift as well as attenuation. An exception is the type of attenuation known as 'rotary vane', which has negligible phase shift. Some attenuator manufacturers have achieved fairly constant phase shift for microwave coaxial structures. Often the level set attenuator in the reference arm is of lower quality with large phase shift, but the operator encounters little difficulty because the reference level is seldom changed. The reader can appreciate the frustration that would occur if for every change in the main attenuator setting an accompanying phase shift were to occur.

Small phase differences between the reference and resonator arms result in ESR resonance lines of the proper shape (i.e., no dispersive admixture, which may surprise

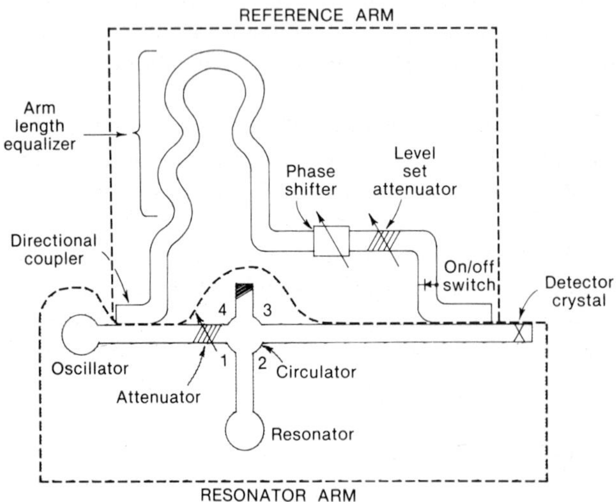

Figure 22. Block diagram of an ESR bridge, showing the components of the reference and resonator arms.

some persons) but of somewhat reduced intensity. This becomes a source of quantitative error.

If the reference and cavity arm pathlengths differ by an odd number of half wavelengths, an unstable operating condition can occur when the cavity arm power is less than or equal to the reference arm power. The condition is recognized on a mode sweep oscilloscope display by a resonator 'dip' of the opposite sense, leading to 'anti-stabilization' by the automatic frequency control (AFC). The spectrometer manufacturer normally organizes the bridge so that this condition cannot be reached. If no attention is made to this detail, the phase shifter in the reference arm must have a range of 360° or more, and the operator must become proficient in noticing the difference between odd and even number of half wavelength path differences. The on/off switch in the reference arm is a convenient aid in setting the phase shift properly.

While various automatic schemes for precise closed-loop control of the reference phase are possible [285], this deluxe feature is not usually present. Manual adjustment is accomplished by very carefully adjusting the phase for a maximum in detector crystal current (assuming that the resonator is well matched).

Subtle difficulties in operation of the bridge occur if low levels of microwave power reach the detector crystal by paths other than the reference arm or the cavity arm. One possible route is directly from arm 1 to arm 3 of the circulator. For this reason, four-port circulators of high quality are used in commercial bridges.

One of us (J.S.H.) has described an approach to the detection of dispersion, using a reference arm bridge with AFC locking to the resonator [286]. It was pointed out that since dispersion is a frequency shift of the resonator that is induced by magnetic resonance, it is necessary that the AFC have negligible loop-gain at the field modulation frequency. AFC information is gathered by satisfying the condition of equal pathlengths in the reference and resonator arms, and signal information is gathered by introducing an additional 90° phase shift. One approach is to introduce an additional path for the AFC with its own detector, the first path being for the ESR signal and the two paths differing by 90°.

Future developments of reference arm bridges include the following:

(1) Increased use of octave bandwidth bridges. In our laboratory reference arm bridges operating from 1 to 2 GHz and from 2 to 4 GHz are in regular use. Increased availability of octave bandwidth microwave components in coaxial configuration make this development possible.

(2) Use of field effect transistor (FET) amplifiers for two purposes: amplification of the oscillator output to one or more watts, and also for amplification of the ESR microwave signal using a very low noise FET. Care must be taken in protection of detector diodes, in design of switches to introduce the amplifiers into the circuit, and in maintaining good operator discipline in use of the bridge.

(3) Increased use of digitally controlled components not only for convenience but also to compensate for variations in oscillator power output with microwave frequency, microwave phase with attenuation, and microwave attenuation with microwave frequency.

(4) Increased use of the so-called three-arm bridge [287,288] in which a pump arm is used in addition to the resonator and reference arms, and microwave power suitable for pulse ESR or ELDOR is introduced to the resonator through this arm.

(b) Sensitivity

Sensitivity involves noise and signal. These should be considered separately, and this is our procedure here.

Noise can be divided into three categories: (1) radiofrequency interference, microphonics, unstable main power lines and other disturbances in the environment; (2) oscillator source noise, both AM and FM; (3) the overall receiver noise figure. In well-designed spectrometer systems, the third category is always the limiting noise source.

Most ESR experiments are performed, at least in chemistry and biology, using 100 kHz magnetic field modulation. It was originally introduced as a simultaneous attack on all three noise sources. Detectors in use in the early 1960s exhibited noise that had a $1/f$ dependence and became negligible at frequencies above 100 kHz, oscillator noise falls off rapidly from the carrier and is low at 100 kHz, and environmental noise, both electrical and acoustic, is low at 100 kHz. Detector noise in modern detectors of the back diode or Schottky diode types reaches its minimum level at frequencies of a few hundred Hertz, so this is no longer a compelling rationale.

The noise, both AM and FM, of microwave oscillators used in ESR bridges is equal to or worse than in early years. There has been little improvement in tube oscillators and solid state oscillators have somewhat worse noise in our experience. It is important to recognize that both oscillator AM and FM noise are enhanced in a microwave bridge of the reference arm type. Wilmshurst [12] discusses AM noise enhancement, deriving the following equation for the noise voltage

$$\frac{V(\text{AM noise})}{V(\text{incident})} \propto Q \frac{\delta\omega}{\omega_0} \tag{31}$$

Hyde [289] derived a similar expression for FM noise when detecting absorption,

$$\frac{V(\text{FM noise})}{V(\text{incident})} \propto \frac{V(\text{incident})}{V(\text{reference})} \frac{Q^2(\delta\omega)^2}{\omega_0^2} \tag{32}$$

where $(\delta\omega)$ corresponds to noise at the field modulation frequency and V refers to microwave voltages. When detecting dispersion,

$$\frac{V(\text{FM noise})}{V(\text{incident})} \propto Q \frac{\delta\omega}{\omega_0} \tag{33}$$

The equations show that use of resonators with low Q is advantageous in decreasing the sensitivity of the system to source noise.

This brief discussion leads to the conclusion that the main reason in modern spectrometers for use of 100 kHz field modulation is to operate at a frequency that is free of environmental instabilities and noise. This remains a compelling reason.

We consider next the subject of signal intensity. Some time ago Hyde [290] introduced a system of classification of ESR samples. In this system, all samples were divided into one of eight categories depending on yes or no answers to these three questions: does the sample exhibit dielectric loss (and one usually thinks of water when answering 'yes'); is the sample limited in availability or size because of cost or rarity; and does the sample exhibit microwave power saturation at microwave magnetic field intensities available in a particular spectrometer (see Table 4). (The question of 'limitedness' can be additionally complicated because samples can be limited in one dimension and unlimited in the other two, as is the case for a thin film, or limited in two dimensions and unlimited in the third, as is the case for a filament, fiber or needle type geometry.) One then takes cognizance of Feher's formula for the signal intensity V [284].

$$V \propto Q\eta\chi\sqrt{P_0} \tag{34}$$

where P_0 is the incident power, η is the filling factor and χ the radiofrequency susceptibility (which can contain P_0 if power saturation occurs). A particular resonator geometry can then be compared with another geometry by evaluating this equation for the eight classes of samples or one might compare two scaled but otherwise identical resonators at two microwave frequencies. In these comparisons one normally drops χ from consideration, arguing either that the sample is unsaturated or that it is brought in both resonators to the same level of saturation by adjusting P_0.

An alternative and simpler approach is to consider just two classes of samples, viz, (1) nonsaturable, low loss, limited (which in practice, often means a single crystal speck

TABLE 4
The eight classes of samples, with examples of each class

Class	Saturable	Unlimited	Dielectric loss	Examples
1.	yes	yes	yes	many systems of biological interest
2.	yes	yes	no	paramagnetic defects in diamond
3.	yes	no	yes	radicals with well-resolved hyperfine structure formed by dissolving aromatic molecules in concentrated H_2SO_4
4.	yes	no	no	color centers in alkali halides
5.	no	yes	yes	many systems of biological interest
6.	no	yes	no	photoexcited triplet state in single crystals of naphthalene in durene
7.	no	no	yes	aqueous solution of Mn^{2+} or Co^{2+}
8.	no	no	no	solutions of DPPH in benzene

of DPPH (2,2-diphenyl-1-picrylhydrazyl)) and (2) saturable, high loss, unlimited (which in practice, often means an aqueous solution of a nitroxide radical spin label). For category 1, sensitivity is expressed as the minimum detectable number of spins extrapolated to $P_0 = 0.2$ W for a 1 G linewidth. For category 2 one can use any amount of sample and any power to obtain the best possible signal. Sensitivity is expressed as minimum detectable concentration of spins in moles per liter.

The usual ESR laboratory acquires a variety of resonators over a period of time. The spectroscopist generally has little control over the noise in the spectrometer

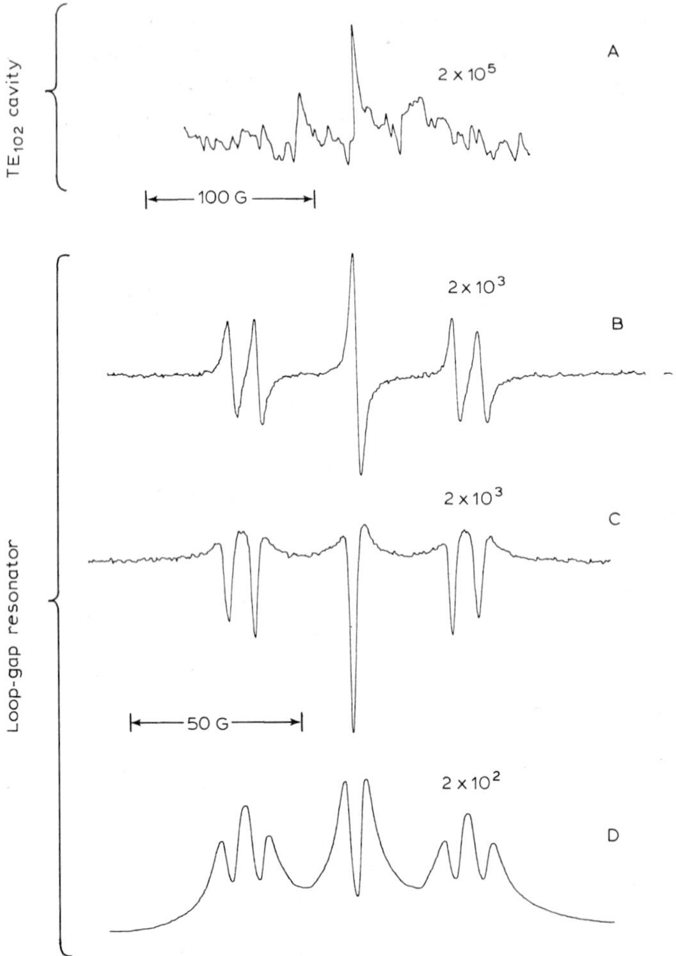

Figure 23. Comparison of ESR signals from a single crystal of synthetic diamond in the TE_{102} cavity and in the loop-gap resonator. Spectra were obtained at room temperature with 2 G field modulation. Spectrometer gains are indicated. A, Absorption signal, 1 mW, 9.3 GHz, 1 second time constant. B, Absorption signal, 2 μW, 8.8 GHz, 0.25 second time constant. C, Dispersion signal, other conditions same as in B. D, Dispersion signal, 1 mW, 8.8 GHz, 0.25 second time constant. From [53], with permission.

system, but he has considerable control over the signal level. He should always optimize signal intensity by careful consideration of the terms in Feher's formula for the sample of interest, taking into account the geometries of the available resonators.

Attention is called to a recent paper from our laboratory illustrating the use of signal and noise analysis as outlined here, and making an additional point [53]. If Q is sufficiently low to avoid demodulation of oscillator FM noise when tuned to dispersion, Eqn. 33 above, and if H_1^2 at the sample is sufficiently high, improved signal-to-noise ratio can be achieved by observing the dispersion under saturating conditions. An example is shown in Figure 23, taken from this paper. Figure 23D is a so-called adiabatic rapid passage display. Although the physics of this display is complicated, it can often be used for purely spectroscopic purposes, without paying particular concern to the physics.

(c) Resonators

It is our thesis that the loop-gap lumped circuit resonator introduced recently by us will eventually supplant microwave cavity resonators in ESR spectroscopy except for a few specialized applications [53,291–293]. Figure 24 (from Ref. 291) shows this resonator. In a sense, this is a hybrid structure midway between low-frequency lumped circuits where a capacitor and an inductor are connected by a transmission line, and high-frequency distributed circuit cavity resonators where the electric and magnetic

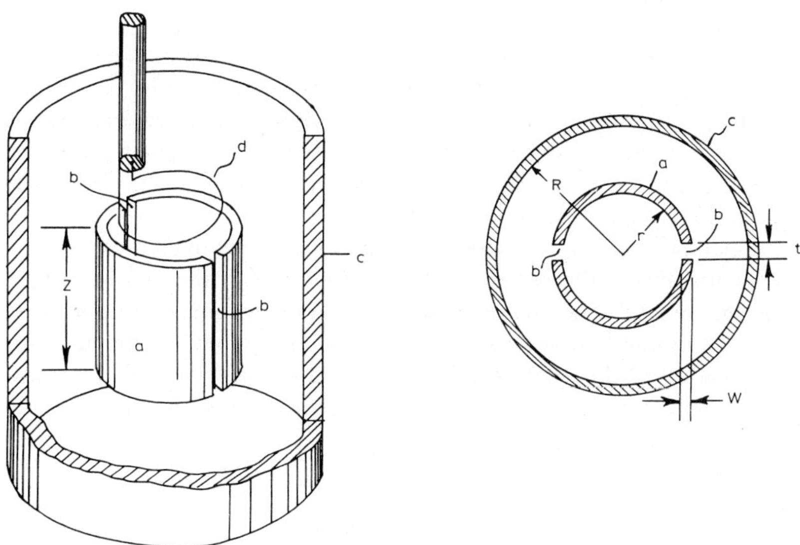

Figure 24. The loop-gap resonator showing the principle components (a, loop; b, gaps; c, shield; d, inductive coupler) and the critical dimensions (Z, resonator length; r, resonator radius; R, shield radius; t, gap separation; W, gap width). The sample is inserted into the loop a through the coupler, d. The microwave magnetic field in the loop is parallel to the axis of the loop. From [291], with permission.

fields share the same space. In the loop-gap resonator the magnetic field lies largely in the loop and the electric field in the gap(s), but, because there is no transmission line, Maxwell's field equations must be satisfied at the interface of the loop and gap (namely E and H must be perpendicular to each other).

To a fair approximation, this resonator is an infinite one-turn solenoid, and in infinite solenoids the magnetic field is uniform.

Because the Q is relatively low, of the order of 1000 compared with 7000 for a TE_{102} cavity, enhancement of both AM and FM noise originating in the oscillator is reduced as predicted by Eqns. 31–33 above.

Because the structure is small compared to the wavelength, whereas a cavity resonator has dimensions similar to a wavelength, the energy density is extremely high – more than 100 times in some geometries compared with a conventional TE_{102} cavity resonator operating at the same frequency and same input power.

Of the eight classes of samples introduced in the preceding section, comparison of the loop-gap resonator with the TE_{102} cavity resonator leads to the following conclusions. The loop gap is markedly superior for the four classes of limited samples, and for the two unlimited classes that are additionally 'nonsaturable'. A more careful comparison is required for the two categories 'non-limited, high loss, saturable' and 'non-limited, low loss, saturable'.

This last category, 'non-limited, low loss, saturable', is suitable for gas phase ESR or for liquid phase experiments in very low loss solvents such as heptane or benzene. In principle, best performance is obtained by filling the resonator with sample and employing a resonator of the highest possible Q, such as one oscillating in the cylindrical TE_{011} mode.

The category of 'non-limited, high loss, saturable' is problematical. The present experience is that the sensitivity is about the same expressed on a molarity basis when comparing the TE_{102} and loop-gap resonator, even though 20 times less material is required in the latter. This perspective may shift to favor the loop-gap resonator in the future by increased use of the dispersion mode at high powers, as mentioned in the preceding section and described in Ref. 53.

Several additional points can be made. (1) The TE_{102} rectangular cavity is a superior geometry for optical illumination. (2) A special case of the limited sample category is rapid mixing and flow experiments, where large amounts of reactants are often consumed. The loop-gap resonator requires about 100 times less material, for comparable data quality. (3) Because of the very high radiofrequency field intensity, certain new experiments requiring such intensity become feasible in the loop-gap resonator.

(d) Field modulation

Choice of magnetic field modulation frequency is based on the following considerations: (1) minimization of noise by converting resonance information to a relatively noise-free frequency, (2) experiments where the field modulation is itself in the spin Hamiltonian and is a crucial physical parameter such as saturation transfer spectroscopy [49,56,57], and (3) possible spectral distortions from modulation sidebands. Sometimes experimentalists propose other types of modulation in ESR spectroscopy, such as microwave amplitude or microwave frequency modulation. These are almost always poor ideas because they violate the 'transfer of modulation' principle: namely, that modulation should be transferred to the microwave carrier only when magnetic resonance occurs.

There is an intrinsic experimental problem when using magnetic field modulation in ESR spectroscopy that is associated with spurious resonance-like modulations transferred to the microwave carrier and arising from modulations of the resonator microwave characteristics. This phenomenon, perhaps apocryphally, was called 'potato' by F. Bloch because of the resemblance when the effect is displayed as a lissajous figure.

Potato of the first kind involves acoustic coupling to the resonator of vibrations in the field modulation coils that arise from forces determined by the cross product of current and magnetic field. It is controlled by mechanical damping and acoustic isolation.

Potato of the second kind arises from forces determined by the cross product of the magnetic field with eddy currents induced in metallic walls of the resonator by the modulation field. It is controlled by interrupting eddy current paths and using resonators with wall thicknesses less than one skin depth thickness at the field modulation frequency.

Both kinds of potato can cause changes in resonator Q, so-called resistive potato, and in resonator frequency, so-called dispersive potato. Potato is seen at both first and second harmonics of the modulation. Potato effects are thought to be exacerbated if the resonator dimensions are comparable to the acoustic wavelength in the resonator material at the field modulation frequency, leading to acoustic standing waves. Generally, potato is manifested in a spectrum by a sloping baseline. Sometimes mechanical instabilities, perhaps associated with metallic burrs or plating defects, can lead to excess noise that has its origin in forces generated by the simultaneous presence of a strong magnetic field and currents.

Mechanical displacements associated with potato are in the micron range. Quantitation of potato effects is difficult, and no perfect solution to potato problems exists. Every ESR spectroscopist is aware of the problem, and partial solutions can often be found based on the conceptual models outlined in this section.

(e) Accessories

Electron spin resonance spectroscopy generally involves obtaining spectra as a function of some third variable, such as temperature, light intensity, light wavelength, pH, voltage on an electrochemical cell, mixing chamber flow rates, viscosity, solvent, time, etc., etc. Accessories are absolutely essential to ESR spectroscopy. Almost always a commercial accessory is purchased initially, or a published design copied. And almost always, as the spectroscopist's experience increases, he designs his own version of the requisite accessory.

(f) ENDOR, ELDOR, time domain ESR and multifrequency ESR

The reader is referred to the book by Kevan and Kispert [19] for a thorough discussion of ENDOR and ELDOR, to the book by Kevan and Schwartz [20] for a discussion of time domain ESR, and to the review by Hyde and Froncisz [242] of multifrequency ESR.

Electron nuclear double resonance (ENDOR) is the detection of nuclear magnetic resonance by way of the effect of nuclear resonance on the ESR signal of a paramagnet that is coupled to the nucleus. It improves the effective resolution, generally at the expense of lower signal-to-noise ratio. Its primary utility is in resolution improvement, although information on relaxation processes can be obtained. ENDOR can be performed on single crystals, unordered solids, and liquids. One sometimes speaks of self ENDOR, which would be ENDOR on a transition metal nucleus, or of ligand ENDOR, which would be on the nucleus of an ion that exhibits both dipolar and covalent coupling to a paramagnet.

One key aspect of ENDOR spectroscopy is the nuclear relaxation time, which is generally governed by the dipolar coupling between nucleus and electron. Another key aspect is the ENDOR enhancement factor, as discussed by Geschwind [294]. The radiofrequency frequency field as experienced by the nucleus is enhanced by the ratio of the nuclear hyperfine field to the nuclear Zeeman interaction. Still another point is the 'selection of orientation' concept introduced by Rist and Hyde [276]. In ENDOR of unordered solids, the ESR resonance condition 'selects' molecules in a particular orientation, leading to 'single crystal' type ENDOR. Triple resonance is also possible, irradiating simultaneously two nuclear transitions, as shown by Mobius et al. [295].

There are many modulation methods used in ENDOR. Since the details of relaxation processes are seldom understood, the usual experimental approach is unfortunately to search rather blindly for experimental conditions that yield good signals.

Electron-electron double resonance (ELDOR) [296] is the direct observation at one part of an ESR spectrum of the effect of intense radiation with microwaves at another part. Therefore, it measures transverse relaxation, and its utility has been more to study relaxation processes rather than to obtain improved spectral resolution (although the latter is possible). ELDOR requires a special microwave geometry that permits irradiating the sample with two different microwave frequencies. Two displays are common: field-swept ELDOR where both frequencies are fixed at a difference that corresponds to a hyperfine interval and the magnetic field is swept, and frequency-swept ELDOR, which is performed at constant field and observing frequency, sweeping the pumping frequency. Other displays are possible. ELDOR can be both continuous wave (CW) and pulse. Experiments have been published on single crystals, unordered solids and liquids. A key aspect of ELDOR in solid phases involves so-called 'forbidden' transitions that share one energy level with an allowed transition. In liquids, nuclear relaxation induced by rotational modulation of the electron-nuclear dipolar interaction is an important mechanism.

Heisenberg exchange, ω(Hex), is another mechanism of importance, yielding information on translational diffusion and bimolecular collision rates. The ELDOR effect from exchange is approximately equal to $\omega(\text{Hex})T_1$, where $\omega(\text{Hex})$ is the Heisenberg existence frequency and T_1 is the electron spin-lattice relaxation time, whereas effects of Heisenberg exchange on ordinary ESR linewidths are approximately $\omega(\text{Hex})T_2^*$, where $(\gamma T_2^*)^{-1}$ is a characteristic ESR envelope (inhomogeneous) width. Since $T_2^* \ll T_1$, ELDOR can be used to study much smaller Heisenberg exchange frequencies, $\omega(\text{Hex})$, than in ordinary ESR spectroscopy. Heisenberg exchange can be divided into homospecies and heterospecies exchange, depending on whether or not the pumped and observed species are identical.

A very important ELDOR mechanism is diffusion of saturation because of very slow rotational diffusion of nearly immobilized species. This is the basis of saturation transfer spectroscopy [57].

Pulse or time domain ESR can be divided into two categories: the transient response of spin systems to abrupt or step changes in resonant condition and the transient response to sequence of pulses [20]. The step response, as in saturation recovery, is used commonly to measure T_1; and the pulse response, as in 90–180° spin echo, to measure T_2.

Very severe experimental problems have restricted the growth of time domain ESR, but these problems are now being solved. Many interesting experiments, often without clear analogy to NMR experience, can be expected in the future. One example from our laboratory was the discovery that Heisenberg exchange between spin labels and molecular oxygen is an apparent T_1 mechanism for the spin label [40]. Thus, T_1 measurements on spin labels in membranes gave useful information on oxygen transport.

Multifrequency ESR implies examining the same sample under the same conditions at two or more microwave frequencies. This has long been a recommended, but nevertheless neglected, procedure. It is an extremely critical test of one's theoretical model or simulation. A primary rationale is to exploit forbidden ESR transitions involving either nuclear state mixing, which occurs when the nuclear Zeeman or nuclear quadrupole coupling is comparable to the hyperfine coupling, or electron state mixing, which occurs when nuclear hyperfine and electron Zeeman interactions are comparable. A further rationale lies in the subtleties of the field to frequency transformation. It is now technically possible to create, with a computer, frequency-swept displays from a series of field-swept displays obtained at monitonically stepped microwave frequencies.

In our laboratory microwave bridges at 35, 8.8 to 9.6, 1 to 2 and 2 to 4 GHz are in regular usage, and at the time of preparation of this article, bridges at 0.5 to 1 and 4 to 8 GHz were under construction, illustrating the importance the authors place on multifrequency capability.

(g) ESR and computers

The enormous diversity of ESR applications, the relatively small size of the commercial ESR market, and the great variety of computers has made standardization impossible. We have little to suggest to the reader other than to say the obvious:

computer-aided ESR spectroscopy is important and growing. Since the scientist is on his own, some computer expertise is necessary. The main subdivisions of the field are as follows:

(1) Spectrometer control. We prefer dual control wherein the instrument can alternatively be addressed from the front panel or a keyboard. Spectrometer control by computer permits the convenient acquisition of huge amounts of data by systematic variation of experimental parameters.

(2) Data massaging. By this we mean the application of a filter based on existing knowledge to enhance the quality of unknown information. Primary examples are signal averaging or smoothing.

(3) Data transformation. By this we mean the application of a transform for which a reverse transform exists including integrals, derivatives and Fourier transforms.

(4) Spectral simulation. Simulations based on appropriate spin Hamiltonians of experimental spectra.

(5) Library comparison. The computer should have in mass storage various libraries of experimental and computed reference spectra with appropriate scaling capability for comparison with unknown spectra.

(6) Mass storage with preprocessing. As spectrometer control by computer becomes more common, large amounts of data will be generated. Examples might be CW saturation curves as a function of temperature or pulse ESR response as a function of magnetic field (conceivably a 1000×1000 block of information). Increasingly large storage and some preprocessing of data will be required.

References

1 Specialist Periodical Reports (1973–1981) The Chemical Society, London.

2 Carrington, A. and McLachlan, A.D. (1967) Introduction to Magnetic Resonance, Harper & Row, New York.

3 Ayscough, P.B. (1967) Electron Spin Resonance in Chemistry, Methuen, London.

4 Alger, R.S. (1968) Electron Paramagnetic Resonance: Techniques and Application, Wiley-Interscience, New York.

5 Swartz, H.M., Bolton, J.R. and Borg, D.C. (eds.) (1972) Biological Applications of Electron Spin Resonance, Wiley-Interscience, New York.

6 Wertz, J.E. and Bolton, J.R. (1972) Electron Spin Resonance: Elementary Theory and Practical Applications, McGraw-Hill, New York.

7 Atherton, N.M. (1973) Electron Spin Resonance, Halsted Press, New York.

8 Gordy, W. (1980) Theory and Applications of Electron Spin Resonance, Wiley, New York.

9 Atkins, P.W. and Symons, M.C.R. (1967) The Structure of Inorganic Radicals, Elsevier, Amsterdam.

10 Symons, M.C.R. (1978) Chemical and Biochemical Aspects of Electron Spin Resonance Spectroscopy, Halsted Press, New York.

11 Poole, C.P., Jr. (1983) Electron Spin Resonance, 2nd edn., Wiley-Interscience, New York.

12 Wilmshurst, T.H. (1967) Electron Spin Resonance Spectrometers, Hilger, London.

13 Knowles, P.F., Marsh, D. and Rattle, H.W.E. (1976) Magnetic Resonance of Biomolecules, Wiley-Interscience, New York.

14 Schlicter, C.P. (1963) Magnetic Resonance, Harper & Row, New York.

15 Pake, G.E. and Estle, T.L. (1973) The Physical Principles of Electron Paramagnetic Resonance, 2nd edn., Benjamin, Reading.

16 Molin, Yu. N., Salikhov, K.M. and Zamaraev, K.I. (1980) Spin Exchange: Principles and Applications in Chemistry and Biology, Springer-Verlag, Berlin.
17 Feher, G. (1970) Electron Paramagnetic Resonance with Applications to Selected Problems in Biology, Gordon and Breach, New York.
18 Abragam, A. and Bleaney, B. (1970) Electron Paramagnetic Resonance of Transition Ions, Oxford University Press, Oxford.
19 Kevan, L. and Kispert, L.D. (1976) Electron Spin Double Resonance Spectroscopy, Wiley-Interscience, New York.
20 Kevan, L. and Schwartz, R.N. (eds.) (1979) Time Domain Electron Spin Resonance, Wiley-Interscience, New York.
21 Schweiger, A. (1982) Electron Nuclear Double Resonance of Transition Metal Complexes with Organic Ligands, Springer-Verlag, New York.
22 Spin Labeling, Theory and Applications (1976) (Berliner, L.J., ed.), Academic Press, New York. Spin Labeling II, Theory and Applications (1979) (Berliner, L.J., ed.), Academic Press, New York.
23 Volodarsky, L.B., Grigor'ev, I.A. and Sagdeev, R.Z. (1980) in Biological Magnetic Resonance, 2 (Berliner, L.J., ed.), Plenum Press, New York, pp. 169–241.
24 Likhtenshtein, G.I. (1976) Spin Labeling in Molecular Biology, Wiley, New York.
25 Jost, P.C. and Griffith, O.H. (1978) Methods Enzymol. 49G, 369–418.
26 Berliner, L.J. (1978) Methods Enzymol. 49G, 418–480.
27 Morrisett, J.D. (1976) in Spin Labeling Theory and Applications (Berliner, L.J., ed.), Academic Press, New York, pp. 273–338.
28 Seelig, J. (1970) J. Am. Chem. Soc. 92, 3881–3887.
29 Nakaie, C.R., Goissis, G., Schreier, S. and Pawa, A.C.M. (1981) Braz. J. Med. Biol. Res. 14, 173–180.
30 Smith, I.C.P., Schreier-Muccilo, S. and Marsh, D. (1976) in Free Radicals in Biology, Vol. 1 (Pryor, W.A., ed.), Academic Press, New York, pp. 149–197.
31 Forrester, A.R. (1979) in Landolt-Bornstein New Series (Hellwege, K.-H., ed.), Group II, Vol. 19, Part c. 1, Springer-Verlag, Berlin, pp. 192–1066.
32 Pasenkiewicz-Gierula, M., Hyde, J.S. and Pilbrow, J.R. (1983) J. Magn. Reson. 55, 255–265.
33 Hyde, J.S. and Rao, K.V.S. (1980) J. Magn. Reson. 38, 313–317.
34 Briere, R., Lemaire, H., Rassat, A., Rey, P. and Rousseau, A. (1967) Bull. Soc. Chim. Fr., 4479–4484.
35 Briere, R., Lemaire, H., Rassat, A. and Dunand, J.-J. (1970) Bull. Soc. Chim. Fr., 4220–4226.
36 Kreilick, R.W. (1967) J. Chem. Phys. 46, 4260-4264.
37 Kreilick, R.W. (1973) in Advances in Magnetic Resonance, 6 (Waugh, J.S., ed.), Academic Press, New York, pp. 141–181.
38 Eaton, S.S., VanWilligan, H., Heinig, M.H. and Eaton, G.R. (1980) J. Magn. Reson. 38, 325–330.
39 Hyde, J.S. and Subczynski, W.K. (1984). J. Magn. Reson. 56, 125–130.
40 Lai, C.-S., Hopwood, L.E., Hyde, J.S. and Lukiewicz, S. (1982) Proc. Natl. Acad. Sci. U.S.A. 79, 1166–1170.
41 Percival P.W. and Hyde, J.S. (1976) J. Magn. Reson. 23, 249–257.
42 Kusumi, A., Subczynski, W.K. and Hyde, J.S. (1982) Proc. Natl. Acad. Sci. U.S.A. 79, 1854–1858.
43 Hyde, J.S. and Sarna, T. (1978) J. Chem. Phys. 68, 4439–4447.
44 Subczynski, W.K. and Hyde, J.S. (1981) Biochim. Biophys. Acta 643, 283–291.
45 Popp, C.A. and Hyde, J.S. (1982) Proc. Natl. Acad. Sci. U.S.A. 79, 2559–2563.
46 Stone, T.J., Buckman, T., Nordio, P.L. and McConnell, H.M. (1965) Proc. Natl. Acad. Sci. U.S.A. 54, 1010–1017.
47 Goldman, S.A., Bruno, G.V., Polnaszek, C.F. and Freed, J.H. (1972) J. Chem. Phys. 56, 716–735.
48 Freed, J.H. (1972) in Electron Spin Relaxation in Liquids (Muus, L.T. and Atkins, P.W., eds.), Plenum Press, New York, pp. 165–192.
49 Hyde, J.S. and Dalton, L.R. (1979) in Spin Labeling II, Theory and Applications (Berliner, L.J., ed.), Academic Press, New York, pp. 1–70.
50 Hyde, J.S. and Hyde, D.A. (1981) J. Magn. Reson. 43, 137–140.
51 Lajzerowicz-Bonneteau, J. (1976) in Spin Labeling, Theory and Applications (Berliner, L.J., ed.), Academic Press, New York, pp. 239–250.

52 Piette, L. and Hsia, J.C. (1979) in Spin Labeling II, Theory and Applications (Berliner, L.J., ed.), Academic Press, New York, pp. 247–290.
53 Hyde, J.S., Froncisz, W. and Kusumi, A. (1982) Rev. Sci. Instrum. 53, 1934–1937.
54 Tao, T. (1969) Biopolymers 8, 609–632.
55 Broido, M.S., Belsky, I. and Meirovich, E.J. (1982) J. Phys. Chem. 86, 4197–4205.
56 Hyde, J.S. (1978) Methods Enzymol. 49G, 480–511.
57 Hyde, J.S. and Thomas, D.D. (1980) Annu. Rev. Phys. Chem. 31, 293–317.
58 Backer, J.M., Budker, V.G., Eremenko, S.I. and Molin, Y.N. (1977) Biochim. Biophys. Acta 460, 152–156.
59 Hyde, J.S., Swartz, H.M. and Antholine, W.E. (1979) in Spin Labeling II, Theory and Applications (Berliner, L.J., ed.), Academic Press, New York, pp. 71–114.
60 Leigh, J.S., Jr. (1970) J. Chem. Phys. 52, 2608–2612.
61 Eaton, S.S. and Eaton, G.R. (1978) Coord. Chem. Rev. 26, 207–262.
62 Hyde, J.S. and Rao, K.V.S. (1978) J. Magn. Reson. 29, 509–516.
63 Borg, D.C. (1972) in Biological Applications of Electron Spin Resonance (Swartz, H.M., Bolton, J.R. and Borg, D.C., eds.), Wiley-Interscience, New York, pp. 265–350.
64 Borg, D.C. (1976) in Free Radicals in Biology, Vol. 1 (Pryor, W.A., ed.), Academic Press, New York, pp. 69–147.
65 Yamazaki, I. (1977) in Free Radicals in Biology, Vol. 3 (Pryor, W.A., ed.), Academic Press, New York, pp. 183–218.
66 Mason, R.P. (1982) in Free Radicals in Biology, Vol. 5 (Pryor, W.A., ed.), Academic Press, New York, pp. 161–222.
67 McCay, P.B., Noguchi, T., Fong, K.-L., Lai, E.K. and Poyer, J.L. (1980) in Free Radicals in Biology, Vol. 4 (Pryor, W.A., ed.), Academic Press, New York, pp. 155–186.
68 Sealy, R.C., Felix, C.C., Hyde, J.S. and Swartz, H.M. (1980) in Free Radicals in Biology, Vol. 4 (Pryor, W.A., ed.), Academic Press, New York, pp. 209–259.
69 Janzen, E.G. (1980) in Free Radicals in Biology, Vol. 4 (Pryor, W.A., ed.), Academic Press, New York, pp. 114–154.
70 Mason, R.P. (1979) in Reviews of Biochemical Toxicology, Vol. 1 (Hodgson, E., Bend, J.R. and Philpot, R.M., eds.), Elsevier/North Holland, New York, pp. 151–200.
71 Mason, R.P. and Chignell, C.F. (1982) Pharmacol. Rev. 33, 189–211.
72 Kalyanaraman, B. (1982) in Reviews of Biochemical Toxicology, Vol. 4 (Hodgson, E., Bend, J.R. and Philpot, R.M., eds.), Elsevier/North Holland, New York, pp. 73–139.
73 Perkins, M.J. (1980) Adv. Phys. Org. Chem. 17, 1–64.
74 Buettner, G.R. (1982) in Superoxide Dismutase, Vol. II (Oberley, L.W., ed.), CRC Press, Boca Raton, pp. 63–81.
75 Symons, M.C.R. (1969) J. Am. Chem. Soc. 91, 5924.
76 Symons, M.C.R. (1974) J. Chem. Soc. Perkin Trans. II, 1618–1620.
77 Pryor, W.A. (1966) Free Radicals, McGraw Hill, New York, p. 9.
78 Bielski, B.H.J., Richter, H.W. and Chan, P.C. (1975) Ann. N. Y. Acad. Sci. 258, 231.
79 Trifunac, A.D., Norris, J.R. and Lawler, R.G. (1979) J. Chem. Phys. 71, 4380–4390.
80 Beinert, H. (1972) in Biological Applications of Electron Spin Resonance (Swartz, H.M., Bolton, J.R. and Borg, D.C., eds.), Wiley-Interscience, New York, pp. 351–410.
81 Janzen, E.G. (1971) Accounts Chem. Res. 4, 31–40.
82 Perkins, M.J. (1970) in Essays in Free Radical Chemistry, Spec. Publ. No. 24, Chemical Society, London, pp. 97–115.
83 Kalyanaraman, B. and Sealy, R.C. (1982) Biochem. Biophys. Res. Commun. 106, 1119–1125.
84 Misra, H.P. and Fridovich, I. (1972) J. Biol. Chem. 247, 3170–3175.
85 Adams, M., Blois, M.S. and Sands, R.H. (1958) J. Chem. Phys. 28, 774–776.
86 Perez-Reyes, E. and Mason, R.P. (1981) J. Biol. Chem. 256, 2427–2432.
87 Felix, C.C. and Sealy, R.C. (1981) J. Am. Chem. Soc. 103, 2831–2836.
88 Felix, C.C., Reszka, K. and Sealy, R.C. (1983) Photochem. Photobiol. 37, 141–147.

89 Mauzerall, D. and Feher, G. (1964) Biochim. Biophys. Acta 79, 430–432; Biochim. Biophys. Acta 88, 658–660.
90 Novak, M., Miller, A., Bruice, T.C and Tollin, G. (1980) J. Am. Chem. Soc. 102, 1465–1467.
91 Borg, D.C., Schaich, K.M., Elmore, J.J. Jr. and Bell, J.A. (1978) Photochem. Photobiol. 28, 887–907.
92 Bishop, C.A. and Tong, L.K.J. (1965) J. Am. Chem. Soc. 87, 501–505.
93 Stone, T.J. and Waters, W.A. (1965) J. Chem. Soc., 1488–1494.
94 Borg, D.C. and Elmore, J.J. (1967) in Magnetic Resonance in Biological Systems, (Berliner, L.J. and Reuben, J., eds.), Pergamon Press, Oxford, pp. 341–349.
95 Lambeth, D.O. and Palmer, G. (1973) J. Biol. Chem. 248, 6095–6103.
96 McMillan, M. and Norman, R.O.C. (1968) J. Chem. Soc. (B), 590–597.
97 Hemmerich, P., Dervartanian, D.V., Veeger, C. and VanVoorst, J.D.W. (1963) Biochim. Biophys. Acta 77, 504–506.
98 Felix, C.C. and Sealy, R.C. (1982) J. Am. Chem. Soc. 104, 1555–1560.
99 Hemmerich, P. (1964) Helv. Chim. Acta 47, 464–475.
100 Massey, V. and Palmer, G. (1966) Biochemistry 5, 3181–3189.
101 Massey, V. and Hemmerich, P. (1978) Biochemistry 17, 1–16.
102 Moan, J. and Wold, E. (1979) Nature 279, 450–451.
103 Moan, J. (1980) Acta Chem. Scand. B34, 519–521.
104 Adams, R.N. (1969) Electrochemistry at Solid Electrodes, Dekker, New York.
105 Fischer, H. and Hellwege, K.-H. (eds.) (1977–1979) Landbolt-Bornstein New Series, Group II, Vol. 9, Springer-Verlag, Berlin.
106 Saunders, B.C. (1982) in Inorganic Biochemistry, Vol. 2. (Eichborn, G.L., ed.), Elsevier, New York, pp. 988–1021.
107 Hager, L.P., Doubek, D.L., Silverstein, R.M., Hargis, J.H. and Martin, J.C. (1972) J. Am. Chem. Soc. 94, 4364.
108 Roberts, J.E., Hoffman, B.M., Rutter, R. and Hager, L.P. (1981) J. Am. Chem. Soc. 103, 7654–7656.
109 Schulz, C.E., Devaney, P.W., Winkler, H., Debrunner, P.G., Doan, N., Chiang, R., Rutter, R. and Hager, L.P. (1979) FEBS Lett. 103, 102–105.
110 Roberts, J.E., Hoffman, B.M., Rutter, R. and Hager, L.P. (1981) J. Biol. Chem. 256, 2118–2121.
111 Kneko, Y., Tamura, M. and Yamazaki, I. (1980) Biochemistry 19, 5795–5799.
112 Rutter, R. and Hager, L.P. (1982) J. Biol. Chem. 257, 7958–7961.
113 Aasa, R., Vanngard, T. and Dunford, H.B. (1975) Biochim. Biophys. Acta 391, 259–264.
114 Yonetani, T., Schleyer, H. and Ehrenberg, A. (1966) J. Biol. Chem. 241, 3240–3243.
115 Peisach, J., Blumberg, W.E., Wittenberg, B.A. and Wittenberg, J.B. (1968) J. Biol. Chem. 243, 1871–1880.
116 Hoffman, B.M., Roberts, J.E., Brown, T.G., Kang, C.H. and Margoliash, E. (1979) Proc. Natl. Acad. Sci. U.S.A. 76, 6132–6136.
117 Gilbert, B.C., Hodgeman, D.K.C. and Norman, R.O.C. (1973) J. Chem. Soc. Perkin Trans. II, 1748–1752.
118 King, N.K., Looney, F.D. and Winfield, M.E. (1964) Biochim. Biophys. Acta 88, 235–236.
119 Butler, J., Land, E.J., Prutz, W.A. and Swallow, A.J. (1982) Biochim. Biophys. Acta 705, 150–162.
120 Yamazaki, I., Mason, H.S. and Piette, L. (1960) J. Biol. Chem. 235, 2444–2449; (1961) Biochim. Biophys. Acta 50, 62–69.
121 Ohnishi, T., Yamazaki, H., Iyanagi, T., Nakamura, T. and Yamazaki, I. (1969) Biochim. Biophys. Acta 172, 357.
122 Shiga, T. and Imizumi, K. (1975) Arch. Biochem. Biophys. 167, 469–479.
123 Blumberg, W., Goldstein, M., Lanber, E. and Peisach, J. (1965) Biochim. Biophys. Acta 99, 187.
124 Skotland, T. and Ljones, T. (1980) Biochim. Biophys. Acta 630, 30–35.
125 Mason, R.P., Kalyanaraman, B., Tainer, B.E. and Eling, T.E. (1980) J. Biol. Chem. 255, 5019–5022.
126 DeGroot, J.J.M.C., Garssen, G.J., Vliegenthart, J.F.G. and Boldingh, J. (1975) Biochim. Biophys. Acta 337, 71–79.
127 Kalyanaraman, B., Mason, R.P., Tainer, B. and Eling, T.E. (1982) J. Biol. Chem. 257, 4764–4768.

128 Nakamura, T. (1960) Biochem. Biophys. Res. Commun. 2, 111–113.
129 Buettner, G.R. and Coffman, R.E. (1977) Biochim. Biophys. Acta 480, 495–505.
130 Boas, J.F., Hicks, P.R., Pilbrow, J.R. and Smith, T.D. (1978) JCS Faraday Trans. II 74, 417–431.
131 Wallis, O.C., Bray, R.C., Gutteridge, S. and Hollaway, M.R. (1982) Eur. J. Biochem. 125, 299–303.
132 Finlay, T.H., Valinsky, J., Sato, K. and Abeles, R.H. (1972) J. Biol. Chem. 247, 4297.
133 Golding, B.T. and Radom, L. (1976) J. Am. Chem. Soc. 98, 6331–6338.
134 Gilbert, B.C., Larkin, J.P. and Norman, R.O.C. (1972) J. Chem. Soc. Perkin Trans. 2, 794–802.
135 Ehrenberg, A. and Reichard, P. (1972) J. Biol. Chem. 247, 3485–3488.
136 Barlow, T., Eliasson, R., Platz, A., Reichard, P. and Sjoberg, B.-M. (1983) Proc. Natl. Acad. Sci. U.S.A. 80, 1492–1495.
137 Sjoberg, B.-M., Reichard, P., Graslund, A. and Ehrenberg, A. (1978) J. Biol. Chem. 253, 6863–6865.
138 Sjoberg, B.-M., Loehr, T.M. and Sanders-Loehr, J. (1982) Biochemistry 21, 96–102.
139 Palmer, G., Muller, F. and Massey, V. (1971) in Flavins and Flavoproteins (Kamin, H., ed.), University Park Press, Baltimore, pp. 123–140.
140 Ehrenberg, A., Ericksson, L.E.G. and Hyde, J.S. (1971) in Flavins and Flavoproteins (Kamin, H., ed.), University Park Press, Baltimore, pp. 141–152.
141 Ehrenberg, A., Muller, F. and Hemmerich, P. (1967) Eur. J. Biochem. 2, 286–293.
142 Walker, W.H., Salach, J., Gutman, M., Singer, T.P., Hyde, J.S. and Ehrenberg, A. (1969) FEBS Lett. 5, 237–240.
143 Ohnishi, T., King, T.E., Salerno, J.C., Blum, H., Bowyer, J.R. and Maida, T. (1981) J. Biol. Chem. 256, 5577–5582.
144 McKean, M.C., Sealy, R.C. and Frerman, F.E. (1982) in Flavins and Flavoproteins (Massey, V. and Williams, C.H., eds.), Elsevier, New York, pp. 614–617.
145 Ericksson, L.E.G. and Ehrenberg, A. (1973) Biochim. Biophys. Acta 295, 57–66.
146 Beinert, H. and Hemmerich, P. (1965) Biochem. Biophys. Res. Commun. 18, 212–220.
147 Yasukochi, Y., Peterson, J.A. and Masters, B.S. (1979) J. Biol. Chem. 254, 7097–7104.
148 Iyanagi, T. and Yamazaki, I. (1969) Biochim. Biophys. Acta 172, 370.
149 Rosen, H. and Klebanof, S.J. (1979) J. Clin. Invest. 64, 1725.
150 Green, M.R., Hill, H.A.O., Okolow–Zubkowska, M.J. and Segal, A.W. (1979) FEBS Lett. 100, 23–26.
151 Ruzika, F.J., Beinert, H., Schepler, K.L., Dunham, W.R. and Sands, R.H. (1975) Proc. Natl. Acad. Sci. U.S.A. 72, 2886–2890.
152 Ruuge, E.K. and Konstantinov, A.A. (1976) Biofizika 21, 586–587.
153 Salerno, J.C. and Ohnishi, T. (1980) Biochem. J. 192, 769–781.
154 Hales, B.J. (1975) J. Am. Chem. Soc. 97, 5993–5997.
155 Hales, B.J. and Case, E.E. (1981) Biochim. Biophys. Acta 637, 291–302.
156 Hyde, J.S. and Pilbrow, J.R. (1980) J. Magn. Reson. 41, 447–457.
157 Fridovich, I. (1970) J. Biol. Chem. 245, 4053–4057.
158 Nakamura, S. and Yamazaki, I. (1969) Biochim. Biophys. Acta 189, 29–37.
159 Nakamura, S. and Yamasaki, I. (1973) Biochim. Biophys. Acta 327, 247–256.
160 Knowles, P.F., Gibson, J.F., Pick, F.M. and Bray, R.C. (1969) Biochem. J. 111, 53–58.
161 Mason, H.S., Yamazaki, I. and Spencer, E. (1961) Biochem. Biophys. Res. Commun. 4, 236–238.
162 Uemura, T., Shimazu, T., Miura, R. and Yamano, T. (1980) Biochem. Biophys. Res. Commun. 93, 1074–1081.
163 Josephy, P.D., Eling, T.E. and Mason, R.P. (1983) J. Biol. Chem. 258, 5561–5569.
164 Sealy, R.C., Hyde, J.S., Felix, C.C., Menon, I.A., Prota, G., Swartz, H.M., Persad, S. and Haberman, H.F. (1982) Proc. Natl. Acad. Sci. U.S.A. 79, 2885–2889.
165 Chio, S.-S., Hyde, J.S. and Sealy, R.C. (1982) Arch. Biochem. Biophys. 215, 100–106.
166 Chio, S.-S., Hyde, J.S. and Sealy, R.C. (1980) Arch. Biochem. Biophys. 199, 133–139.
167 Sarna, T. and Sealy, R.C. (1984) Photochem. Photobiol. 39, 69–74.
168 Laroff, G.P., Fessenden, R.W. and Schuler, R.H. (1972) J. Am. Chem. Soc. 90, 9062.
169 Dodd, N.J.F. and Swartz, H.M. (1980) Br. J. Cancer 42, 349–351.
170 Swartz, H.M. (1979) in Submolecular Biology and Cancer, Ciba Foundation Series 67, 107–130.

171 Piette, L.H., Bulow, G. and Yamazaki, I. (1964) Biochim. Biophys. Acta 88, 120–129.
172 Josephy, P.D., Eling, T. and Mason, R.P. (1982) J. Biol. Chem. 257, 3669–3675.
173 Sayo, H. and Hosokawa, M. (1980) Chem. Pharm. Bull 28, 683–685.
174 Griffin, B.W. and Ting, P.L. (1978) Biochemistry 17, 2206–2211.
175 Mottley, C., Mason, R.P., Chignell, C.F., Sivarajah, K. and Eling, T.E. (1982) J. Biol. Chem. 257, 5050–5055.
176 Hill, H.A.O. and Thornalley, P.J. (1981) FEBS Lett. 125, 235–238.
177 Rauckman, E.J., Rosen, G.M. and Kitchell, B.B. (1979) Mol. Pharmacol. 15, 131–137.
178 Oritz de Montellano, P.R., Augosto, O., Viola, F. and Kanze, K.L. (1983) J. Biol. Chem. 258, 8623–8629.
179 Griffin, B.W., Marth, C., Yasukochi, Y. and Masters, B.S. (1980) Arch. Biochem. Biophys. 205, 543–553.
180 Mason, R.P. and Holtzman, J.L. (1975) Biochemistry 14, 1626–1632.
181 Perez-Reyes, E., Kalyanaraman, B. and Mason, R.P. (1980) Mol. Pharmacol. 17, 239–244.
182 Mason, R.P. and Holtzman, J.L. (1975) Biochem. Biophys. Res. Commun. 67, 1267–1274.
183 Kalyanaraman, B., Perez-Reyes, E., Mason, R.P., Peterson, F.J. and Holtzman, J.L. (1979) Mol. Pharmacol. 16, 1059–1064.
184 Kalyanaraman, B., Perez-Reyes, E. and Mason, R.P. (1980) Biochim. Biophys. Acta 630, 119–130.
185 Sato, S., Iwaizumi, M., Handa, K. and Tamura, Y. (1977) Gann 68, 603–608.
186 Bachur, N.R., Gordon, S.L. and Gee, M.V. (1977) Mol. Pharmacol. 13, 901–910.
187 Sealy, R.C., Swartz, H.M. and Olive, P.L. (1978) Biochem. Biophys. Res. Commun. 82, 680–684.
188 Poyer, J.L., Floyd, R.A., McCay, P.B., Janzen, E.G. and Davis, E.R. (1978) Biochim. Biophys. Acta 539, 402–409.
189 Kalyanaraman, B., Mason, R.P., Perez-Reyes, E., Chignel, C.F., Wolfe, C.R. and Philpot, R.M. (1979) Biochem. Biophys. Res. Commun. 89, 1065–1072.
190 Harrelson, W.G., Jr. and Mason, R.P. (1982) Mol. Pharmacol. 22, 239–242.
191 Pan, S.-S. and Bachur, N.R. (1980) Mol. Pharmacol. 17, 95–99.
192 Vanngard, T. (1972) in Biolgical Applications of Electron Spin Resonance (Swartz, H.M., Bolton, J.R., and Borg, D.C., eds.), Wiley-Interscience, New York, pp. 411–447.
193 Low, W. (1960) Paramagnetic Resonance in Solids, Academic Press, New York.
194 Yen, T.F. (ed.) (1969) Electron Spin Resonance of Metal Complexes, Plenum, New York.
195 McGarvey, B.R. (1966) in Transition Metal Chemistry, Vol. 3 (Carlin, R.L., ed.), Dekker, New York, pp. 89–201.
196 Brill, A.S. (1977) Transition Metals in Biochemistry, Molecular Biology, Biochemistry and Biophysics (Kleinzeller, A., Springer, G.F. and Wittman, H.G., eds.), Vol. 26, Springer-Verlag, New York, pp. 1–86.
197 Umezawa, H., Maeda, K., Takeuchi, T. and Okami, J. (1966) Antibiot. Ser. A. 29, 200–209.
198 Sausville, E.A., Peisach, J. and Horwitz, S.B. (1978) Biochemistry 17, 2740–2745.
199 Sausville, E.A., Stein, R.W., Peisach, J. and Horwitz, S.B. (1978) Biochemistry 17, 2746–2754.
200 Burger, R.M., Peisach, J., Blumberg, W.E. and Horwitz, S.B. (1979) J. Biol. Chem. 254, 10906–10912.
201 Burger, R.M., Peisach, J. and Horwitz, S.B. (1981) J. Biol. Chem. 256, 11636–11644.
202 Sugiura, Y. and Kikuchi, T. (1978) J. Antibiot. 31, 1310–1312.
203 Oberley, L.W. and Buettner, G.R. (1979) FEBS Lett. 97, 47–49.
204 Solaiman, D., Rao, E.A., Petering, D.H., Sealy, R.C. and Antholine, W.E. (1979) Int. J. Radiat. Oncol. Biol. Phys. 5, 1519–1521.
205 Dabrowiak, J.C. (1980) J. Inorg. Biochem. 13, 317–337.
206 Sugiura, Y. (1979) Biochem. Biophys. Res. Commun. 87, 643–648.
207 Sugiura, Y. (1979) Biochem. Biophys. Res. Commun. 88, 913–918.
208 Sugiura, Y. (1980) J. Am. Chem. Soc. 102, 5208–5215.
209 Antholine, W.E. and Petering, D.H. (1979) Biochem. Biophys. Res. Commun. 91, 528–533.
210 Blumberg, W.E. and Peisach, J. (1971) in Probes of Structure and Function of Macromolecules and Membranes, Vol. 2 (Chance, B., Yonetani, T. and Mildvan, A.S., eds.), Academic Press, New York, pp. 215–227.
211 Sugiura, Y. (1980) J. Am. Chem. Soc. 102, 5216–5221.

212 Wagner, G.C., Gunsalus, I.C., Wang, M.-Y.R. and Hoffman, B.M. (1981) J. Biol. Chem. 256, 6266–6273.
213 Bossu, F.P. and Margerum, D.W. (1976) J. Am. Chem. Soc. 98, 4003–4004.
214 Sugiura, Y. and Mino, Y. (1979) Inorg. Chem. 18, 1336–1339.
215 Kruger, H.J., Huynh, B.H., Ljungdahl, P.O., Xavier, A.V., DerVartonian, D.V., Moura, I., Peck Jr., H.D., Teixeira, M., Moura, J.J.G. and LeGall, J. (1982) J. Biol. Chem. 257, 14620–14623.
216 Spiro, T.G. (ed.) (1981) Copper Proteins, Wiley-Interscience, New York, and other references therein.
217 Peisach, J. and Blumberg, W.E. (1974) Arch. Biochem. Biophys. 165, 691–708.
218 Peisach, J. Stern, J.O. and Blumberg, W.E. (1973) Drug Metab. Dispos. 1, 45–61.
219 Barber, M.J., Salerno, J.C. and Siegel, L.M. (1982) Biochemistry 21, 1648–1656.
220 Blum, H. and Poole, R.K. (1982) Biochem. Biophys. Res. Commun. 107, 903–909.
221 Burgess, B.K., Stiefel, E.I. and Newton, W.E. (1980) J. Biol. Chem. 255, 353–356.
222 Moura, J.J.G., Moura, I., Kent, T.A., Lipscomb, J.D., Huynh, B.H., LeGall, J., Xavier, A.V. and Munck, E. (1982) J. Biol. Chem. 257, 6259–6267.
223 Slappendel, S., Veldink, G.A., Vliegenthart, J.F.G., Aasa, R. and Malmstrom, B.G. (1981) Biochim. Biophys. Acta 667, 77–86.
224 Antonaitis, B.C. and Aisen, P. (1982) J. Biol. Chem. 257, 1855–1859.
225 Ecker, D.J., Lancaster Jr., J.R. and Emery, T. (1982) J. Biol. Chem. 257, 8623–8626.
226 Palmer, G. (1979) in The Porphyrins, Vol. IV (Dolphin, D., ed.), Academic Press, New York, pp. 313–354.
227 Lin, W.C. (1979) in The Porphyrins, Vol. IV (Dolphin, D., ed.), Academic Press, New York, pp. 355–378.
228 Louro, S.R.W. and Bemski, G. (1977) J. Magn. Reson. 28, 427–431.
229 Rakhit, G. and Sarkar, B. (1981) J. Inorg. Biochem. 15, 233–241.
230 Sarkar, B. (1981) in Metal Ions in Biological Systems, Vol. 12 (Sigel, H., ed.), Marcel Dekker, New York, pp. 233–282.
231 Rifkind, J.M. (1981) in Metal Ions in Biological Systems, Vol. 12 (Sigel, H., ed.), Marcel Dekker, New York, pp. 191–232.
232 Iyer, K.S., Lau, J., Laurie, S.H. and Sarkar, B. (1978) Biochem. J. 169, 61–69.
233 Brown, C.E., Antholine, W.E. and Froncisz, W. (1980) J. Chem. Soc. Dalton, 590–596.
234 Boas, J.F., Pilbrow, J.R., Hartzell, C.R. and Smith, T.D. (1969) J. Chem. Soc. (A) 572–577.
235 Boas, J.F., Pilbrow, J.R. and Smith, T.D. (1978) in Biological Magnetic Resonance (Berliner, L.J. and Reuben, J., eds.), Plenum, New York, pp. 277–342.
236 Hyde, J.S., Swartz, H.M. and Antholine, W.E. (1979) in Spin Labeling II (Berliner, L.J., ed.), Academic Press, New York, pp. 71–113.
237 Saryan, L.A., Mailer, K., Krishnamurti, C., Antholine, W.E. and Petering, D.H. (1981) Biochem. Pharmacol. 30, 1595–1604.
238 Antholine, W.E. and Taketa, F. (1984) J. Inorg. Biochem. 20, 69–78.
239 Minkel, D.T., Saryan, L.A. and Petering, D.H. (1978) Cancer Res. 38, 124–129.
240 Antholine, W.E., Knight, J., Whelan, H. and Petering, D.H. (1977) Mol. Pharmacol. 13, 89–98.
241 Antholine, W.E. and Taketa, F. (1982) J. Inorg. Biochem. 16, 145–154.
242 Hyde, J.S. and Froncisz, W. (1982) Annu. Rev. Biophys. Bioeng. 11, 391–417.
243 Antholine, W.E., Basosi, R., Hyde, J.S. and Petering, D.H. (1984) Inorg. Chem. 23, 3543–3548.
244 Basosi, R., Niccolai, N. and Rossi, C. (1978) Biophys. Chem. 8, 61–69.
245 Webb, M.R., Ash, D.E., Leyh, T.S., Trentham, D.R. and Reed, G.H. (1982) J. Biol. Chem. 257, 3068–3072.
246 Eccleston, J.F., Webb, M.R., Ash, D.E. and Reed, G.H. (1981) J. Biol. Chem. 256, 10774–10777.
247 Schramm, V.L. and Reed, G.H. (1980) J. Biol. Chem. 255, 5795–5801.
248 Jemiolo, D.K. and Grisham, C.M. (1982) J. Biol. Chem. 257, 8148–8152.
249 Ash, D.E. and Schramm, V.L. (1982) J. Biol. Chem. 257, 9261–9264.
250 Chasteen, N.D., Berliner, L.J. and Reuben, J. (1981) Biological Magnetic Resonance, Vol. 3, Plenum, New York.
251 Chasteen, N.D. and Theil, E.C. (1982) J. Biol. Chem. 257, 7627–7677.
252 Sakurai, H., Goda, T., Shimomura, S. and Yoshimura, T. (1982) Biochem. Biophys. Res. Commun. 104, 1421–1426.

253 Post, R.L., Hunt, D.P., Walderhaug, M.O., Perkins, R.C., Park, J.H. and Beth, A.H. (1979) in Na, K-ATPase Structure and Kinetics (Skow, J.C. and Norby, J.G., eds.), Academic Press, London, pp. 389–404.
254 Cantley, L.C., Josephson, L., Warner, R., Yanagisawa, M., Lechene, C. and Guidotti, G. (1977) J. Biol. Chem. 252, 7421–7423.
255 Macara, I.G., Kustin, K. and Cantley, L.C. (1980) Biochim. Biophys. Acta 629, 95–106.
256 Valentine, J.S., Pantoliano, M.W., McDonnell, P.J., Burger, A.R. and Lippard, S.J. (1979) Proc. Natl. Acad. Sci. U.S.A. 76, 4245–4249.
257 Himmelwright, R.S., Eickman, N.C., LuBien, C.D. and Solomon, E.I. (1980) J. Am. Chem. Soc. 102, 5378–5388.
258 Himmelwright, R.S., Eickman, N.C., LuBien, C.D., Lerch, K. and Solomon, E.I. (1980) J. Am. Chem. Soc. 102, 7339–7344.
259 Markham, G.D. (1981) J. Biol. Chem. 256, 1903–1909.
260 Antholine, W.E., Basosi, R., Hyde, J.S. and Taketa, F. (1984) J. Inorg. Chem. 21, 125–136.
261 Froncisz, W. and Hyde, J.S. (1980) J. Chem. Phys. 73, 3123–3131.
262 Froncisz, W. and Aisen, P. (1982) Biochim. Biophys. Acta 700, 55–58.
263 Rakhit, G., Antholine, W.E., Froncisz, W., Hyde, J.S., Pilbrow, J.R., Sinclair, G.R. and Sarkar, B. (1985) J. Inorg. Biochem., in press.
264 Antholine, W.E., Hyde, J.S., Sealy, R.C. and Petering, D.H. (1984) J. Biol. Chem. 259, 4435–4440.
265 Froncisz, W., Scholes, C.P., Hyde, J.S., Wei, Y., King, T.E., Shaw, R.W. and Beinert, H. (1979) J. Biol. Chem. 254, 7482–7484.
266 Taketa, F. and Antholine, W.E. (1982) J. Inorg. Biochem. 17, 109–120.
267 Mims, W.B. and Peisach, J. (1981) in Biological Magnetic Resonance, Vol. 3 (Berliner, L.J. and Reuben, J., eds.), Plenum, New York, pp. 213–263.
268 Burger, R.M., Adler, A.D., Horwitz, J.B., Mims, W.B. and Peisach, J. (1981) Biochemistry 20, 1701–1704.
269 Zweier, J., Aisen, P., Peisach, J. and Mims, W.B. (1979) J. Biol. Chem. 254, 3512–3515.
270 Mims, W.B. and Peisach, J. (1978) J. Chem. Phys. 69, 4921–4930.
271 Fee, J.A., Peisach, J. and Mims, W.B. (1981) J. Biol. Chem. 256, 1910–1914.
272 Hyde, J.S. (1974) Annu. Rev. Phys. Chem. 25, 407–434.
273 Norris, J.R., Scheer, H. and Katz, J.J. (1979) in The Porphyrins, Vol. IV (Dolphin, D., ed.), Academic Press, New York, pp. 159–197.
274 Scholes, C.P. (1979) in Multiple Electron Resonance Spectroscopy (Dorio, M.M. and Freed, J.H., eds.), Plenum, New York, pp. 297–329.
275 Sands, R.H. (1979) in Multiple Electron Resonance Spectroscopy (Dorio, M.M. and Freed, J.H., eds.), Plenum, New York, pp. 331–374.
276 Rist, G.H. and Hyde, J.S. (1970) J. Chem. Phys. 52, 4633–4643.
277 Rist, G.H., Hyde, J.S. and Vanngard, T. (1970) Proc. Natl. Acad. Sci. U.S.A. 67, 79–86.
278 Roberts, J.E., Brown, T.G., Hoffman, B.M. and Peisach, J. (1980) J. Am. Chem. Soc. 102, 825–829.
279 Van Camp, H.L., Wei, Y.H., Scholes, C.P. and King, T.E. (1980) Biochim. Biophys. Acta 537, 238–247.
280 Hoffman, B.M., Roberts, J.E., Swanson, M., Speck, S.H. and Margoliash, E. (1980) Proc. Natl. Acad. Sci. U.S.A. 77, 1452–1457.
281 Feher, G., Isaacson, R.A. and Scholes, C.P. (1973) Ann. N. Y. Acad. Sci. 222, 86–101.
282 Scholes, C.P., Isaacson, R.A., Yonetani, T. and Feher, G. (1973) Biochim. Biophys. Acta 322, 457–462.
283 Mulks, C.F., Scholes, C.P., Dickinson, L.C. and Lapidot, A. (1979) J. Am. Chem. Soc. 101, 1645–1654.
284 Feher, G. (1959) Bell Systems Tech. J. 36, 449–484.
285 Praddaude, H.D. (1967) Rev. Sci. Instrum. 38, 339–347.
286 Hyde, J.S. (1963) U.S. Patent 3,350,633.
287 Huisjen, M. and Hyde, J.S. (1974) Rev. Sci. Instrum. 45, 669–675.
288 Percival, P. and Hyde, J.S. (1975) Rev. Sci. Instrum. 46, 1522–1529.
289 Hyde, J.S. (1962) Evaluation of an EPR Superheterodyne Spectrometer, Varian Associates, 7th Annual NMR-EPR Workshop.

290 Hyde, J.S. (1965) Signal Amplitudes in Electron Paramagnetic Resonance, Varian Associates, Technical Information Bulletin, Fall 1965.
291 Froncisz, W. and Hyde, J.S. (1982) J. Magn. Reson. 47, 515–521.
292 Thomas, D.D., Wendt, C.H., Froncisz, W. and Hyde, J.S. (1983) Biophys. J. 43, 131–135.
293 Wood, R.L., Froncisz, W. and Hyde, J.S. (1984) J. Magn. Reson. 58, 243–253.
294 Geschwind, S. (1967) in Hyperfine Interactions (Freeman, A.J. and Frankel, R.B., eds.), Academic Press, New York, pp. 225–286.
295 Mobius, K., Frohling, W., Lendzlan, F., Lubitz, W., Plato, M. and Winscom, C.J. (1982) J. Phys. Chem. 86, 4491–4507.
296 Hyde, J.S., Chien, J.C.W. and Freed, J.H. (1968) J. Chem. Phys. 48, 4211–4226.

Neuberger/Van Deenen (eds.) Modern Physical Methods in Biochemistry, Part A

CHAPTER 3

Mass spectrometry

J.C. TABET and M. FÉTIZON

Ecole Polytechnique, Laboratoire de Synthèse Organique, 91128 Palaiseau Cedex, France

1. General

During the last 15 years, mass spectrometry has undergone a veritable explosion of its possibilities, both at the level of new ionization techniques and the development of peripheral methods. These peripheral methods, leading to a better rationalization of mass spectrometry, were introduced only during the use of electron impact ionization.

These methods became necessary for analytical applications, especially in the field of biological and natural molecules, where it was necessary to examine very small diluted samples.

Two different aspects appeared in this evolution: the determination of the structure of organic compounds, and the quantitative study of mixtures of compounds with known structures.

The behavior of organic compounds under electron impact allowed the interpretation of ion formation at the source. Isotopic labeling was primarily utilized in this approach (^{2}H and ^{13}C). Thus, fragmentation mechanisms of a very general nature, and even in certain cases highly specific, could be elucidated in an extremely empirical fashion.

Several works have appeared on these topics [1–4], the best known being those of Djerassi and co-workers [5] and Waller and co-worker [6] in the field of biochemistry.

Alongside this empirical approach, more basic research led to the discovery of the main kinetic and thermodynamic factors which govern competitions among the various decomposition reactions, based on a theory introduced by Rosenstock et al. [7] and subsequently improved [8]. This theoretical approach led certain workers to perform determination at low electron energy (10–15 eV) to obtain utilizable data, for example, during isotopic assays [9] on specific centers.

(a) Peripheral techniques in mass spectrometry

An understanding of these mechanisms enables the second aspect of mass spectrometry to be approached with more confidence. One of the results of this approach led chemists to introduce functional groups [10–12] in order to induce certain specific fragmentations, which may considerably simplify the spectra. This type of utilization furnishes structural data on the environment of the exchanged group.

The presence of easily detected and intense fragmentations also led to the possibility of studying complex mixtures quantitatively with techniques such as selected ion monitoring (SIM) and multiple ion detection (MID), which are used extensively especially when the mass spectrometer is coupled to a gas chromatograph. In their review, Lehman and Schulten [13] showed the various possibilities of these methods when mass spectrometry is applied quantitatively to biochemistry or medicine. It should be noted that these detection modes are valid even at high resolution (>10,000) [14,15].

In the fields of pharmacology, toxicology and medicine, Jellum [16] showed how GC/MS opened the door to new possibilities. His exhaustive work describes the problems associated with GC/MS, as well as those involved with the processing of the data obtained. The use of stable isotopes for these studies is a non-negligible contribution, as shown by Krahmer and McCloskey [17] during the development of this methodology. Klein and Klein [18,19] performed a veritable census of labeled molecules so far prepared for these analyses.

During the last few years, a considerable number of reviews (prior to 1981) have appeared on the subject of GC/MS [20] and also on the new possibilities offered by LC/MS [21,22], which seems to have a promising future. Computer-assisted data processing has enabled workers to study complex problems, even when only a very small sample is available. The rapid repetition of such measurements leads to sufficient precision, even at high resolution [23].

It is difficult to give an exhaustive list of the various possibilities which simplify analytical and structural MS studies, in particular in the fields of natural and biological products. We wish to stress, however, the particular importance of computer-assisted data processing, used to determine the structure of these compounds in a mixture and even in trace amounts [24].

These advantages are due to: the creation of actual spectrum libraries [25,26]; and the development of Dendral type [27] 'artificial intelligence' programs.

Concomitant with these peripherals was the development of new types of analyzers for ions produced in the source (other than magnetic and electric analyzers), leading to the construction of high-performance quadrupoles. These less costly instruments have greater flexibility, especially when coupled with gas [28] and liquid [29] chromatography. They lead not only to very rapid recording*, but also to the generation of simultaneous positive and negative ion spectra, as introduced by Hunt and co-workers [30].

Although these techniques were developed partially as a result of improvements of the electron impact techniques, new ionization methods were developed simultaneously. These new methods appeared when the discipline of mass spectrometry

---

* Recently, a new generation of laminated magnetic analyzers has led to faster scanning rates.

entered the study of large, fragile high-molecular-mass biological molecules ($M_r = 1000$–5000 amu and more) with sufficient specificity. The main difficulty was to obtain abundant molecular ions with good resolution in the absence of pyrolysis of these thermally unstable molecules. These new ionization techniques obviated the need for modifying these compounds to render them more volatile; the disadvantage of this type of method was very often the exaggerated increase of the mass of the molecular ions.

(b) Chemical ionization (CI)

(i) Positive CI

The physical phenomenon utilized for the first time by Field [31a] and Munson [31b] is as old as the universe itself. In the MS source, a gas plasma is produced at a pressure of 0.1–1 Torr (in electron impact, this pressure is of the order of 10^{-6}–10^{-7} Torr). If the reagent gas is methane, CH_5^+, $C_2H_5^+$, etc., are produced after reaction. These ions have been detected in the gas plasmas surrounding Jupiter and Saturn [32] and those which compose certain stars.

These plasmas are used to produce two types of ion-molecule reactions.

(i-a) Protonation reactions (and the formation of adducts). These are true acid-base reactions in the sense of Brönsted:

$$BH^+ + A \rightleftharpoons B \ldots \overset{+}{H} \ldots A \rightarrow B + HA^+$$

The reagent gas BH^+ can lead to an adduct ion $[A + BH]^+$, the abundance of which is directly related to source temperature and pressure conditions and to the standard enthalpy of the reaction. Then, protonated molecular ions (MH^+) are often produced.

The internal energy of MH^+ ions is sensitive to thermal conditions and to the difference in proton affinity between the gas, B, and the compound studied, A. The choice among the following reagent principal gases: CH_5^+, H_3O^+, $iC_4H_9^+$, and NH_4^+, listed in order of increasing proton affinity, leads to either a large number of fragment ions (and low peak intensity of MH^+) with CH_5^+ or, in constrast, to abundant protonated molecules with NH_4^+.

In addition to the interest of obtaining MH^+ ions, it should be noted that in general fragment ions are characterized by an odd number of electrons and are formed by the elimination of neutral molecules, either by simple rupture or by hydrogen transfer.

The fragmentations are very often complementary to those produced in EI and may be of considerable importance for the determination of complex structures, especially peptides [33] (Fig. 1), carbohydrates [34], etc.

The use of ND_4^+ (or of D_3O^+) leads to the determination of the number of acidic protons (–OH, –NH, etc.) on the protonated molecule and occasionally on the fragment ions [35].

(i-b) Adduct ion formation reactions and their decompositions. Ion-molecule reactions which lead to electrophilic attachment reactions are very interesting because

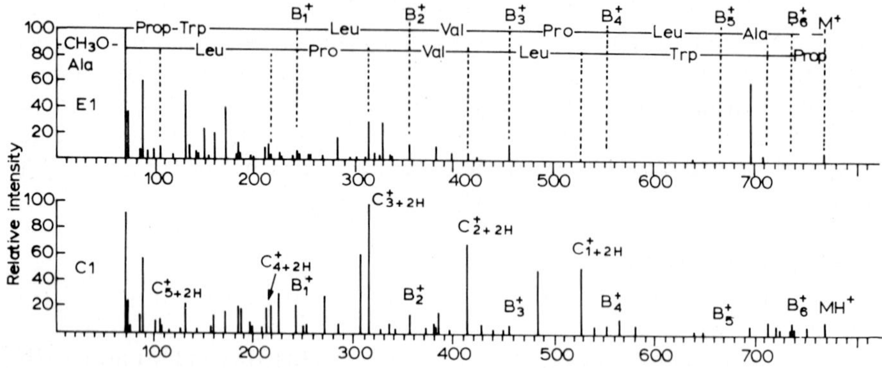

Figure 1. The EI and CI mass spectra of EtCo-Trp-Leu-Val-Pro-Leu-Ala-OMe [33].

they present several analogies with condensed phase reactions and generate rich structural data.

Hunt and Ryan [36] chose NO^+ as the reagent gas for these reactions, which enabled them to characterize several classes of compounds (Table 1).

The absence of $[M–H]^+$ ions in the case of tertiary alcohols is consistent with the supposition that the extracted hydrogen was bound to the functionalized carbon, a situation which is also found for aldehydes. These are true oxidation reactions produced with NO^+.

The use of ammonia generates other types of data. When the adduct ion $[M + NH_4]^+$ is stable* the measurement of the ratio $MH^+/M + NH_4]^+$ gives direct access to the distinction between diastereomeric diols, for example, the cyclopentane and cyclohexane diols [37,38]. This behavior is due to slight variations of proton affinity, in these cases certainly resulting from the possibility of forming hydrogen bonds (Table 2).

TABLE 1
Behavior of oxygenated compounds in a high-pressure NO^+ plasma [36]

	Alcohols			Carbonyls		Acids
	n	*sec-*	*tert-*	aldehyde	ketone	
$[M + NO]^+$	–	–	–	×	×	×
$[M–H + NO]^+$	×	×	–	–	–	–
$[M–H]^+$	×	×	–	×	–	–
$[M–OH]^+$	–	×	×	–	–	×

The ions observed in the CI mass spectra are indicated by ×.

* This often requires measuring with the lowest possible source temperature (100–130°C).

TABLE 2
Partial NH_4^+/CI mass spectra of diastereomer cyclohexane diols (1–3 and 1–4) [37]

Isomers	$[M+NH_4^+]$	$[MH^+]$
cis (1–3)	29	51
trans (1–3)	54	2.2
cis (1–4)	15	64
trans (1–4)	19	0.5

Intensities are in percentage of substrate ions, $\%\Sigma 80$.

The $[M+NH_4]^+$ ions, produced from ketones and aldehydes, lead protonated Schiff bases to form $[M+NH_4–H_2O]^+$ (noted as $[M_sH]^+$) in the gas phase [39] (Fig. 2).

In a different manner, protonated amines are obtained, probably from adductions of alcohols according to a nucleophilic substitution reaction which could be demonstrated to be stereospecific with inversion of configuration (Fig. 3) [40].

The $[M_sH]^+$ ions thus produced are located by a molecular peak shifted by 1 amu in the mass spectrum.

The proton affinities of these compounds are highly sensitive to the environment of the functional group. This leads to the distinction of diastereoisomers comprising only one alcohol function (Fig. 4).

The knowledge of such a mechanism, close to SN_2, makes it possible to differentiate enantiomers, as well as to assay them by the use of an 'ad hoc' chiral reagent gas

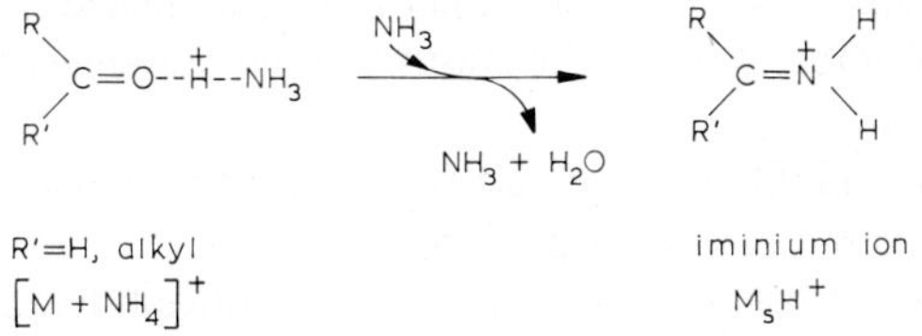

R'=H, alkyl
$[M+NH_4]^+$

iminium ion
M_sH^+

Figure 2. Formation of iminium ion by ion molecule reaction with NH_3 from adduct ions [39].

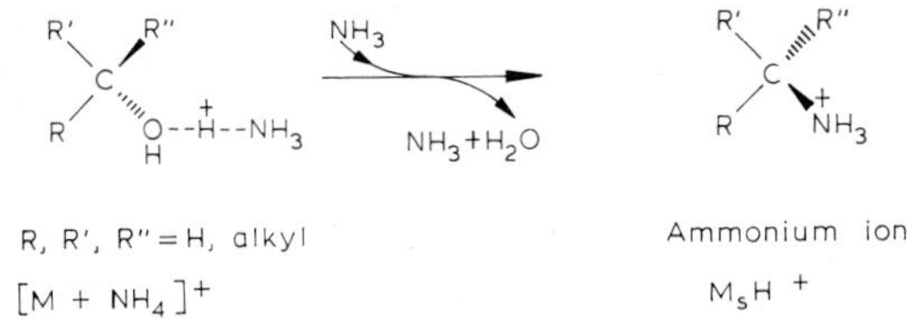

R, R', R''=H, alkyl
$[M+NH_4]^+$

Ammonium ion
M_sH^+

Figure 3. Bimolecular nucleophilic substitution by NH_3 [40b].

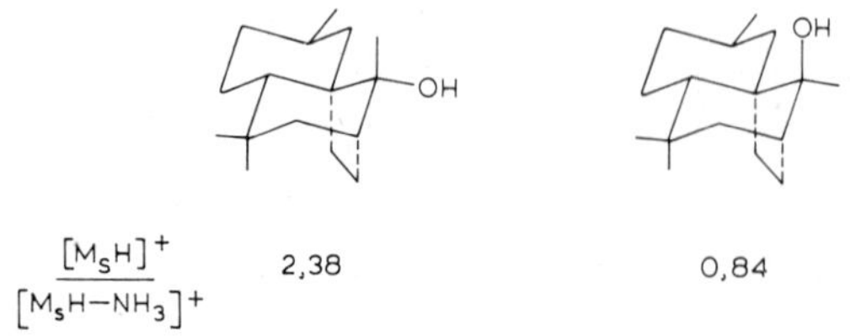

$\frac{[M_sH]^+}{[M_sH-NH_3]^+}$	2,38	0,84

Figure 4. Evolution of ratio $[M_sH]^+$ (m/z 236)/$[M_sH-NH_3]^+$ (m/z 219) according to initial alcohol configuration [41].

which, in the gas phase, no longer generates antipodes, but distinguishes a mixture of diastereomers after nucleophilic substitution [41].

(i-c) Charge-exchange reactions. All the previous reactions are produced with a gas plasma, leading either to the protonation of functional groups or to the formation of adduct ions. Chemical ionization is also characterized by the possibility of obtaining charge-exchange reactions in the course of ion-molecule reactions [42]: $A^{+\cdot} + M \rightarrow A + M^{+\cdot}$.

Again, in this case, there is a wide choice of reagent gases: on one hand, the inert gases (He, Ne, Ar, Xe, Kr) and, on the other hand, di- or tri-atomic gases, such as N_2, NO, CO, CO_2, N_2O, etc. Charge-transfer reactions generate $M^{+\cdot}$ ions – the same species produced in EI – with the difference that their decomposition may be reduced considerably, especially when the reagent gas has a low recombination energy. In general, the internal energy of $M^{+\cdot}$ ions is lower and often corresponds to an energy window which is narrower than that of EI, and their fragmentation is sensitive to thermal effects. The decomposition mechanisms remain analogous to EI.

Mixtures of gases, such as $(NO + N_2)$ [43], where NO is highly diluted ($= 5\%$), may be used. Considering recombination energies, the advantage of this mixture is that it produces $-NO^{+\cdot}$ which, after reaction, leads to $M^{+\cdot}$. Isomers may be distinguished by the use of such mixtures [44], the double interest being the simultaneous presence of fragment ions and intense molecular ions.

Finally, the use of mixtures such as $(Ar + H_2O)$ leads to additional data: certain are analogous to EI, the others are due to chemical ionization [45].

Greathead and Jennings [46] utilized ion-molecule reactions produced by a mixture of N_2/CS_2/methyl vinyl ether on alkenes to localize the double bonds (Fig. 5).

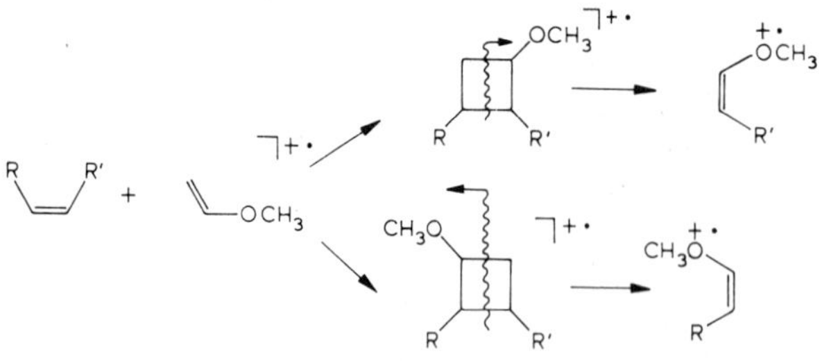

Figure 5. Ion-molecule reaction between methyl vinyl ether ion and alkene molecule [46].

Chemical reactions in the gas phase generally are often very sensitive to stereochemical effects, leading to the generation of very specific structural data.

Chemical ionization coupled with the different peripheral methods enumerated above widens its possibilities, even if sensitivity remains lower than that obtained with electron impact. Among the review articles appearing in this field, we may cite those of Fales [47], Field [48], Harrison [50], Hunt and Sethi [51], Munson [52] and Jennings [53], which cover the various aspects of these techniques.

(ii) Negative chemical ionization

There is currently an increasing tendency towards negative chemical ionization, as a result of its high sensitivity. The minor role that negative ions had for a long time is due above all to the low yield of their production under usual low pressure conditions, as utilized in electron-impact ionization. Indeed, it is necessary to obtain electrons with low kinetic energies (< 10 eV), which is very difficult under the conventional condiiions of electron impact. The appearance of high-pressure sources has generated a new interest in negative ions [54], by the possibility of preparing very 'slow electrons'.

There are several types of reaction produced: resonance electron capture, 0 eV $AB + e \rightarrow AB^{-\cdot}$; dissociative electron capture, 0–15 eV $AB + e \rightarrow A^{\cdot} + B^{-}$ (or $A^{-} + B^{\cdot}$); ion-pair dissociation, > 15 eV $AB + e \rightarrow A^{+} + B^{-} + \bar{e}$ (or $A^{-} + B^{+} + e$).

Thus, molecular ions, $M^{-\cdot}$, can be formed by the reaction of thermal electrons. These molecular ions are sufficiently stable so that they can be demonstrated in negative mode spectra. Different gases may be used, depending on the data sought. Reviews on these various aspects have been published by Dillard [55a,b], Hunt and co-workers [56a,b], Field [57a,b], Harrison [58], Jennings [59a–c] and Bowie and co-worker [60a,b]. The ions thus prepared may be studied either with conventional electric and magnetic field instruments or with quadrupole analyzers. In the context of the utilization of quadrupole instruments, the particularly astute set-up published by Hunt and Sethi [56b] should be noted: this leads to the simultaneous recording by pulses of positive and negative ion CI spectra, a process termed PPNICI. In order to generate all these different types of ions, it is possible in this case to choose different gas mixtures (Table 3). A large quantity of complementary data can thus be obtained.

TABLE 3

Mixture of reagent gases to obtain PPNICI spectra [56b]

Gas	PICI	NICI
Ne (or Ar)	Charge transfer with N_2^+ (formation of $M^{+\cdot}$ and fragment ions)	Electron capture (formation of $M^{-\cdot}$ and fragment ions)
CH_4 (or iC_4H_{10})	Protonation with CH_5^+ (MH^+ and fragment ions)	Proton abstraction ($[M-H]^-$ only)
$CH_4 + CH_3ONO$		Proton abstraction and electron capture ($[M-H]^-$ and fragment ions)

Jennings [59a] has shown that other mixtures, such as N_2 (or CH_4)* $+N_2O$ to form $O^{-\cdot}$, may be used. These ions abstract H or H_2 according to the position of the double bond in alkene compounds. With this reagent it is possible to demonstrate if the α carbon(s) of a nitrile or ketone function is (are) secondary (or not). In this case, an intense $[M-H_2]^{-\cdot}$ ion appears.

Field [57b] shows theoretical aspects and the applications of the reactivity of differents anions, including OH^-, on certain organic (acids, ketones, alcohols) and biological (amino acids, steroids) molecules. He stressed the sensitivity of the method, which can be augmented further by introducing functional groups, e.g., the pentafluorobenzoyl group can be introduced in amines and phenols [30a,b], resulting in sensitivities in the fg range with SIM (or MID) detection, with a good signal/noise ratio, as shown by Hunt and co-workers [56a,b] in the case of compounds of the amphetamine $\Delta^{1,6}$-tetrahydrocarbinol type.

Selva et al. [61] were able to study metabolites of *Aspergillus wentii* Whehmer in negative mode (Fig. 6).

Roy et al. [62] studied the behavior of 35 different steroids with OH^- and explained the role of molecular structure for favoring certain reactions, such as losses of H_2 and H_2O, enabling the environment of the functional groups present to be determined.

In studies of pollution problems, Hass et al. [63] utilized negative ions for detecting very low quantities of dioxin (dimethoxane).

Other workers [64] utilized the NICI mode for studying melatonin (*N*-acetyl methoxy tryptamine) derivatized with pentafluoropropionic anhydride and showed that this method had a great sensitivity.

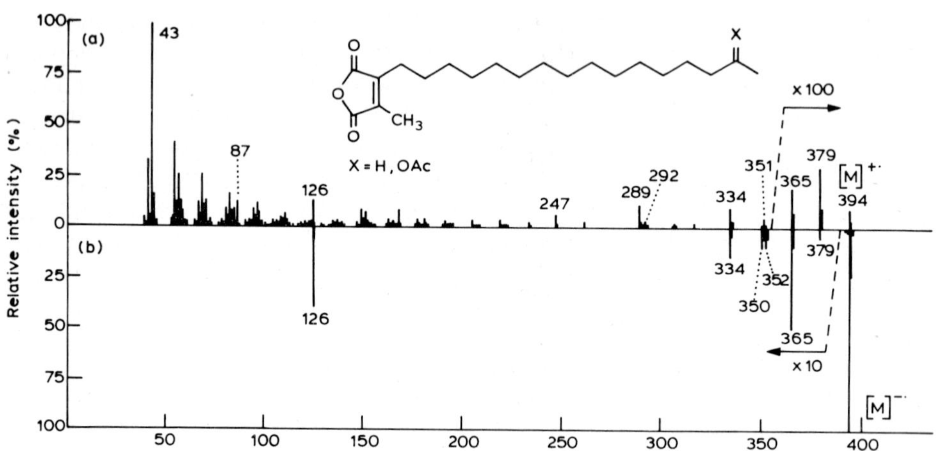

Figure 6. (a) Positive EI spectra and (b) negative electron capture mass spectra of a metabolite of *Aspergillus wentii* [61].

---

*Under certain conditions, the $CH_4 + N_2O$ mixture may yield OH^- ions, leading to $[M-H]^-$ ions by deprotonation.

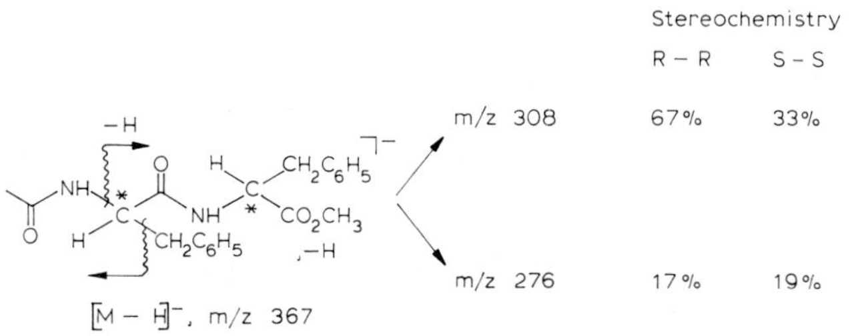

Stereochemistry

	R – R	S – S
m/z 308	67%	33%
m/z 276	17%	19%

[M – H]$^-$, m/z 367

Figure 7. Relative loss of CH_3CONH_2 and $CH_2C_6H_5$ from [M–H]$^-$ molecular anion produced by deprotonation from diastereomeric dipeptide [65].

Negative ionization is also very sensitive to stereochemical effects. Winkler and Stahl [38] were thus able to differentiate diastereomeric diols. More recently, we [65] were able to distinguish diastereomeric dipeptides R–R and R–S by the variations in the abundance of losses of benzyl radical and of acetamide (Fig. 7).

This is very difficult in PICI, since determinations must be performed with NH_4^+ at low source temperature (100–120°C).

(c) Chemical ionization at atmospheric pressure (API)

A slightly different chemical ionization technique was introduced by Horning et al. [66]: atmospheric pressure ionization (API), in which the reagent gas plasma (N_2, O_2, H_2O) is used at a pressure of 1 atm. These gases are ionized with β irradiation (slow electrons) arising from the decomposition of ^{63}Ni. Ionization may also be performed with discharges. Although the reactions in this plasma are very complex, the number of collisions is very high, leading to a better sensitivity in comparison to conventional CI at 1 Torr. This method was used with both positive and negative ionizations for detecting amino acids and their derivatives in biological mixtures [67].

(d) Thermal desorption

(i) Flash desorption

The utilization of the previous techniques is limited by the necessity of evaporating compounds with a very low vapor pressure which are also often thermo-labile. Nevertheless, the vaporization at high temperature is often more rapid than the decomposition reactions of the ions produced after passage into the vapor phase (reverse of the situation at lower temperatures). Thus, Anderson et al. [68] introduced the 'flash desorption technique', in which the sample is brought from 25 to 1000°C in less than 0.5 second and is ionized by electron impact. This technique appears to be much more efficient than the conventional technique. It is sufficiently powerful to obtain the vaporization of a polypeptide so as to obtain intense protonated molecules, MH^+, and identifiable fragmentations, leading to the peptide sequence.

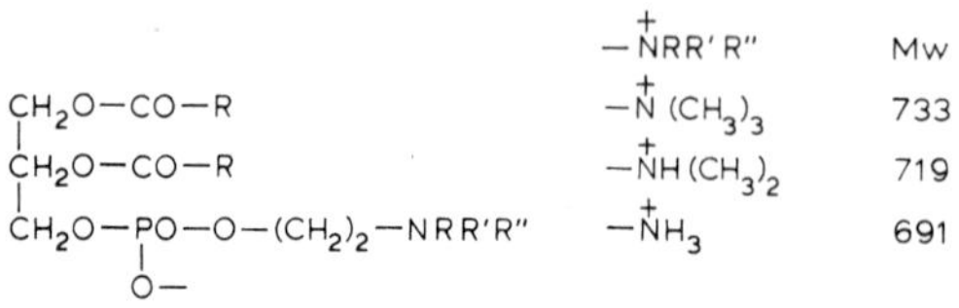

	$-\overset{+}{N}RR'R''$	Mw
$CH_2O-CO-R$	$-\overset{+}{N}(CH_3)_3$	733
$CH_2O-CO-R$	$-\overset{+}{N}H(CH_3)_2$	719
$CH_2O-PO(O^-)-O-(CH_2)_2-NRR'R''$	$-\overset{+}{N}H_3$	691

Figure 8. Derivatives of –CO–R (palmitoyl group) leading to $M^{+\cdot}$ and MH^+ ions [72a].

(ii) Desorption by 'electron (or ion) beam' technique

Reed and Reid [69] observed that if a sample is deposited on the edges of a tip, which is then placed in proximity to an electron beam, a desorption phenomenon occurs, even if the compound is not volatile.

This led to the development of 'in beam' desorption techniques: in an electron beam [70] (EI 'in beam'); or, as reported by Baldwin and McLafferty [71], in an ion beam, e.g., from a protonating plasma (CI 'in beam').

More recently, a French group [72a] utilized a slightly different holder. Rather than using a quartz tip, a gold wire was chosen as desorption support. Thus, the spectra of certain phospholipids could be obtained (Fig. 8).

Ohashi et al. [72b] utilized in-beam EI to study *N*-carbobenzoxy derivatives of an oligopeptide (pentapeptide) composed of leucine and isoleucine. The spectra were characterized by the abundant $[M+H]^+$ ions (Table 4).

They also obtained the sequence of *z*-pentapeptides (carbamic acid) as well as the protonated molecules of these nonvolatile compounds. The presence of leucine in these compounds could be differentiated from that of isoleucine to some extent, but reproducibility of measurements remains a problem.

TABLE 4

The relative abundances of $(M+H)^+$ and M^+ ions in the 'in-beam EI' and conventional EI spectra of several *N*-carbobenzoxyoligopeptides [72b]

z-Peptide	Molecular weight	In-beam base peak	EI[a]		
			MH^+	Base peak	$M^{+\cdot}$ or MH^+
Leu-Leu	378	176	16	91	<1
Leu-Ileu	378	176	4	91	<1
Ileu-Ileu	378	176	10	91	2 (MH^+)
Leu-Leu-Leu	491	91	28	91	1 (MH^+)
Leu-Ileu-Leu	491	91	20	91	2 (MH^+)
Leu-Leu-Leu-Leu-Leu	717	474	22	86	–
Leu-Ileu-Leu-Ileu-Leu	717	474	18	86	–

[a]Values are percentages.

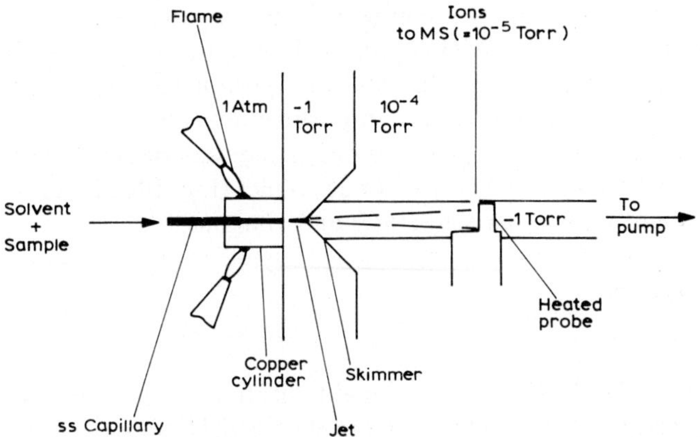

Figure 9. Principles of Vestal's LC/LS interface [73].

(iii) Formation and ionization of aerosols

Recently, Blaskley et al. [73,74] have developed a new soft ionization method, introduced in order to be compatible with liquid chromatography. This technique allows the study of mixtures of compounds which are difficult to volatilize and which are thermally labile.

The emerging, partially vaporized*, liquid phase of the chromatography is

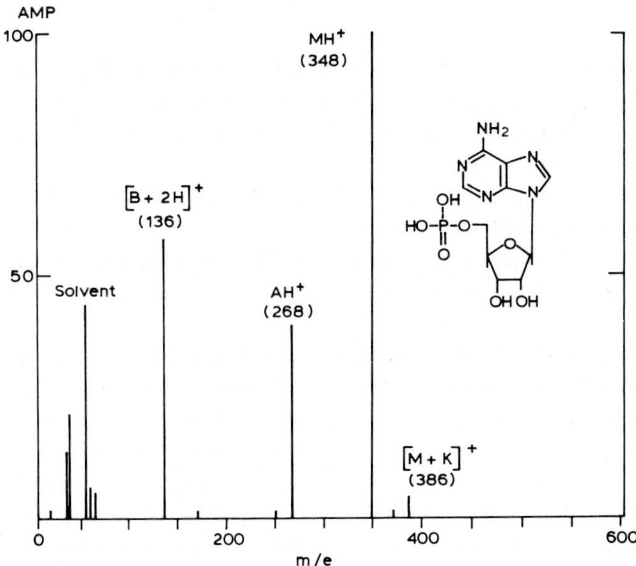

Figure 10. Conventional mass spectrum, obtained under 'Vestal's conditions', of adenosine [73].

* This vaporization is produced in a furnace heated with an oxy-hydrogen torch.

transformed into an aerosol containing the component to be analyzed. After undergoing an adiabatic expansion, the aerosol is projected onto a target heated to 250°C, installed in a CI-type source (Fig. 9). In this way, protonated (MH^+) or cationized $[M + cation]^+$ molecules were observed in positive ionization. Negative ionization can also be used. Initial results are quite promising and were obtained with different classes of biological molecules, such as mononucleotides (Fig. 10), dinucleotides, tri- and pentapeptides, antibiotics, carbohydrates and vitamins.

(e) Field ionization and desorption

(i) Field ionization (FI)

In addition to the above methods utilizing conventional ionization modes, the field ionization technique has appeared [75]. The very intense electric field (about 1 V/Å), produced by an electrode*, results in the ionization of molecules in the gas phase. This soft ionization technique is often used competitively with CI, since it does not pollute the source and may yield sufficiently reproducible results. The transit time of ions in the source is on the order of 10^{-14} to 10^{-12} second. The radical molecular ions ($M^{+\cdot}$) produced are characterized by a low internal energy, and thus can be detected easily. As a result of dispersion within the source, however, sensitivity is about two orders of magnitude lower than that of EI. As in the case of EI, the fragments produced by FI can furnish interesting structural data on carbohydrates, amino acids, peptides and cardenolides [76].

(ii) Field desorption (FD)

Another method, developed to a much greater extent, is that of field desorption. This method involves the desorption of the sample which is deposited on the filament, activated as above*, and which emerges in ionic form. Again, internal energy remains very low and is highly sensitive to thermal effects.

When studying mixtures, it is necessary to increase the emitter current slowly, in order to induce the sequential desorption of the components of the mixture. Thus, in addition to these molecular species, there also appear decomposition ions (probably by pyrolysis) of the molecular species (protonated, cationized or not) which were emitted initially.

Molecular ions formed under low pressure may have an odd number of electrons ($M^{+\cdot}$) or may be protonated (MH^+). There are several possibilities of detecting them, including a search for doubly charged ions (M^{2+}), which often appear and which may be used as a molecular test. It is possible [77] to use cationization when it is difficult to obtain molecular ions and $[M + cation]^+$ ions may thus appear, as well as doubly charged ions, such as $[M + Na]^{2+\cdot}$. The choice of an alkali cation leads to charge localization on the metal, thus reducing the

---

*The electrode may be activated by the development of fine carbon microneedles under electric discharges, which improve efficiency of desorption.

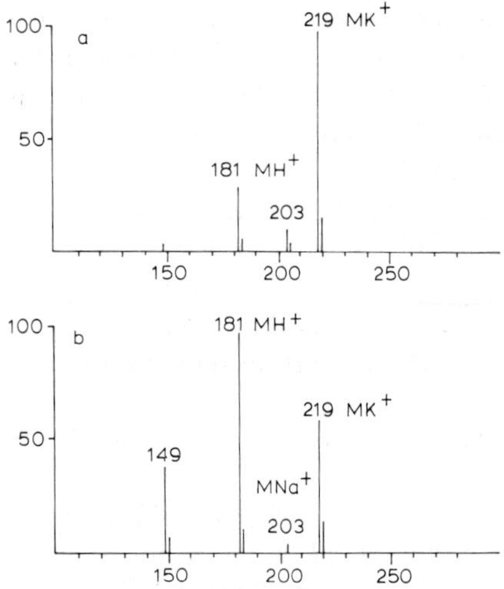

Figure 11. Field desorption spectra of (a) D-glucose and (b) D-fructose, at 7 mA wire current.

possibility of fragmentation [78]. Fragments eventually produced can retain the alkali cation.

This method, introduced by Beckey [79], has been successfully utilized by different groups studying high-molecular-mass molecules which are difficult to volatilize. A number of reviews exists [80–82], including Schulten's most recent one [83].

The field of application of this method is quite large, especially at the level of all classes of biological macromolecules, e.g., polysaccharides, antibiotics, natural metal complexes, etc. This is shown by the observation of ions with masses of 4000 units produced by peptides composed of 29 amino acid residues [84]. The method seems to be sensitive in the ng and pg range of sample detection.

Deutsch [85] has shown that the $[M+K]^+$ ions produced from isomeric monosaccharides decompose differently, leading to their distinction. We have obtained analogous results (Fig. 11).

In order to increase the possibilities of this method, thermal effects may be controlled, in order to lead to the pyrolysis of products (by Curie Point), and thus to furnish a 'fingerprint' of the substances, e.g., DNA [86]. High resolution may play an important role in these studies; the quantities of product detected remain on the order of 50 ng in the best cases.

One of the limitations of this method is due above all to the mediocre reproducibility of the results, partially resulting from emitter activation conditions.

(iii) Desorption by chemical ionization (DCI)

Hunt et al. [87] recently introduced the technique of desorption by chemical ionization (DCI), using a quadrupolar instrument. The chemical ionization of the sample apparently occurs on the surface of the emitter and the thermal effect may lead to the desorption of ions. The gases used to produce the reagent plasma are the same as those used in conventional CI, and the use of NH_4^+ as reagent leads to the formation of high abundance $[M + NH_4]^+$ ions, showing the slight energy excess of the adduct ions so formed [88].

In 1979, Arpino and Devant [89] published a review of this technique, which should develop rapidly as a result of its simplicity; no emitter activation is required. The sensitivity of this technique (<1 µg) should undergo considerable improvement.

The DCI spectra of amino acids, vitamins and trisaccharides are obtained easily, without the necessity of creating derivatives. Antibiotics can be studied by this method (Fig. 12).

Rapp et al. [90] presented a comparison between different ionization methods for the various classes of compound (Table 5).

These results are sufficiently convincing to demonstrate the usefulness of DCI as an analytical method, especially since the same authors [91] obtained protonated molecular ions and the adducts of triglycerides (trimyristin, *m/z* 742; triolein, *m/z* 905; trierucin, *m/z* 1080) in large abundance.

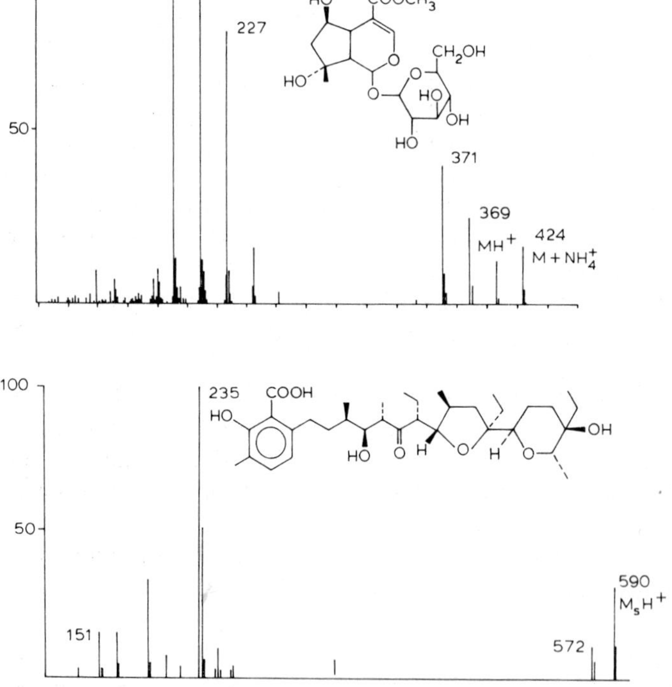

Figure 12. DCI/NH_4^+ spectra of ionophore antibiotic (lower) and iridöid (upper).

TABLE 5
Comparison of various soft ionization methods: CI, FD and DCI, with NH_3 as reagent gas [90]

Compound	Molecular weight	Ionization methods	m/z (% relative intensity)
Trehalose	(342)	CI NH_3	360(20), 198(48), 180(100) 162(18)
		DCI NH_3	702(14), 360(100), 198(18) 180(55), 162(20)
Raffinose	(504)	CI NH_3	342(3), 324(1), 306(4), 198(11), 180(100), 162(21), 144(17)
		DCI NH_3	522(1), 504(1), 402(2), 360(4), 342(9), 198(24), 180(100), 162(18)
Naringin	(580)	FD (MeOH) 21 mA	581(5), 603(100), 619(13)
		CI NH_3	419(2), 326(6), 273(100), 180(14)
		DCI NH_3	598(43), 581(8), 435(5), 418(19), 326(100), 273(77), 180(10), 164(23), 146(16)
Riboflavin	(376)	FD	377(54), 376(100), 285(7), 256(4), 242(7)
		CI NH_3	377(6), 257(84), 243(100), 225(60), 199(21), 137(4)
		DCI NH_3	377(23), 260(30), 257(24), 243(7), 198(6), 137(100)
Cascaroside A	(580)	FD (MeOH) 16 mA	418(58), 280(100), 256(52)
		CI NH_3	436(4), 419(7), 281(100), 257(14), 198(11), 180(40), 162(8)
		DCI NH_3	596(1), 436(9), 198(14), 180(100), 162(24)

(f) Other types of desorption

As indicated in the review by Daves [81], there are numerous methods of desorption from surfaces, leading to the vaporization of fragile, high-molecular-mass molecules.

(i) ^{252}Cf plasma desorption [92] (PDMS)

This method was developed by MacFarlane [93], who, in order to induce desorption, used high kinetic energy fragments (^{105}Tc, 100 MeV) produced by nuclear fission of ^{252}californium. Different ionization modes may occur: electron transfer ($M^{+\cdot}$ and $M^{-\cdot}$) [94], proton transfer ($[M+H]^+$ and $[M-H]^-$) and, finally, cationization ($[M+Na]^+$). Figure 13 shows the comparison of spectra obtained under PDMS-conditioned laser desorption.

The possibilities of this type of desorption are rather extensive, since it can be used for the detection of the molecular ions of various compounds, such as antibiotics (hedomycin, erythromycin, etc.), vitamins (B_{12}), amino acids, peptides (β-endorphin,

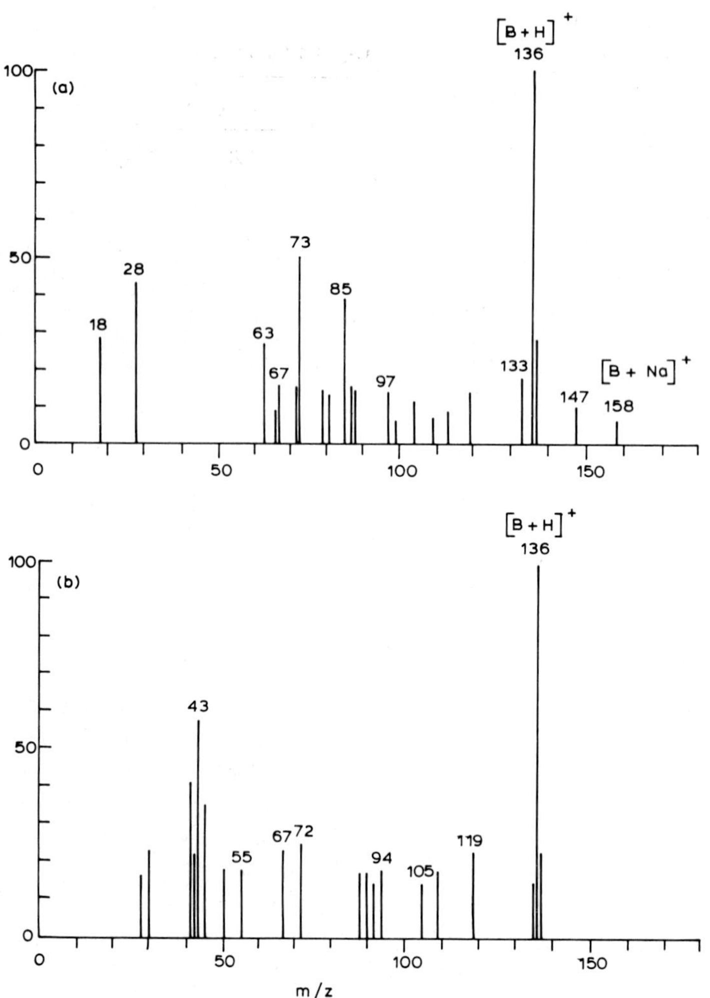

Figure 13. Comparison of mass spectra of adenine under (a) PDMS and (b) LDMS conditions [100].

3487 ± 1.06 amu), nucleotides, etc. In addition to obtaining the molecular mass of an oligonucleotide (4741.5 ± 0.7 amu), MacFarlane and co-workers [95] succeeded in determining the sequence of the compound by breaking each phosphotriester bond. Now, the Swedish group has obtained high molecular weights (> 20 000 u) using the PDMS technique.

(ii) Laser-induced desorption (LDMS)

This technique was introduced by Posthumus et al. [96] in 1978. The ions can be produced in an EI source (LDEIMS). The laser beam is obtained with either a pulsed CO_2 or ruby neodymium type YAG laser. Here again, the technique is useful for labile biological molecules and yields molecular ionic species with very low internal energies. This method is applicable to the study of nucleosides, nucleotides, amino acids, peptides and carbohydrates [97].

More recently, Cotter and co-worker [98] used a CI source to study with a greater sensitivity desorbed neutrals from glucuronic conjugated steroids and bile acids. Hunt et al. [99] utilized the same ionization method (CI/Ar) and showed the presence of 50 ng of strychnine. Studying trisaccharides they observed cationized ions, such as $[M+Na]^+$, $[M+K]^+$ and $[M+Cs]^+$, in their respective mass spectra.

Peptides were also investigated: thus, the molecular peak at *m/z* 555 was obtained corresponding to Tyr-Gly-Gly-Phe-Leu peptide and fragment ions corresponding to peptide sequence.

More recently, Schueler and Kreuger [100] compared these two latter desorption methods, PDMS (or FFID) and LDMS (or LID), in a study of nucleoside compounds and their corresponding bases (Fig. 13).

Common fragmentations appeared in each of the spectra, especially in those of adenosine, recorded in negative ionization mode.

In general, the results obtained with this method are comparable to the methods presented in this section (in-beam CI, FD, PDMS, etc.) [101]. Here again, this mode is consistent with LC/MS coupling, as developed by Hardin and Vestal [97] and utilized for the first time in studies of sucrose, the peptide Gly-Trp, guanosine, Na_2ATP, etc.

(iii) Desorption by ionic bombardment (SIMS)

This method is already relatively old and has been used for the past several years in organic chemistry and biochemistry. High kinetic energy (several keV) primary ions, e.g., Ar^+, bombard a surface on which the sample has been deposited. Under these conditions, ions are extracted from the surface and can be analyzed. Benninghoven and co-workers [102] presented a number of examples: carbohydrates, alkaloids, amino acids (and derivatives) and peptides. As with the other methods, both positive and negative ionization modes are possible (Fig. 14). More recently, the same author [103] demonstrated the possibility of studying non-volatile nucleic acids and compared the results obtained with the other desorption methods. Sensitivity limits are on the order of ng.

We have summarized in the previous pages the recent developments (prior to 1980*) of new soft ionization techniques and of various peripherals. These aspects often responded to the need to obtain molecular ions, in order to determine the molecular masses of these often thermo-labile biological molecules, leading to the development of soft ionization methods; and to obtain sufficient vaporization (or desorption) of the sample studied. Among other methods, field desorption (and DCI), the use of lasers and radioactive decomposition of ^{252}Cf have been introduced.

---

*A useful method was introduced by Barber [260] in 1981; this is fast atom bombardment (FAB), comparable to SIMS, using atoms with 10 KeV as kinetic energy and a liquid matrix. However, this report had been prepared before the method was introduced.

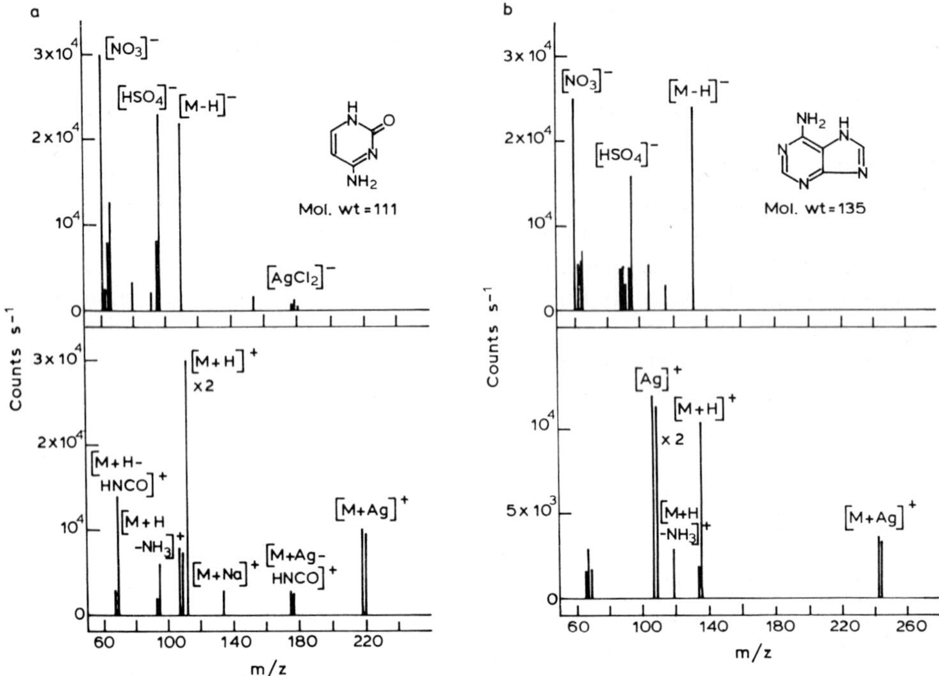

Figure 14. Positive and negative SIMS of pyrimidine bases [103].

Mass spectrometry may be characterized by successful results in the field of biochemistry; this success is due above all to the development of GC/MS (and LC/MS), combined with a computer for rapid data processing. The use of mass fragmentometry (SIM, MID) has resulted in assays of very small quantities in mixtures (currently on the order of a pg, and even a fg, scale).

Although these techniques are widely used and are of great interest for biochemists, several limitations may appear and condition the use of new methodologies for studying mixtures: 1, the necessity of modifying labile and polar compounds in order to render them more volatile or for inducing specific fragmentations for structural determinations; 2, the necessity of separating all the components of a mixture, without degradations or 'losses' during GC/MS separation: this often occurs under high resolution conditions to obtain a better specificity; 3, the necessity of separation times, which may reach 1–2 hours; 4, GC/MS techniques cannot always be used in parallel with certain ionization conditions, e.g., with FD, PDMS and LDMS limitations appear; 5, the necessity for specific structural studies, such as the differentiation of diastereomers. When the ions chosen are produced from various precursors, this approach can be erroneous.

Clearly, it is very difficult to obtain an exhaustive list of limitations, which in fact may only apply to individual cases. Nevertheless, mass spectrometry offers other possibilities which enable one to obtain, in addition to the same results as with GC/MS, qualitative or quantitative data on mixtures, thus being an alternative for circumventing the limitations cited above.

2. *Ion metastable studies and MS/MS methodology*

This analytical field no longer utilizes fragment ions formed in the source, but rather ions characterized with a greater lifetime ($>10^{-6}$ second) and which decompose in the field free regions (FFR)* of the mass spectrometer. These ions are termed metastable ions and are quite useful for demonstrating the existence of characteristic unimolecular (or bimolecular) decompositions.

The use of these ions enables one to answer the preceding questions and we will examine below the new methodology required; however, first we will propose several direct applications of this method (even in the case of mixtures): the origin of fragment ions, thus enabling veritable 'family trees' of these ions to be produced; the structure of molecules and the generation of 'fingerprints' for their respective identification; isotopic assay when the molecular peak is too small, or 'polluted' by the presence of $[M-H]^+$ ions; the localization of isotopes (^2H, ^{15}N, ^{13}C, ^{18}O, etc.); the study of fragmentations of labeled molecules as pure ions in a mixture of more or less labeled molecules; the study in a mixture of molecules that are difficult to separate, where molecular or quasi-molecular ions may be obtained by field desorption, and the study of their characteristic decompositions to identify them without derivatization, even if certain molecules (of different molecular ions) generate the same fragment ions.

The demonstration of impurities, as well as the determination of their structure, is a direct result. MID and SIM techniques may be applied with this method. The presence of different compounds with the same molecular weight may constitute a limitation of the method.

The use of high resolution and collisional methods will be useful tools for rendering such studies possible.

(a) Detections of metastable ions

Jennings and co-workers [104] and Beynon and co-workers [105], as well as McLafferty and co-workers [106], have contributed significantly to the development of these methods for detecting uni- and bimolecular reactions. Their reviews (or books) have demonstrated the value of these techniques. Several applications have been examined by Schlunegger [107]. In addition, the development of collisional reactions (CA or CAD**) (bimolecular decompositions) in the field free regions by

---

*Only ions formed in these regions can be easily detected. We will study in particular the ions produced in the first and second FFR.

**CA, collision-activation; CAD, collisionally activated decomposition.

Jennings and co-workers [108], McLafferty and co-workers [109], Beynon, Cooks and co-workers [108c,110] has shown new possibilities for studying the ion structures; the application of this method, by Maquestiau et al. [111], Levsen and Schwarz [112], etc., among others, contributes valuable data.

A large number of techniques for observing only these ions has been introduced in the absence of possible interference with the ion beam produced in the source*. These methods differ by the type of field variation and by the geometry of the double focusing mass spectrometer utilized.

These methods are based on the necessity to detect the ions formed external to the source, which thus have a kinetic energy lower than eV_0 (V_0 being the initial accelerating voltage of ions from the source). This is done either by increasing V_0, by decreasing the electric field of the energy filter or by varying both E and B (or E and V).

Field scan mode: conventional geometry, HV scan (generation of 'parent' ions or precursor ions); IKE (generation of all the metastable transitions); reversed geometry, MIKE (generation of 'daughter' ions in the second FFR).

Simultaneous variation of magnetic and electric fields (linked scan methods, conventional or reversed geometry): B/E, constant 'parent' spectrum; E^2/V, simulated MIKE; B^2/E, constant 'daughter' spectrum; $(B/E)\,(1-E/E_0)^{1/2}$, constant 'neutral fragment' spectrum.

(i) Methods involving the variation of one field

The two oldest methods, high voltage (HV) scan and ion kinetic energy (IKE) were introduced on conventional geometry mass spectrometers of the Nier-Johnson (or Mattauch-Herzog) type. They enable the demonstration of transitions, e.g., of the type: $m_1^+ \rightarrow m_{1,1}^+ + n$, either by determining the possible parents of the $m_{1,1}^+$ ion (HV scan), or by obtaining the daughter of the m_1^+ ion itself (IKE and MIKE), these ions being produced in the first FFR (conventional geometry) or in the second FFR (reversed geometry).

The kinetic energy of the $m_{1,1}^+$ ion formed in the field free region preceding the electric sector, when it is produced from m_1^+, corresponds to:

$$\frac{1}{2} m_{1,1} v^2 = \frac{m_{1,1}}{m_1} eV_0 \tag{1}$$

thus rendering problematic the existence of these $m_{1,1}^+$ ions from the various fields, E_0 and $B_{m_{1,1}}$.

If a fraction of the $m_{1,1}^+$ ions formed within the source has an energy, eV_0, then it can emerge from the electric field E_0 (constant) according to a certain radius r_e:

$$eE_0 = \frac{m_{1,1} v^2}{r_e} \tag{2}$$

---

* There occasionally appear artifacts which, in a large number of cases, can be interpreted [113].

$$r_e = 2\frac{V_0}{E_0}, \tag{3}$$

thus, the ratio V_0/E_0 defines the radius of the electric field. This ion will emerge from the magnetic field for a value $B_{m_{1,1}}$:

$$B_{m_{1,1}} = \frac{1}{r_b}\sqrt{\frac{m_{1,1}}{e}2V_0} \tag{4}$$

(where r_b is the ion trajectory radius in the magnetic field, also imposed by instrument geometry), then the other fraction of the $m_{1,1}^+$ ions produced in the first FFR will no longer be able to leave the fields: electric (with a value of E_0) and magnetic (with a value of $B_{m_{1,1}}$).

The metastable $m_{1,1}^+$ ions thus formed can in fact no longer respect the value of r_e defined in Eqn. 3, since the combination of Eqns. 1 and 2 leads to the value developed in Eqn. 5:

$$r'_e = 2\frac{m_{1,1}}{m_1} \times \frac{V_0}{E_0} \quad \text{with} \quad r'_e \neq r_e \tag{5}$$

Two solutions are offered to obtain the same value for these ratios: increase the accelerating voltage of ions from the source so that the kinetic energy of the $m_{1,1}^+$ ions produced in the first FFR becomes equal to eV_0; decrease the value E_0 of the energy filter so that ions with a kinetic energy $< eV_0$ can still leave ($V_0 = \text{constant}$) the field E.

(i-a) Variation of accelerating voltage (HV scan or defocused metastable scanning). Jennings [114] and Barber and Elliott [115] developed this technique, which enables all the m_i^+ parents* of the $m_{i,1}^+$ ion to be localized. It is sufficient to adjust with E_0 the magnetic field to $B_{m_{i,1}}$ so as to obtain the signal $m_{i,1}/e$, corresponding to ions produced in the source, and to defocus the source by increasing the ion acceleration voltage, which leads to the disappearance of the signal $m_{i,1}$ of ions produced within the source; it will reappear as metastable $m_{i,1}^+$ ion only when Eqn. 6 is respected:

$$\frac{V_i}{V_0} = \frac{m_i}{m_{i,1}} \tag{6}$$

Using the ratio V_i/V_0, it is thus possible to calculate the value of the m_1^+ ion, parent of the $m_{i,1}^+$ ion (produced in the first FFR).

The disadvantage of this method is related above all to the ratio V_i/V_0, which limits the mass ratio $m_i/m_{i,1}$ (3 to 4). In addition, defocusing the source can lead to an attenuation of sensitivity.

---

* The m_i^+ ions are considered here as parents and the daughter ions are $m_{i,j}^+$.

Nevertheless, this method of ion precursor search has extensive applications. Several of these aspects have been shown by Gallegos [116], who studied terpanes and steranes in samples of Green River shale under electron-impact ionization conditions. All the compounds generated characteristic ions (base peaks) at *m/z* 191 and 217, arising, respectively, from the molecular ions of terpanes and steranes (Fig. 15), as shown by the study of the precursors of these ions (Fig. 16).

Using these results, Gallegos studied the molecular $M^{+\cdot}$ ions, precursors of the *m/z* 191 ions in the mixture. The $M^{+\cdot}$ ions were thus detected and were characterized, by numbers of carbon atoms, as C_{31}, C_{30}, C_{29}, C_{21} and C_{20} (majority).

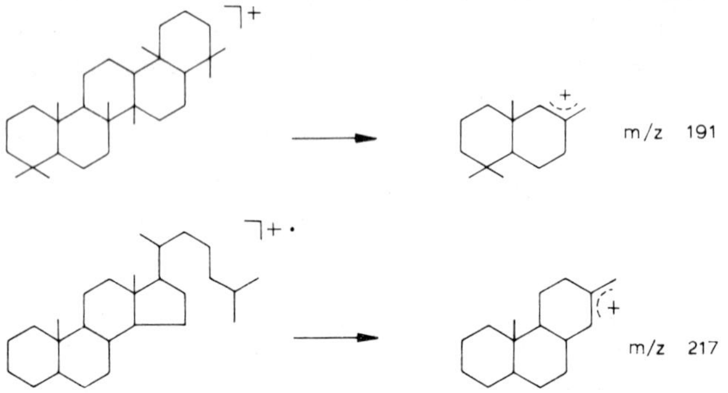

Figure 15. Formation of *m/z* 191 and *m/z* 217 ion for terpanes and steranes, respectively, ionized in EI [116].

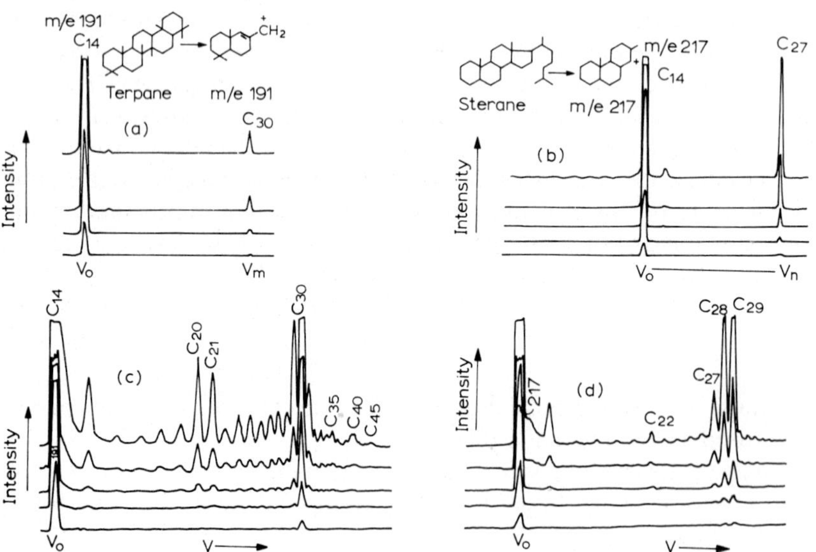

Figure 16. HV scan spectra of *m/z* 191 (a) and *m/z* 217 ions (b) produced by decomposition of terpanes and steranes. Precursors of *m/z* 191 (c) and *m/z* 217 (d) ions detected in the EI mass spectra of 'Green River Shale' mixture [116].

TABLE 6
Comparison of GC/MS and HV scan method results* of terpane and sterane analysis in the saturated portion of Green River Shale [116]

Terpanes	GC/MS	HV scan method	Steranes	GC/MS	HV scan method
C_{31}	1.0	1.1	C_{29}	10.3	10.1
C_{30}	13.6	13.2	C_{28}	7.6	6.3
C_{29}	2.3	2.2	C_{27}	1.9	1.8
C_{21}	0.4	0.4	C_{22}		0.6
C_{20}	1.4	1.7	Total	19.8	18.8
Total	18.7	18.6			

*Relative abundances.

Similarly, the origin of the m/z 217 ions in the same mixtures showed $M^{+\cdot}$ ions with C_{22}, C_{27}, C_{28}, and C_{29}. The only comment here is that numerous isomers exist for each family of terpanes and steranes, but they are not detailed in this work. In addition, the comparison of the results obtained with GC/MS and with HV scan (Table 6) demonstrate a satisfying analogy.

This technique also furnishes quantitative possibilities when assaying mixtures. The use of a computer furnishes the values of metastable transitions, as well as the abundance of the ions [117].

(i-b) Variation of the electric field (IKE technique).* Contrary to the above technique, in this case the accelerating voltage V_0 value is constant and the electric field E is varied. Under these conditions, Eqns. 3 and 5 are identical to Eqn. 7:

$$E_1 = \frac{m_{i,1} E_0}{m_i} \tag{7}$$

and all the $m_{i,1}^+$ ions produced in the first FFR can emerge from the energy filter; however, they do not have the required kinetic energy, eV_0, to leave the magnetic field at the value of $B_{m_{i,1}}$, and thus cannot reach the collector.

For this reason the magnetic field has no use in this case; a deflector is installed at the exit of the electric field and all the ions which emerge at the value E_0 correspond to the totality of ions (of all masses) which are formed in the source. By decreasing E_i as soon as an $m_i^+ \rightarrow m_{i,1}^+$ transition appears, a signal is thus recorded. The results for a given signal, however, may be ambiguous and lead to several solutions for each signal:

$$\frac{m_{i,1}}{m_i} = \frac{m_{2,1}}{m_2} = \frac{m_{3,1}}{m_3} = \frac{m_{4,1}}{m_4} = \ldots\ldots \tag{8}$$

* IKE, ion kinetic energy.

These energy spectra have been extensively utilized. Termed IKE spectra, they are veritable 'fingerprints' of compounds [105b,c,118,119a,b], thus leading to the otherwise impossible characterization of isomeric compounds.

Isomers occasionally generate identical IKE spectra [120,121]. For instance, the various methyl indole isomers cannot be distinguished. This is explained by the possibility of isomerization of the $M^{+\cdot}$ molecular ions (produced in EI at 70 eV) into the same isomeric ion. This disadvantage may be eliminated by using other ionization techniques which can be softer than electron impact.

Gallegos [116] published an example of the differentiation among alkene isomers. In the case of mixtures, this technique may be combined with GC, as in GC/MS. GC/IKE leads to a more satisfactory characterization of a mixture when GC/MS is insufficient.

*(i-c) MIKE (or DADI) technique**. This method is a derivative of the previous technique, since it still involves energy spectra. The principle of this method is to obtain the IKE spectra of a unique chosen ion, thus avoiding any ambiguities involving the metastable transition.

Various groups, such as those of Beynon, Cooks, etc., were thus led to build instruments with reversed geometry, in which the magnetic field is located between the source and the electric field [105a,122,123]. The magnetic field, which in the case of IKE was not used, is used in the present case to select a molecular ion (or a fragment ion) and the uni- (or bi-) molecular decompositions of this ion are then studied in the second FFR (after the magnetic field). Fragment ions are analyzed by their respective kinetic energy by the variation of electric field.

In this way, all the decomposition ions, $m_{i,j}^{+}$, of the m_i^{+} parent ion are obtained. It should be noted here that if a metastable transition ($m_2^{+} \rightarrow m_{2,1}^{+}$) also exists in the first FFR such that it verifies Eqn. 9,

$$\frac{m_{2,1}}{m_2} = m_{1,1} \tag{9}$$

then the $m_{2,1}^{+}$ ion, which could eventually fragment in the second FFR, will generate artifacts already cited [113].

The measurement of MIKE spectra to elucidate ion structures and to study fragmentation mechanisms remains the method of choice, providing they lead to numerous fragmentations, which is not always the case when soft methods, such as CI, FD, etc., are utilized. It is necessary to use an ionization which transmits a higher energy to the molecular ions. A larger number of fragments can be observed for the ions which decompose in the second FFR.

Using chemical ionization, it is difficult to use the unimolecular decomposition spectra of the MH^{+} ion to distinguish cyclohexanone from its isomers, 2-methyl- and 3-methyl-cyclopentanone, since they lead uniquely to the elimination of a molecule of

---

* MIKE, mass analyzed ion kinetic energy; DADI, direct analysis of daughter ions.

TABLE 7
MIKE spectra* of isomeric $[C_6H_{10}O]^{+\cdot}$ (m/z 98) molecular ions produced under EI conditions [124]

m/z	Cyclohexanone	Methyl-2-cyclopentanone	Methyl-3-cyclopentanone
43	1.3	1.2	0.5
55	0.5	0.4	1.6
56	1.6	1.4	1.0
69	4.2	3.2	24.2
70	4.6	4.4	29.3
80	66.6	71.9	1.2
83	14.8	12.06	32.9
97	4.7	3.3	1.1

*Relative abundances of metastable daughter ions.

water. Using electron-impact ionization, however, the $M^{+\cdot}$ molecular ions lead to different MIKE spectra in the case of 3-methyl-cyclopentanone and cyclohexanone, and to similar spectra in the case of the latter and 2-methyl-cyclopentanone (Table 7) [124].

This result may be explained by the possible molecular isomerization of the molecular ions of these two isomers (Fig. 17), which may occur as a result of the particularly long lifetime ($10^{-5.5}$ second $<\tau<10^{-5}$ second) of these ions.

In the case of bicyclic ketones, such as ethyl hydrindanones (a and b) [125], such a partial molecular isomerization does not prevent their identification (Fig. 18), since their MIKE spectra are quite different (Fig. 19).

The $[M–C_2H_4]^{+\cdot}$ fragment ions arising uniquely from a type b form may be isomerized.

These results show the caution that must be exercised in special cases, in terms of the conclusions to be drawn from identical MIKE spectra.

Using this methodology, Beynon and co-workers [126,127] showed how the use of MIKE spectra provided the possibility of obtaining 'family trees' for the ions,

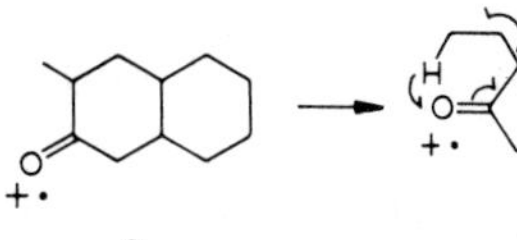

Figure 17. Isomerization of cyclohexanone [124].

Figure 18. Molecular isomerization prior to the C_2H_3 loss, and partial isomerization of fragment $[M–C_2H_4]^{+\cdot}$ ions [125].

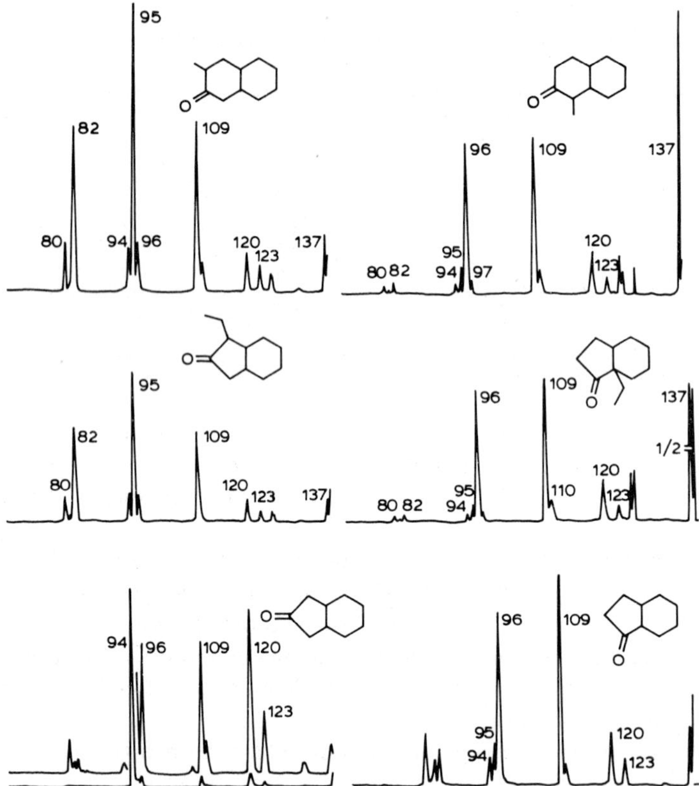

Figure 19. Comparison of MIKE spectra of $[M–C_2H_4]^{+\cdot}$ fragment ion from isomeric ethyl hydrindanone, methyl decalone and the molecular ions, $M^{+\cdot}$, formed from isomeric hydrindanones [125].

which can elucidate unknown ion structures. The association of high-resolution measurements with this technique is a very useful methodology.

Considering the previous remarks, the use of these techniques remains interesting, even in the case of complex molecules, e.g., the interpretation of the biotin methyl ester spectrum [128] (Fig. 20).

It was thus possible to show that the loss of 60 amu concerned the elimination of urea ($(NH_2)_2CO$), to yield the *m/z* 198 ion, and that the loss of 83 amu was due to the elimination of the $.CH_2COOCH_3$ radical, to give rise to formation of the *m/z* 185 ion from the $M^{+\cdot}$ molecular ion.

Thus, the generation of 'family trees' of ions may be very useful in biochemistry, especially when labeled molecules are studied [128c].

The practical applications of these metastable spectra will be discussed below, but we should first indicate an inherent disadvantage associated with HV scan and MIKE. Peak widths are directly related to reaction mechanisms in mass spectrometry: they may be very useful for more theoretical studies, since they enable one to obtain

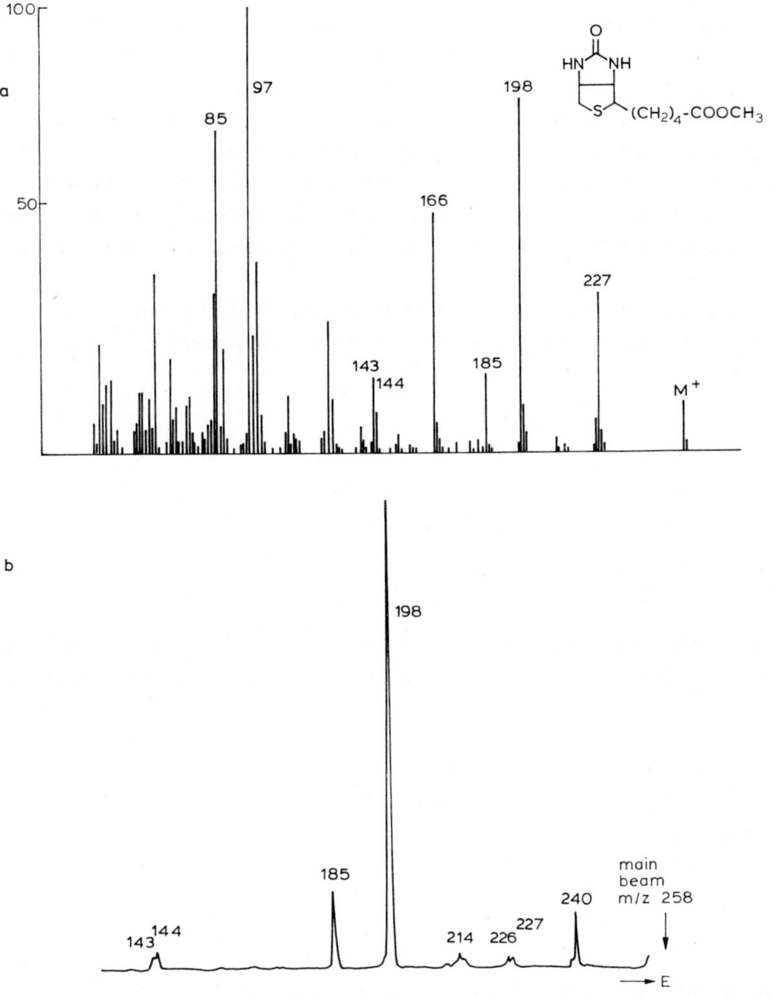

Figure 20. (a) Conventional EI mass spectrum of biotin methyl ester and (b) MIKE spectrum of its molecular ion, *m/z* 258 [128].

information on the transition states of ions during their decompositions. Inversely, this leads to poor resolution of energy spectra, which renders difficult the determination (± 1 amu) of the masses of fragment ions thus produced as soon as masses exceed 200 amu. Maquestiau et al. [129] showed that this limit – even if it can be improved – will always exist, even if ion acceleration is increased to 10 kV and more.

(ii) Linked scan methods

An alternative has been introduced by instrument designers by varying both fields at the same time. These methods are termed 'linked scan' and enable one to study only the decomposition of ions in the first FFR according to a sufficient resolution.

The work of Boyd and Beynon [130] and Lacey and MacDonald [131] indicates all possible combinations for detecting metastable ions produced in the first FFR, either by varying one field or by varying both fields at the same time.

(ii-a) E^2/V linked scan (simulated MIKE). The first of these methods to be introduced was developed on a mass spectrometer with conventional geometry (MS 50 AEI/Kratos), as a result of the work of Evans and Graham [132] and Weston et al. [133]. It involves varying both V and E in a constant E^2/V ratio for a magnetic field value of B_{m_1} and then determining all the $m_{1,j}^+$ daughters of the m_1^+ ion formed in the first FFR.

If the $m_{1,j}^+$ ions are to emerge from the magnetic field at the value B_{m_1}, the accelerating voltage must be increased (V_j)*, thus enabling movement quantity to be identified, and thus

$$\sqrt{\frac{m_{1,j}^2}{em_1} 2V_j} = \sqrt{\frac{m_1}{e} 2V_0} \tag{10}$$

$$V_j = \frac{m_1^2}{m_{1,j}^2} V_0 \tag{11}$$

which requires an increase of the electric field (E_j)* so that these $m_{1,j}^+$ ions can leave again. Under these conditions, the ratio E/E_0 of Eqn. 7 is inverted to yield Eqn. 12

$$E_j = \frac{m_1}{m_{1,j}} E_0 \tag{12}$$

and Eqn. 5 becomes Eqn. 13:

$$E_j = 2\frac{m_1}{m_{1,j}} \frac{V_0}{r_e} \tag{13}$$

Under these conditions, there is an identity between the ratios:

$$\frac{E_j^2}{V_j} = \frac{E_0^2}{V_0} \tag{14}$$

at the value $4V_0/r_e^2$. Thus, for the field set at B_{m_1}, all the $m_{1,j}^+$ metastable daughters will emerge each time Eqn. 14 is applied.

It should be noted that varying V and E enables metastable transitions to be analyzed with good resolution, as shown by Lacey and MacDonald [134], who studied the decomposition of the cyclohexanone molecular ion (m/z 98). They observed

---

* V_j and E_j represent field values for which the $m_{1,j}^+$ ions produced in the first metastable region may leave the magnetic field at the value B_{m_1} (constant).

isobaric metastable ions such as: $(M–C_2H_5)^+$ or $(M–HCO)^+$ and $(M–C_2H_4)^{+\cdot}$ or $(M–CO)^{+\cdot}$.

The limitation for measuring mass ratios, introduced by the limitation of the ratio V_j/V_0 – which can change only from 3 to 4 – led designers to introduce the following methods, as published by Boyd and Beynon [130] and Jennings and co-workers [135].

Thus, by maintaining V_0 constant and varying E and B in various manners, it is possible to obtain: all the daughters; all the parents of a chosen ion; all the ions leading to the loss of constant neutral (or radical).

These methods are related to the necessity of respecting the following relationships, unless the $m_{1,j}^+$ ions produced by the decomposition of m_1^+ reach the collector

$$E_j = 2\frac{m_{1,j}}{m_1}\frac{V_0}{r_e} \tag{15}$$

$$B = \frac{1}{r_b}\sqrt{\frac{m_{1,j}^2}{em_1}2V_0} \tag{16}$$

when the m_1^+ ions have been subjected to an accelerating voltage of V_0, maintained constant in this case.

(ii-b) B/E linked scan method (daughter $m_{1,j}^+$ ions of m_1^+) [136]. The identity of the ratios B_{m_1}/E_0 and B/E_j with the value (Eqn. 17):

$$\frac{B}{E_j} = \frac{r_e}{r_b}\sqrt{\frac{m_1}{e}\frac{1}{2V_0}} \tag{17}$$

shows that the $m_{1,j}^+$ ion will reach the collector if the values of the electric and magnetic fields are sufficiently attenuated so that the ratio B/E remains identical to B_{m_1}/E_0. The knowledge of the value of E_j so that a signal appears leads to the determination of the value of $m_{1,j}^+$ with the relationship $m_1(E_j/E_0)$.

This analytical mode has a resolution sufficient for the fragment ions. For instance, the B/E linked scan spectrum of the unimolecular decomposition of the MH^+ ion (m/z 259) of biotin methyl ester, produced by CI ammonia, has a good resolution (<1 u). The m/z 227, 241 and 243 ions are due to the elimination of neutral fragments (CH_3OH, H_2O and CH_4), which characterize CI mass spectra (Fig. 21).

The major disadvantage of this method, which presents no mass limitations, is the presence of numerous inherent 'artifacts' [137], e.g., in the B/E spectra of the decomposition of partially labeled molecules.

These artifacts can nonetheless be identified in this case by slightly varying either the accelerating voltage or the electric field energy value in order to shift the center of mass. Tulp et al. [138] show an example of this in the case of the B/E spectra of molecular ions having several chlorine atoms (2′,5′-dichloro-3-methoxybiphenyl), where a large number of artifacts is present due to great abundance of ^{37}Cl.

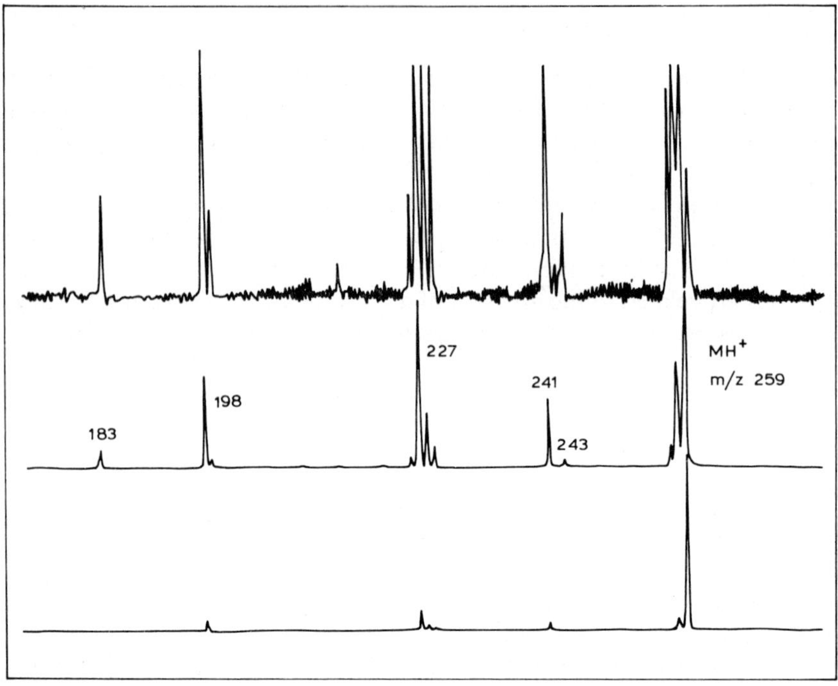

Figure 21. B/E linked scan spectra of MH^+ ions produced from biotin methyl ester under CI/NH_4^+ conditions.

The B/E linked scan method may be applied with different ionization techniques, such as EI, CI, FD, SIMS, FAB, LDMS, under positive or negative mode [139]. The success enjoyed by this type of technique is related above all to the large number of detection possibilities and to the reproducibility of spectra, due to the development of thermostatically controlled Hall probes.

The $m_i/m_{i,1}$ ratio for a given metastable transition does not appear to be limited. Warburton et al. [140a,b,c] were able to generate a veritable 'family tree', starting with the m/z 1165 ion, belonging to the EI spectrum of oil of fomblin, a polyfluorinated polyether:

$$[(\mathrm{O}-\underset{\displaystyle |}{\overset{\displaystyle \mathrm{CF_3}}{\mathrm{CF}}}-\mathrm{CF_2}-\mathrm{O}-\overset{\displaystyle \mathrm{CF_3}}{\mathrm{CF}}-\mathrm{CF})_{2x}-\mathrm{O}-(\mathrm{CF_2})_y]_z$$

The decomposition of this polymeric ion leads to losses of 66, 116 and 169 amu (or more).

In addition, resolution is sufficient, since it enables one to differentiate unambiguously the $C_5H_9]^+$, $C_5H_{10}]^{+\cdot}$ and $C_5H_{11}]^+$ ions produced from a C_{28} hydrocarbon such as octacosane [141].

Lacey and MacDonald [142] determined metastable peak widths by setting the B/E ratio so as to be on the $m_i^+ \rightarrow m_{i,1}^+$ transition and then scanned the accelerating voltage (V) to describe both the width and the form of the peak.

The relative heights of these peaks have a physical significance, as shown by Bruins et al. [135b]. They compared the ratios of the peaks $(114^{+\cdot}/113^+) = 2.65$ and $(100^+/99^+) = 1.61$ of the B/E spectrum and of isotopic ion ($^{13}C^{12}C_9H_{16}$) (m/z 143) and found that they were close to the ratios calculated with the hypothesis of a statistical loss of ^{13}C atoms. Furthermore, no mass discrimination was encountered.

As a result of the reproducibility conditions of these spectra, it is possible to use them for the differentiation of closely related isomers. Thus, Boyd and co-workers [143] were able to characterize polyaromatic hydrocarbon isomers of molecular mass m/z 228 by measuring the ratios of losses of $H^{\cdot}$ and H_2 from $M^{+\cdot}$ (Table 8).

The value of such a method is shown by these results, especially since it is applicable with any instrument geometry and in spite of two major weaknesses: the presence of artifacts; and the poor resolution of the parent ion.

(ii-c) B^2/E linked scan method (precursors of $m_{i,1}^+$ ions decomposing in the first FFR). In a reciprocal manner, an analogous identity exists for the ratios of $B^2_{m_{i,1}}/E_0$ and B^2/E:

$$\frac{B^2}{E} = \frac{r_e}{r_b^2} \times \frac{m_{i,1}}{e} \tag{18}$$

The $B^2_{m_{i,1}}/E_0$ ratio is set so that the $m_{i,1}^+$ ions produced in the source reach the collector. The electric and magnetic fields are scanned with Eqn. 18 constant.

Under these conditions, the $m_{i,1}^+$ ions, produced this time in the first FFR during the metastable decomposition of m_i^+ for instance, will generate a signal. Each peak will have a corresponding different peak whose mass is determined by the relationship $(m_{i,1}E_0/E_i)$.

The resolution of the signals obtained is lower in this case in comparison to the B/E technique. The peaks are wide [144], enabling one to obtain valuable energy release data. Nevertheless, the resolution of the main beam is sufficient so that artifacts due to the presence of the natural isotope ^{13}C are avoided.

As an example, the $C_7H_{15}^+$ ion of n-hexadecane has different precursors, but none of

TABLE 8
Intensity ratios in relation to the main beam $M^{+\cdot}$ (reproducibility of ratios on the order of $\pm 5\%$) for various polyaromatic isomer compounds [143]

	$M^{+\cdot}/[M-H]^+$	$[M-H]^+/[M-H_2]^{+\cdot}$
Chrysene	78	7.2
1,2-Benzanthracene	128	2.4
2,3-Benzanthracene	189	1.5
Triphenylene	63	20

them corresponds to a ^{13}C isotopic ion. Thus, the resolution of the principal ion can be better than in B/E linked scan [145].

The determination of ion precursors is extremely interesting, since it complements information obtained in B/E mode. For example, it was previously shown that the m/z 185 and 198 ions were produced by the $M^{+\cdot}$ (or MH^+) ions of biotin. It is possible, however, that they are also produced by decomposition of other ions, but the B^2/E spectra of the m/z 185 and 198 ions (Fig. 22a and b) confirm that the molecular ion was the unique precursor ion.

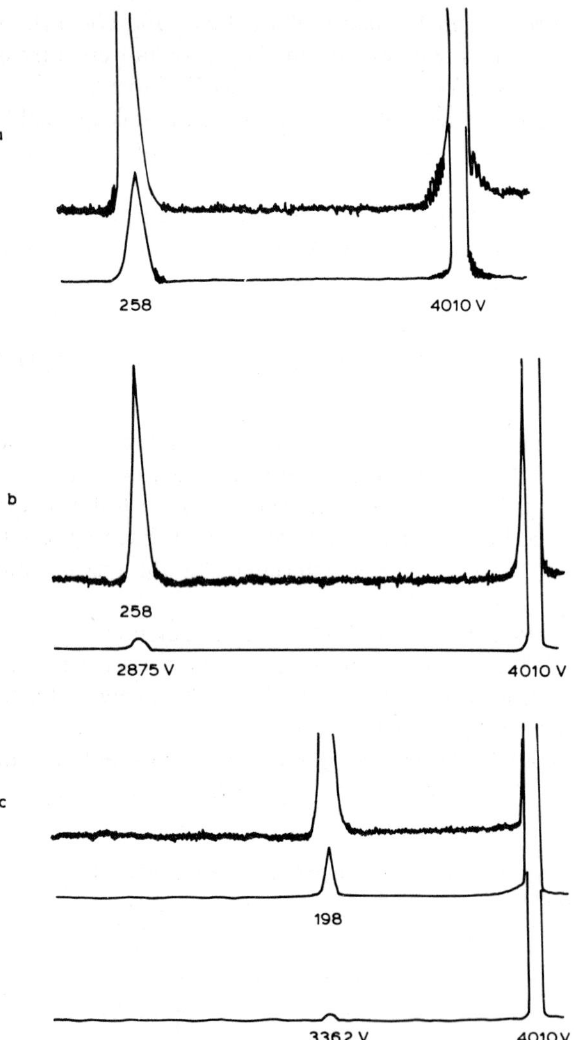

Figure 22. B^2/E linked scan spectra of m/z 198 (a), m/z 185 (b) and m/z 166 (c) ions produced by decomposition of biotin methyl ester in the source under EI conditions.

It was also shown in this way that the unique precursor of the m/z 166 ion (intense peak of the conventional EI spectrum) was produced from the m/z 198 ion, having a precise structure, as indicated by the study of deuterated derivatives.

(ii-d) $B/E\sqrt{1-(E/E_0)}$ linked scan spectra. These spectra are termed 'constant neutral fragment spectra' and were introduced first by Lacey and MacDonald [146a,b] and more recently by Haddon [147] and Shushan and Boyd [148]. They express the 'fictional' ratio B_n/E_0, where B_n corresponds to $\frac{1}{r_b}\sqrt{\frac{n}{e} \times 2V_0}$, as a function of the field values B, E and E_0, chosen for a neutral n.

For a set of metastable transitions eliminating a fragment n, (molecule or radical) Eqns. 17 and 18 enable the relationship $m_i^+ \rightarrow m_{i,1}^+ + n$ to be expressed.

$$n = \frac{B^2}{E^2} \times \frac{r^2}{r_e^2} \times 2V_0\left(1 - \frac{E}{E_0}\right) \tag{19}$$

Given

$$\frac{1}{r_b^2} \times \frac{n}{e} \times 2V_0 = \frac{B^2}{E^2} \times \frac{1}{r_e^2} \times 4V_0\left(1 - \frac{E}{E_0}\right) \tag{20}$$

However,

$$\frac{4V_0^2}{r_e^2} = E_0^2 \tag{21}$$

from which

$$\frac{1}{r_b}\sqrt{\frac{n}{e} \times 2V_0} = \frac{B}{E} \times E_0 \times \left(1 - \frac{E}{E_0}\right)^{1/2} \tag{22}$$

Again, given

$$B_n = \frac{B}{(E/E_0)}\sqrt{1-(E/E_0)} \tag{23}$$

Thus, having chosen the constant neutral fragment n, defining all the transitions:

$$m_i^+ \rightarrow m_{i,1}^+ + n$$

It is possible to adjust the magnetic field on n with E_0 and to vary B and E so as to have Eqn. 23 remain consistently equal to $B_n E/E_0$.

By measuring E/E_0, each signal obtained allows the determination of the transition $m_i^+ \rightarrow m_{i,1}^+ + n$.

Zakett et al. [149] demonstrated a direct application of this method to analyse a complex mixture of various acetylated phenols. They were studied under CI conditions and protonated molecules MH^+ were observed. In addition, these ionic species eliminated ketene molecule according to the mechanism shown in Figure 23.

The conventional, CI spectrum was complex; however, it may be clarified by studying the neutral fragment 42 amu loss as shown in Figure 24.

Thus, it was possible to identify the different phenols in the mixture which were substituted with either the $CH_3^{\cdot}$ or $C_2H_5^{\cdot}$ radicals or with chlorine or bromine (Table 9).

This method is quite interesting for studying complex mixtures. We believe that it may also be very effective in the elucidation of the sequence of large polypeptides.

Indeed, all the A_i type ions produced by direct decomposition of the protonated peptides produced in PICI can eliminate a molecule of CO (Fig. 25).

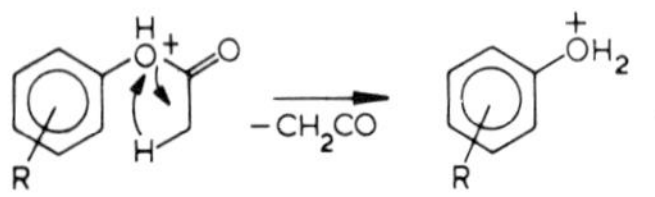

Figure 23. Loss of ketene from protonated acetate of phenols [149] under CI conditions.

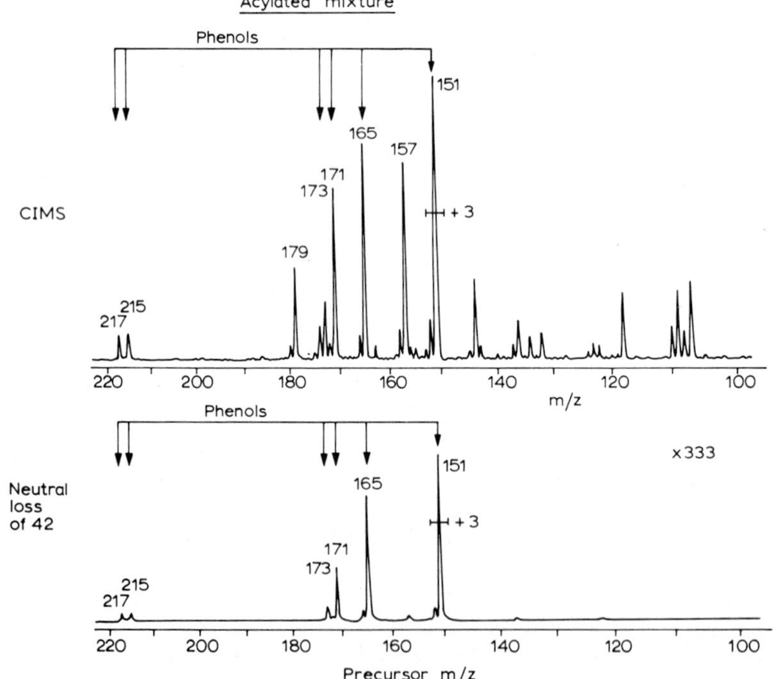

Figure 24. CI mass spectrum and $B/E\sqrt{1-(E/E_0)}$ linked scan spectrum of neutral loss of ketene (42 amu) for a mixture of various acetylated phenols [149].

TABLE 9
Mixture of various acetylated phenols in the source analysed under chemical ionization [149]

MH^+ (*m/z*)	Phenols
151	*o*-cresol
165	*p*-ethyl phenol
171	*p*-chlorophenol
173	
215	*p*-bromophenol
217	

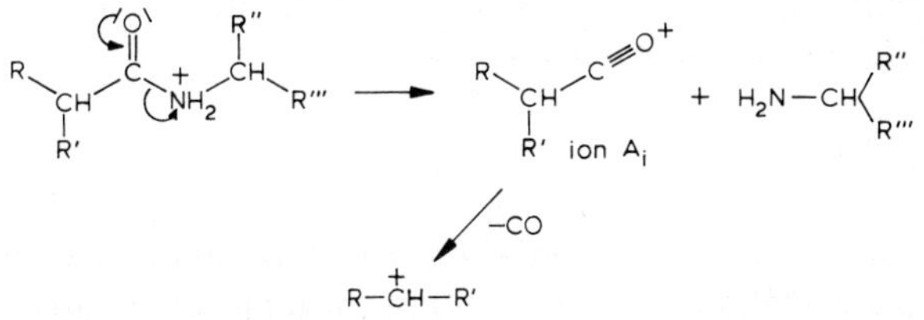

Figure 25. Decomposition of protonated molecules produced from peptides.

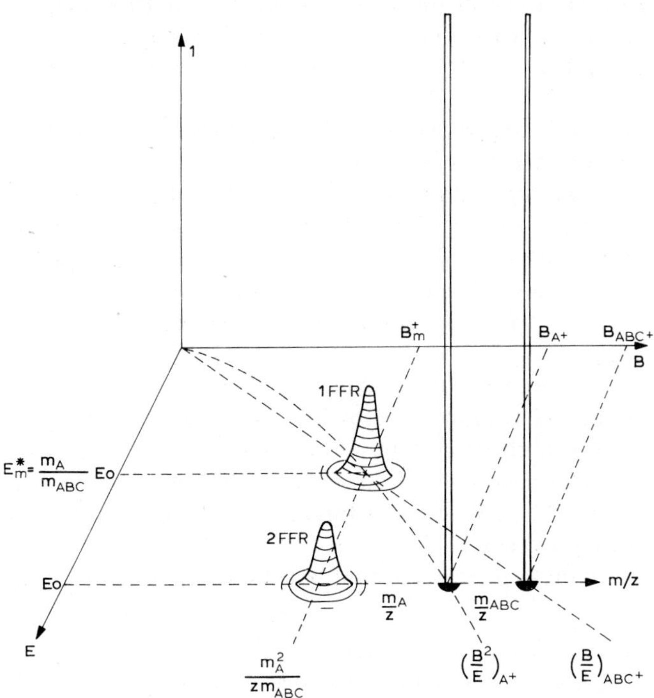

Figure 26. Various linked scan techniques for determination of neutral loss, precursor and daughter ions.

Thus, it is possible to trace back to all the ions leading to this type of elimination, which is not always obvious with conventional spectra. Initial results of this method are encouraging.

Haddon [147] recently provided an example similar to our proposition. He sought all the ions produced by *n*-hexadecane (*m/z* 266), which decomposes by eliminating $C_4H_6 (m_n = 54)$, $C_4H_7 (m_n = 55)$, $C_4H_9 (m_n = 57)$, etc. The results are sufficiently convincing so that such a method can be applied in practice, in particular for oleonalic acid.

The three linked scan methods described are summarized in Figure 26, where the conventional spectrum of fragmentations produced in the source is located for $E = E_0$ at the same time as those produced in the second FFR of a conventional instrument just before the magnetic field*. The fragmentations in the first FFR are related only for the previously determined values of B and E.

(b) Collisionally activated fragmentations

In parallel to the development of the above methods for detecting spontaneous decompositions of ions with a longer lifetime than those fragmenting in the source, decomposition reactions induced by collisions in a cell where gas pressure is much higher than that in the analyzers were developed. The gas may be nitrogen, helium, etc. The collision cell is located either in the first FFR or second FFR and the respective detection methods are linked scan (B/E or B^2/E) and MIKE.

The spectra obtained are called CA (collision activation) or CAD (collision-activated dissociation).

McLafferty and co-workers [150,151] were the first to apply these reactions to the determination of ion structures in organic chemistry.

The $M^{+\cdot}$ ions which are not decomposed in the source (or before the collision cell) receive energy 'packages' during the collisions. Then they fragment very rapidly in this zone, according to a mechanism requiring high activation energies, as simple ruptures. Unimolecular decompositions, on the other hand, are unfavorable for these simple ruptures and favor fragmentations which involve rearrangements characterized by lower activation energies (Fig. 27).

Levsen and Schwarz [112] showed the possibility of comparing ion structures with different origins by studying the collision spectra. Thus, their induced decompositions are related uniquely to their structure and no longer to their initial internal energy. As a first approximation for complex molecules, we may assume that identical (or very similar) spectra correspond to identical structures (or to a mixture of identical structures which undergo interconversion).

MIKE/CAD spectra, in addition to their interest for structural determinations, are

---

*In a reversed geometry instrument, it is not possible to observe such metastable transitions in the conventional spectrum, since the ions produced in the first FFR cannot emerge from the electric field at the value E_0.

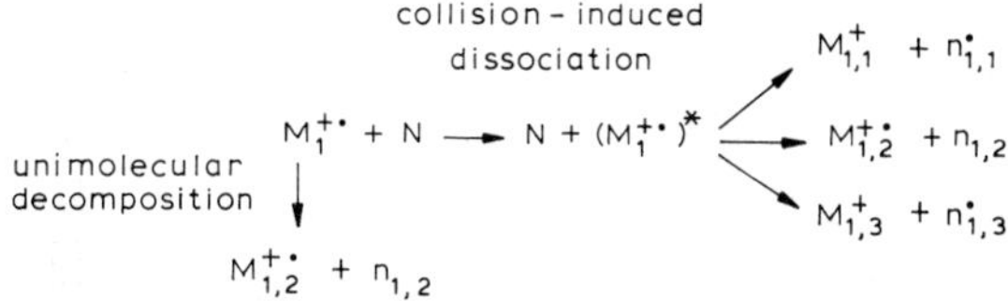

Figure 27. Unimolecular decompositions (MIKE, *B/E*) and decompositions induced by collision gas (MIKE/CAD and *B/E*/CAD).

a useful tool for analytical applications. This results from the increased number of ionic species produced by collisions (10 to 1000 times greater, in abundance, than those formed during unimolecular decompositions).

This situation is even more true as the molecular (or quasimolecular) ions arise under conditions of soft ionizations (chemical ionization*, field ionization, field desorption, FAB, SIMS, etc.).

As an example, it is difficult to distinguish the structures of the protonated molecules MH^+ produced from isomeric compounds with the elemental formula $C_6H_{10}O$ (cyclohexanone, cyclohexane oxide and 1,4-cyclohexane ether) [152] on the basis of their respective unimolecular decomposition spectra. Indeed, only the loss of water was observed in the MIKE spectra of these products. Collision-induced decomposition spectra (MIKE/CAD), on the other hand, lead to an easy structural distinction (Fig. 28).

The non-negligible disadvantage of this mode in analytical applications should not be forgotten: the considerable widening of some peaks which are already wide in spontaneous decomposition MIKE spectra.

As an example, the MIKE spectrum of the m/z 135 ion (MH^+) of 3-phenylpropene oxide (measured under CI/$tC_4H_9^+$ conditions) [153] presents a base peak at m/z 117 (elimination of a water molecule) which is particularly wide**, corresponding to energy release of about 440 meV.

Its pentadeuterated derivative, phenyl-(D_5)propene oxide, under the same conditions, leads to the elimination of one molecule of deuterated (or not) water. The unambiguous attribution of the signal obtained requires precise measurements of centroids for these cluster peaks.

Under collisional conditions, this situation is aggravated, since these peaks are considerably widened. In spite of this drawback, the flexibility of this method renders it extremely useful.

Indeed, it is possible to obtain MIKE/CAD spectra of isobaric ions [154a,b]. For example, after elimination of 43 amu fragment, cyclohexanone leads within the

---

* The use of protonating gases with high proton affinity, such as NH_3 or $tC_4H_9^+$, and of gases producing charge transfers (low recombination energies), such as argon or nitrogen, generally leads to such ions.

** These experimental results are interpreted on the basis of an exchange of internal energy (vibrational origin) into kinetic energy during fragmentation.

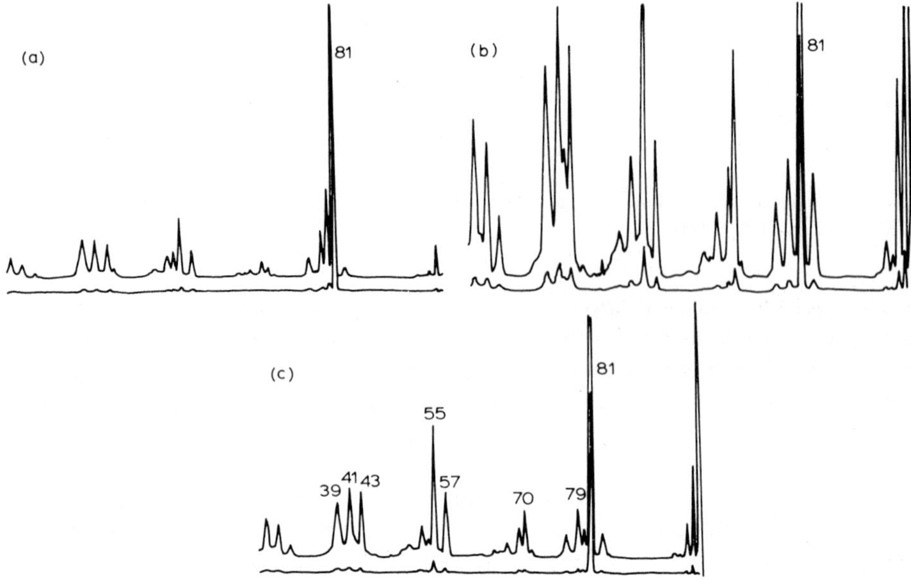

Figure 28. MIKE/CAD spectra of various protonated molecules of 1,4-cyclohexane ether (a), cyclohexanone (b) and cyclohexane oxide (c) produced under tC_4H_9/CI [152].

source to the formation of an *m/z* 55 ion, composed of a mixture of isobaric ions: $[C_3H_3O]^+$ and $[C_4H_7]^+$. The use of a resolution on the order of 3000 to 3500 for the magnetic field allows the separation of these isobaric ions, and thus enables the study of each of these ion structures.

The MIKE/CAD spectra of each ion are sufficiently different so that we may admit the separation of these ions, as shown by the non-existence of ions arising from the loss of 4, 5 or 6 hydrogen atoms, as well as that of the *m/z* 39 ion in the MIKE/CAD spectrum of the isobaric ion $C_3H_3O]^+$. Inversely, the very low intensity of the *m/z* 41 peak in the MIKE/CAD spectrum of the $C_4H_7]^+$ ion is proof of the good separation of these isobaric ions (Fig. 29).

In order to increase the possibility of collisional activation, the current trend is to develop this mode (on universally accessible conventional geometry instruments), utilizing the detection of decomposition produced in the first FFR by the *B/E* linked scan method. Nevertheless, although the resolution of fragment ions is better in this mode, that of precursor ion resolution remains mediocre and does not lead to as good a separation of isobaric ions as in the case of MIKE mode*.

Various groups, including those of Jennings [155], Burlingame [156], etc., have considerably developed the applications of this method.

---

*The MIKE and MIKE/CAD) method is used on a ZAB 2F reversed geometry instrument, designed by VG Micromass Instruments.

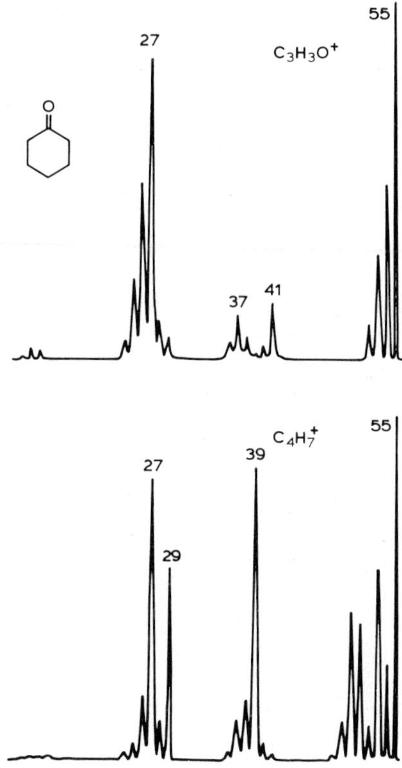

Figure 29. MIKE/CAD spectra of $C_3H_3O^-$ and $C_4H_7^+$ isobaric ions (m/z 55) produced from cyclohexanone under EI conditions [154a].

Burlingame's group installed the collision cell in the EI source located just at the exit of the field desorption source on a modified AEI MS 902 instrument (conventional geometry). This installation enabled them to study the decompositions of high-molecular-weight molecules (>600), as we will show later.

In addition, ions decomposed by collisions generally do not have the time to undergo transpositions (as for those produced within the source), thus avoiding isomerization of the molecular ion (when this occurs) if the lifetime of the ion is too long.

Mass resolution of the *B/E* technique is better than 1 amu, which in this case leads to the unambiguous study of decompositions induced by collisions of labeled molecules contrary to the MIKE technique, which generated peaks too wide, especially for high-mass molecules.

Harris et al. [157] showed that the same methodology could be applied to *B/E* spectra to determine the structure of ions. The only information which is no longer available is the width of decomposition peaks.

TABLE 10
Partial collisionally activation B/E linked scan spectra of various $C_5H_8^+$ isomers ions formed in EI [157]

m/z	Isoprene	Piperylene	Cyclopentene	Allene	Limonene
67	(230)	(325)	(600)	(345)	(230)
66	(83)	(133)	(278)	(59)	(70)
65	(92)	(153)	(220)	(63)	(74)
64	15	23	38	7	9
63	45	74	99	29	39
62	25	40	58	18	23
61	13	19	30	11	11

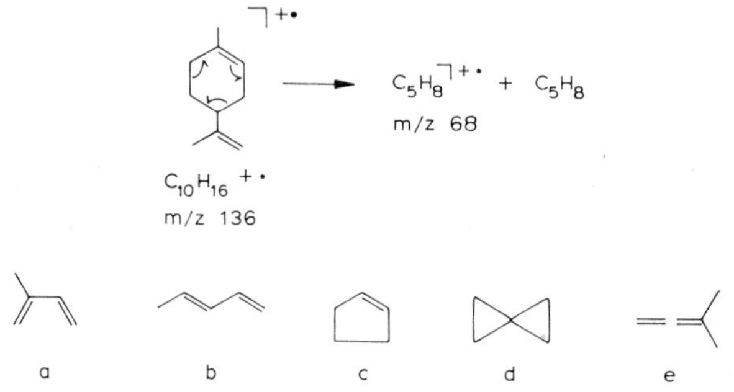

$C_{10}H_{16}^{+\cdot}$ m/z 136 → $C_5H_8^{+\cdot}$ + C_5H_8 m/z 68

a b c d e

Figure 30. Various isomers of $[C_5H_8]^{+\cdot}$ studied [157].

It was thus shown that the $C_5H_8]^{+\cdot}$ ions produced by the decomposition of limonene have a similar structure to that of isoprene (a) and slightly different from those of the isomers (b, c, d and e) (Table 10, Fig. 30).

Finally, as a last example of the utilization of these methods, we cite the use of the $E/2$ spectrum of collisionally produced doubly charged ions:

$$M^+ + N \rightarrow M^{2+} + N + \bar{e}$$

Zakett et al. [158] used these reactions to demonstrate the protonated molecule MH^+ of nitrogen-containing compounds in a complex mixture, such as coal liquid.

With the direct CI spectrum of such a mixture, it is difficult to determine which peaks correspond to MH^+ ions for nitrogen-containing compounds (Fig. 31).

As shown below, the operation involves the induction of collisions in the second FFR so as to produce doubly charged ions which emerge from the electric field for a constant value of $E_0/2$. The nitrogen-containing MH^+ ions produce doubly charged ions more easily and only these can emerge from $E_0/2$ (Fig. 32).

In order to obtain a spectrum, it is necessary merely to vary the magnetic field under collisional conditions, with $E_0/2$ as an energy filter.

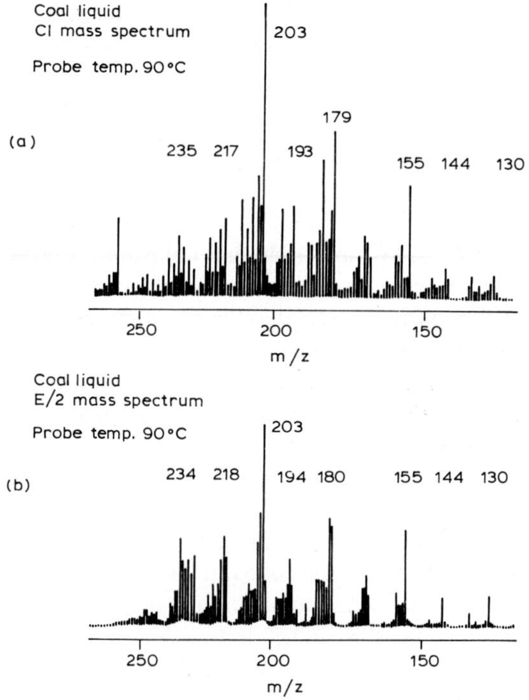

Figure 31. Comparison of $CI/tC_4H_9^+$ mass spectrum of coal liquid (a) and the $E/2$ mass spectrum (b) recorded under the same conditions [158].

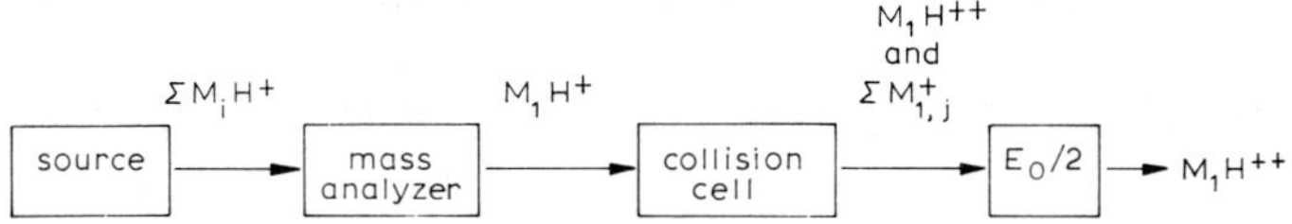

Figure 32. Principle of $E_0/2$ spectra.

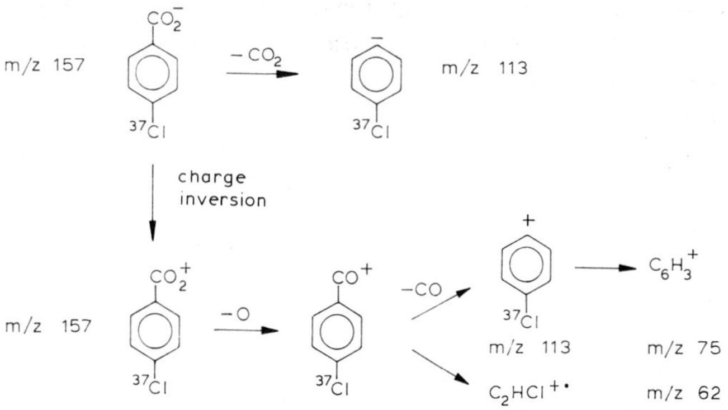

m/z 157 CO_2^- ^{37}Cl — $-CO_2$ → ^{37}Cl m/z 113

charge inversion

m/z 157 CO_2^+ ^{37}Cl — $-O$ → CO^+ ^{37}Cl — $-CO$ → ^{37}Cl (+) m/z 113 → $C_6H_3^+$ m/z 75

→ $C_2HCl^{+\bullet}$ m/z 62

Figure 33. Comparison of decompositions of molecular anion and molecular ion [164].

The $E_0/2$ spectrum thus presents MH^+ ions at m/z 234, 218, 204, 194, 180, 144, 130, etc., undoubtedly corresponding to quasi-molecular ions of mononitrogen compounds (Fig. 33).

(c) Special case of negative ions

(i) IKE spectra

The field of application of the metastable decompositions (whether collision-induced or not) of positive ions is very important. Nevertheless, Hunt and Sethi [51] showed the value of studying the behavior of negative ions: they very often provide complementary ion structural data, with a better sensitivity.

Bowie and Hart [159] used the IKE technique to study the decomposition of negative molecular ions $M^{-\cdot}$ (produced by electron capture) into fragment ions according to mechanisms of simple rupture and of rearrangements.

The following compounds: *o*-nitroacetanilide and phenyl-(*p*-nitrophenyl) acetate lead to the elimination of radicals, such as (a) $NO_2^{\cdot}$ (and also $^{\cdot}NHCOCH_3$ for acetanilide) directly from the initial negative molecular ion; (b) $NO^{\cdot}$, $OH^{\cdot}$, $(NO + CH_2CO)$ after molecular ion isomerization. Under collision conditions in the FFR, direct cleavages occur without isomerization of the molecular ion and lead to the formation of $[M-NO_2^{\cdot}]^-$, $[M-O-Ph]^-$ and $[M-PhCOO]^-$ ions.

Measurements are performed with a conventional geometry instrument and the reactions occurred in the first FFR (unimolecular or collisional-induced decompositions). The value of the transitions is determined by measuring the E/E_0 ratio, which yields m_2/m_1 (m_1 and m_2 unknown); each term can be determined as soon as the ion beam which emerges from $(-E)$ leaves the magnetic field at the value m_2/m_1. Thus, m_1 and m_2 can be calculated.

Other types of reactions have been studied in collisional conditions, especially charge-inversion reactions.

(ii) MIKE spectra and charge inversion reactions induced by collisions

Bowie and co-worker [160,161] used the MIKE technique to measure such spectra in a study of compounds which led to practically no fragmentations. The charge inversion produced during collisions can be utilized to obtain a larger number of types of fragment ions:

$$\begin{array}{c} M^{-\cdot} + N \rightarrow (M^{+\cdot}) + N + 2e \\ \swarrow \quad \downarrow \quad \searrow \\ M_1^+ + r_1^{\cdot} \quad M_2^+ + r_2^{\cdot} \quad M_3^+ + r_3^{\cdot} \end{array}$$

The energy spectrum is recorded by inverting the electric field $(-E) \rightarrow (+E)$. Thus, the positive ions may once again emerge from the field and be recorded.

Other application examples are encountered, such as the comparison of the behavior of benzoate type anions and cations during decomposition [162], the retro

Diels-Alder rearrangement from flavone and flavolone [160,161] and, finally, molecular transpositions occurring in sulfur-containing compounds [163].

Cooks and co-workers [164,165] studied the decompositions of nitrated derivatives under NCI conditions, with $Cl^-(CH_2Cl_2 + iC_4H_{10})$ as reagent gas.

The ions formed by Cl^- attachment decompose to form $[M–H]^+$ ion during an elimination of HCl:

$$M + Cl^- \leftrightarrows [M + Cl]^- \rightarrow [M–H]^- + HCl \overset{He}{\rightrightarrows} [M–H]^+ + 2e$$

Then the $[M–H]^-$ ions may undergo a collisional charge inversion.

This reaction is used especially when the $[M–H]^-$ decomposition spectrum contains no fragments.

A typical example is furnished by the study of *p*-chlorobenzoic acid. Indeed, the *m/z* 157 ion eliminates only CO_2 to produce the *m/z* 113 negative ion. On the other hand, charge inversion generates a larger number of fragment ions (Figs. 33 and 34).

This example also shows the possibility to select specifically the acid with only isotope ^{37}Cl, thus avoiding interference with other peaks. This method is shown to be particularly useful and sensitive in the case of mixtures.

It should also be noted that negative CI by OH^- leads to the distinction of diastereomers, especially diols, and bifunctional cyclic compounds by utilizing metastable decomposition spectra – both unimolecular and collisional induced – from deprotonated $[M–H]^-$ molecules.

The latter lead to the $[M–H_3]^-$ and $[M–H–H_2O]^-$ ions in different relative abundances according to the diastereomer diol studied [37,38,166].

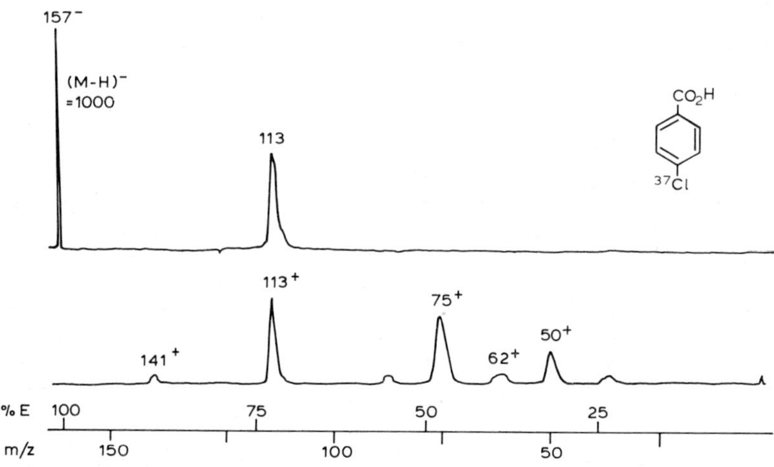

Figure 34. MIKE/CAD spectra showing positively and negatively charged products both arising from the $[M–H]^-$ anion of *p*-chlorobenzoic acid [164].

(d) Use of computers for processing unimolecular and collisional-induced decomposition spectra

Data processing leads to a greater flexibility in the exploitation of these spectra. In the case of mixtures, it is important to be able to pilot the analyzers of mass spectrometer with sufficient precision, especially when using various linked scan modes.

In cooperation with instrument designers, the group of Hass [167] showed the advantages of data acquisition techniques, especially to automate: the adjustment of the main beam studied; scanning of the different fields and detection of multiple metastable ions; the acquisition of metastable spectra; the various calculations of exact mass and energy liberated during the formation of metastable ions; the different automatic processes of the reversed geometry mass spectrometer, such as ZAB 2F.

This program contains all the possibilities of rapid and sensitive analysis of metastable spectra, regardless of the mode employed, with averages of MIKE (or MIKE/CAD) spectra.

Several groups have developed programs for processing the spectra of metastable ion decompositions. We may cite the Lausanne group [168], which used a PDP 11 computer to obtain the fragment ion abundances in MIKE/CAD spectra and the separation of uni- and bimolecular decompositions during collisions*, but also an accurate approach to top peak, and thus the mass of the fragment ion, leading to the detection of observed 'artifacts'. Beynon and co-workers [169] utilized a PDP 8 to measure the widths of metastable peaks. Mandli et al. [170] combined a computer with a conventional geometry instrument for the automatic acquisition of metastable data in the first FFR as well as for determining energy release and energy distribution of ions.

Finally, Haddon [171] introduced programs to use linked scan methods. He

TABLE 11

Fragment ions detected in the B/E linked scan spectrum of *n*-dotriacontane molecular ion (molecular weight, 450.5166) [171]

Neutral lost	m/z		Error (amu)
	Calculated	Measured	
C_4H_9	393.4506	393.4460	+0.054
C_7H_{15}	351.3611	351.3990	−0.038
C_9H_{19}	323.3138	323.3677	−0.054
$C_{11}H_{11}$	295.2775	295.3364	−0.058
$C_{13}H_{31}$	239.1940	239.2738	−0.079
$C_{18}H_{37}$	197.1559	197.2269	−0.071

*It is often necessary to separate ions produced in the cell (pure CAD spectrum) from those occurring outside the cell (by spontaneous decompositions or by collision resulting from gas leaks in the analyzer). The pure MIKE/CAD spectrum is shifted by applying a negative voltage on the cell.

obtained a resolution greater than 500 for metastable fragment ions, such as he showed in the case of dotriacontane (Table 11).

Powers et al. [172] developed a program which, among other features, is characterized by the rapid analysis of metastable spectra detected in the first FFR in B/E, B^2/E and $(B/E)(1 - E/E_0)^{1/2}$ modes. The microprocessor employed also enabled SIM and MID (on a large number of different ions) modes to be used for the same analyses with rapidity. VG Micromass Instruments currently market such a computer, for example, the PDP 8 and PDP 11/23, leading to total piloting of the mass spectrometer.

(e) New generation of mass spectrometers for MS/MS techniques

(i) Magnet and electric analyzer instrument as tandems

The creation of new generations of instruments, in addition to conventional mass spectrometers (as E/B), became necessary as a result of the importance of detecting reactions occurring outside the source.

It is true that the success of the MIKE method is taken for granted. It was necessary to design new inverted geometry instruments (B/E), such as the ZAB 2F (VG Micromass) and the MAT 312 (Varian). In addition, other types of 'tandem' (MS/MS) instruments have appeared.

Although there are relatively few of these instruments, we may cite results obtained by McLafferty and co-workers [173] and by Maquestiau et al. [174]. In particular, the instrument (built by the latter authors) is composed of three fields: $E/B/E$.

Russel et al. [175] also designed an instrument with analogous configuration ($E/B/E$) with a slightly different conception, giving a resolving power of about 15 000. This instrument has exceptional performance for high-resolution analyses and for the different modes of detecting metastable ions. Beynon and colleagues [176] modified a CEC 21 110.

An MS 50 (AEI Kratos), transformed into triple field ($E/B/E$) [177], results in a very high sensitivity, on the order of 1 part per trillion (ppt). The tetrachlorodibenzo-*p*-dioxin (TCDD) studied is introduced using a direct probe and the losses of $CO^{35}Cl$ and $CO^{37}Cl$ are selected (Fig. 35).

In these conditions, 50 pg can be detected with a precision of $\pm 20\%$. A complete MIKE/CAD spectrum can be recorded with about 500 pg of mixture. Here again, the results are comparable to HR/MS*, but the measurements are rapid and more specific.

The main interest of the four field ($E/B/E/B$) tandem instrument of McLafferty and co-workers [178] is, above all, at the level of analytical possibilities: among others by

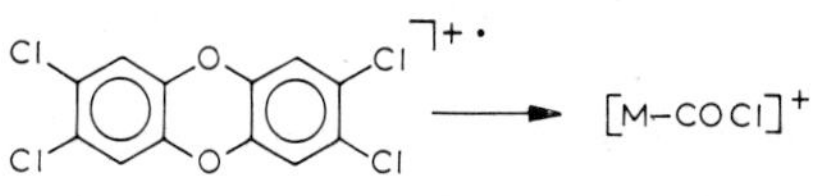

$[M-COCl]^+$

Figure 35. Decomposition of TCDD molecular ion produced under EI conditions [177].

---

*HR/MS, high-resolution mass spectrometry.

the high mass resolution of first MS (from 1000 to 50 000) and of second MS, which may vary from 100 to 10 000. It enabled the separation of the isobaric $C_6H_5N^{+\cdot}$ (91.0422) and $C_7H_7^+$ (91.0547) ions (mass resolution: 5000). This instrument is characterized by a 25 kV ion acceleration voltage to increase sensitivity and energy resolution*.

Boerboom and co-workers [179] improved energy resolution by reaccelerating the ions after the first magnetic field of the tandem, resulting in improved ion transmission, and thus sensitivity. This was done by utilizing a configuration in which several quadrupoles ($B/Q/Q/Q/B/Q$) were introduced between two magnetic fields. This also led to the rapid and simultaneous detections of various ions.

(ii) Triple quadrupole instruments

Recently, the development of quadrupole instruments has given rise to the introduction of another class of instrument which is more flexible and less costly. These instruments are termed triple quadrupole ($Q/Q/Q$), in which the collision cell is located in the second quadrupole, in which no mass separation occurs. These instruments were introduced by Yost, Enke and co-workers [180].

Collision reactions are as efficient (or improved) as those produced in magnetic and electric field instruments (at several kV) [180c] in spite of the relatively low acceleration (on the order of 1 to 100 eV) of the ions produced in this type of instrument. Boitnott et al. [181] showed that the cross-section was increased by a factor of about 20 in these instruments. Hunt et al. [182,183] indicated the possibilities offered for studying negative ions, since collisions can be 50–100 times more efficient than those produced in double focusing instruments. Giordani et al. [184] showed applications in the quantitative analysis of biological compounds and in the elucidation of polyphenol structures. In particular, they used the constant neutral fragment spectra. They showed the presence of aromatic acids in samples of human urine. After the first quadrupole (Q_1), they isolated the $[M-H]^-$ ions produced under NICI(N_2O/CH_4) conditions, which collisionally eliminate one molecule of CO_2 (44 amu) in the second quadrupole (Q_2). The third quadrupole is used to detect fragment $[M-H-CO_2]^-$ ions (Fig. 36).

This method of parent detection leading to the same neutral fragments with a $Q/Q/Q$ instrument is interesting, since the other ions remain transparent to the fast analysis and a very good specificity is obtained in comparison to GC/MS.

Finally, this system was used to study mixtures of polyaromatic compounds. A Townsend CI discharge source with N_2O_4 as reagent gas improves the production of $[M-H+NO_2]^{+\cdot}$ ions, which fragment in the second quadrupole with elevated efficiency to eliminate, among others, NO_2 radical. Thus, all the molecular ions of the polyaromatic compounds were obtained by studying the NO_2 (46 amu) constant neutral spectrum.

Zakett and Cooks [186] studied mixtures obtained from SRC II refined coal by analysing constant neutral fragment spectra under collision conditions. To increase

* Recently, VG Micromass Instruments designed a new four field tandem ($B/E/B/E$).

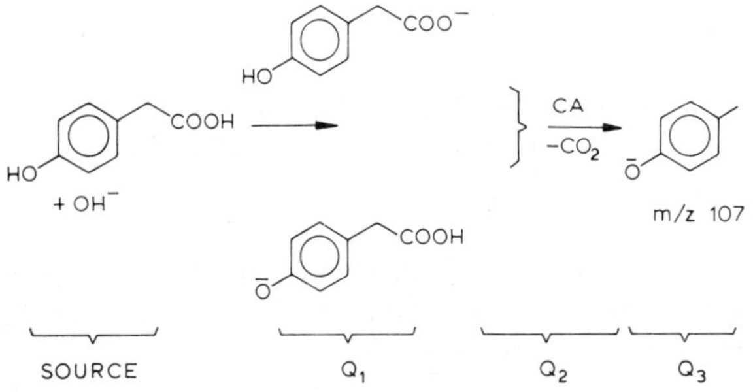

Figure 36. Neutral and ionic species analyzed in the triple quadrupole instrument.

the sensitivity of the method, the components of the mixture (phenol, isoquinoline, indanol) first were chemically modified. In this case, they also demonstrated a better resolution of spectra in comparison to MIKE spectra.

A simplified 'multiquadrupole' type of instrument was introduced by Siegel [185]. This is a double quadrupole analyzer which analyzes primary ion (or precursors) and fragment ions. In place of the second normal quadrupole (for *Q*/*Q*/*Q* instruments), a ferrite ceramic collision cell is installed which overlaps with the two quadrupoles. The first and second quadrupoles analyzed, respectively, the primary and fragment ions. The collisions occur in the intermediate region.

It is possible to accelerate (or decelerate) the ions in the collision cell*.

Furthermore, this generation of instrument may be favorably coupled with GC, leading to rapid measurements**: 0.2 u/s in the primary ion selector and 100 u/s in the second quadrupole.

Nevertheless, an inherent limitation in this type of instrument is related to the relatively limited resolution, in comparison to magnetic and electric sector tandem instruments.

(iii) Hybrid instruments

This family of instruments should nevertheless be developed, especially for fast routine analyses.

To improve the mass resolution, Glish et al. [188] designed a *B*/*Q*/*Q* type instrument. The advantage of this type of instrument is related to a better mass resolution, while retaining the quadrupole system, which requires low kinetic energy. This instrument may be utilized for determining the structure of ions. They also showed applications for studying complex biological mixtures and mixtures of isomers***.

---

* The possibility of producing ion-molecule reactions in the second region has also been shown.

** Several applications using this instrument have been shown [187].

***Now, *B*/*E*/*Q*/*Q* and *E*/*B*/*Q*/*Q* instruments have been introduced.

(f) A new methodology for the study of mixtures: MS/MS

In analogy with the techniques of coupling GC and LC to a mass spectrometer for the identification of the components of a mixture, McLafferty and co-workers [106a,173] designated as MS/MS the method of detecting fragmentations outside the source, as discussed previously.

The basic difference between GC (or LC)/MS and MS/MS is the first analytical step: in one case separation occurs before ionization while in the second it occurs afterwards. Figure 37 shows the principles of the two methods.

Thus, the mixture $\sum_i M_i$ analyzed is first ionized to obtain a set of molecular ions (and eventually fragmentation ions), which are then separated by the magnetic analyzer, which plays the role of a particularly efficient chromatographic system.

After choosing an $M_1^{+\cdot}$ mixture of ions, its structure is studied by its $\sum_j m_{1,j}^+$ fragmentations (spontaneous or collisionally induced) with the second mass spectrometer.

The energy spectrum recorded is thus used as a fingerprint to identify the molecular ion studied. In the case of GC (or LC)/MS, the traditional spectrum is used to determine the structure of the ion studied.

Concerning the 'linked scan' detection technique, the methodology remains the same, since in this case an ion may also be studied independently of other ions in the source (the decompositions measured are those produced in the first FFR).

In the context of applications, Cooks and co-workers [158,189] first showed the value of each of these methodologies, which are perfectly complementary. It is obvious that MS/MS will never replace GC (or LC)/MS. Harvan et al. [190] explained that the use of GC/MIKES (or GC/MS/MS) may be a powerful tool for studying mixtures of isomers and will be more efficient than measurements performed at high resolution (GC/MS/HR).

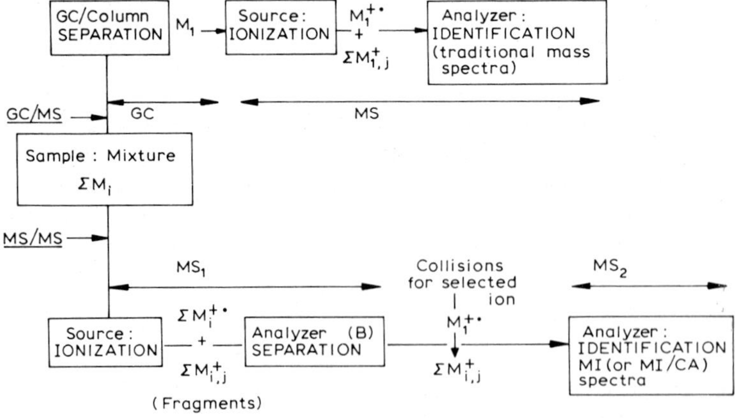

Figure 37. Comparison of GC/MS and MS/MS methodologies.

Furthermore, McLafferty [173b] confirmed future development and added LC/MS/MS as a possible technique. Each of these methods may have its own characteristics.

In order to show a direct application of this methodology, we cite the example of the detection of 150 ng of biotin in a mixture from the biosynthetic reaction from dethiobiotin [128].

The CID spectrum of this *m/z* 259 ion in the complex mixture is sufficiently close to that produced by an authentic biotin sample that we may conclude that the *m/z* 259 ion in the mixture is the protonated molecular ion of biotin (Fig. 38).

This example of application, combined with those we will develop below, naturally leads to the comparison of GC (LC)/MS and MS/MS – in studies of mixtures – through their proper characteristics.

Table 12 shows a few of the advantages and disadvantages of each method.

This table gives rise to several remarks: substances are normally modified to render them more volatile. This process may also be used to induce fragmentations and, in the case of MS/MS, it may become necessary. In a mixture in which one wishes to identify each component, MS/MS requires the knowledge of the spectrum of each

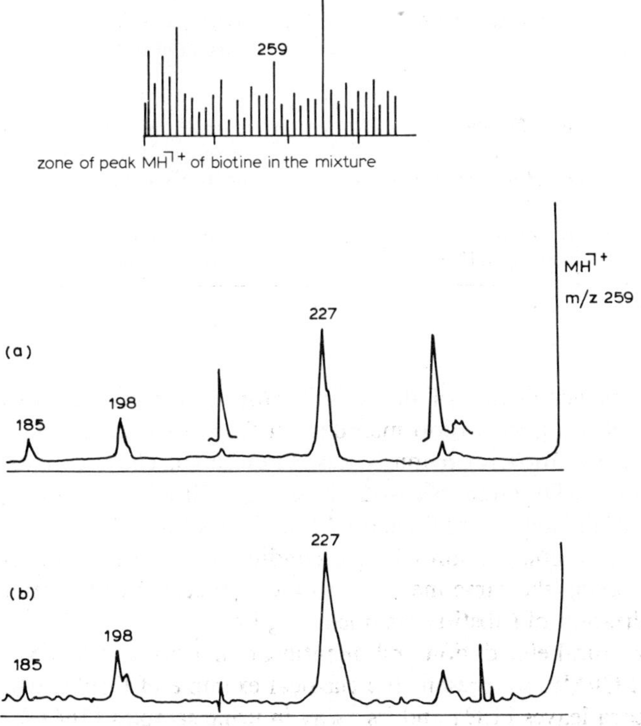

Figure 38. Analysis by MIKE/CAD mode of ion *m/z* 259 (MH^+) produced from an authentic sample of biotin methyl ester (in CI/NH_4^+) (a) and from the unknown mixture (b).

TABLE 12
Comparison of GC/MS and MS/MS techniques

	GC/MS	MS/MS
Extraction and modification	often necessary	not useful
Thermolabile samples	delicate separations	no difficulty
Separation times	columns: 10–120 minutes	magnetic field: several seconds
Ionization methods	few difficulties for ionization using desorptions	no limitations
Structural analysis		
Separation of high-mass compounds	sometimes difficult if column efficiency not sufficient (derivatization used)	always possible within mass limits of instrument
Separation of products of same mass		
Isomers	possible	delicate but possible
Isobars	sometimes require high resolution	HR: 5000, tandem (15 000)
Separation of isotopes	rarely possible	technique of choice
Localization of isotopes	possible	partial labelling migration, but possible
Identification of molecular sequences	sometimes delicate	technique of choice, since fingerprint is obtained
Quantifications		
Assay with MID/SID	no difficulty	possible; renders the method as sure as in high resolution
Introduction of background, sample loss and memory effect	disadvantage of interface	no memory effect
Sensitivity limit	fg range	pg and lower
Reproducibility	correct ($<1\%$)	several percent

component of the mixture, as well as that of all molecular (or quasimolecular) ions, leading to the importance of using ionization methods. In the search for a unique component in a mixture, the situation is obviously less complicated. The use of soft ionization methods (CI and FD) often necessitates using collisionally induced fragmentations as a result of the low internal energy of the ions studied.

The above remarks lead to a comparison of these conditions at the level of the separation of compounds having the same mass (or not) for structure identification and, more so, for the localization of substituents (including isotopes).

In order to show the structural elucidation and quantification aspects of MS/MS, routinely used in GC (or LC)/MS, we present the classical example of Cooks, who demonstrated cocaine in coca leaves [189] and its assay in urine samples [164].

The presence of cocaine molecular ion is detected with chemical ionization for coca leaves. Under collisions, this MH^+ ion decomposes by elimination of benzoic acid.

This loss leads to the most abundant transition from the MH^+ ion (m/z 304). The signal obtained is sufficiently intense that as little as 2 ng of sample need be used. This procedure is quite close to that of selected ion monitoring (SIM). The signal/noise ratio is on the order of 5 for 600 pg of cocaine.

The sensitivity can be improved considerably if – rather than recording peak intensity during electric field scanning (MIKE spectra after setting on the main beam) – the electric field is set on the transition, e.g., $304^+ \rightarrow 182^+$ and peak area is recorded as a function of time.

It is also possible to detect several metastable ions simultaneously. The ions are detected and analyzed by varying the accelerating voltage in order to select precursor ions. This is done with the variation of electric field voltage for the selection of the fragment ions chosen. The method allows an assay of complex mixtures, leading to identical intense ions, as for example the measurement of the ratio between cocaine and a minor component, such as cinnamoylcocaine, in coca leaf samples (Fig. 39).

A better signal/noise ratio is obtained by heating the sample rapidly. Furthermore, linearity for the assay is good.

In this case, there is an approximately 30% error for 1 ng of cocaine. In a more general case, Cooks [164] demonstrated the presence of drugs in 1 ml of urine: 40 ng of phenobarbital, 450 ng of aspirin, 80 ng of caffeine, 260 ng of phenacetin, 60 ng of mexaline and 25 ng of morphine.

Fraisse and Maquin [191] utilized B/E linked scan to demonstrate traces – on the fg level – of hydroxyphenyl 2-butanone, the natural flavour of strawberry.

McLafferty [173b] showed the possibility of detecting the presence of thiophene, THF and *n*-propylbenzene at 25, 50 and 500 ppm, respectively, in gasoline, which is a particularly complex mixture. Hass and co-workers [190,192] showed how GC/MS can be complemented by MS/MS, and thus lead to an approach for determining complex structures, such as that of humic acid. In addition, they demonstrated the presence of very small quantities of tetrachlorodibenzo-*p*-dimethoxane (TCDD) (>5 pg) in methyl stearate samples.

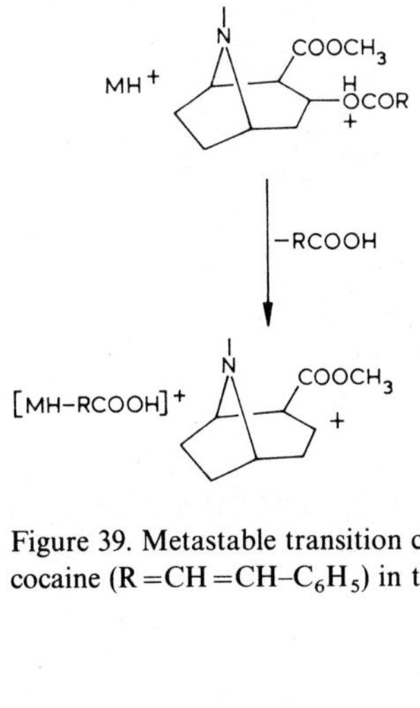

Figure 39. Metastable transition chosen to perform quantifications of cocaine ($R = C_6H_5$) and cinnamoylcocaine ($R = CH{=}CH{-}C_6H_5$) in the mixture [189].

Gross and co-workers [193] recently used the HV scan mode for the identification of TCDD by direct probe introduction at the 1 ppt level. The results obtained with this method were compared to those furnished by GC/HRMS. The metastable transitions: $320^{+} \rightarrow 257^{+} + CO^{35}Cl$ and $322^{+} \rightarrow 259^{+} + CO^{35}Cl$, were chosen as specific reactions when polychlorinated biphenyl (PCB) is present. Interferences appear in GC/HRMS, which is not the case when the metastable ion method is utilized (Fig. 40).

Among all the aspects of MS/MS, isotopic assay should be mentioned. If the molecular ion is sufficiently abundant [194] then the conventional mass spectrum can be used to analyse the labeled molecules. However, if an isomeric labeled mixture is present it becomes difficult to attribute total deuterium for each of these species, as well as localization. For instance, the classical method cannot be used for the study of the mixture d_0, d_1 and d_2 of phenyl-2 ethanol *a* and phenyl-1 ethanol *b*. The decomposition ions in the source are not sufficiently specific (randomization of deuterium). The *m/z* 92 and *m/z* 107 ions are characteristic decompositions of *a* and *b* compounds (Fig. 41).

A calibration curve can be used to determine the *a/b* ratio by measuring the ratio of abundances of ions: (92 + 93 + 94)/(107 + 108 + 109). The localization of D in each isomer is possible by measuring the shift of the *m/z* 92 ion (for *a*) and of the *m/z* 107 ion

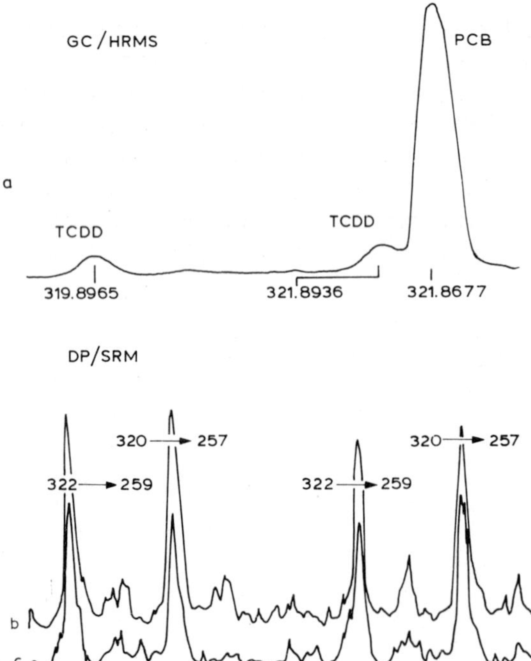

Figure 40. Comparison of analysis of standard solutions containing 200 pg of TCDD and 200 ng of 'Arachlor 1260' using (a) both the GC/HRMS and DP/SRM (direct probe/specific reaction monitoring) for (b) this mixture and (c) a sample containing only 200 ng of TCDD [193].

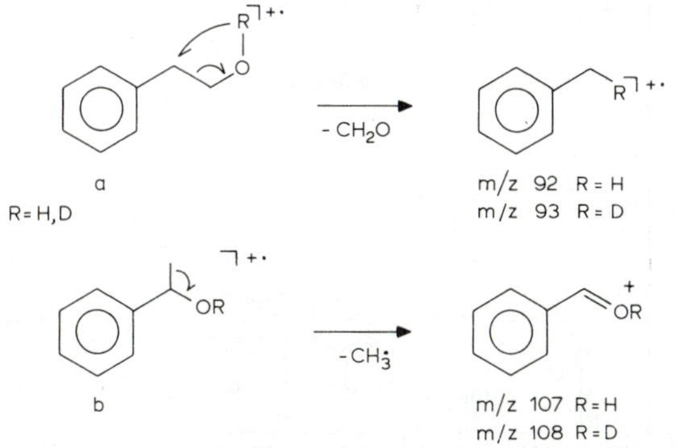

- CH_2O

a

R = H,D

m/z 92 R = H
m/z 93 R = D

$-CH_3^{\bullet}$

b

m/z 107 R = H
m/z 108 R = D

Figure 41. Characteristic decompositions of a and b in first FFR. Only these ions are observed in, respectively, the B/E spectra of a and b isomeric compounds.

(for *b*) for each labeled isomer. Then it becomes possible to determine the labeling distribution of compounds in the mixture.

Isotopic assays, especially for deuterated compounds, are generally possible in conventional spectra, except in certain cases in which the abundance of the molecular ion is very low or when it is 'contaminated' by the presence of satellites, such as MH^+ (or $M–H]^+$).

Finally, it has been possible to localize the labeling on the biotin skeleton after extraction of biotin from biological mixture, and consequently to understand its biosynthesis better [128].

The situation is less complicated when MS/MS is used. It is occasionally possible, however, to perform isotopic assays on fragment ions, providing their mechanism of formation is known and one is sure that no specific exchange of H or D atoms occurs in the positions assayed. It is sufficient to know a fragment ion (no longer labeled) which is produced at least from the molecular ion. The search for these precursors with the HV scan and B^2/E linked scan methods leads to the direct determination of isotope distribution (sensitivity and reproducibility are good).

These examples show the value of these highly specific techniques when studying traces of compounds in complex mixtures in the fields of organic chemistry and biochemistry. The diversity of applications of MS/MS makes it an alternative to conventional methods. These techniques are constantly evolving. We will now present several examples of application, by class of compounds, in order to show the scope of possibilities offered by MS/MS.

3. *Applications*

(a) Analysis of steroid compounds

The various analytical aspects of MS/MS were first investigated with steroids. The identification of compounds, even in complex mixtures, has been performed up to the

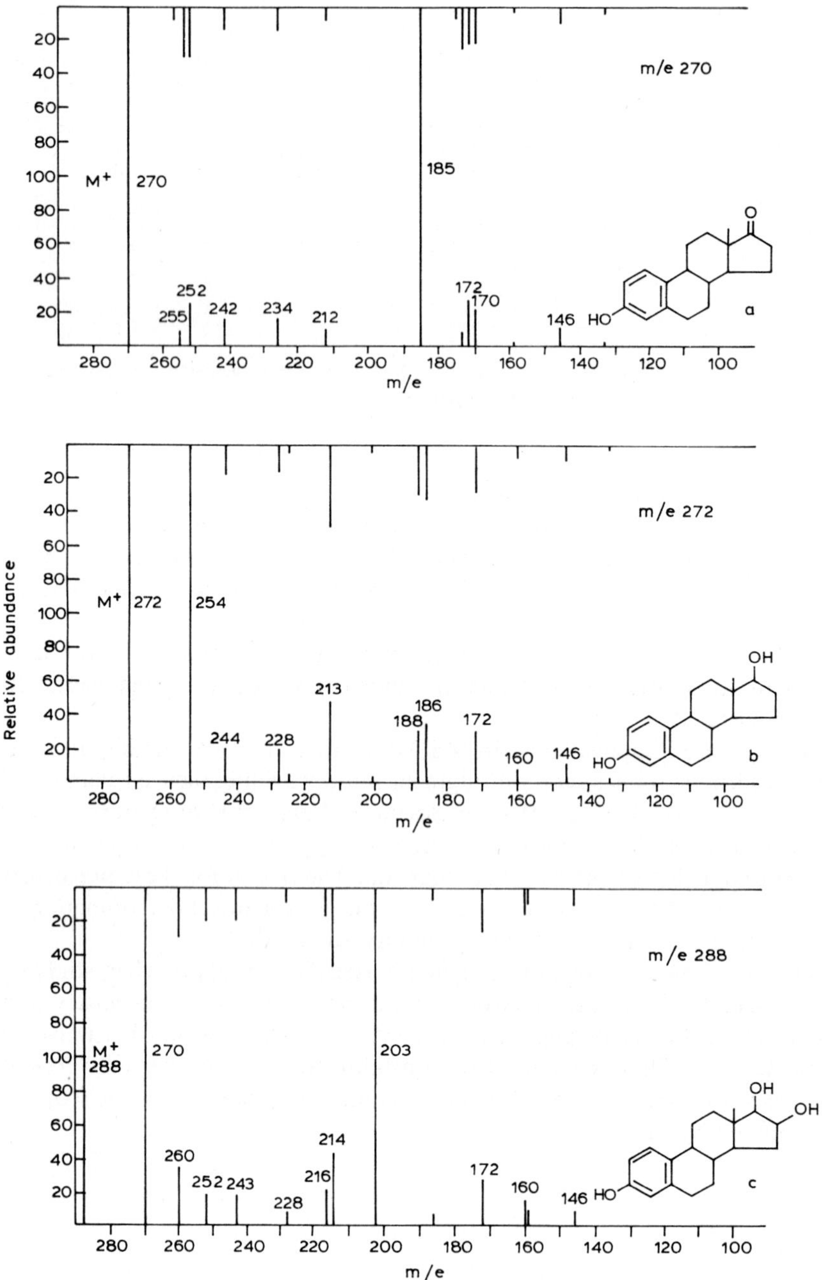

Figure 42. Comparison of MIKE spectra of ions *m/z* 270, 272 and 288 produced in EI from mixture of estrone (a), estradiol (b) and estriol (c) (as 'reflected' spectra) with the same respective ions formed for authentic samples of these compounds [196].

level of the 'stereochemical aspect,' as well as the quantitative determination of trace amounts in urine (or blood) samples.

The two aspects of uni- and bimolecular decompositions of ions are used with their respective specificity. These decompositions are detected with both MIKE and linked scan methods.

In a study of the structure of cholesterol, McLafferty et al. [195] chose to induce the fragmentations by collisions on the molecular ion peak produced by EI. They demonstrated the presence of different groups, such as $-CH_3$, $-OH$ ..., on the steroid skeleton studied.

The following year, Djerassi and co-workers [196] utilized unimolecular decompositions to show the possibility of identifying estrone derivatives (Fig. 42), estrone (a), estradiol (b) and estriol (c).

Figure 42 shows the presence of each of these three compounds in the artificial mixture by measuring the MIKE unimolecular decomposition spectra of the $M^{+\cdot}$ ion.

The spectra of pure products (reflected spectra) are practically identical to those obtained in the mixture, verifying the specificity of the method.

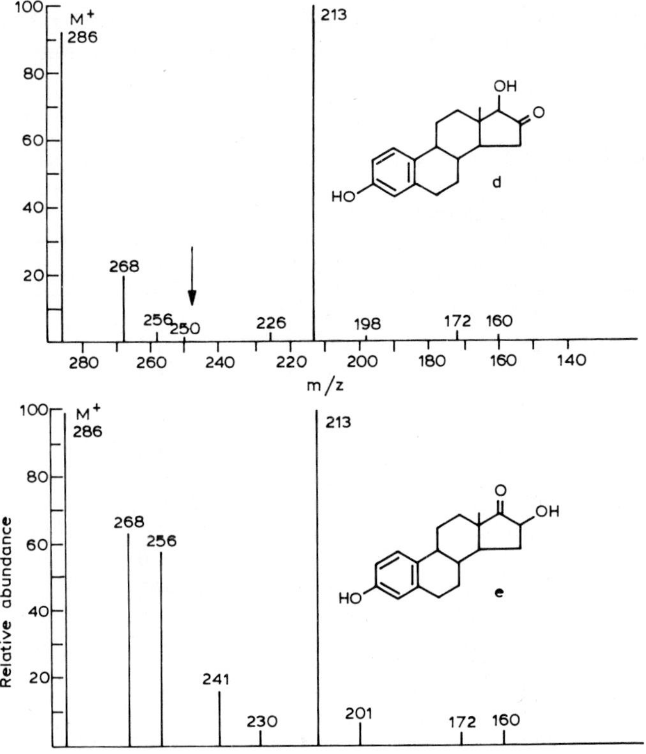

Figure 43. MIKE spectra of isomeric molecular ions m/z 286 for 16-keto estradiol (d) and 16-hydroxy-estrone (e) [196].

It should be noted that the molecular ion of estriol (c, *m/z* 288) eliminates a water molecule to yield the $[M_c\text{–}H_2O]^{+\cdot}$ abundant ion (*m/z* 270) with the same *m/z* as the estrone molecular ion. However, the observed differences between these two spectra indicate that the *m/z* 270 ions decompose by various pathways according to the structure of their precursor ions.

It should be noted that these ions would no longer have the same internal energies and would generate somewhat different ion abundances in their respective MIKE spectrum.

The MIKE spectra of isomers d and e are each characterized by considerable differences between the *m/z* 268, 258 and 250 peaks (Fig. 43).

Previously, we also showed the advantage of this method for studying the decomposition of ions entirely labeled in a mixture of insufficiently deuterated molecular ions as clusters (p. 201). It should not be forgotten, however, that when this method is applied to fragment ions, it may lead to false data if non-specific H (or D) exchanges occur.

In order to obtain more intense molecular peaks, Levsen and co-workers [197] chose the field ionization (FI) method, much 'softer' than electron impact ionization, and utilized collision-induced decompositions rather than unimolecular decompositions. They were thus able to identify estrone (*m/z* 270) and progesterone (*m/z* 314) in an artificial mixture (Table 13). The FI/MIKE/CAD spectra of the molecular ions are

TABLE 13
Analysis of steroid mixtures using MIKE/CAD spectra under FI conditions (FI/CAD spectra) [197]

Estrone (*m/z* 270)		Progesterone (*m/z* 314)		
m/z	FI/CAD	*m/z*	FI/CAD	EI
41	2–3	42	5–2	
55	2–2	55	0–9	10
65	1–6	65	1–1	3–2
77	4–1	77	1–7	7–4
91	3–3	84	0–6	2–2
97	1–8	91	4–3	9–5
107	4–2	93	4–2	6–1
115	6–3	105	4–6	5–5
132	12	124	20	15
146	19	135	0–8	3–0
159	15	147	1–5	4–5
172	0–9	159	1–1	1–9
185	13	173	1–1	2–3
199	2–3	191	19	2–7
213	7–7	229	2–0	8–7
226	1–7	244	0–5	4–1
242	2–5	272	27	11
		296	1–3	0–8
		299	3–9	1–6

characterized by ions similar to those observed in the EI mass spectra of authentic products. The comparison of these spectra is a practical means of verifying the value of the method.

Baczybskyj and Duchamp [198] studied the fragmentation of the calusterone molecular ion (Fig. 43) without collisions. An IBM 1800 computer was used to calculate the mass of fragment ions. High-resolution measurements complemented the interpretation of the experimental results obtained. The fragmentation mechanisms are characteristic of the A and B rings of calusterone (Fig. 44).

Horvath and Ambrus [199] studied the common loss of carbons at positions 16 and 17 (regardless of the substituents) in the norethisterone family in norgestrel and in norethynodrel. Although these characteristic eliminations were observed for many derivatives, the authors showed the specific migration of the hydrogen atom of 17-OH (Fig. 45). This migration occurred specifically for the Δ^4-3-keto derivatives.

The *m/z* 231 ion is shifted to *m/z* 232 if the –OH group is labeled.

As shown by HV scan spectra, the only origin for this ion is the molecular ion. Under these conditions, the reverse of squalene cyclization in sterol biosynthesis could be produced and followed by the migration of H (17-OH) during decomposition, to yield the *m/z* 110, 162 and 231 ions, such as Figure 45 indicates.

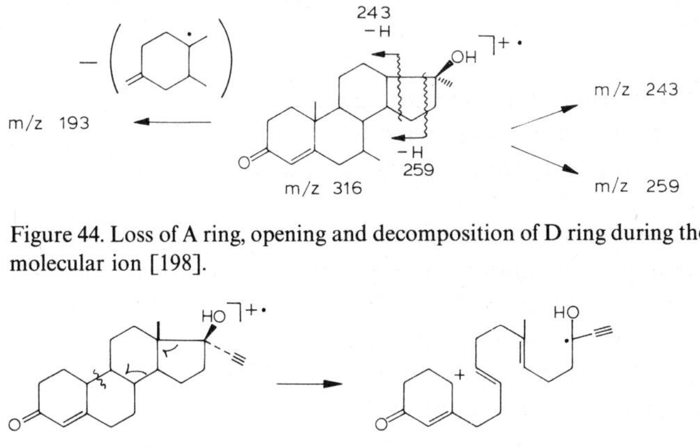

Figure 44. Loss of A ring, opening and decomposition of D ring during the unimolecular decompositions of molecular ion [198].

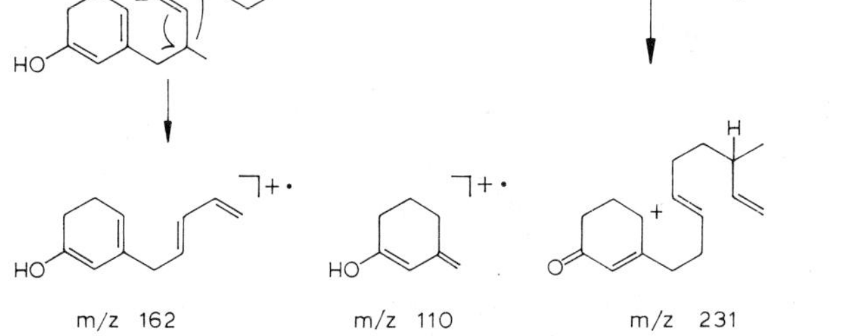

Figure 45. Unimolecular decompositions of 17-hydroxy-19-nor-17α-pregn-4-en-20-yne-3-one detected in the MIKE spectrum of the molecular ion [199].

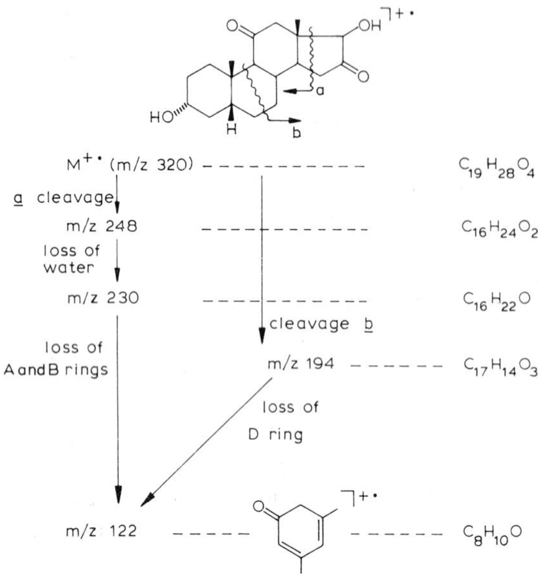

Figure 46. Consecutive unimolecular decompositions of molecular ion of 3α-17β-dihydroxy-5β-androstane-11,16α ions in the second FFR [200].

Another example shows the possibility of constructing a 'family tree' of fragment ions. Schlunegger [200] was able to explain the formation of the *m/z* 122 ion (Fig. 46) during the fragmentation of 3α,17β-dihydroxy-5β-androstane-11,16-diones with the MIKE method.

These examples, while not being the most recent, show the possibilities offered by the metastable detection techniques for understanding fragmentation mechanisms.

Brown and Djerassi [201] performed detailed studies of the decomposition of Δ^4-3-keto steroids under EI conditions in order to use them for the elucidation of the structures of new steroids isolated from marine organisms, and especially to locate functions in the adrenal and diverse sex hormone skeleton.

The search for the precursors of abundant fragment ions in the EI spectra of different Δ^4-3-keto steroids leads to the almost complete interpretation of the spectra (Table 14). Figure 47 summarizes these fragmentations, largely due to rings A and B cleavage.

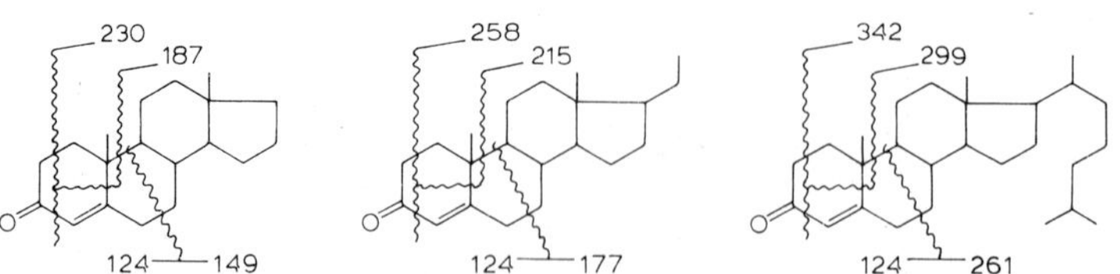

Figure 47. Fragmentations of various Δ^4-3-keto steroids under EI conditions [201].

TABLE 14
Precursors of fragment ions formed in EI from various Δ^4-3-keto steroids by HV scan spectra [201]

Compounds	Daughter ions (*m/z*) as main beam	Precursor ions (*m/z*)
Δ^4-Cholestene-3-one	342, 261, 124,	384 ($M^{+\cdot}$)
	299	384 ($M^{+\cdot}$), 10%
		342 $(M-C_2H_2O)^{+\cdot}$, 90%
Δ^4-Pregnene-3-one	258, 177, 124,	300 ($M^{+\cdot}$)
	215	258 $(M-C_2H_2O]^{+\cdot}$, 90%
		230 10%
Δ^4-Androsten-3-one	230, 149, 124	272 ($M^{+\cdot}$)
	187	230 $[M-C_2H_2O]^{+\cdot}$, 90%
		202 $(M-60)^{+\cdot}$, 10%
$\Delta^{1,4}$-Cholestadiene-3-one	367, 261, 147, 122	282 ($M^{+\cdot}$)

Isotopic labeling was used to define the various hydrogen atom migrations. The results obtained led to the interpretation of the effects of substituents on 'key' positions in the steroid. In addition, the stereochemistry of specific protons led to increased intensity of the *m/z* 124 peak.

The behavior of rings A and B is well known; nevertheless, it is useful to have a method for analyzing the structure of the side chain on various cholestane skeletons under EI conditions.

Wieber and co-workers [202] utilized unimolecular decomposition spectra (as MIKE), as well as high-resolution measurements for this type of structural study. This methodology was applied to the following derivatives: 5α-6-dihydroergosterol (1), cholest-1-ene-3-one (2), methyl trinor-5-cholestane-3-one-24-oate (3), 22β-hydroxycholesterol acetate (4) and 22-ketocholesterol acetate (5) (Fig. 50).

The spectra of these compounds were often complex, but fragmentations were occasionally predominant, which gave rise to formation of $[M-R]^+$ and $[M-(R+42)]^+$ ions, corresponding to the loss of the side chain and the cleavage of D ring (Fig. 48).

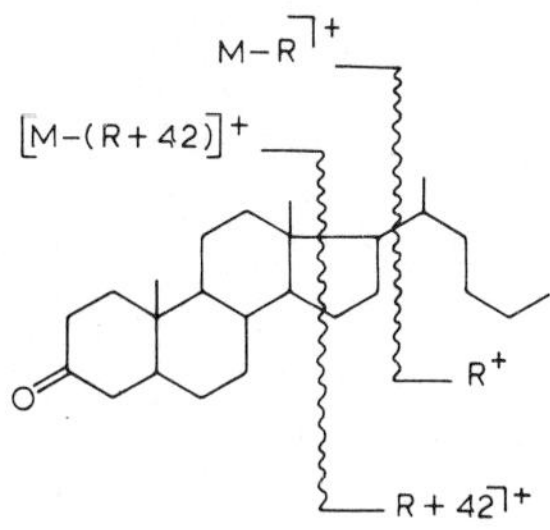

Figure 48. Cleavage of D ring and lateral chain loss during the unimolecular decompositions of the molecular ion $M^{+\cdot}$ [202].

Although the abundances of the $[R]^+$ and $[R+42]^+$ ions were very low in the conventional mass spectra, MS/MS rendered the study of the structures possible.

The interest of choosing the $[R]^+$ and $[R+42]^+$ ions is related to additional data which may be obtained from the MIKE spectra of these ions, as well as confirmation of structures. These must be compatible among themselves, considering that the only difference is due to the presence of C-15, 16 and 17 (forming the D ring) in the $[R+42]^+$ fragment ion.

The 'decomposition maps' obtained show that (after verification of basic decompositions): the $[R+42]^+$ ion yields $[R]^+$ by propene elimination (carbons 15, 16 and 17). This fragmentation is a general occurrence in the compounds studied; the loss of CH_3OH from $[R+42]^+$ and $R]^+$ ions (*m/z* 125 and 83), as well as the formation of the *m/z* 59 ion $[COOCH_3]^+$, are characteristic of the presence of methyl esters and shed light on the structure of the side chain; ethylene eliminations (such as C_2H_4, C_3H_6 and C_4H_8) indicate the presence of the hydrocarbon skeleton; finally, the $[M–R]^+$ and $[M–(R+42)]^+$ ions obviously do not lead to R^+ ions (Fig. 49).

The 'artifacts' that may arise can be detected by: their presence at non-whole mass numbers; the form of these peaks; the impossibility of interpreting these peaks, accounting for R^+ (or $R+42]^+$); finally, the absence of these signals in conventional spectra.

Thus, the structures of various side chains on steroid skeletons could be identified, as Figure 50 shows.

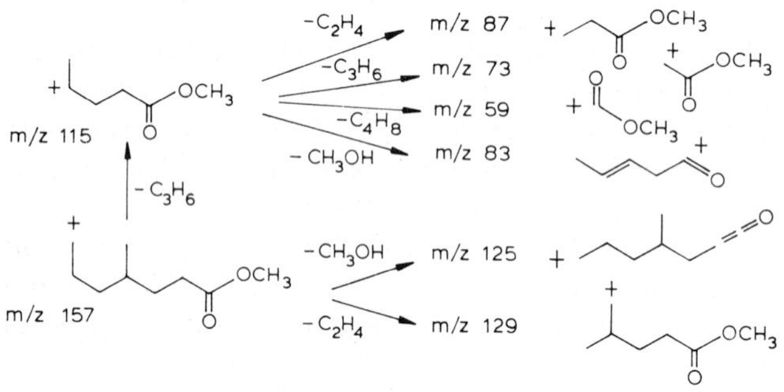

Figure 49. Decomposition of R^+ and $R+42]^+$ ions [202].

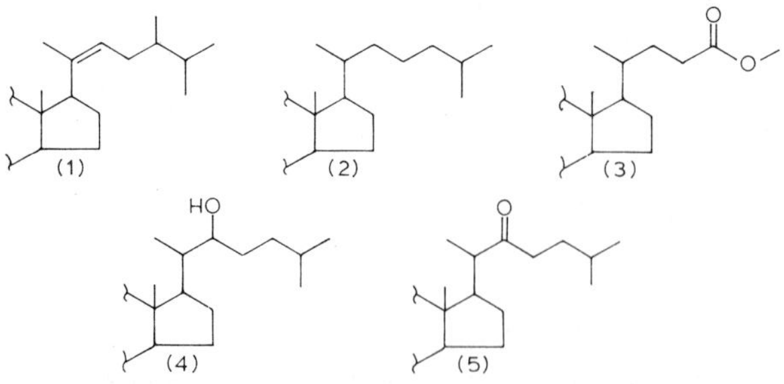

Figure 50. Structures of side chains identified using the MIKE technique.

TABLE 15
Stereochemistry effect of ring junction A/B observed in HV scan spectra for $M^{+\cdot} \rightarrow [M - CH_3]^+$ transition [203]

Steroid compounds	$[m^*]^+/[M\text{–}CH_3]^+$
5α-Androstane-3-one	0.08
5β-Androstane-3-one	0.17
5α-Androstane-17-one	0.23
5β-Androstane-17-one	0.30
5α-Androstane-3,17-dione	0.16
5β-Androstane-3,17-dione	0.25
5α-Androstane-3,11,17-trione	0.10
5β-Androstane-3,11,17-trione	0.28

These examples show that the structure of the carbon skeleton can be studied with metastable decompositions. In addition, the stereochemistry of functional groups can play a non-negligible role in the orientation of fragmentations.

Zaretskii [203] showed that the geometry at the ring A/B junction also led to different rates of decomposition. In particular, this is the case of the elimination of the methyl radial, as expressed by the ratio $m^*/(M\text{–}Me)$ (m^* being related to the $M^{+\cdot} \rightarrow [M\text{–}Me^{\cdot}]^+$ transition detected in the first field free region by HV scan mode) (Table 15).

Although the situation is less clear-cut in the case of the hydrocarbons themselves, the differences are greater in oxygenated compounds, and thus enable *cis* and *trans* isomers to be distinguished (rings A and B).

More recently, the same author [204] studied the MIKE spectra of molecular ions produced by electron impact to distinguish *cis* and *trans* configurations of steroid hydrocarbon ring junctions (mono-, di-, tricarbonylated). It was shown that the loss of ring A is favored when the A/B junction is *cis*. Concerning the ketones, the losses of $.CH_3$, H_2O and ring A are more abundant for *cis* geometry.

Differences in the *cis/trans* positions of the B/C and C/D ring junctions can also be recognized. Thus, all combinations of the skeletons shown in Figure 51 could be distinguished.

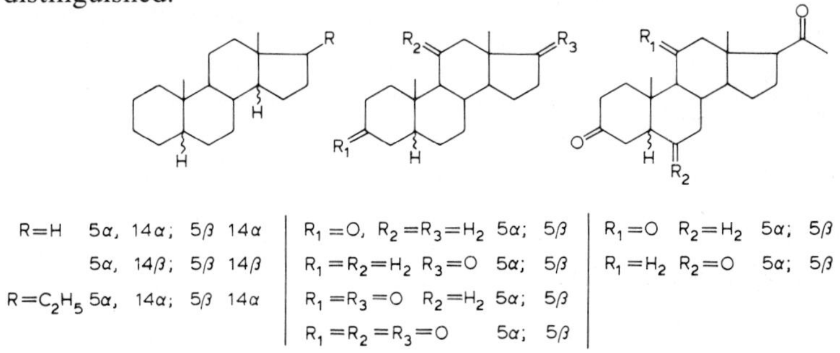

R=H 5α, 14α; 5β 14α
5α, 14β; 5β 14β
$R=C_2H_5$ 5α, 14α; 5β 14α

$R_1=O$, $R_2=R_3=H_2$ 5α; 5β
$R_1=R_2=H_2$ $R_3=O$ 5α; 5β
$R_1=R_3=O$ $R_2=H_2$ 5α; 5β
$R_1=R_2=R_3=O$ 5α; 5β

$R_1=O$ $R_2=H_2$ 5α; 5β
$R_1=H_2$ $R_2=O$ 5α; 5β

Figure 51. Different isomer families distinguished by the HV scan method.

Djerassi and co-workers [205] contributed additional data by studying the role of the unusual stereochemistry of certain carbon atoms at the junction of the ring: 8α, 9β or 14β. This work was performed by measuring HV scan spectra (of the most abundant ions) and high resolution spectra, and the fragmentations of deuterated compounds.

In particular, the following conventional spectra (Fig. 52) indicate the facility for obtaining *m/z* 193, 246 and 289 ions from 14β stereomeric compounds. Other ions at lower abundances were also characteristic at *m/z* 219, 229 and 342.

Table 16 gives the precursors of these intense ions. They were measured with HV scan mode for the 8α and 8β stereomers of 5α,14β-cholestane-3,11-dione.

Thus, in addition to characterizing compounds with 14α and 14β stereochemistry by measuring conventional mass spectra, it is possible with the 14β series to differentiate those with 8β configuration from those with 8α, as shown in the table. In

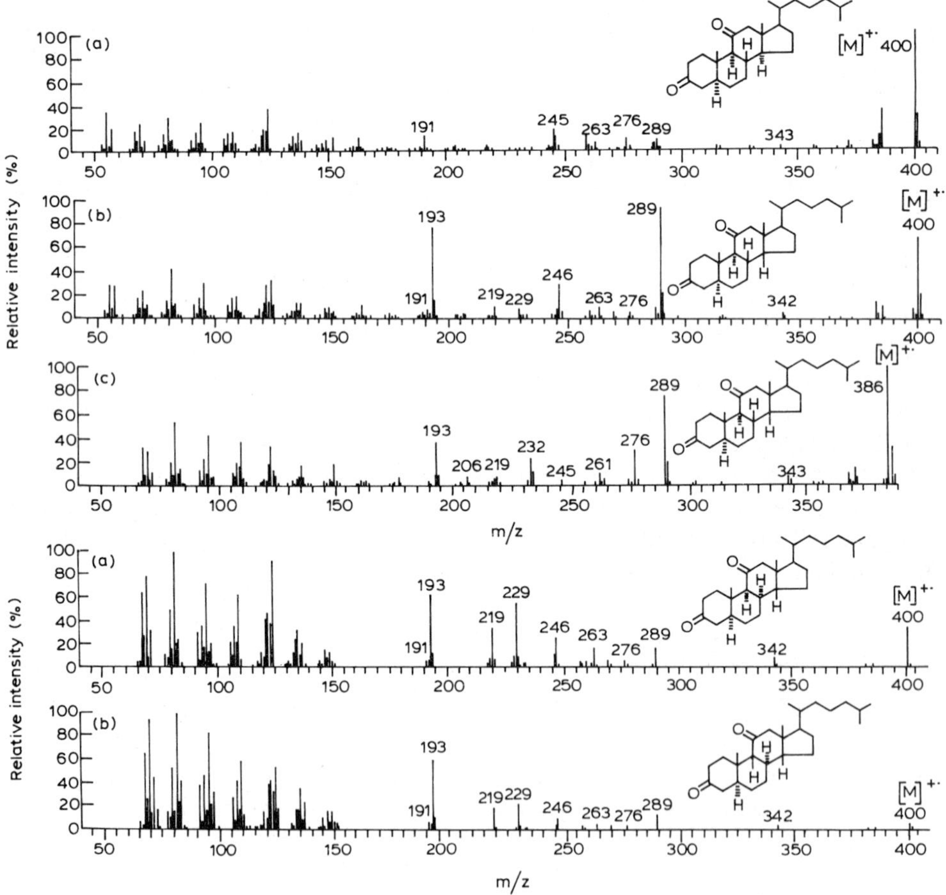

Figure 52. Conventional mass spectra of 14β (and 14α) keto and diketo steroids [205].

TABLE 16
Precursors of abundant ions observed in EI mass spectra of 5α,14β-cholestane-3-11-diones (8α and 8β) (HV scan spectra) [205]

Daughter ions (*m*/*z*)	Precursor ions (*m*/*z*)	Abundances	
		8β	8α
342, 289, 246	400	100	100
263	400	54	56
	382	31	22
	291	15	22
245	400	61	59
	287	9	–
	263	22	41
229	400	–	6
	382	5	–
	342	73	90
	287	8	–
	247	13	–
219	400	67	67
	382	7	7
	342	26	26
193	400	53	18
	342	7	44
	289	31	17
	263	4	21

particular, the abundances of the precursors of the *m*/*z* 193 ion, such as the *m*/*z* 400, 342, 289 and 263 ions, are sensitive to the α or β stereochemistry of C-8.

The ratios m^*_{400}/m^*_{342} and m^*_{289}/m^*_{263} for the 8α derivative are 0.41 and 0.81, whereas in the 8β isomer they are 7.6 and 7.75.

Returning to the more general study of 14β compounds, it is seen that this stereochemistry plays a double role during fragmentations: (i) the possible migration of H on C-14 only when stereochemistry is 14β; (ii) ring conformation by this stereochemistry, favoring hydrogen atom migrations without involving the H at C-14.

As an example, the *m*/*z* ion ($C_{24}H_{38}O^{+\cdot}$) is produced specifically by the migration of the hydrogen in position 14β. The reaction is favored to a greater extent if the proton in position 8 is β (Fig. 53) and is produced directly from the *m*/*z* 400 molecular ion (Table 16).

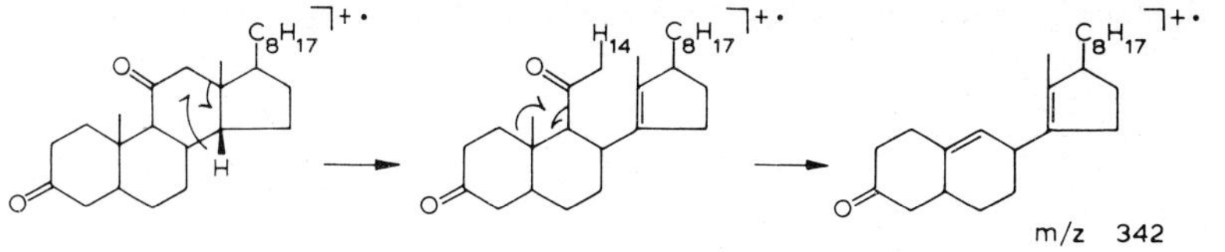

Figure 53. Migration of H in position 14β and loss of acetone after methyl migration [205].

The mechanism of acetone loss is perhaps not concerted, but rather is a surprising two-step reaction: $-CH_3^{\cdot}$ and $-COCH_3^{\cdot}$ (or the reverse). In this case, and taking the results of metastable spectra into account, it must be admitted that these consecutive cleavages are very fast in order to be produced in the first field free region, and thus the m/z 400 ion would be only a 'grandparent' ion.

The second effect is manifested by the formation of the m/z 193 ion which is also sensitive to C-14 stereochemistry, but which nonetheless is not produced by the migration of the hydrogen on this atom. Stereochemistry participates directly on the carbon ring, thus facilitating the formation of this m/z 193 ion (Fig. 54) via a similar McLafferty rearrangement.

The other consequence is related to the fact that the ring is apparently in twist conformation, generating a partial flexibility; the 1β hydrogen is thus very close to the carbonyl in position 11, favoring the initial step of ring B cleavage to lead to the m/z 289 ion (Fig. 55).

Other compounds, such as ketols (11-keto-3-ol), as well as 11-monoketo, characterized by this β stereochemistry (for C-14), have a similar behavior concerning the formation of these ions. In addition to the interest in distinguishing ring junction geometries, Gaskell and co-workers [206a,b] studied the behavior of α and β stereochemistry of the di-OTBDMS (*t*-butyldimethylsilyl) groups in derivatives of androstane-3,17-diol (Fig. 56).

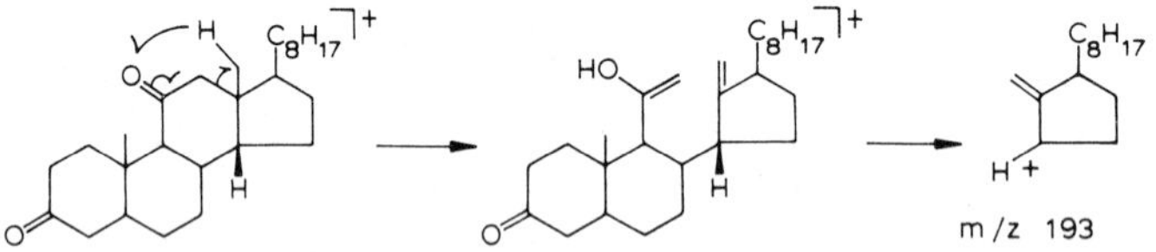

Figure 54. H transfer and A, B, C ring eliminations from 14β diketo steroids [205].

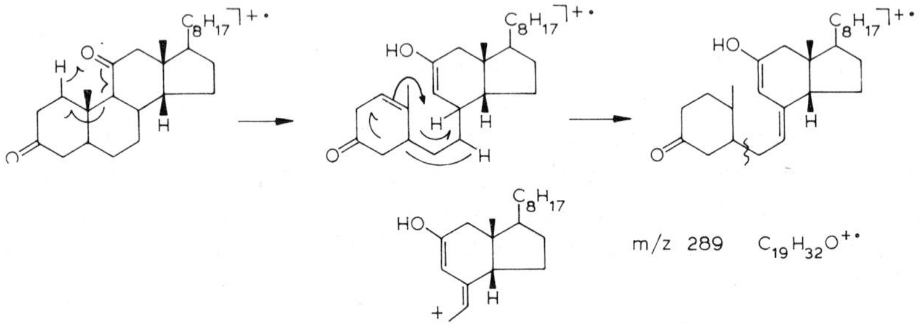

Figure 55. H_2 double transfers and allylic cleavage to give rise to formation of m/z 289 ions [205].

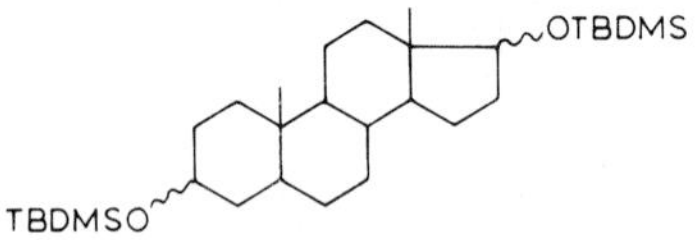

Figure 56. Structure of di-OTBDMS 3-17-androstane [206].

TABLE 17 [206]
(a) Conventional spectra of *t*-butyldimethylsilyl ethers of isomeric androstanediols

Steroid	Mass spectrum (70 eV)					
	m/z: 463	387	373	345	331	255
1, 5α-A-3α,17α-diol TBDMS	47	7	1	–	45	100
2, 5α-A-3α,17β-diol TBDMS	100	35	2	–	18	87
3, 5α-A-3β,17α-diol TBDMS	100	8	–	–	6	66
4, 5α-A-3β,17β-diol TBDMS	100	11	–	–	4	34
5, 5β-A-3α,17α-diol TBDMS	43	6	11	6	24	100
6, 5β-A-3α,17β-diol TBDMS	50	7	1	–	7	96
7, 5β-A-3β,17α-diol TBDMS	53	7	15	4	47	100
8, 5β-A-3β,17β-diol TBDMS	100	9	–	–	12	56

(b) *B/E* spectra of *m/z* 463 derived from *t*-butyldimethylsilyl ethers of isomeric androstanediols

	Daughter ions				
Steroid	*m/z*: 387	373	345	331	255
1, 5α-A-3α,17α-diol TBDMS	35	13	4	100	20
2, 5α-A-3α,17β-diol TBDMS	100	6	1	45	24
3, 5α-A-3β,17α-diol TBDMS	100	3	1	41	70
4, 5α-A-3β,17β-diol TBDMS	100	5	1	11	16
5, 5β-A-3α,17α-diol TBDMS	24	97	24	100	28
6, 5β-A-3α,17β-diol TBDMS	100	9	2	30	40
7, 5β-A-3β,17α-diol TBDMS	21	99	14	100	28
8, 5β-A-3β,17β-diol TBDMS	100	13	2	84	21

The study of low-energy (20 eV) electron impact spectra gives an idea of the stereochemistry effects in the formation of abundant ions (*m/z* 463, 387, 331 and 255), as shown in Table 17. According to the analysis of high-resolution spectra, these ions correspond respectively to $[M{-}C_4H_9]^+$, $[M{-}(But + HMe_2SiOH]^+$, $[M{-}(But + ButMe_2SiOH)]^+$ and $[M{-}(But + ButMe_2SiOH{-}HMe_2SiOH)]^+$. The study of the B/E and B^2/E spectra enables the 'family tree' of the principal ions to be obtained (Fig. 57), including the *m/z* 255 ion*.

This is one of the first examples of the migration of hydrogen and the alkyl group during the elimination of Me_2SiH, MetButSiH and Me_3SiOH from the $[M{-}tBut^{\cdot}]^+$ ion (*m/z* 463). In this context, the abundance of the trimethylsilyl cation (*m/z* 73) is not negligible, thus proving that this type of elimination is a one-step reaction.

In addition, although the low-energy conventional spectra indicate several differences in the abundant ions, the ions resulting from the unimolecular decompositions of the abundant *m/z* 463 ion are more sensitive, and lead to an unambiguous distinction between stereomers (Table 17).

*The *m/z* 255 ion is produced from $[M{-}C_4H_9]^+$ by fast consecutive decompositions in the first FFR, via *m/z* 387, 373, 345 and 331 ions.

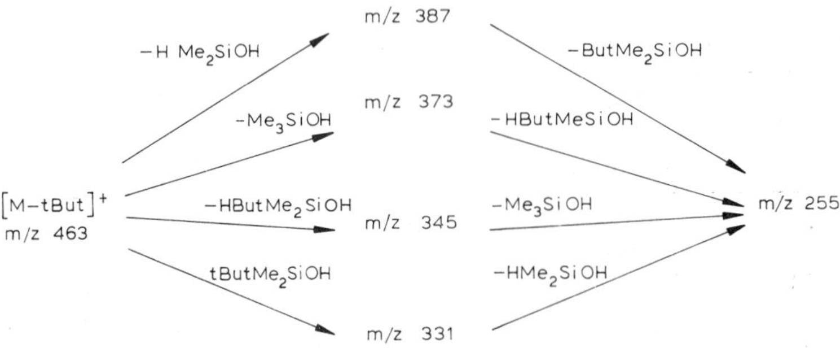

Figure 57. Various origins of *m/z* 255 ion produced during consecutive decompositions in the first FFR.

It should be noted, as shown in Table 17b, that it is possible to determine the α or β stereochemistry of C-5, especially by using the *B/E* linked scan spectra of the *m/z* 463 ion. These variations show the extent to which the stability of this intermediate ion (*m/z* 463) is sensitive to stereochemistry.

The main advantage of this method is the identification of these stereomers in mixtures, in which impurities may generate interferences.

The different methods of unimolecular decompositions under electron impact, HV scan, MIKE, and *B/E* and B^2/E linked scan, have been widely used for the determination of molecular structures and stereochemistry. Their identification in mixtures is now possible.

Gaskell and co-workers [207a,b] attempted to perform quantitative determintions with the steroid series. They used *B/E* linked scan with SIM system coupled with GC/MS.

The presence of isomeric (or isobaric) ions could eventually represent a limitation of the linked scan method when working on a unique transition. This is why coupling with GC is interesting.

The advantage of GC/MS is the capacity to separate these isomeric (or isobaric) compounds. Its disadvantage is, among others, to introduce coeluting compounds, leading to uncertainties and to a reduced sensitivity (in the case where high resolution becomes necessary); this no longer occurs when MS/MS is utilized.

The combination of the two methods thus leads to a very high specificity (comparable to or even greater than that obtained at high resolution of 12 000), an opinion which is shared by Hass and co-workers [192].

Gaskell and co-workers [208a,b] developed a technique for assaying endogenous steroids in blood plasma samples after first creating the TBDMS ether derivative.

The traditional spectrum of this derivative in these conditions is characterized by an intense peak at *m/z* 347 (M–*t*But]$^+$) and the *B/E* linked scan spectrum of this ion demonstrates the elimination of $H(CH_3)_2SiOH$, to yield a very intense peak (=10% of the main beam) at *m/z* 271 (M–*t*But–$H(CH_3)_2SiOH]^+$).

The analysis is thus performed on the *m/z* 347–*m/z* 271 transition, produced in the first FFR. Figure 58 shows the signal obtained from 100 pg of derivatized 5α-DHT

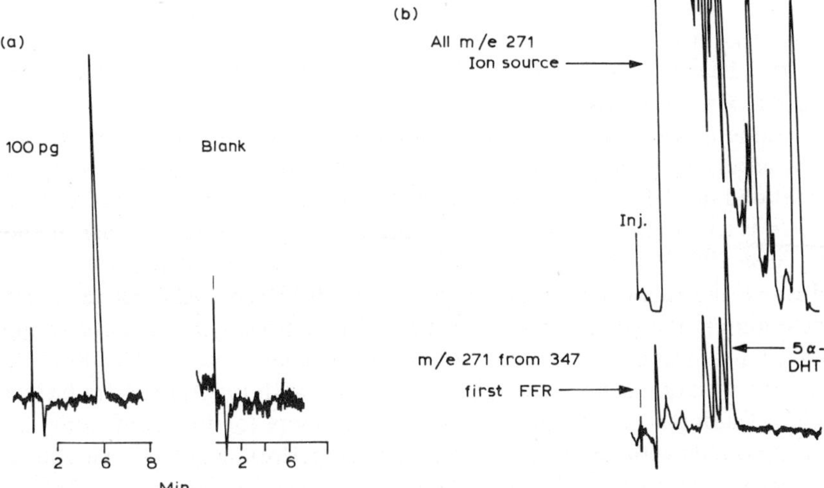

Figure 58. (a) Sensitivity of metastable peak monitoring for authentic sample of 5-DHT. (b) Comparison of conventional low resolution with SIM for 5-DHT and its biological isomers in human blood plasma and metastable peak monitoring for m/z 347 → m/z 271 transition [208].

with a reproducibility on the order of 10% [208a]. The detection threshold is about 20 pg.

The m/z 271 peak is produced in the source from series of isomers and other compounds, which renders any conventional measurement mode impossible. In B/E linked scan, however, the situation is greatly simplified, since only four signals remain which correspond to the different isomers, which are known. Multiple ion detection (MID) will render this technique even finer (Fig. 58b).

The use of labeled internal standards, such as 16,16,17d_3-testosterone for testosterone assays in human blood, has interesting possibilities. In this case, the steroids are transformed into methyloxime and TBDMS ether. Thus, less 1 ng of testosterone per ml (1 ppb) can be detected. Using GC/MS with metastable peak monitoring, a linear calibration curve has been obtained.

In the context of this method, we should stress the specific nature and the disappearance of all foreign peaks (low probability of encountering the same transitions). The recent work of Gaskell and co-workers [208b] demonstrates a detection threshold of 20–50 pg in mixtures of diastereomers of TBDMS ethers.

The development of MIKE/CAD or B/E/CAD linked scan methods will undoubtedly result in greater possibilities than those offered by methods of detecting metastable ions, especially at the level of sensitivity.

Maquestiau et al. [209] identified fractions in complex mixtures of free marine sterols with MIKE/CAD of molecular ions formed in EI. The method was tested on six samples from corals and sponges. For example, the EI conventional mass spectrum of the fraction isolated from *Eunicella stricta* is especially complex: high-mass peaks are located at m/z 428, 426, 414, 412, 400, 398, 396, 386, 384 and 382. They may correspond to various molecular ions. As stated above, it is possible that lower mass peaks are partially due to eliminations from heavier ions (m/z 400 $\xrightarrow{-H_2O}$ m/z 382, for example), and thus when analyzing the MIKE/CAD spectrum of the m/z 382 ion the method may eventually lose its specificity to some extent. Nevertheless, considering that veritable fingerprints are obtained, it is necessary merely to seek several characteristic peaks and to compare their relative abundance. If their ratio remains constant, one may affirm the presence of a given sterol (it is true that the GC/MS/MS method used by Millington, Gaskell and co-workers [206a–208b] is an alternative for circumventing this difficulty).

As stated above, the analysis of certain of these peaks in MIKE/CAD spectra is very useful for determining the structure of molecules [209], especially for demonstrating the presence of unsaturations, of cyclopropenes on side chains, etc. (Table 18).

In practice, the situation is not as simple as that, since MIKE/CAD spectra have a low resolution, leading to a few difficulties: when isomers (or fragment) ions are present (or compounds with new structures); and when performing certain identifications. Nevertheless, the objective can be achieved, as shown by this review [209].

Thus, in the spectrum of a sample isolated from *Alcyonum digitatum*, molecular ions

TABLE 18

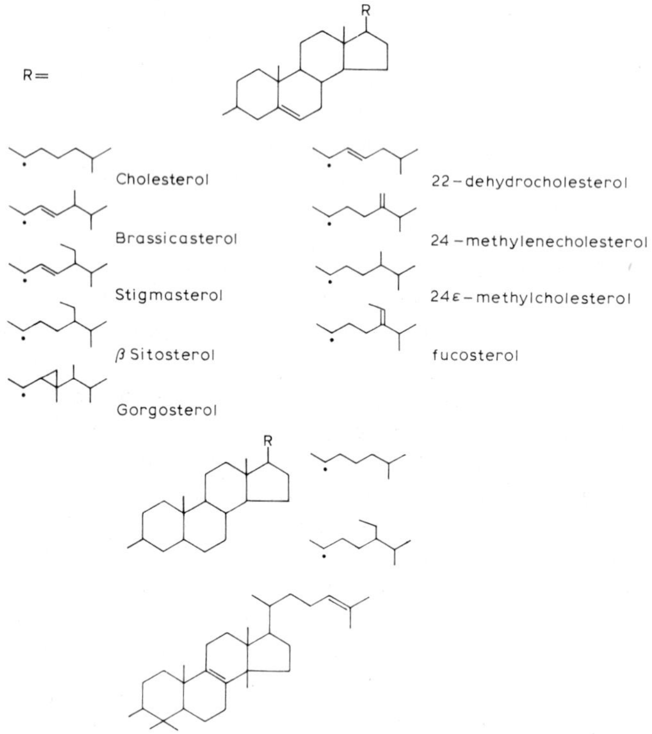

R=

R

Cholesterol

22-dehydrocholesterol

Brassicasterol

24-methylenecholesterol

Stigmasterol

24ε-methylcholesterol

β Sitosterol

fucosterol

Gorgosterol

R

with the same m/z are encountered: at m/z 398, $M^{+\cdot}$ of brassicasterol and $[M-H_2O]^{+\cdot}$ produced by β-sitosterol ($M^{+\cdot}$, m/z 414) are both present.

The study of six different extract mixtures has been performed. There is no ambiguity concerning the structure of the various compounds in each mixture, but the same study conducted uniquely with GC/MS cannot furnish the same specificity for the identification of these marine sterols.

When MIKE/CAD is applied to the study of molecular ions produced by EI, it is possible to detect them specifically and with a very low detection threshold. The limitations of this methodology are greatly reduced when CI is used, since molecular ions (protonated or not) which form may be very abundant.

Cooks and co-workers [210] identified each individual protonated molecule MH^+ from various steroids contained in biological matrices by studying their MIKE/CAD spectra. Collisions are necessary in this ionization mode in order to increase the number of characteristic daughter ions. In certain cases, however, spontaneous decompositions are sufficiently numerous, especially in the first FFR.

The study of MIKE/CAD spectra of protonated molecules of various compounds characterized by a Δ^4-3-keto system (such as testosterone, corticosterone, norgesterel, etc.) leads to the demonstration of peaks with non-negligible intensities at m/z 147, 135, 123, 109 and 97, with ratios that are very close. Certain fragment ions, including the $[MH-CH_3OH]^+$ ion, produced from hydroxycortisone (with the same Δ^4-3-keto sequence) also generate the same characteristic ions, and thus a certain amount of caution is required.

It is possible to distinguish these different compounds by measuring the abundances of ions formed by the elimination of small molecules (H_2O, CH_3OH, $2H_2O$, etc.) which are variable as a function of their precursors.

Protonated molecules from steroids with a phenolic A ring may be characterized similarly by the presence of intense peaks at m/z 157, 135 and 107, as well as $[MH-H_2O]^+$ in their MIKE/CAD spectra.

Protonated isomeric molecules, such as testosterone and dehydroepiandrosterone, are characterized in the MIKE/CAD spectra by peak widths, but also by intensities, which vary as a function of collision gas pressure.

Mixtures of the latter isomers which are not normally encountered in the urine of women, except in cases of ovarian tumors, could be studied. This was done by demonstrating uniquely dehydroepiandrosterone by the study of the spontaneous loss of water which is not observed in practice during the unimolecular decomposition of testosterone.

It is possible to determine the presence of testosterone – 100 times lower than the dehydro compounds – in 2 μl of urine, i.e., 100 pg of free testosterone mixed with 100 ng of the dehydroepiandrosterone can be detected. These assays were made possible by the use of a calibration curve [210].

Dunholke et al. [211] studied collision-induced decompositions of MH^+ ions produced by CI with B/E linked scan mode. It was shown that these ions, in spite of collisions in the first FFR, generated only a limited number of fragment ions, which nonetheless permitted their identification. In the same report, the authors similarly sought the presence of certain prostaglandins (prostaglandin-2) which under the same conditions generate a larger number of fragment ions. Quantitative determinations can also be performed without great difficulty, thus offering new and highly specific possibilties for analzying drugs and modifications.

(b) Analysis of peptide compounds

The technique of MS/MS is also very promising for the elucidation of the structures of peptides. A large number of examples shows the various applications of this technique, according to a very practical methodology which inevitably leads to the identification of the species studied, the component amino acid residues and the peptide sequence, even in a mixture.

The presence of metastable peaks in conventional spectra has occasionally been utilized. The ions detected are those which are produced just before the magnetic field in conventional single or conventional double focusing (as *E–B* configuration) instruments.

Sun and Lovins [212] studied the elimination of neutral fragments, obtained from amino acids liberated during the Edman degradation, which are transformed into derivatives of methyl- (or phenyl-) thiohydantoin (Fig. 59).

The data in Table 19 show that, for example, it is possible to distinguish leucine from isoleucine by their methylthiohydantoin derivatives (Fig. 60), as a result of the respective losses of $C_3H_7^{\cdot}$ and $C_2H_5^{\cdot}$ from their molecular ions.

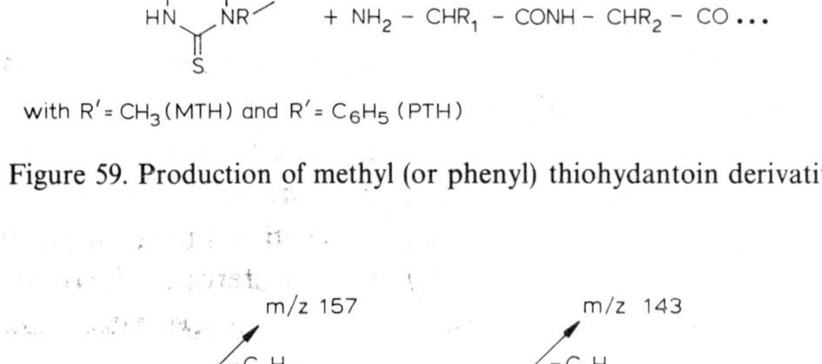

R'–N=C=S + NH$_2$ – CHR – CO – NH – CHR$_1$ – CO – NH – CHR$_2$ – CO ...

R' – NH – C(=S) – NHCHR – CO|NH – CHR$_1$ – CO – NH – CHR$_2$ – CO ... (H$^+$)

(thiohydantoin: R, O, HN, NR', S) + NH$_2$ – CHR$_1$ – CONH – CHR$_2$ – CO ...

with R' = CH$_3$ (MTH) and R' = C$_6$H$_5$ (PTH)

Figure 59. Production of methyl (or phenyl) thiohydantoin derivatives from polypeptides.

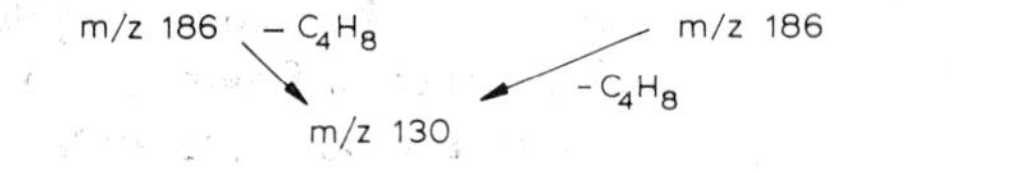

ISOLEUCINE (MTH)$^{+\cdot}$ m/z 186 → –C$_2$H$_5$ → m/z 157; – C$_4$H$_8$ → m/z 130

LEUCINE (MTH)$^{+\cdot}$ m/z 186 → –C$_3$H$_7$ → m/z 143; – C$_4$H$_8$ → m/z 130

Figure 60. Characteristic metastable decompositions of isoleucine and leucine methyl hydantoins.

TABLE 19
Metastable transitions observed for amino acid derivatives

Amino acid	Methylthiohydantoin derivatives			Phenylthiohydantoin derivatives		
	Parent	Daughter	Losses	Parent	Daughter	Losses
Alanine	144	116	CO	207	192	CH_3
Asparagine	187	170	NH_3	249	232	NH_3
	170	142	CO	232	204	CO
	187	143	$CONH_2$	249	205	$CONH_2$
				93	66	HCN
Glutamine	201	184	NH_3	263	246	NH_3
	184	142	C_3H_6	204	203	H
	156	56	C_3H_2ONS	246	204	C_3H_6
				135	77	NCS
				77	51	C_2H_2
Glycine	130	102	CO	192	191	H
				192	164	CO
				164	137	HCN
				135	77	NCS
				77	51	C_2H_2
Isoleucine	186	130	C_4H_8			
	186	157	C_2H_5			
Leucine	186	143	C_3H_7	248	192	C_4H_8
	186	130	C_4H_8	248	205	C_3H_7
				248	77	$C_7H_{11}ON_2S$
				135	77	NCS
				77	51	C_2H_2
Methionine	204	156	CH_3SH	266	192	$CH_3SCH{=}CH_2$
	204	130	C_3H_6S	266	205	CH_2SCH_2
				192	191	H
				266	218	CH_3SH
Phenylalanine	220	91	$C_4H_5ON_2S$	282	91	$C_9H_7ON_2S$
	91	65	C_2H_2	191	136	CH:NCO
				282	250	S
Proline	170	142	CO	69	68	H
				232	135	C_5H_7ON
				232	231	H
Threonine	286	157	$C_4H_5ON_2S$	236	192	CH_3CHO
	286	130	$C_6H_8ON_2S$			
	156	56	C_3H_2ONS			
Tryptophan	259	130	$C_4H_5ON_2S$	130	103	HCN
	130	103	HCN	321	130	$C_9H_7ON_2S$
	103	77	C_2H_2			
Valine	172	130	C_3H_6	234	205	C_2H_5
				234	192	C_3H_6
				234	233	H
				192	191	H
				192	120	CH_2NHCS

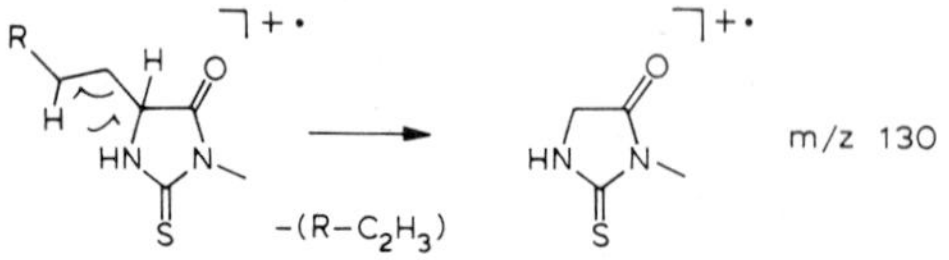

Figure 61. Formation of *m/z* 130 ion from methylthiohydantoin derivatives under EI conditions [212].

The peak at *m/z* 130, which is often encountered, corresponds to the elimination of the side chain of the derivatized amino acid (Fig. 61).

When using conventional geometry instruments, HV scan method combined with high resolution may occasionally be effective. In this way, Das and co-workers [213a,b] studied the components of pithamycolide fractions of extracts from *Pithomyces chartarum* cultures. The elucidation of structure of dipeptide has been done by this method. The measurements of the main peaks obtained in EI under high-resolution conditions furnished the results shown in Table 20, and these enabled a true 'family tree' to be generated when combined with the determination of their precursor ions (by HV scan) (Table 21).

Considering the knowledge of the elemental composition of the molecular ion $M^{+\cdot}$, as well as the other data presented here, the authors [213a,b] were able to propose the following structure for this type of peptide: [cyclo(L-*N*-methyl-alanyl-L-valyl-D-3-oxo-3-phenyl-propionyl-L-2-oxo-3-methyl-butyryl)] (Fig. 62).

TABLE 20
High resolution of abundant peaks produced in EI mass spectra [213]

Accurate masses (*m/z*)	Corresponding ions	δ (ppm)
552, 2477	$C_{30}H_{36}N_2O_8^{+\cdot}$	<1
465, 2153	$C_{27}H_{31}NO_6^{+\cdot}$	<1
332, 1499	$C_{18}H_{22}NO_5^+$	<1
316, 1544	$C_{18}H_{22}NO_4^+$	2
231, 1013	$C_{14}H_{15}O_3^+$	3.5
131, 0494	$C_9H_7O^+$	2

TABLE 21
Precursor ions of main fragment ions observed in the EI mass spectrum of depsipeptide (HV scan method) [213]

Ion (*m/z*)	Neutral loss	Precursors (*m/z*)
465	alanine	552, 510
332	phenyl propanoid	552, 465, 424, 404, 376
316	phenyl propanoid	552, 465, 404, 388, 361, 332
231	*N*-methyl alanine	380, 316, 288, 259
131	*N*-valeric acid	285, 231, 203

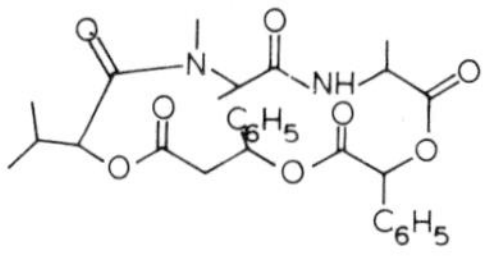

Figure 62. Proposed structure for the cyclic depsipeptide isolated in *Pithomyces chartarum*.

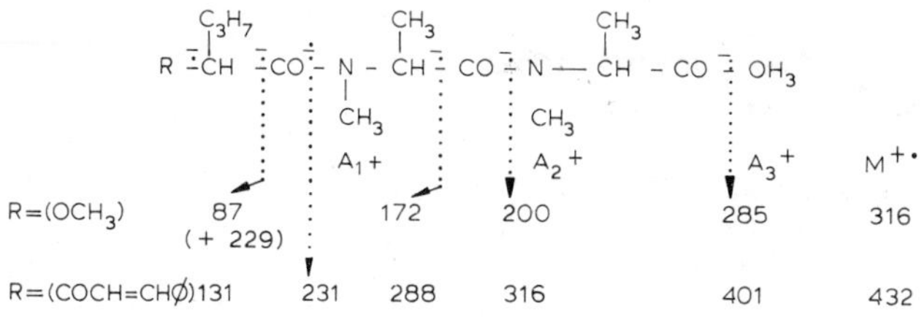

Figure 63. Fragment ions produced under EI conditions.

In addition to this peptide, more volatile peptides with molecular weights of m/z 316 and 432 could be detected using their characteristic fragmentations (Fig. 63).

The presence of another cyclic depsipeptide, a *Pithomyces maydicus* metabolite, could be proven with an analogous methodology [213a]. This technique could be improved by the use of either MIKE or B/E linked scan techniques. The former has been very useful for determining peptide* structures, as shown by Schlunegger and co-workers [214a,b] in several reviews. At the beginning of these studies, only the unimolecular decompositions of molecular ions produced in the second FFR were considered (Fig. 64).

The relatively low abundance of the A_i–CO]$^+$ ions is to be noted. This ion must be produced by the consecutive decompositions of the $M^{+\cdot}$ and $A_i]^+$ ions in the second FFR.

In addition, only few ions of the A_i^+ type are produced during the decomposition of this hexapeptide, thus limiting the direct study to $M^{+\cdot}$, the A_1^+, A_2^+ and A_3^+ ions not being observed.

An alternative may be proposed in this case, consisting of: determining the highest mass fragment produced by decomposition of $M^{+\cdot}$; setting the instrument on this daughter ion produced in the source, performing the MIKE spectrum and screening for the high-mass fragment ion which is produced directly; by successive repetitions of the above, the peptide sequence can be determined under EI conditions.

This method is quite useful when distinguishing isomeric peptides, i.e., with the same amino acids but in a different sequence (Fig. 65). a, AcN-Val-Ala-Leu–COOMe; b, AcN-Leu-Ala-Val–COOMe.

---

* One notes the $A_i]^+$ ion, produced by the cleavage of the peptide bond and which retains the charge on the carbonyl (–C≡O$^+$). The A_i–CO]$^+$ ions could be produced by elimination of CO from $A_i]^+$ or by direct cleavage.

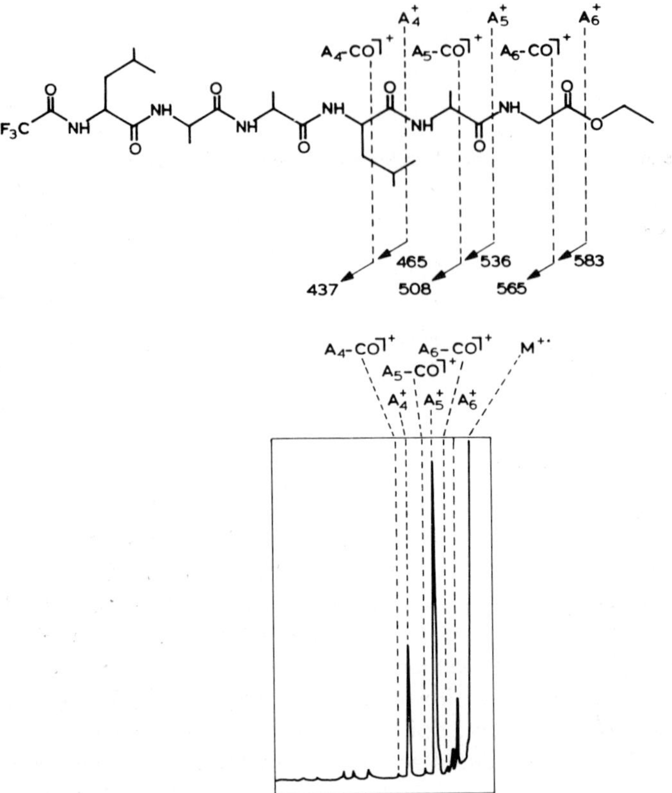

Figure 64. Unimolecular decompositions (MIKE mode) of $M^{+\cdot}$ ion and formation of various A_i^+ and $[A_i\text{–}CO]^+$ ions from $M^{+\cdot}$ [214a,b].

The above tripeptides, differing in the terminal residues (Val and Leu), thus exhibit different transitions during the decomposition of the $[M\text{–}COOCH_3]^{\cdot}$ (m/z 298) ion (Fig. 65).

It can be seen that the first ionic species leads to the m/z 213 ion (loss of the Leu fragment), while the second eliminates Val, to generate the m/z 227 ion (Fig. 65).

In the case of a mixture of two to three oligopeptides, the sequence of each can be determined [215] with this methodology.

Ronje and Grutzmacher [216] studied systematically various polypeptide sequences using MIKE spectra from derivative (labeled or not) peptides and mixtures under EI and CI conditions.

The same investigation can be performed with B/E linked scan spectra. In these spectra the same metastable transitions are encountered; however, a variation of ion abundances is observed [214a].

These detection techniques (especially collision-induced) are very sensitive; ions are occasionally produced, formed by elimination of low-mass neutrals, and which are difficult to explain with classical mechanisms.

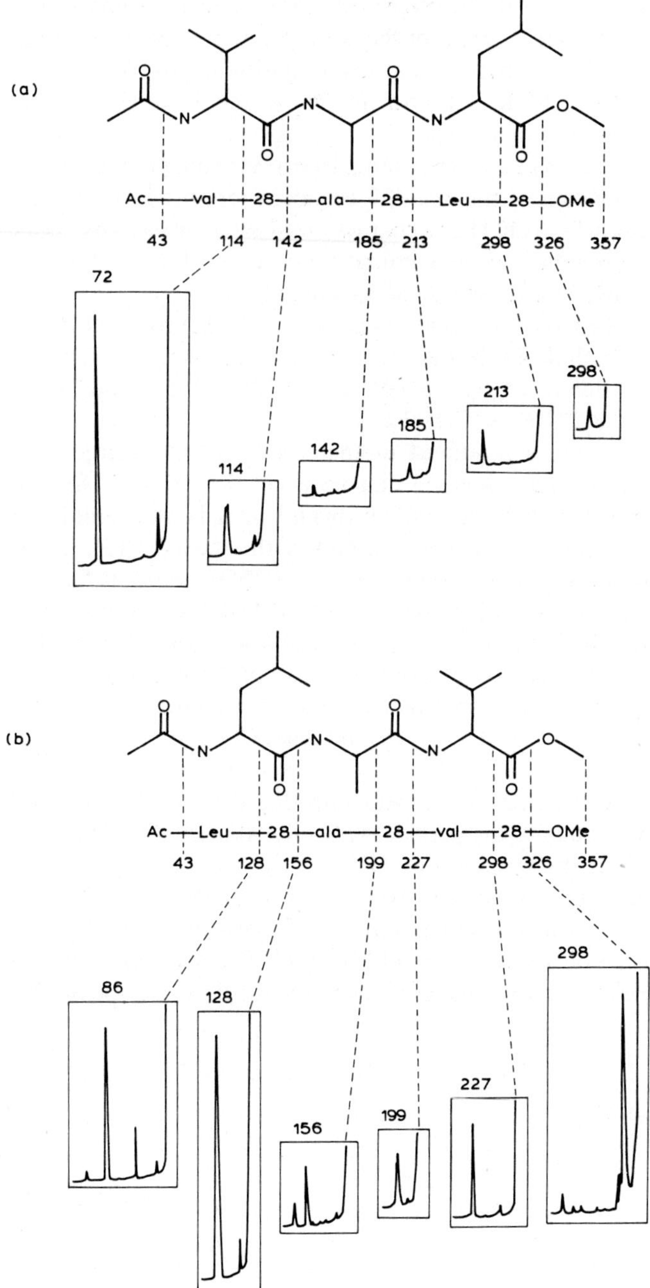

Figure 65. (a) MIKE spectra of corresponding fragment ions formed in the source for (a) AcN-Val-Ala-Leu–COOMe and (b) AcN-Leu-Ala-Val–COOMe isomeric peptides.

In particular, when the abundance of the precursor is low, 'artifact' signals may appear which are difficult to attribute. Certain of them, however, may be interpreted, especially those which may be due to unimolecular decompositions produced: in the first FFR, for signals obtained in MIKE spectra; in the second FFR when these artifacts are present in B/E linked scan spectra.

Schlunnegger et al. [214a] cited an interesting example corresponding to the second case. The B/E linked scan spectrum of the m/z 227 fragment ion, produced by the decomposition of AcN-Leu-Ala-Leu–OCH_3, is characterized by a very intense peak at m/z 188 (base peak), corresponding to the elimination of a 39 u fragment, which is impossible when the structure of this peptide is considered.

When this ion characterized by the 'apparent mass' of m/z 188 is produced in the first FFR, it will emerge from both fields (in the case of reversed geometry instrument) [214c]: for a magnetic field value of B^*_m, i.e., at $m^* = (188^2/227) = 155.7$; and for an electric sector value of $E = (188/227)E_0 = 0.828\ E_0$.

Furthermore, if an m/z 156 ion is produced in the source it will emerge at a magnetic field value close to (or even equal to) that necessary for the m/z 188 ion, produced from m/z 227 in the first FFR, especially if this m^* peak is wide; and if the m/z 156 ion decomposes, eliminating a molecule of CO in the second FFR, it will produce the m/z 128 ion, which will emerge only if $E = (128/156)E_0 = 0.820\ E_0$.

Thus, these two decompositions (first FFR, hypothetical metastable transition, m/z 277 $\xrightarrow{-39}$ m/z 188; the second FFR, m/z 156 $\xrightarrow{-28}$ m/z 128) will require close electric and magnetic field values in order to emerge and only the most probable transition, i.e., in this case the loss of 28 u from the m/z 156 fragment, must be retained [217]. Thus, among the fragmentations, we note that the m/z 156 ion loses CO in the second FFR (Fig. 66).

The use of collision-induced decompositions was introduced by McLafferty and co-workers in 1974 [218] in a study of the peptide AcN-Gly-Ala-Leu–OCH_3 and its fragment ions produced by electron impact (Fig. 67).

Thus, there is no difficulty in determining the sequence of the AcN-Gly-Ala-Leu–OCH_3 peptide. In a more thorough report of this work [218], the authors presented all the aspects offered by the use of MIKE/CAD spectra, including the study of a mixture of oligopeptides with sufficient sensitivity; the decreased possibility

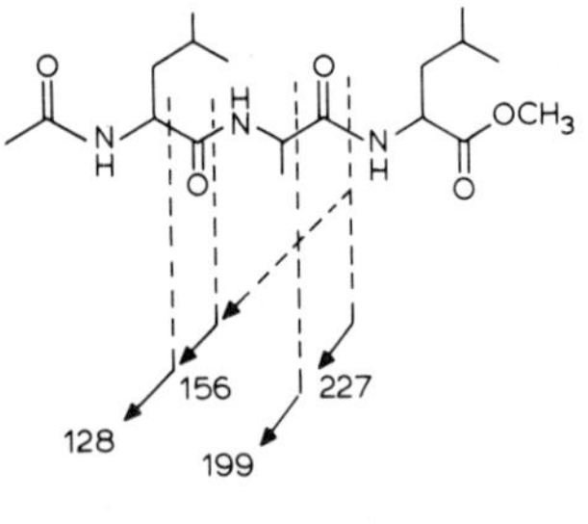

Figure 66. Decompositions in the first and second FFR which give rise eventually to formation of artefact peaks in the B/E spectrum.

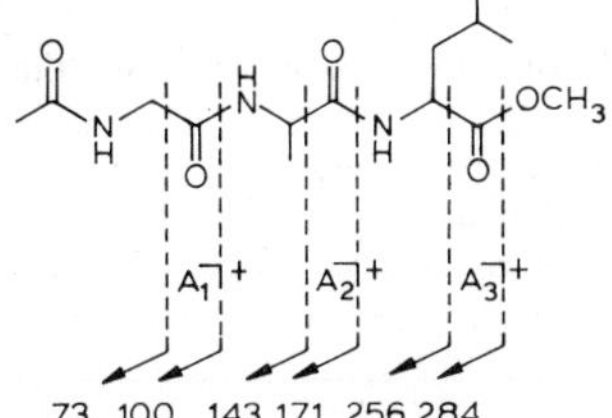

Fig. 67. Collision-induced dissociations of NAc-Gly-Ala-Leu–OCH_3 in the second FFR.

of molecular isomerizations; the creation of a larger number of daughter ions from simple cleavage.

The major disadvantage of MIKE/CAD spectra, however, is related to the widening of metastable peaks, which reduces energy resolution.

The isomeric and labeled Ac-Ile-Leu–OCD_3 and Ac-Leu-Ile–OCD_3, lead to practically identical conventional EI spectra (with the exception of several low intensity peaks, e.g., *m/z* 61 and 69). The most abundant ions are *m/z* 103, 145 (base peak) and 173, with the following structures: *m/z* 103, $CD_3\overset{+}{N}H{=}CH(C_4H_9)$; *m/z* 145, $CH_3CO\overset{+}{N}(CD_3){=}CH(C_4H_9)$; *m/z* 173, $CH_3CON(CD_3){-}CH(C_4H_9){-}CO^+$.

All three include the $C_4H_9^{\cdot}$ radical and can decompose spontaneously (or collisionally) to eliminate either the primary ($-CH_2-CH(CH_3)_2$) or secondary ($-CH(CH_3)-CH_2-CH_3$) isobutyl radical. Consequently, several differences are observed in the MIKE/CAD spectra. However, the greatest differences appear in the decomposition spectra of *m/z* 103 ion (Table 22), which is consistently present when the polypeptide has an Ile or Leu residue in terminal position.

The *m/z* 69 and 61 ions are among the fragment ions characterized by an elevated abundance sensitive to the terminal Ile or Leu structure. These ions are produced by competitive mechanisms (Fig. 68) and lead to the distinction between Leu and Ile derivatives.

These mechanisms are favored to such an extent that the conventional mass spectra of Leu and Ile compounds are characterized by, respectively, the intense metastable peaks at *m/z* 36.1 (103 → 61) and *m/z* 46.3 (103 → 69)*.

Furthermore, the intensity of the peak at *m/z* 103 in the conventional mass spectra is sensitive to the N-terminal (for Leu or Ile) or C-terminal (for Ile or Leu) position. In the latter case, the abundance of this ion is attenuated considerably.

If amino acid isomers can be distinguished, the operation remains possible when differentiating isomeric fragment ions (*m/z* 303: Leu-Leu]$^+$ and Ile-Ile]$^+$), produced for example in the decomposition of polypeptides such as Leu-Leu-Leu and Ile-Ile-Ile.

The unimolecular decomposition spectra of these isomeric *m/z* 303 ions present, in

*The apparent mass corresponds to (m_2^2/m_1) for the transition $m_1^+ - m_2^+ + n$.

TABLE 22
MIKE and MIKE/CAD spectra of ion ($CD_3\overset{+}{N}H{=}CH(C_4H_9)$) m/z 103 produced from peptides where the Leu (or Ileu) group is present in various positions [218]

	LeuGlyGly	GlyGlyLeu	LeuIle	IleLeu
Precursor ions	100[a]	8[a]	77[a]	72[a]
Unimolecular metastable spectra				
75	0.6	2	0.3	7
69	2	4	4	58
61	66	60	68	8
47	31	33	27	10
35	0.4	0.9	0.6	15
32			0.1	2
Collisional activation spectra				
87	13	9	14	13
75	1		1	11
74	4	6	5	33
73	10	15	9	31
69	6	1	6	46
61	109	106	108	18
59	13	13	13	5
47	20	19	22	20
45	11	14	11	10
43	26	23	27	5
41	13	11	12	12
39	4	2	4	5
35	1	2	1	7
32	4	2	3	3
29	6	8	6	6

[a]Abundance of the m/z 103 ion in the conventional mass spectra.

one case, m/z 274 and 173 and, in the other, m/z 260 and 173. This is sufficient for the identification of the fragment ions Ile-Ile]$^+$ and Leu-Leu]$^+$.

The presence of the abundant common ion m/z 173 enables the N-terminal Leu (or Ile) to be demonstrated. The presence of m/z 260 and 274, corresponding to the

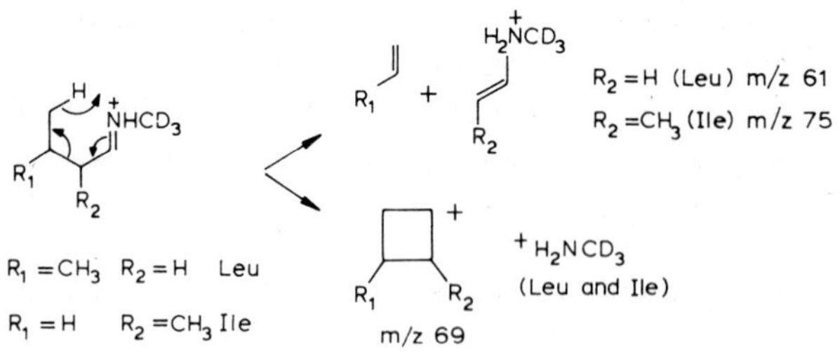

R_2 = H (Leu) m/z 61
R_2 = CH_3 (Ile) m/z 75

R_1 = CH_3 R_2 = H Leu
R_1 = H R_2 = CH_3 Ile

m/z 69

$^+H_2NCD_3$ (Leu and Ile)

Figure 68. Origins of m/z 69 and 61 fragment ions which are formed by rearrangement mechanisms in the FFR [218].

eliminations of the propyl and ethyl radicals, leads to the conclusion that Leu is present in the first case and Ile in the second case.

In the case of the isomeric dipeptides Leu-Ile and Ile-Leu themselves, the above sequences are again encountered during the spontaneous decompositions of the $M^{+\cdot}$ (*m/z* 337) ion, i.e., in addition to the *m/z* 173 ions, *m/z* 308 $[M–C_2H_5]^+$ and *m/z* 294 $[M–C_3H_7]^+$ are present. The differences between the latter two remain minor, however, so that there is no ambiguity.

The fragmentation which is determinant for the identification of Leu or Ile corresponds to that leading to the *m/z* 206 ion, which is 20 times more abundant in the MIKE spectrum of the $[\text{Leu-Ile}]^+$ ion than in that of $[\text{Ile-Leu}]^+$. This is due to the secondary or tertiary character of the hydrogen atom, which migrates during the elimination of the neutral fragment (Fig. 69).

In other words, the formation of the *m/z* 206 ion will be favored when R_2 is a methyl group, which is the case when Ile is terminal.

These examples of this methodology show the usefulness of MS/MS, especially in the case of mixtures: isomeric linkages can be identified; peptide sequences (during either spontaneous or collisional-induced decompositions) can be determined.

One of the major drawbacks of this methodology is related to the electron impact ionization technique itself. It often leads to very low intensity molecular ions, which eliminates any degree of sensitivity with MIKE. Nevertheless, the EI method may be interesting when studying fragment ions ($[A_i]^+$ or $[A_i–CO]^+$) which, in contrast, are highly abundant.

The use of softer ionization methods, such as positive and negative CI (during electron capture), provides additional and complementary data.

Hunt et al. [30a] utilized this mixed mode for the first time: PPNICI (pulsed positive negative ion chemical ionization mass spectrometry) with quadrupolar instruments for the study of peptide sequences. The reagent gas chosen was a mixture of CH_4 and CH_3ONO. The resulting plasma was composed of CH_5^+ and CH_3O^- (among others) and led to the positive $[MH]^+$ and negative $[M–H]^-$ ions, respectively.

Under these conditions, the PPNICI spectrum of the permethylated tetrapeptide AcN-Met-Gly-Met-Met could be obtained in the same scan (Fig. 70).

Conventional spectra are characterized by the presence of the fragment $[A_i]^+$, $[A_i]^-$ and $Z_jH_2]^+$ ions, produced as shown in Figure 71 [219].

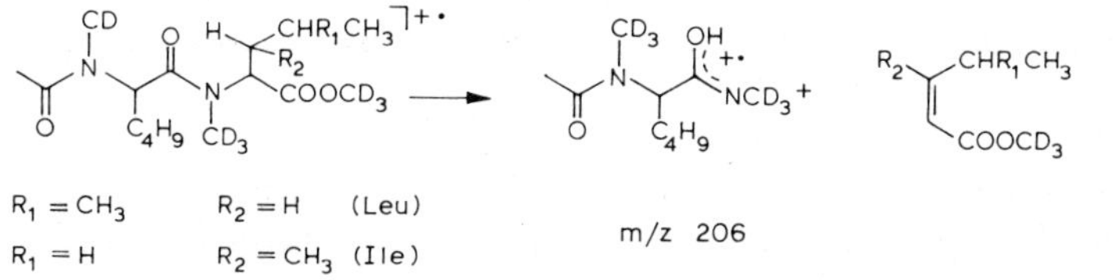

Figure 69. H migration leading to production of *m/z* 206 ion independently to the R_1 and R_2 terminal groups.

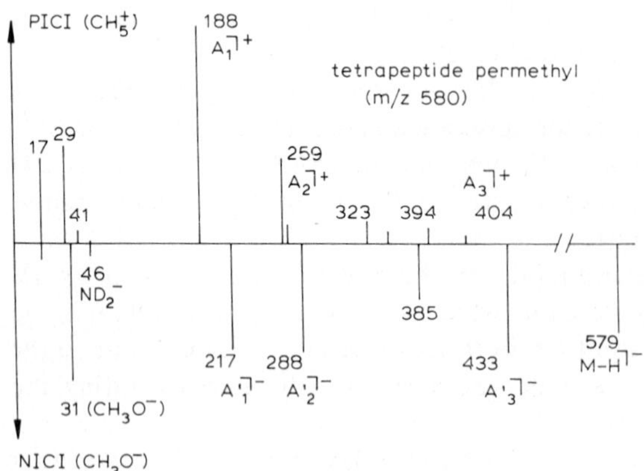

Figure 70. PPNICI spectrum of permethylated tetrapeptide/AcN-Met-Gly-Met-Met- [30a].

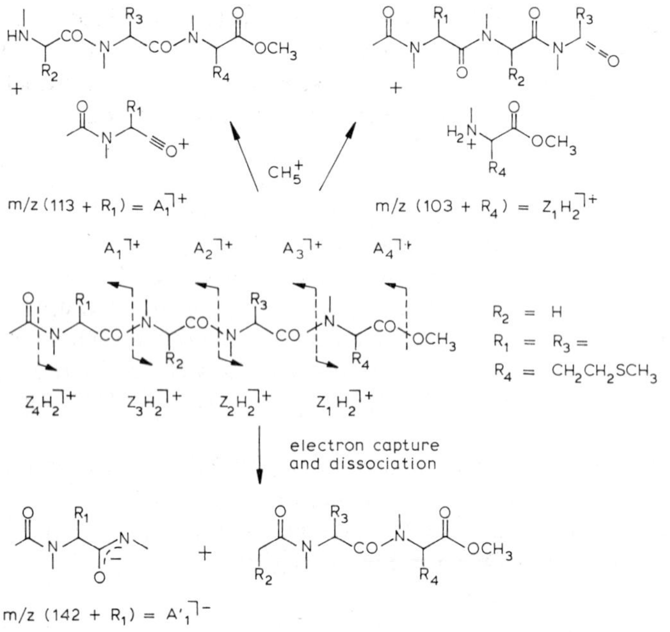

Figure 71. Origins and structures of $A_i]^+$, $A'_i]^-$ and $Z_jH_2]^+$ produced by decomposition of tetrapeptide AcN-Met-Gly-Met-Met.

The use of the negative ion mode led to a 10–1000-fold increase of peak intensities. Additional data were thus obtained, leading to the identification of the peptide sequence with no difficulty.

The same authors showed [220] that sequences of apolipoprotein B (ApoB) of low-density lipoprotein (LDL) and of sickle-cell hemoglobin (HbS) could be determined under PICI conditions by studying two groups of ions, such as $A_i]^+$ and $Z_j]^+$, after performing several chemical modifications. It was thus possible to identify the N-terminal sequences.

These determinations were first attempted in PICI/CAD [221] and the polypeptides were modified as follows: 1, $Ac_2O(d_6/d_0)$; 2, $Na^+CH_2^-SOCH_3$; 3, MeI; 4, H_2O, $CHCL_3$.

The N-terminal residue thus contained the $-COCD_3$ (and $-COCH_3$) substituent and the other residues were permethylated (d_3/d_0-*N*-acetyl-*N*-*O*-permethyl polypeptide).

Under positive CI conditions with isobutane (or methane), these derivatives generate conventional spectra characterized by doublets separated by 2 amu and by isolated peaks.

The first signals correspond to the fragment ions $\left[A\left(\frac{d_0}{d_3}\right)_i\right]^+$ which retain the N-terminal residue, while the second set of singlet peaks is formed by the protonated $Z_jH_2]^+$ ions, which possess the C-terminal residues. Thus, each of these fragment ions* could be analyzed in CAD mode with a triple quadrupole instrument (Fig. 72).

For instance, the $A_{d_03}]^+$ and $A_{d_33}]^+$ doublets can decompose to form either the $\left[A\left(\frac{d_0}{d_3}\right)_2\right]^+$ ions via direct cleavage or the $A_{2,3}]^+$ and $A_{2,2}]^+$ ions during hydrogen transfer followed by peptide bond cleavage (Fig. 73), and the $Z_jH_2]^+$ singlets can yield the different $Z_{j-1}H_2]^+$ ions and the corresponding ammonium derivatives (Fig. 74).

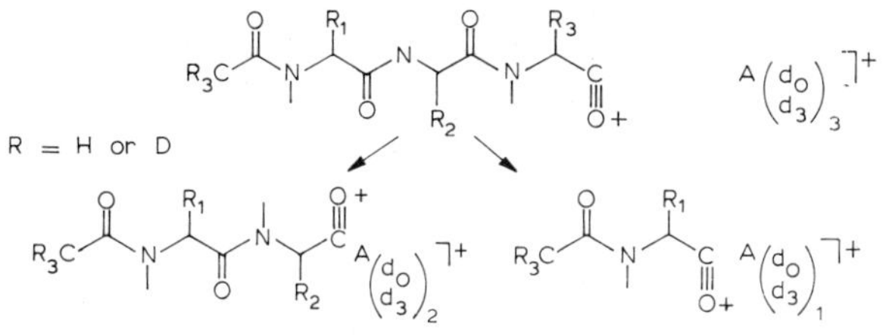

Figure 72. Formation of $A_{\left(\substack{d_0\\d_3}\right)^2}]^+$ and $A_{\left(\substack{d_0\\d_3}\right)^1}]^+$ ions by decomposition of fragment $[A_{\left(\substack{d_0\\d_3}\right)^3}]^+$ ions.

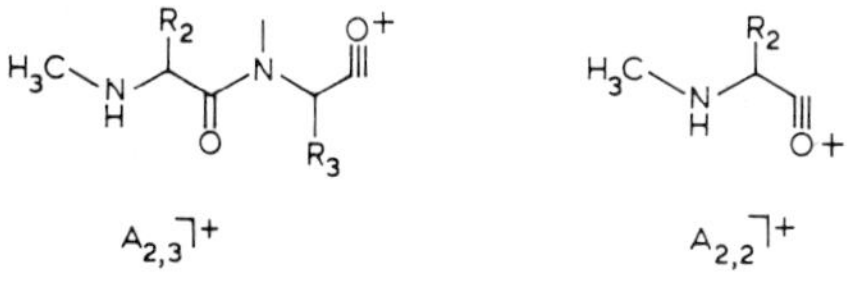

Figure 73. Structure of $A_{2,3}]^+$ and $A_{2,2}]^+$ ions.

---

*The ion chosen can be characterized by the $[A_{d_3i}]^+$ (or $[A_{d_0i}]^+$) or $[Z_jH_2]^+$ structure.

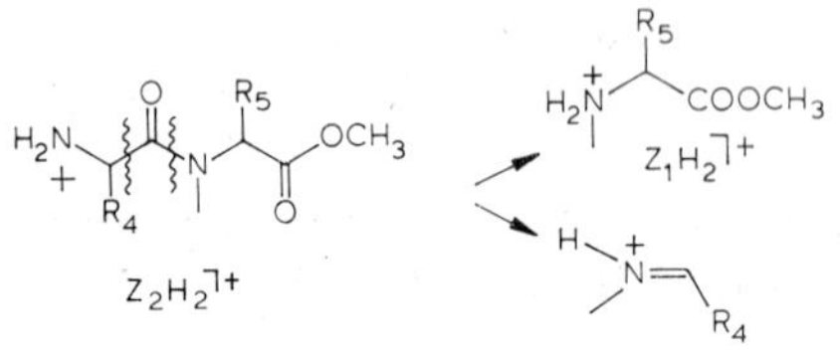

Figure 74. Formation of $[Z_1H_2]^+$ and imminium ions from $[Z_2H_2]^+$ precursor ion (in the case of derivatized pentapeptide).

In order to be efficient, the determination of the sequences of neuropeptides [221] requires a high sensitivity as a result of the small quantities of samples analyzed (on the pmol level).

After dividing the sample in two parts, the following methodology is employed: 1, incubate one half with dipeptidylaminopeptidase I and IV (DAP I/IV) to form a mixture of dipeptides; 2, perform an Edman degradation of the second half to separate N-terminal amino acids, followed by a DAP I/IV digestion, leading to the creation of a second set of dipeptides.

These dipeptides can be hydrolyzed into amino acids which are acetylated with acetic anhydride d_0 and which are added to a standard mixture of d_0–d_6 acetylated amino acids. The d_0/d_6 ratio determined from doublets thus provides the amino acid composition.

The dipeptides can also be transformed into d_3,d_0-1-1-acyl-dipeptides (with a mixture of d_0 and d_6 acetic anhydride (1:1) in methanol). Free carboxyl groups are esterified into pentafluorobenzyl esters (with fluorobenzyl bromide in acetonitrile and

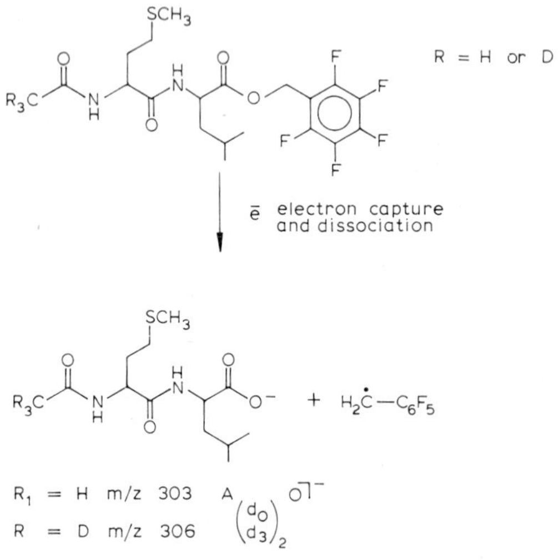

Figure 75. Formation of carboxylate anion $A_{\binom{d_0}{d_3}_2}O]^-$ by loss of radical functional group $[\dot{C}H_2\text{–}C_6F_5]$ under NICI conditions [221].

diisopropylethylamine). The advantage of introducing this type of functional group is both to increase the efficiency of electron capture and to induce characteristic fragmentations after having formed carboxylate ions immediately by dissociative electron capture and loss of $^{\cdot}CH_2C_6F_5$ (Fig. 75).

Collision-induced decompositions of the $RCOO^-$ ion characterize the dipeptide $A_{\binom{d_0}{d_3}_2}O^-$, leading to the identification of the C-terminal residue (Fig. 76).

If the sequence is inverted, i.e., Leu-Met–CH_2–C_6H_5, the collisionally obtained fragment ion would have been localized at *m/z* 148 $[H_2N–CH(C_3SH_7)–COO]^-$.

After analyzing all dipeptide spectra, the polypeptide sequence is identified not only unambiguously, but also with high sensitivity (with quantities close to 1 pmol for each dipeptide).

The importance of these NICI (negative ion chemical ionization) spectra of modified polypeptides led Howe and co-workers [222] to study the spectra of the free compounds. Thus, with OH^- as reagent ion with a mixture of three tripeptides, in addition to fragment ions, $[M–H]^-$ molecular ions are produced in sufficient abundance (Fig. 77) to be analyzed by collisions (MIKE/CAD spectrum).

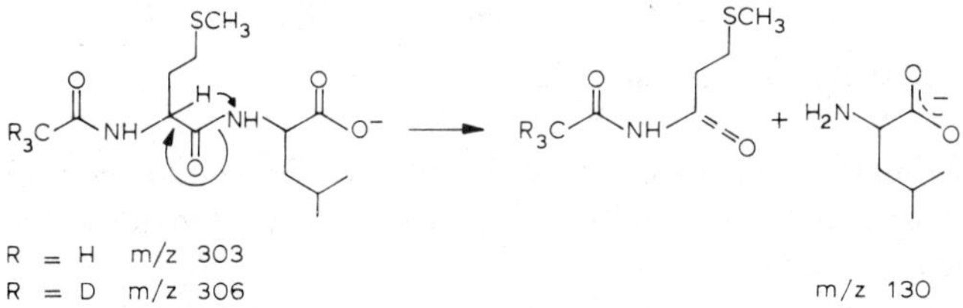

R = H m/z 303
R = D m/z 306
m/z 130

Figure 76. Determination of terminal acid group from carboxylate anion previously formed under NICI conditions [221].

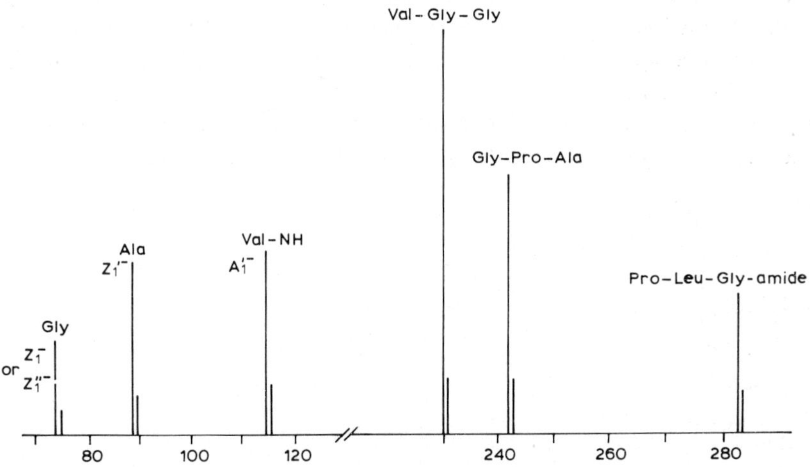

Figure 77. OH^-/NICI mass spectrum of tripeptide mixture [222].

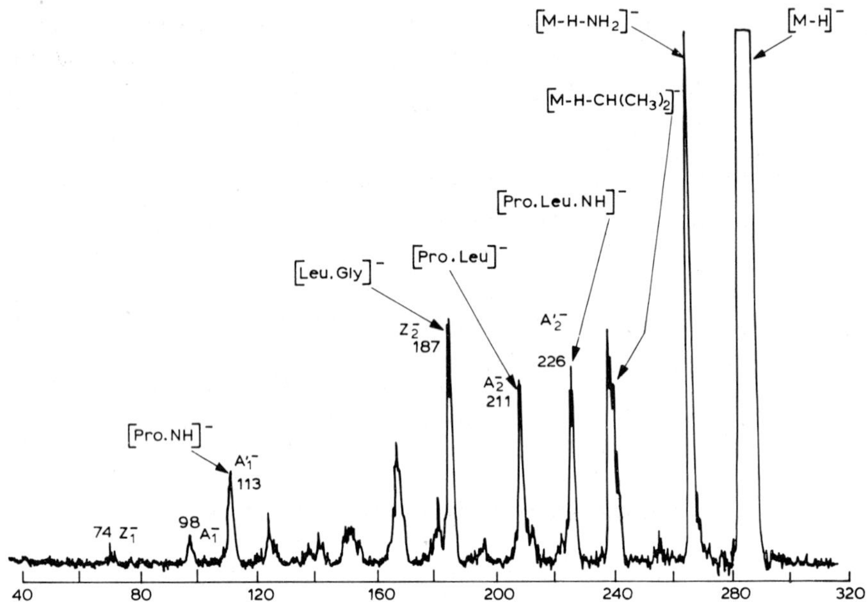

Figure 78. MIKE/CAD spectrum of $[M–H]^-$ (m/z 284) anion [222].

For example, the deprotonated molecule $[M–H]^-$ of the Pro-Leu-Gly–OH tripeptide (m/z 284) decomposes collisionally with argon and provides the MIKE/CAD spectrum shown in Figure 78.

The majority of the ions thus formed can be interpreted and the entire 'puzzle' can be reconstructed, leading to the peptide sequence (Fig. 79).

These results are sufficiently encouraging for studying mixtures of non-derivatized peptides with the following advantages: normal increase of the abundance of the deprotonated molecule $[M–H]^-$, which affects sensitivity directly; no useless increase of the molecular weights of peptides, which had an unfavorable effect on the resolution of MIKE spectra.

Howe and co-workers [222] also recommended the use of laser radiation as an alternative to gas collisions, thus increasing the sensitivity of the method.

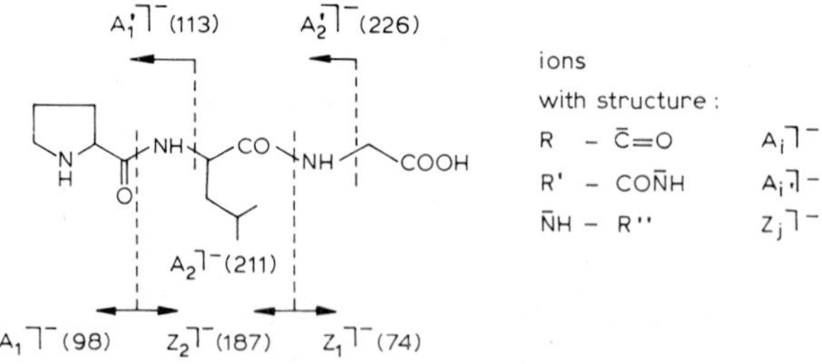

Figure 79. Main A_i^- and Z_j^- ions observed in the MIKE/CAD spectrum of $[M–H]^-$ deprotonated peptide.

Other soft ionization methods, such as field ionization, may also be chosen. Levsen and Beckey [197a] initially attempted to use decomposition-inducing collisions on $M^{+\cdot}$ molecular ions produced with low internal energy* (thus highly abundant) produced by field ionization.

Collisionally activated decompositions were very close, concerning the ions produced, to those produced at 70 eV under EI.

The conventional nature of these fragmentations leads to an unambiguous interpretation** of the FI/CAD spectrum (Fig. 80) of the permethylated and *N*-acetylated tripeptide.

The *m/z* 100 ion (11%), which is also present in the MIKE/CAD spectrum, is an immonium ion: $(CH_3)\overset{+}{N}H{=}CH(C_4H_9)$, previously encountered.

In the case of compounds which are very labile and difficult to volatilize, field desorption becomes an effective alternative for obtaining highly abundant molecular ions, thus sufficient sensitivity***.

When using field desorption, however, it is occasionally difficult to distinguish the fragment ions from impurities and the products of various reactions (surface reactions, pyrolysis). The study of uni- and bimolecular decomposition spectra is very useful, since it enables one to go beyond the simple identification of artifacts to the elucidation of molecular structures.

In addition to reversed geometry instruments generating MIKE [197b] spectra, the use of conventional configuration mass spectrometers, such as the MS 902 (Kratos), is sufficiently flexible to enable one to obtain first FFR decomposition spectra with *B/E* linked scan. This is an alternative to compensate for the poor resolution of MIKE spectra and also enables polypeptides with higher molecular weights to be studied. Burlingame's group [223] utilized the mixed EI/FD source of the MS 902 (where the voltage applied to the extraction electrodes may be +8.2 and −4.8 kV).

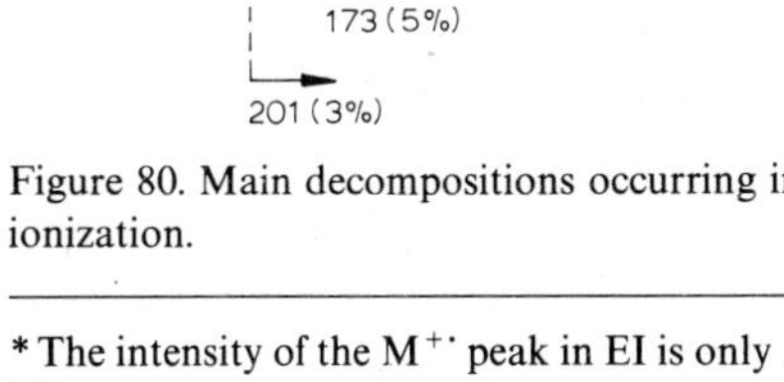

Figure 80. Main decompositions occurring in the second FFR from molecular ion species formed in field ionization.

---

* The intensity of the $M^{+\cdot}$ peak in EI is only 1% that of the base peak, whereas it becomes the base peak in FI.

** The elimination of the leucine side-chain is often encountered. Here, it corresponds to 69% of total ionization (*m/z* 286).

***Now, the Fast Atom Bombardment (FAB) desorption technique associated with the MIKE/CAD technique is preferred, since a constant ion current is obtained.

TABLE 23
Comparison of fragments lost from cyclopeptides a and b observed in the FD/*B*/*E*/CAD spectra [223]

Compound a, FD/CAD (*m*/*z* 573)		Compound b, FD/CAD (*m*/*z* 593)	
m/*z*	Eliminated fragments	*m*/*z*	Eliminated fragments
558	$^{\cdot}CH_3$ (A^+)	550	$^{\cdot}CH(Me)_2$ (C'^+)
529	$^{\cdot}N\ Me_2$ (B^+)	549	$^{\cdot}N\ (Me)_2$ (B'^+)
516	$^{\cdot}CH(Me)C_2H_5$ (C^+)	493	$^{\cdot}CH(NMe_2)CH(Me_2)$ (D'^+)
459	$^{\cdot}CH(NMe_2)$–$CH(Me)C_2H_5$ (D^+)	465	$^{\cdot}COCH(NMe_2)CH(Me_2)$ (F'^+)
443	$^{\cdot}CH_2$–3 indol (E^+)	463	$^{\cdot}CH_2$–3 indol (E'^+)
431	$^{\cdot}CO$–$CH(NMe_2)$–$CH(Me)C_2H_5$ (F^+)	452	$HN^{\cdot}COCH(NMe_2)CHMe_2$ (G'^+)

The EI source precedes the FD source and is used as a collision cell*. The gas has a major role in collisions: collision gas pressure required to attenuate the main beam by 2/3 varies as a function of the products, their molecular weights and the type of gas chosen.

The elucidation of the structure of cyclopeptidic indoles in a mixture could be determined by analyzing the *B*/*E*/CAD spectra of molecular ions. The FD spectra of this mixture indicated the presence of high-molecular-weight ions, including *m*/*z* 593 and 573. It was impossible to separate this mixture, even with liquid chromatography, as a result of the similar structures of these molecules and the mixture had to be studied as it was (about 1 μg of sample). Collisions on these ions led to side-chain cleavage, which led to an interesting approach to molecular structure. Table 23 lists the principal ions observed in *B*/*E*/CAD spectra from the $M^{+\cdot}$ molecular ions produced in FD conditions.

The major fragmentations are represented in Figure 81 for both a and b cyclopeptides.

Baillie et al. [224] presented another example of the characterization of thioethers from acetaminophene, metabolically produced when high doses of analgesics (such as paracetamol) were administered.

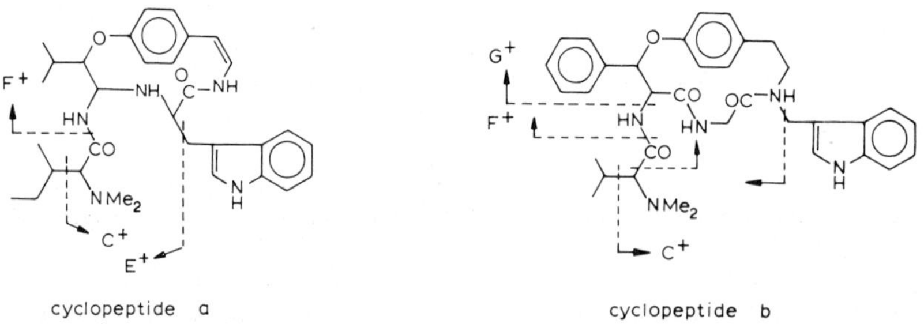

cyclopeptide a

cyclopeptide b

Figure 81. Structure and fragmentation of cyclopeptides a and b [223].

* It is to be borne in mind that it is not always necessary to perform MIKE/CAD spectra, since occasionally spontaneous decompositions are sufficient.

The use of FD/CAD with B/E linked scan mode enabled 200 ng of a compound to be detected, as well as the identification of its structure, although the molecular peak was surrounded by various fragment ions and cationized ions such as the $[M + Na]^+$ ions.

Before concluding this section on peptide sequencing, several additional comments on the contribution of MS/MS to the analysis of amino acids are in order, even though the literature contains relatively few examples of this interesting possibility.

McReynolds and Anbar [225] reported the analysis and assay of ^{15}N-labeled metabolites in very small quantities, which was not possible with conventional methods. The data obtained also enabled ^{15}N to be localized within a molecule. Field ionization was chosen for this study, and led to mass spectra in which the protonated molecule was the base peak in the case of free amino acids, such as glycine, alanine, valine and leucine.

The loss of 46 u, yielding the $[MH\text{–}COOH_2]^+$ ion, i.e., $R\text{–}CH\text{=}\overset{+}{N}H_2$, was the only elimination present characterizing the FI spectra.

In addition, in conditions of collisions of the $[MH]^+$ ion MH, the same loss of HCOOH was encountered, corresponding to the base peak.

In addition to these interesting results, it should be noted that the differentiation of isomers, such as leucine and isoleucine, is also possible (in the same way as in EI/CAD).

These MIKE/CAD spectra are in general characterized by the presence of an m/z 18 ion, which may have originated from either $H_2O^{+\cdot}$ or NH_4^+. The presence of a non-negligible peak at m/z 19 in the MIKE/CAD spectra of the isotopic ion $[MH + 1]^+$ (composed of ^2H, ^{13}C or ^{15}N) indicates that a non-negligible fraction of the m/z 18 ion was due to the presence of NH_4^+.

Calibration curves may be used to estimate the degree of ^{15}N enrichment. The results are sufficiently satisfying to be utilized as an assay method. Nevertheless, for about 0.15 N atoms introduced per molecule, error of 10% remains.

This method can be effective and precise if precursor ions and fragment ions are present in large amounts and are well resolved. Furthermore, if the effect of collisions is to broaden the signal, increasing the ion acceleration voltage could improve resolution, and thus increase the precision of the measurements.

The use of negative ions produced by the electron capture ionization may constitute another alternative for increasing sensitivity. Stapleton and Bowie [226] applied this method to the study of the structures of the nitrobenzoyl derivatives of various natural amino acids. All the species studied led to clear-cut and interpretable fragmentations, enabling the structure of the amino acid in question to be determined.

The mass spectra of the *o*-nitrobenzoyl derivatives may be interpreted according to classical mechanisms (as *ortho* effects). In the case of *m*-nitro compounds, however, only the molecular anion is present in addition to the m/z 46 (NO_2^-) ion. The solution to this problem involves rescanning in positive mode in order to increase the number of fragments*. This operation is not easy and it is more advantageous to measure

---

* To perform these experiments a conventional geometry instrument was used.

the $+E$ spectrum of molecular anion decomposition by simple inversion of the electric fields. Collisions in the first FFR also allow the charge inversion transition: $[M]^{-\cdot} \rightarrow [M]^{+\cdot}$, and thus only the $[M]^{+\cdot}$ ion can pass through the $+E$ field, since there is no loss of kinetic energy. This ion can then partially decompose in the second FFR and the fragment ions can be analyzed by the variation of the magnetic field. The amino acid structure can thus be studied with the $+E$ spectrum obtained.

This method can also be applied to dipeptides. We believe that it can also be used for their identification, even in mixtures. In this context, Cooks and co-workers [227] successfully demonstrated the presence of hippuric acid in certain urine samples. Negative chemical ionization was chosen in order to obtain [M–H] ions, which are intense in the case of acids.

If collision reactions (MIKE/CAD spectra) on this ion are insufficient for the generation of fragment ions, it is then preferable to perform these reactions on the $[M\text{–}H]^+$ ions formed during the reversed charge of $[M\text{–}H]^-$.

This technique renders it possible to demonstrate a small quantity of hippuric acid in as little as 1 μl of a complex mixture such as urine.

Another method can be used, such as study of precursor ion spectra (B^2/E linked scan or HV scan spectra). It may be quite useful in investigations of amino acid mixtures.

Biogenic amines were identified in a mixture following chemical modification to the dansyl derivative (5-(dimethylamino)-1-naphthalenesulfonyl-).

Addeo et al. [228] showed that these compounds have a strong tendency to lead to a very intense fragmentation under electron impact conditions: $(CH_3)_2\text{–}N\text{–}C_{10}H_6SO_2\text{–}N(R_1R_2)]^{+\cdot} \rightarrow (CH_3)_2N\text{–}C_{10}H_7]^+$ *m/z* 171.

The *m/z* 171 ion is consistently present in all the conventional mass spectra of derivatives and so it is sufficient to search its different precursors (as molecular ions) with the HV scan method. The mixture obtained is divided in two parts: one highly volatile, the other relatively involatile.

Each component of the mixture can be analyzed by the metastable spectra without previous separation. Thus the different amino acids (or simpler amines) can be determined. Under EI conditions, the determination of high-molecular-weight precursors of certain fragment ions formed in the first FFR enabled Klein [229] to use the same method to demonstrate the molecular ion at ± 1 u of different 1,2-diacyl glycerylphosphatidylcholines, derived from distearoyl (*m/z* 786.6), 1-stearoyl-2-oleoyl (*m/z* 787.6), etc.

(c) Analysis of polysaccharide and antibiotic compounds

The conventional mass spectra obtained from saccharides transformed into their OTMS and $-OCH_3$ derivatives were interpreted by using labeled derivatives. Systematic studies had never been performed on –OAc derivatives. However, in 1975, Das and Thayumanavan [230] studied various peracetylated disaccharides by their IKE spectra to resolve the ambiguity appearing in the EI mass spectra of the isomeric peracetates of trehalose, sophorose, kojibiose, laminaribiose, maltose, melibiose and gentiobiose, all characterized by the same molecular weights (678).

The difference among these sugars is the disaccharide linkage, which may be $1 \rightarrow 1$, $1 \rightarrow 2$, $1 \rightarrow 3$, $1 \rightarrow 4$ or $1 \rightarrow 6$.

The conventional mass spectra indicate no significant difference between melibiose (1→6) and gentiobiose (1→6), whose structures differ only by the stereochemistry of the disaccharide bond and of one OH group.

Table 24 indicates one possibility for differentiating the various types of glycoside linkages with conventional mass spectra. The relative intensities appear to be highly sensitive to these glycosidic linkages. However, it is difficult to perform any distinction without ambiguity as a result of mass spectrum reproducibility (Table 24). The measurement of their IKE spectra represents an alternative, since they indicate much more spectacular differences (Fig. 82).

TABLE 24

Ratios of characteristic fragment ions measured in the conventional EI mass spectra of various isomers [230]

Ratio of m/z	Sophorose, 1→2	Kojibiose, 1→2	Laminaribiose, 1→3	Maltose, 1→4
331/317	7.5	10.3	3.2	7.4
317/229	1.4	1.5	2.4	2.5
317/245	9.5	9.0	12.0	2.5

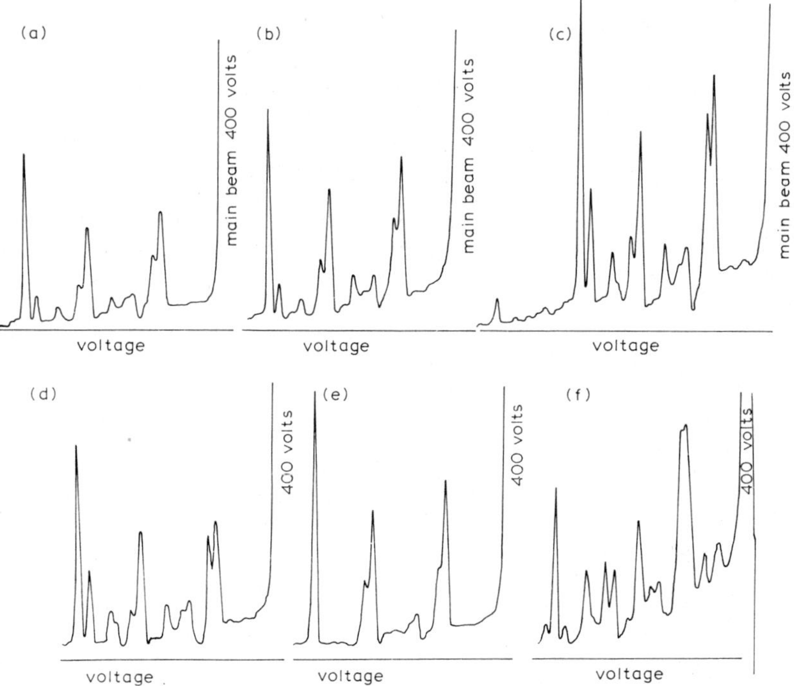

Figure 82. IKE spectra of (a) sophorose (1→2), (b) kojibiose (1→2), (c) maltose (1→4), (d) laminaribiose (1→3), (e) trehalose (1→1) and (f) gentiobiose (1→6) disaccharide isomers [230].

These spectra indeed correspond to veritable 'fingerprints', sensitive to stereochemical differences. Their disadvantage, however, is related above all to the ambiguity of the indicated transitions, since they represent mass ratios m_2/m_1, for which several solutions are possible.

Among the highest mass peaks, certain are due to the loss of acetic acid, and the consecutive decompositions of this $[M\text{–}CH_3COOH]^+$ ion are difficult to study. It may indeed be produced from any modified –OH group, and so this peak may represent a mixture of isomeric structures. The ability to study the behavior of the unique structure molecular ion itself may be more valuable and lead to a greater quantity of data. Chemical ionization, enabling these ions to be studied in their protonated (or not) form, may thus be an alternative to EI.

Milne and co-workers [231] utilized protonation with isobutane reagent and obtained spectra characterized by the presence of highly abundant protonated molecules from aminocyclitol-aminoglucoside antibiotics which were unmodified (Table 25). However, their pattern varies greatly with temperature.

In addition, the high-resolution analyses of EI spectra, as well as metastable spectra, enable one to describe the fragmentations of the low-intensity molecular ion and its fragment ions (Fig. 83).

The identification of methyl hexapyranoside stereomers [232] transformed into trifluoroacetates (TFA) was performed by analyzing the conventional mass spectra of

TABLE 25
Abundances of MH^+ in $CI/iC_uH_9^+$ mass spectra and experimental conditions [231]

	Source temperature (°C)	MH^+ (m/z)	Abundances
Gentamicine C_1	165	478	100
Gentamicine C_2	170	464	100
Gentamicine A	180	469	20
Gentamicine B	185	483	26
Gentamicine X_2	190	483	70

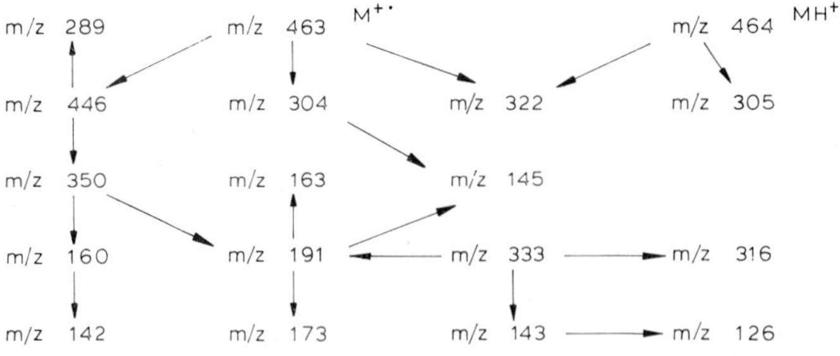

Figure 83. Relation between the main ions observed in the conventional CI mass spectrum using the metastable ions occurring in the first FFR [231].

ions formed in the source. Metastable transitions (first FFR) showed that the consecutive eliminations of the TFA group were competitive, thus leading to the distinction among glucose, galactose and mannose simply by measuring these various eliminations. The loss of TFA and the cleavage of the pyranoside ring are characteristic. It is also possible to differentiate α and β anomers by measuring the intensity of methoxyglycoside elimination.

These relatively old examples of using classic metastable ion spectra for determining molecular structure remain minor. Also, certain aspects, such as the study of mixtures, must be approached with different techniques, such as ionization, even the most recent, or metastable ion detection with B/E (or B^2/E) linked scan and MIKE methods.

Warburton et al. [233] identified the diastereomeric anhydro sugars shown in Figure 84 by comparing the unimolecular decomposition spectra of the $[M\text{–}OH^{\cdot}]^+$ (m/z 115) fragment ion (intense) in B/E linked scan spectra. In comparison to intrasource decomposition spectra – which are identical – the above method leads to characteristic spectra.

The variation of peak intensities at m/z 114 (loss of $H^{\cdot}$) and at m/z 72 (loss of $C_2H_3O^{\cdot}$) is sufficiently sensitive and reproducible to enable each of these isomers to be unambiguously distinguished.

Polysaccharides are often thermo-labile, even when soft ionization methods such as field desorption are utilized. Pyrolysis effects may occur, leading to a low abundant molecular ion, in spite of the slight energy excess. This generates a spectrum which occasionally is complex.

Under these conditions, the unimolecular decomposition spectrum of molecular ions with low internal energy is rich in fragment ions, thus reflecting the facility of their formation, favoring the characterization of the polysaccharides.

For instance, the FD spectrum of permethylated sucrose presents a series of ions at m/z 219, 235, 409 (base peak) and 454 (low intensity molecular peak) [223] (Fig. 85).

The unimolecular decompositions detected in B/E mode are characterized by a large number of consecutive and rapid cleavages in the first FFR, among which are $OCH_3^{\cdot}$ eliminations (Fig. 86).

It should be noted that the inversion of intensities of the m/z 351 peak (base peak in B/E linked scan) and the m/z 409 peak (base peak in the conventional spectrum) reflect the effect of internal energy on the competitive decompositions which are produced from molecular ions with more or less short lifetimes.

All the peaks of the conventional mass spectrum, however, cannot be interpreted. They are likely due to the decompositions of fragment ions, e.g., m/z 409, which lead to ions at

Figure 84. Structures of diastereomeric anhydro sugars.

FD
FD/MI
100
50
219
235
409
M+·
454
200
300
400
500
351
423

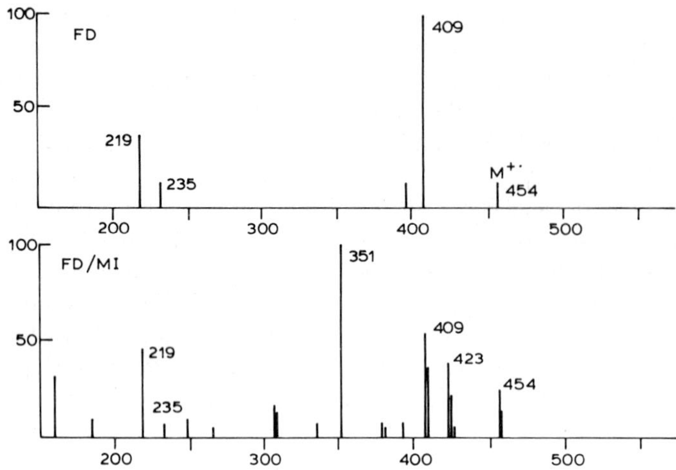

Figure 85. FD and *B*/*E*/FD spectra of molecular ion *m*/*z* 254 from sucrose.

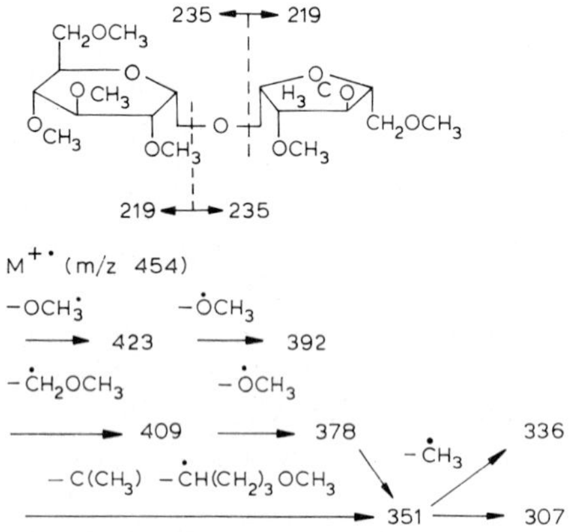

235
219
CH2OCH3
OCH3
CH2OCH3
OCH3
219
235
M+· (m/z 454)
−OCH3
423
392
−CH2OCH3
−OCH3
409
378
336
−CH3
−C(CH3) −CH(CH2)3OCH3
351
307

Figure 86. Unimolecular decompositions of $M^{+\cdot}$ (*m*/*z* 454) in the first FFR (*B*/*E* mode).

m/*z* 219, 378, 351 and 354, as shown by its characteristic unimolecular decomposition spectrum.

The introduction of collisions in the first FFR to analyze the *m*/*z* 409 ions for inducing additional decompositions does not furnish new information in comparison to the conventional spectrum.

This example shows that the use of bimolecular decompositions is not always necessary, especially if the parent ions produced are not stable. The latter generally lead to unimolecular decomposition spectra which contain a large number of ions.

The use of unimolecular decompositions detected in the first FFR with *B*/*E* linked scan may be the determinant for structural studies of high-molecular-weight complex compounds. It was thus shown that the antibiotic 4915.A (molecular weight, 857) [234],

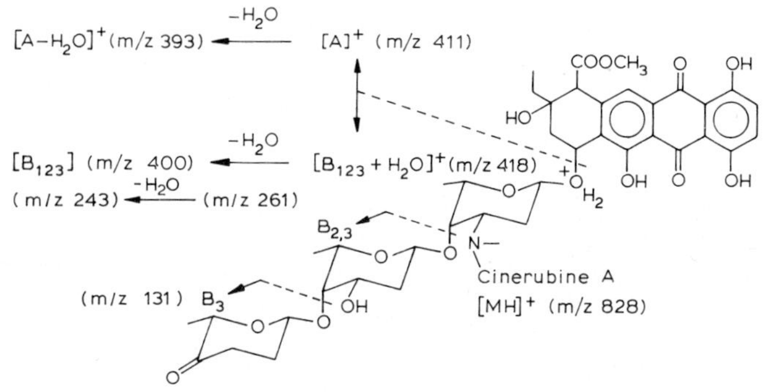

Figure 87. Structure and main decompositions of cinerubin A (m/z 828) under $CI/C_4H_9^+$ conditions.

belonging to the anthracyclin family, was composed of cinerubin A. The B/E spectrum of the protonated molecular ion MH^+ (m/z 828) produced by $CI/tC_4H_9^+$ showed the presence of the aglycone fragment (m/z 411 and 393 ions) and of the heteroside fragment (m/z 400 ion) (Fig. 87).

The systematic metastable study of the m/z 411 and 393 fragment ions, produced from the aglycone fragment, showed the presence of $-OH$, $-COOCH_3$, $-C_2H_5$ and $-CO$ substituents. An analogous study of the m/z 418 and 400 ions led to the construction of a family tree of all the ions, making it possible to determine the heteroside sequence.

It often occurs that collisional spectra may eliminate stereochemical ambiguities. As an example, the disaccharide linkage in a free molecule may be determined by analyzing B/E/CAD spectra of molecular ions produced by field desorption [235]. The disaccharides trehalose ($1 \rightarrow 1$), sophorose ($1 \rightarrow 2$), nigerose ($1 \rightarrow 3$), cellobiose ($1 \rightarrow 4$) and gentiobiose ($1 \rightarrow 6$) generate intense protonated molecular peaks, which enable the collisional-induced decomposition spectra of these ions to be studied with sufficient sensitivity.

These spectra are sensitive to disaccharide stereochemistry structure, and thus contain sufficient differences to result in an unambiguous identification. Table 26 shows all possible combinations.

TABLE 26

Ions selected in the B/E/CAD spectra to distinguish the heterosidic linkages of the various diastereomeric disaccharides

	1→1			
1→2	245/231	1→2		
1→3	289			
	259	289	1→3	
1→4	259	259	259	
				1→4
1→6	259	259	259	275
				289
				273

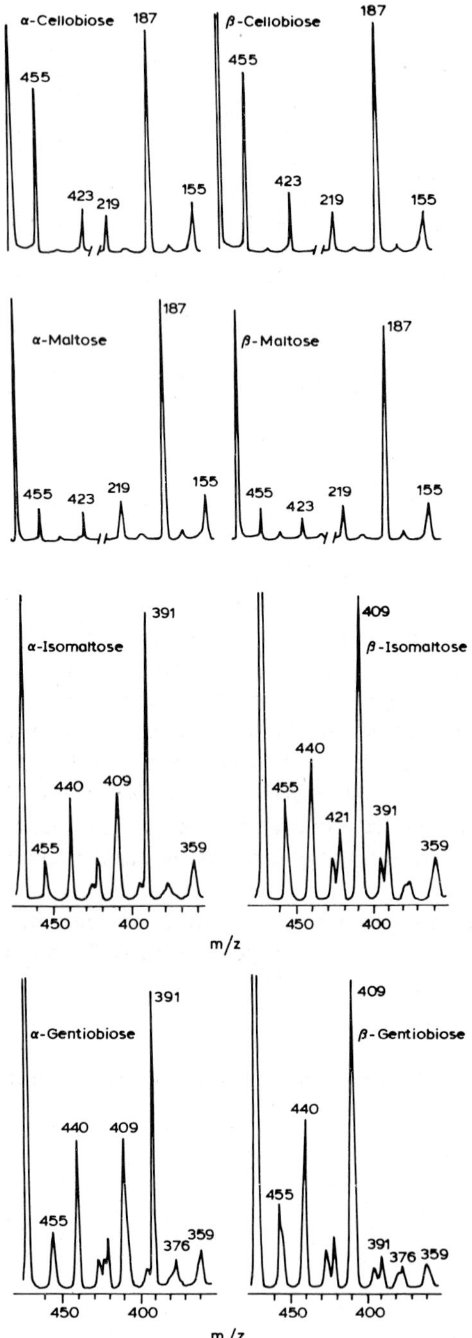

Figure 88. Partial MIKE/CAD spectra of $(M+NH_4)^+$ ions produced from α- and β-cellobiose, α- and β-maltose, α- and β-isomaltose, α- and β-gentiobiose (in CI/NH_4^+) [236].

In addition to distinguishing the disaccharide linkage bond in various stereomeric species, measurements of collisional-induced decompositions (as MIKE/CAD spectra) also enable one to differentiate the diastereomeric disaccharides themselves, providing that protonated molecular ions $[MH]^+$ are not used [236].

The use of ammonia as reagent gas in CI not only increases the abundance of MH^+ ions, but also leads to the formation of adduct $[M+NH_4]^+$ ions.

In contrast to protonated molecules, these $[M+NH_4]^+$ adduct ions enable one to distinguish anomeric disaccharides [236], e.g., α and β isomaltose, and α and β gentiobiose (Fig. 88).

However, the complete work by De Jong et al. [236] showed that in certain cases ambiguities were encountered, as for the anomers of cellobiose and maltose (Fig. 88).

In order to overcome these difficulties, the authors utilized an aqueous solution of 40% trimethylamine and studied the collisional spectra (MIKE/CAD) of the adduct $[M+(CH_3)_3NH]^+$ (*m/z* 514) ions. The results obtained were complementary to the preceding since, although the MIKE/CAD spectra of the $[M+NH_3]^+$ ions produced from a isomaltose and gentiobiose were similar, those obtained with the $M+NH(CH_3)_3]^+$ ions were sufficiently different. The differences between their anomers themselves, however, are no longer evident. Using this methodology, it is possible to differentiate all diastereomers by the analysis of collisional spectra of the adduct ions produced, with either NH_4^+ or $NH(CH_3)_3^+$ as reagent gas (Fig. 89). Other CI conditions are used to distinguish the epimeric isomaltose.

The firm VG Micromass Instruments [237] provided a comparison of the spectra obtained in MIKE and *B/E* linked scan modes with the $[M+NH_4]^+$ ions as main beam, produced by CI with NH_3 from trisaccharides (maltotriose peracetate: $MH]^+$ = 996 u and $M+NH_4]^+$ = 984 u).

The bond ruptures between each of the residues were present in the unimolecular decomposition spectrum of the $[M+NH_4]^+$ ion (*m/z* 984) and were augmented in collisional conditions. *B/E* linked scan spectra provided analogous results, with only quantitative differences in peak intensities.

Burlingame and co-workers [238] published a very interesting example of the fragmentations of a high-molecular-mass Man_xMeMan_y–OCH_3 type polysaccharide, isolated from a *Mycobacterium smegmatis* culture. These compounds play an important role in the regulation of fatty acid synthesis. Although the molecular peak corresponding to the $[M+Na]^+$ cationized molecule can be observed in the FD mass spectrum, decompositions due to the rupture of glycoside bonds occur as soon as the temperature is increased. They may be used to trace back to the structure of the molecule.

In the case of mixtures, molecular ions appear as a function of temperature, which does not simplify their detection. Nevertheless, the presence of doubly (or triply) charged molecular ions (as $[M+2Na]^{2+}$) enables the determination of the molecular species (cationized or protonated). In this case, even when the compounds are of high molecular weight, such as Man_1-Me-Man_9–OCH_3 ($[M+Na]^+$ ion, *m/z* 1801.8), it is possible to study collisional-induced decompositions in *B/E* linked scan. The limitations of this method are related above all to the loss of sensitivity, due to the necessity of using an

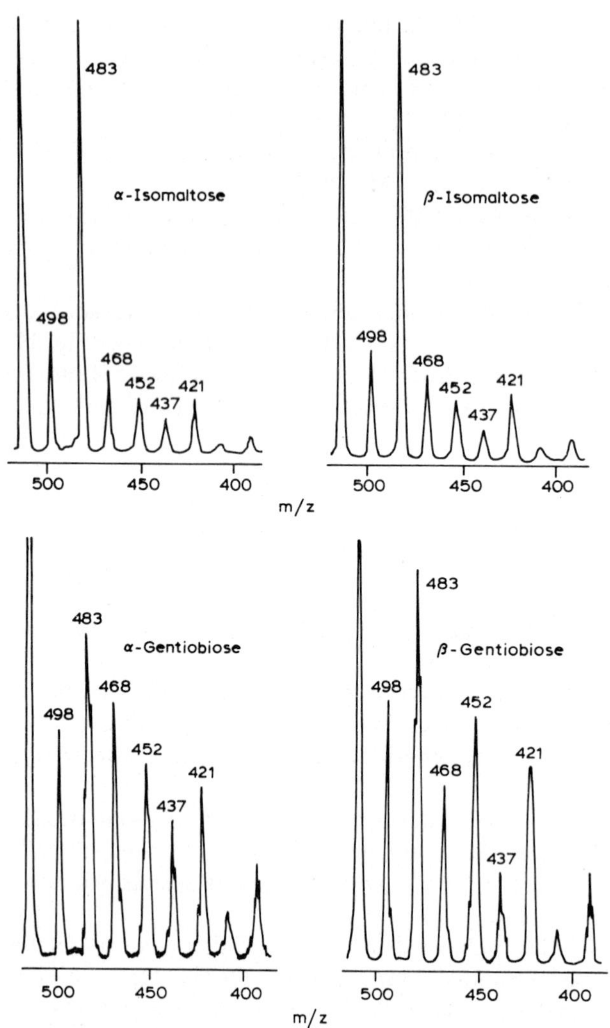

Figure 89. Partial MIKE/CAD spectra $[M + HN(CH_3)_3]^+$ ions produced from α- and β-isomaltose and α- and β-gentiobiose (under $CI/HN(CH_3)_3^+$ conditions) [236].

accelerating voltage of 4 kV*. Nevertheless, it is interesting since it requires no prior chemical modification and only glycoside bonds are broken, thus enabling one to obtain the sequences.

Using FD, it is thus possible to study not only the $M^{+\cdot}$ and MH^+ ions, but also complex ions of the $[M + \text{cation}]^+$ type. Puzo et al. [239] used the presence of these intense peaks to determine the molecular weights of oligosaccharides, although the low abundance of fragment ions always renders the determination of their structures

*Now, higher ion acceleration is used in the new mass spectrometers (8–10 kV).

difficult. Therefore, these authors employed collisional spectra (MIKE/CAD) of 'cationized' molecular ions. They observed that the cation chosen had a strong influence on the decompositions of the [M + cation]$^+$ ion, as well as on its abundance itself. They presented, for example, the analysis of natural mixtures of acetylated trehalones.

Quantitative analyses are possible by the use of SIM (single ion monitoring). Negative ions produced by CI generally provide an increased sensitivity. In particular, the reagent mixture (CH_2Cl_2 + isobutane) easily yields the [M + Cl]$^-$ ions by attachment, which are highly abundant.

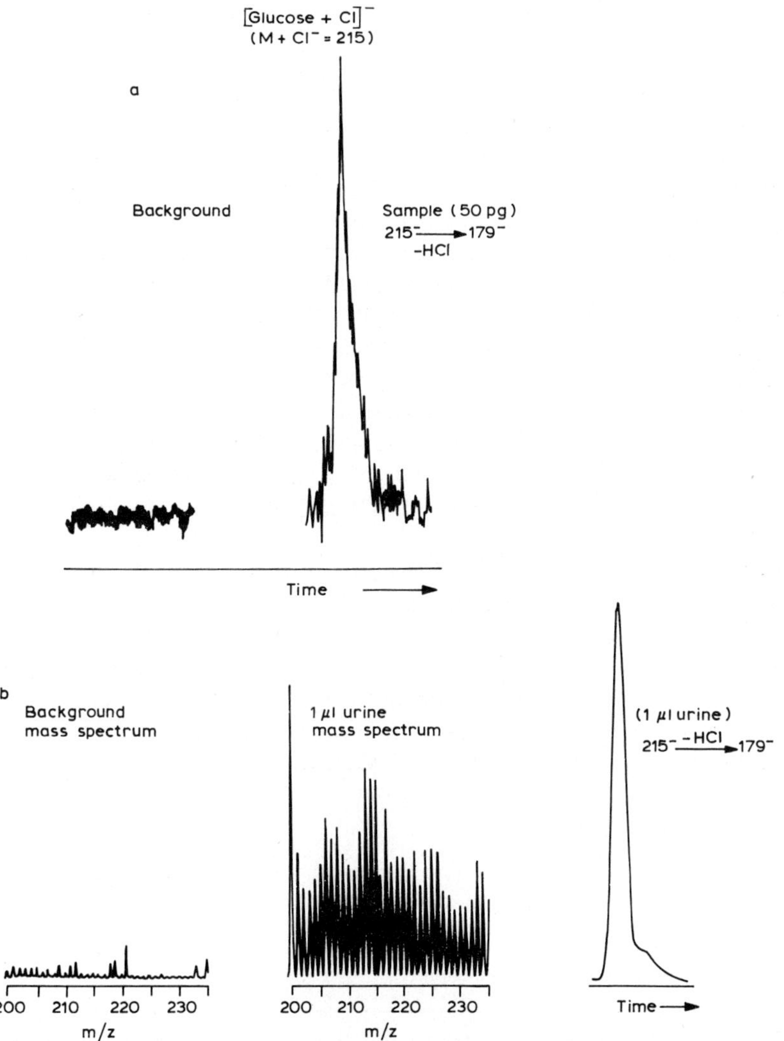

Figure 90. (a) MIKE/CAD/SIM detection of 50 pg of glucose and (b) detection of 100 ng of glucose in 1 µl of urine [240].

These adduct ions decompose preferentially under collisional conditions, to yield a molecule of HCl [240].

In the case of glucose (Fig. 90), this $[M + Cl]^- \rightarrow [M + Cl\text{–}HCl]^-$ transition can be demonstrated in SIM with a sensitivity of 50 pg of glucose and a signal/noise ratio of the order of 10.

The detection of glucose in 1 μl of urine is improbable in conventional spectra. In SIM, however, the $215^- \rightarrow 179^-$ transition is sufficiently intense as to enable the presence of 100 ng of glucose to be estimated.

(d) Analysis of heterocycles and alkaloids

Certain structures of labile alkaloids may often be modified as a result of the various problems of extraction and purification of these plant products.

These complex mixtures may be studied by MS/MS, utilizing the spontaneous (or collision-induced) decompositions in the field free regions. This type of study may be pursued at both the structural and assay levels, without the need for transformation of considerable prior separations of the components of the mixture.

This type of simplification should render chimotaxonomy more accessible, since plant material may be used intact or simply extracted.

Cooks and co-workers [241] analyzed a mixture of volatile phenolics obtained from *Dolichothele uberiformis* by Soxhlet extraction. Ionization method was with EI (or CI).

The conventional EI spectrum of this mixture is characterized by the presence of (a) m/z 165 and m/z 151 ions corresponding to known structures as *N*-methyltyramine and hordenine which are confirmed by MIKE/CAD analysis and (b) m/z 193 and m/z 225 ions of unknown origin. Those at m/z 165 and 151 (Fig. 91) could be identified as *N*-methyltyramine and hordenine by the use of MIKE/CAD spectra. In addition, the m/z 193 and 225 ions are also present.

Their structure could also be defined by studying their respective metastable fragmentations. It should be noted that the m/z 193 ion eliminates a 43-u fragment (in addition to losses of 15 and 17 u), while the m/z 225 ion eliminates a fragment of 58 u (in addition to those of 15 and 28 u). These results can be interpreted, on one hand, according to a 'retro Diels-Alder reaction'* (Fig. 92), and, on the other hand, according to a benzylic cleavage on a non-heterocyclic molecular ion (Fig. 93). Both these fragmentations are used to distinguish the skeletons.

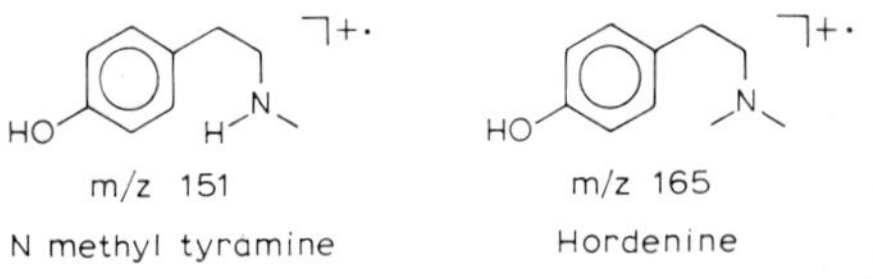

m/z 151 — N methyl tyramine

m/z 165 — Hordenine

Figure 91. Structure of *N*-methyltyramine and hordenine.

---

*It is to be noted that this elimination may be ambiguous, with the consecutive loss of CH_3 and CO (characteristic of methyl-aryl ethers). The intense loss of $H^{\cdot}$, as shown by the spectra of tetrahydroquinoline, confirms the proposed structure.

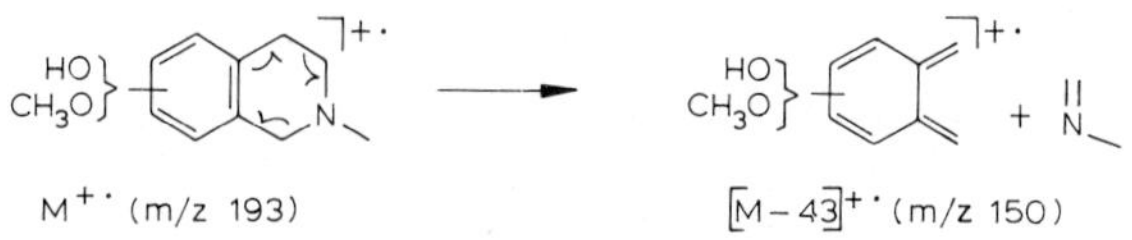

$M^{+\cdot}$ (m/z 193) → $[M-43]^{+\cdot}$ (m/z 150)

Figure 92. Diels-Alder retrogression of isoquinoline derivative.

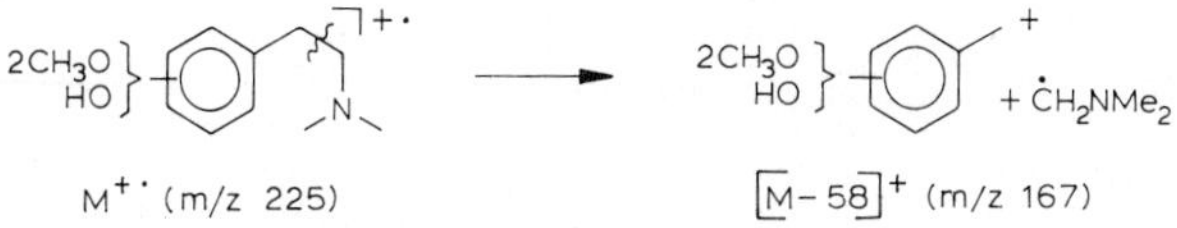

$M^{+\cdot}$ (m/z 225) → $[M-58]^{+}$ (m/z 167) + $\dot{C}H_2NMe_2$

Figure 93. Benzylic cleavage of molecular ion $M^{+\cdot}$.

Chemical ionization with isobutane may be chosen for the analysis of a mixture from another source: *Dolichothele longimamma.* This spectrum is characterized by MH^+ ions, especially an intense peak at *m/z* 166. This ion could be identified as protonated ubine by collisional analysis in MIKE mode (Fig. 94).

The formation of the *m/z* 44, 58, 77 and 91 ions is due, respectively, to $CH_2=$

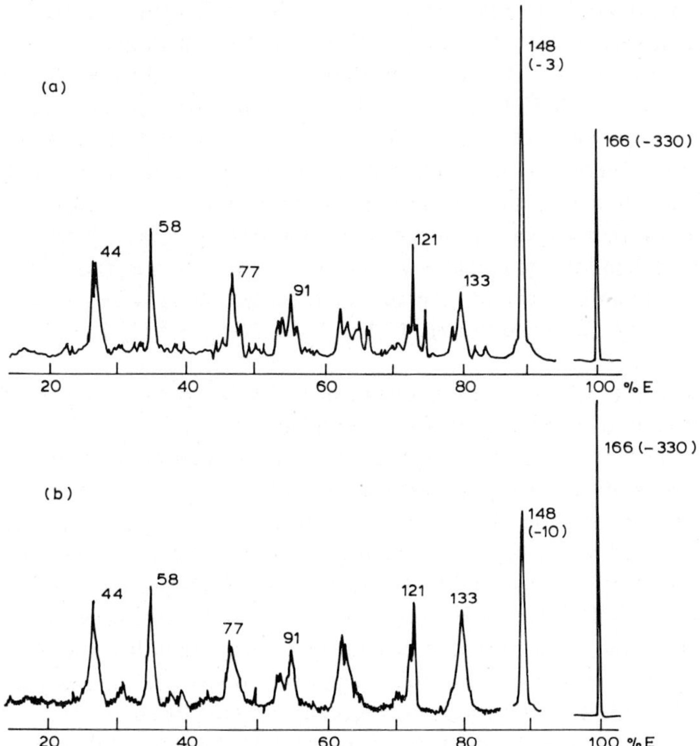

Figure 94. MIKE spectra of *m/z* 166 obtained from (a) a crude '*D. longimamma*' extract and (b) purum ubine hydrochloride [241].

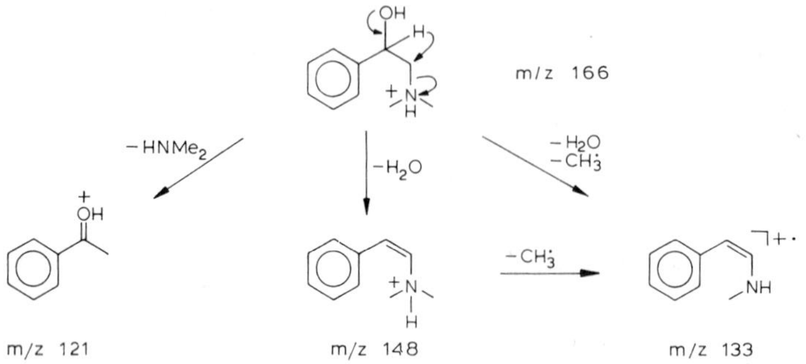

Figure 95. Collisional decomposition of *m/z* 166 ion produced from hordenine isomer.

$\overset{+}{N}H\dot{C}H_3$, $CH_2{=}\overset{+}{N}(CH_3)_2$, $C_6H_5^+$ and $C_7H_7^+$. The *m/z* 148, 133 and 121 ions are more interesting, since they lead to a better definition of the structure of the molecule and also enable it to be distinguished from hordenine (Fig. 95).

It should be noted that these fragmentations are classical, but certain are unusual, such as the loss of $CH_3^{\cdot}$ from MH^+ (or of the *m/z* 148 ion) leading to a radical ion at *m/z* 133.

The identification of alkaloids is thus relatively simple. It is also possible to distinguish their stereochemistry, e.g., the ring junction of indole alkaloids.

Spontaneous decompositions of $M^{+\cdot}$ ions produced by EI (or during charge transfer) lead to characteristic MIKE spectra. The characteristic decomposition chosen is that which gives the retro Diels-Alder, which is particularly sensitive to the *cis* or *trans* ring junction of these alkaloids, as shown by Tamas et al. [242].

Thus, all possibilities of structural elucidation may be shown with MS/MS. Quantitative analysis is another aspect, developed by Cooks and co-workers [189a,c], using CI. In samples of hemlock (*Conium maculatum* L.) they demonstrated the presence of coniine by collisional analysis of protonated molecular ions (*m/z* 128).

This spectrum shows the eliminations of $H^{\cdot}$, C_3H_8, $C_3H_7NH_2$ and $C_4H_9NH_2$ to form carbocations.

The similarity between the MIKE/CAD spectrum of the MH^+ peak (*m/z* 128) of authentic coniine and that of a plant sample are sufficiently convincing that the presence of coniine in the sample can be confirmed. Cocaine and cinnamoylcocaine were both detected in coca leaves using the same methodology.

The MIKE/CAD spectra of the MH^+ peak of both these compounds are characterized by the presence of the same very intense ion at *m/z* 182, corresponding to the loss of C_6H_5COOH in one case and $C_6H_5{-}CH{=}CH{-}COOH$ in the other.

The MH^+ (*m/z* 304) → *m/z* 182, and MH^+ (*m/z* 330) → *m/z* 182 transitions enable cocaine and cinnamoylcocaine, respectively, to be assayed, if they are utilized in MID mode. This method is obviously not applicable to conventional spectra, since an ambiguity remains concerning the origin of this fragment ion (*m/z* 182).

It was thus possible to compare and determine the relative proportions of these two alkaloids in *Erythroxylum coca* Lam. in different plant parts (Tingo Maria) (Table 27) [243].

TABLE 27
Distribution of cocaine and cinnamoylcocaine in *Erythroxylum coca* Lam (from various origins) [243]

Origin	Plant part	% cinnamoylcocaine	Ratio, cocaine/cinnamoylcocaine
Tingo Maria	leaf, powdered (control)	2.3	43
Tingo Maria	leaf, margin, A, B, H, I, K	4.2, 3.6, 2.3, 1.8, 1.1	23, 27, 42, 54, 90
Tingo Maria	leaf, center (with veins), A	6.1	15
Tingo Maria	leaf, center (no veins), G, K	2.6, 1.1	37, 90
Tingo Maria	leaf, stem, A	6.4	15
Tingo Maria	twig, C, D, E	33, 30, 9	2.1, 2.3, 10
Tingo Maria	berry, powdered (control)	51	1.0
Tingo Maria	berry, inside (brown)	68	0.47
Tingo Maria	berry, outside (black)	33	2.0
Tingo Maria	berry, stem	50	1.0

These results show the power of such a method, especially for routine assays of trace amounts. Its overall sensitivity is not lower than that obtained when utilizing conventional mass spectra, since neither extraction nor purification are necessary. In addition, this method is valuable, since no chemical modification is required. It is possible to perform absolute assays by the inclusion of well-chosen internal standards.

Other ionization methods have been employed, including field ionization. McReynolds and Anbar [244a,b] developed a very sensitive FI source.

The effect of collision gas pressure on the sensitivity of the peaks obtained was demonstrated. The relative abundance of these ions can indeed vary by a factor of two, and the pressure chosen was 2.10^{-3} Torr. In the most recent version of their instrument, the authors applied a voltage on the order of 20 kV, thus leading to a much better resolution of the energy spectrum (spherical electric field).

A large number of isobaric compounds, especially those of uric acid, was studied. The spectrum of this compound is distinguished considerably from those of its isobars. Using FI with their instrument, they detected small quantities of caffeine in complex samples (urine). Maquestiau et al. [245] also demonstrated the same compound in tea samples with the same method, but they used CI conditions to produce protonated molecules.

The localization of methyl group on the indolic ring is not always obvious, as reported by Safe et al. [247] using IKE spectra.

Zakett and Cooks [246] rapidly identified the constituents of refined coal liquid. They demonstrated the possibility of distinguishing tetrahydroquinoline from tetrahydroisoquinoline by the intensity of the $MH^{2+\cdot}$ ions produced by charge stripping of MH^+ during collisions [158] (Fig. 96).

The interpretation of the MIKE spectra of heterocyclic compounds may be ambiguous. The study of the decomposition spectra of natural isotopic ions may simplify this situation. Occolowitz et al. [248] studied the structure of the $C_7H_{12}NO_2S^+$ ions (m/z 174) produced during the fragmentations of $M^{+\cdot}$ molecular ions,

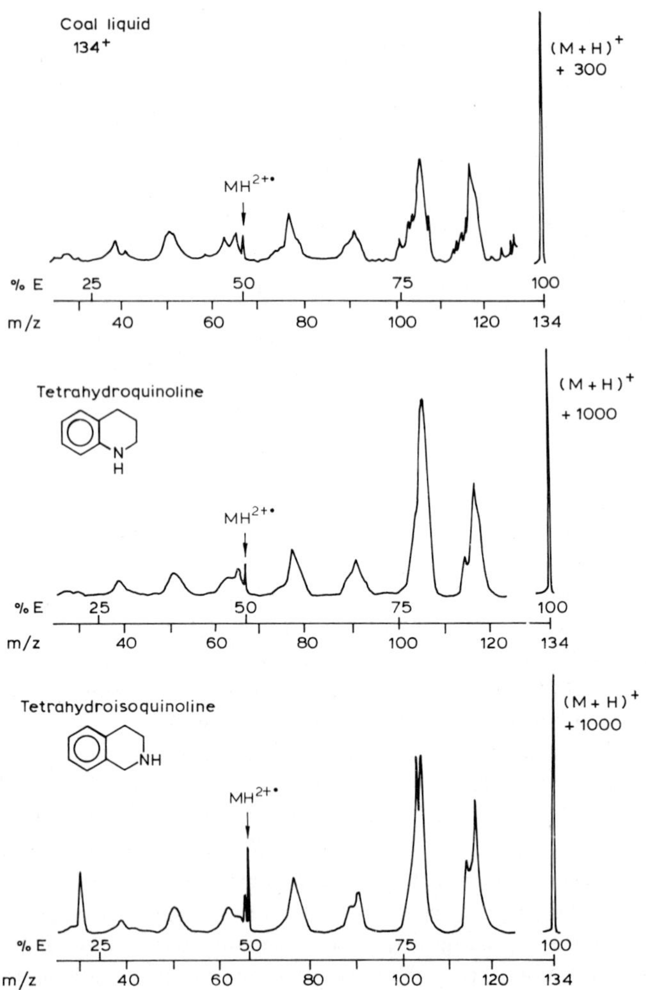

Figure 96. MIKE/CAD spectra of *m/z* 134 ions, differentiating between tetrahydroquinoline and isoquinoline by the intensities of double charge ions [158,246].

formed by EI, of penicillin, dehydrocephalosporin and tetramethylthiazolodine. The collisional-induced decomposition spectra of these isomeric fragment ions are sufficiently different that they can be distinguished (Fig. 97).

Fragment *m/z* 174 ions b and c eliminate 60 u (only b eliminates 32 u), while ion a eliminates only 46 u (Fig. 97).

In order to interpret these 32- and 60-u eliminations, which may correspond to the loss of CH_3OH (or S) and $HCOOCH_3$ (or $S(CH_2)_2$), respectively, the authors recorded the MIKE/CAD spectrum of the natural ^{34}S isotope of the *m/z* 174 ion: $C_{17}H_{12}NO_2{}^{34}S$ (*m/z* 176). The spectrum (Fig. 98) shows a 2-u shift for each of

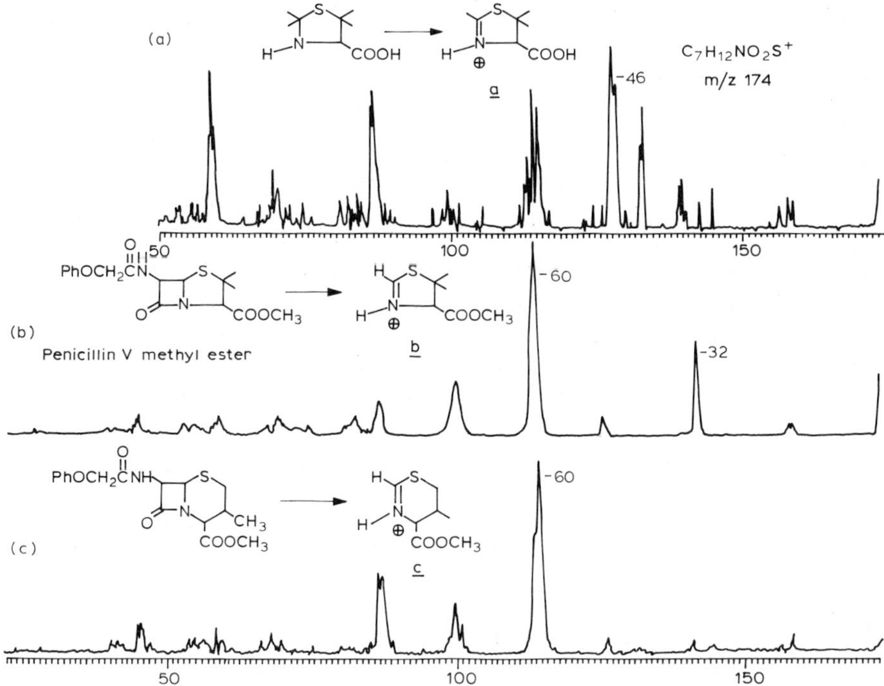

Figure 97. MIKE/CAD spectra of *m/z* 174 ions. These isomer ions are produced from various precursors: tetramethylthiazolidine (a), penicillin (b) and dehydrocephalosporin (c) [248].

these peaks, confirming that sulfur was not eliminated and that only the losses of ethanol and methyl formate remain valid possibilities.

These examples show the applications of MIKE method. In the same way, linked scan techniques may be particularly valuable. These methods are rapid, thus allowing the accumulation of iterative results, since there is a satisfying reproducibility.

Haddon and Molyneux [249] analyzed the pyrrolizidine family of alkaloids, components of certain plant species (*Seneccio*). Their structures are shown in Figure 99. These compounds are related to the toxic principles of plants, such as *S. longilobus* and *S. riddelisi.*

These compounds decompose under both EI and CI to yield a particularly intense *m/z* 120 ion. The high resolution measurements of this ion indicate that the elemental formula is $C_8H_{10}N^+$.

senephylline (SPH) (m/z 333)
retrosine (RET) (m/z 351)
senecionine (SEN) (m/z 335)
riddelliine (RID) (m/z 349)
→ m/z 120

This ion remains the base peak, even in first FFR spectra of the decomposition of the protonated molecular ion, e.g., riddelliine. In fact, this ion is formed totally from

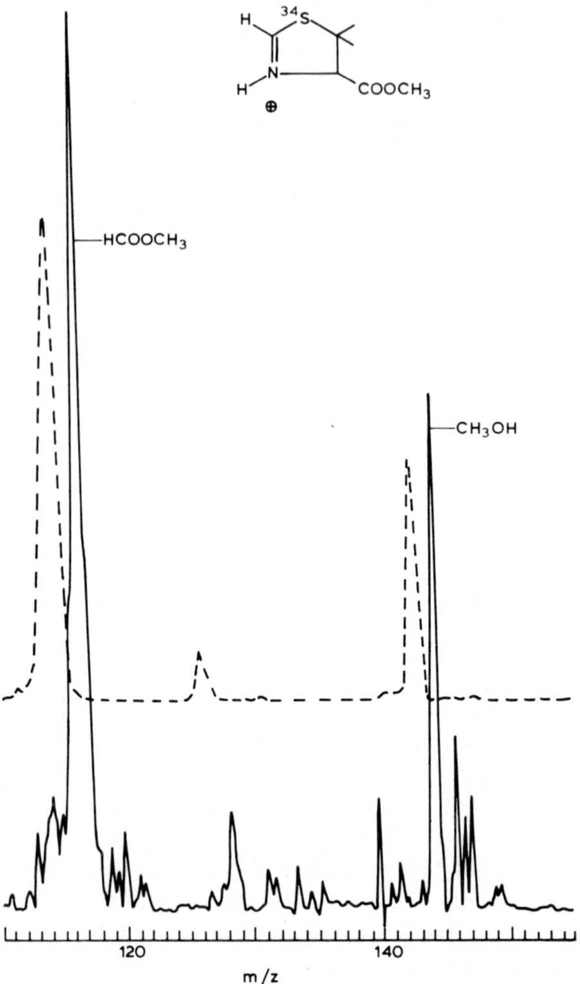

Figure 98. MIKE/CAD spectra of m/z 176 ion (natural ^{34}S isotope of m/z 174 ion) (upper) and m/z 174 from penicillin [248].

the molecular species $MH]^+$ and $[M + C_3H_5]^+$. The B^2/E linked scan spectrum of the m/z 120 ion indeed presents these two possible origins: the protonated molecule MH^+ and the adduct ion $[M + C_3H_5]^+$.

The B^2/E spectra can also be used to identify the components of a mixture. In the case of interest here, although only one alkaloid of this species is present in *S. riddellii*, the same is not true of *S. longilobus*. The B^2/E spectrum of the m/z 120 ion indeed indicates the presence of precursors around m/z 333 (SPH and SEN) ions and m/z 350 (RET and RID) ions.

Quantitative assays of SPH (molecular weight, 351) are possible and an internal

(1)

Seneciphylline
(SPA, MW 333)

(2)

Retrorsine
(CRET, MW 351)

(3)

Senecionine
(SEN, MW 335)

(4)

Riddelliine
(RID, MW 349)

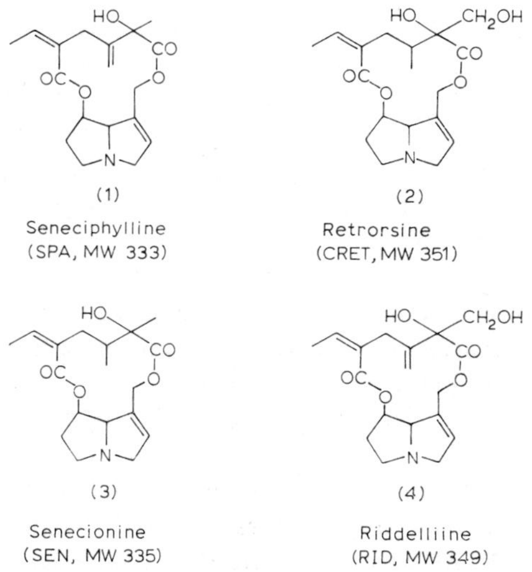

Figure 99. Components of alkaloids in '*Seneccio*' extract [244].

standard can be used, e.g. monocrotalin (molecular weight, 326). Resolution in B^2/E is mediocre and the m/z 326 and 334 (SPH) peaks are poorly separated. Nevertheless the standard compound presents an intense peak at m/z 264 $[M-CO_2-H_2O]^+$ for monocrotalin, which corresponds to fast consecutive decompositions. The 334/264 ion ratio can thus be chosen for the assay and the calibration curve is characterized by good linearity.

The purines and pyrimidines are biochemically important heterocyclic compounds, components of nucleic acids (DNA and RNA). Again, in this case, mass spectrometry may be a valuable tool.

Puzo et al. [250] showed the role of intermediates, e.g., anhydronucleosides, in the synthesis of nucleosides. Their behavior under EI is not very different from that of the nucleosides themselves, rendering interpretation delicate. For instance, the structure of the m/z 153 ion is difficult to prove, and the deuterium labeling and the HV scan spectra lead to an unambiguous structure, as in Fig. 100a.

Although this method is the simplest to use, considering the instrumentation available, it has only a minor role. MIKE and linked scan modes can be very useful for routine determination analysis.

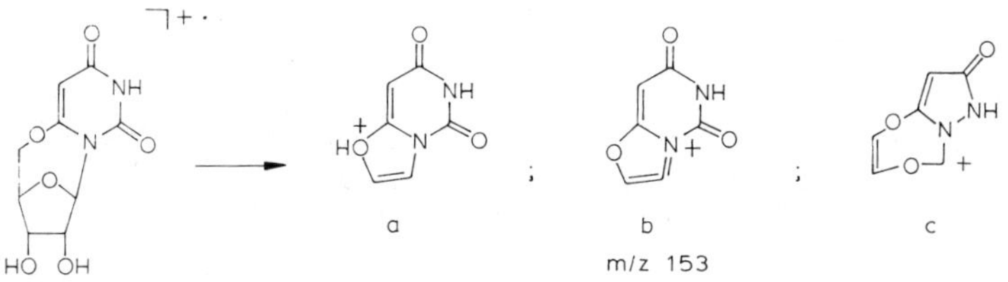

Figure 100. Isomeric structures for m/z 153 ions.

Cooks and co-workers [164,251a,b] studied the structures of modified bases of DNA with a high specificity. They used a direct introduction probe to control sample pyrolysis and measurements were performed in CI with isobutane. The different bases of DNA were thus liberated intact from a salmon sperm DNA sample.

The detection of protonated molecules was performed by searching for different higher mass adducts. High-resolution measurements can be used for the determination of elemental compositions, but this is not sufficient for structural determinations.

For instance, the *m/z* 126 ion as protonated molecule is characterized by a MIKE/CAD spectrum which can be compared to those of authentic modified nucleotides, e.g., 5-methyldeoxycytidine-5′-monophosphate (Fig. 101).

These spectra are sufficiently similar for the identity of these structures to be concluded.

Other protonated molecules may be similarly analyzed. Thus, it was shown that the *m/z* 150 ion did not correspond to N^6-methyladenine, but rather to its isomer, 1-methyladenine, which can be easily identified. At low temperatures, however, other a, b, c and d ions emerge and interfere (Fig. 102).

These peaks disappear at higher temperatures. It was shown that adenine contained about 1% 1-methyladenine, which heretofore had never been detected.

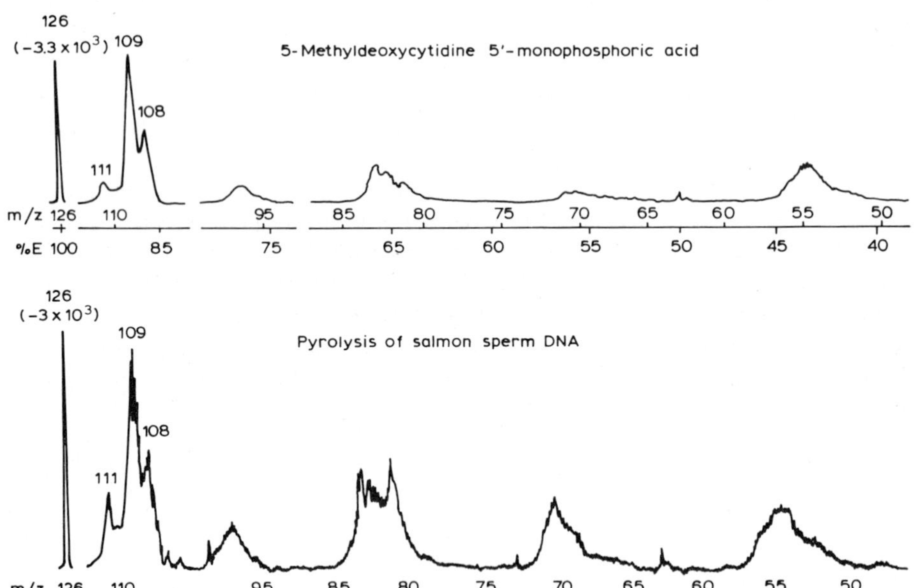

Figure 101. MIKE/CAD spectra of *m/z* 126 ion from salmon DNA (lower) compared to *m/z* 126 ion produced from authentic 5-methyldeoxycytidine sample (upper) [251].

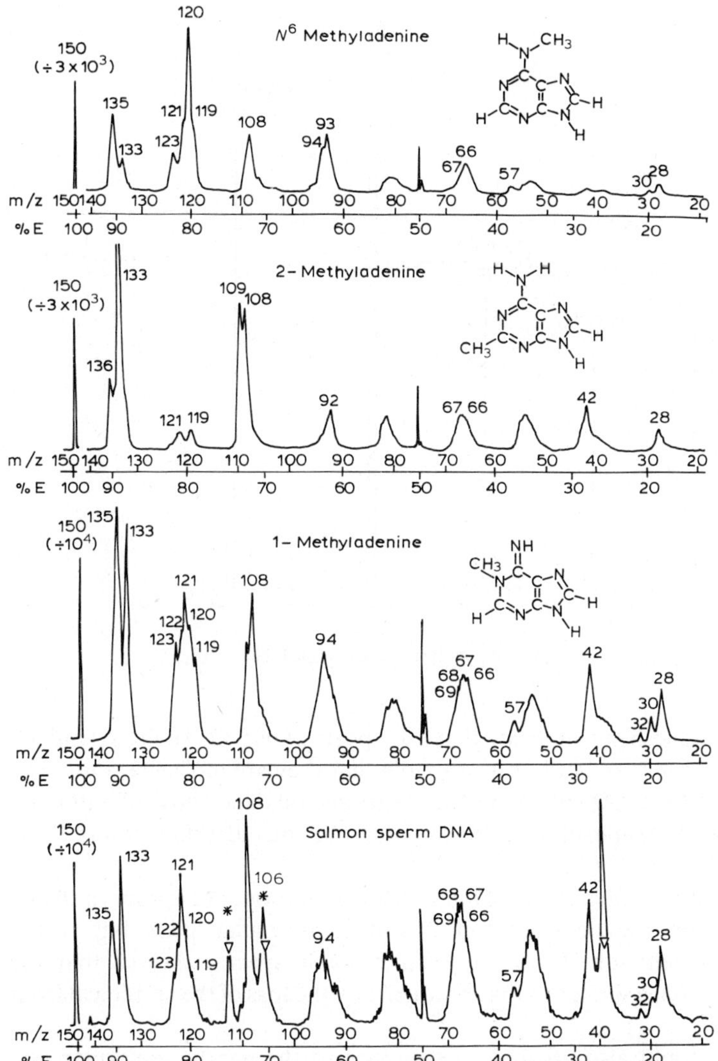

Figure 102. MIKE/CAD spectra of isomer structure m/z 150 ions and presence of ions a, b, c and d formed from impurities in the MIKE/CAD spectrum of m/z 150 ion from salmon sperm DNA (the artefacts are denoted by asterisks in the latter spectrum) [251].

Wiebers and co-workers [252] utilized EI on the fragments produced during the pyrolysis of viral DNA (phage $\lambda\phi$ X 174), which is not present in *Escherichia coli*. Certain unusual nucleosides could be identified and localized by this study.

Conventional mass spectra may be obtained in EI at low energy (14 eV) and can be valuable for locating abundant molecular ions from a mixture. Levsen and Schulten [253] thus studied the mixture furnished by the pyrolysis of DNA (after heating a herring DNA sample to 600°C). The EI spectrum obtained is very complex (Fig. 103).

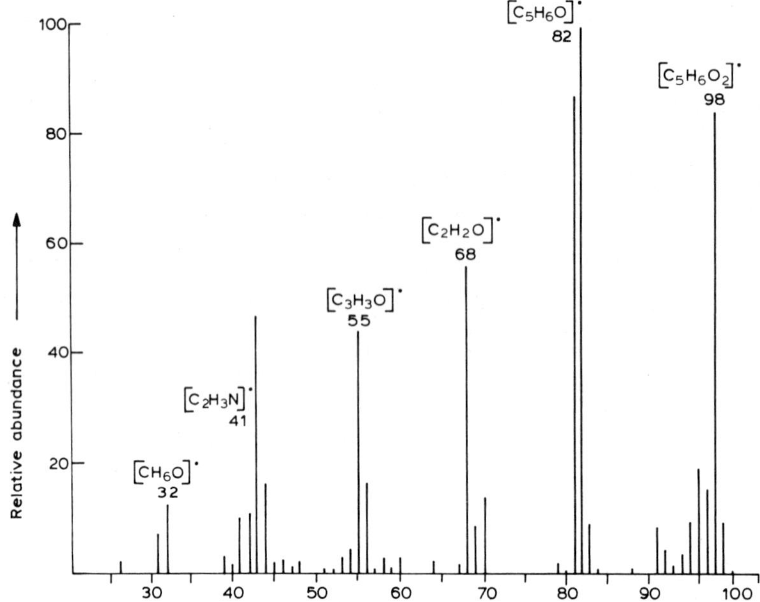

Figure 103. Conventional mass spectra at 14 eV of pyrolysis products of DNA [253].

High-resolution measurements indicate that each peak, at *m/z* 32, 41, 68, 82 and 98, corresponds to radical ions. At 14 eV it is probable that they are molecular ions. The comparison of the MIKE/CAD spectra of these various ions and those of authentic compounds of the same composition led to the unambiguous identification of these ions (Table 28).

The *m/z* 55 and 98 peaks remain unexplained. The former does not appear in the FI spectra, consistent with the belief that it is not a molecular peak but rather corresponds to a fragment ion. The study of metastable spectra showed that the structure of the *m/z* 55 ion is similar to that produced by the loss of the $H^{\cdot}$ radical from propargyl alcohol.

The same authors could also attribute a structure to the *m/z* 98 molecular ion, which corresponds to a 60/40 mixture of molecular ions of α-angelica lactone and

TABLE 28
Structure of ions studied with their elemental composition [253]

m/z	Elemental composition	Structure shown by MIKE/CAD
32	$CH_4O^{+\cdot}$	methanol
41	$C_2H_3N^{+\cdot}$	acetonitrile
68	$C_4H_4O^{+\cdot}$	furane
82	$C_5H_6O^{+\cdot}$	methylfurane

TABLE 29
MIKE/CAD spectra of m/z 98 ions ($C_5H_6O_2^{+\cdot}$) formed by pyrolysis of DNA and other m/z 98 ions produced from α-angelica lactone and furfuryl alcohol [253]

m/z	DNA	α-Angelica lactone (a)	Furfuryl alcohol (b)	60% a, 40% b
15	0.5	0.7	0.2	0.5
27	3.4	4.5	2.6	3.7
29	2.6	1.1	3.7	2.1
31	2.3	0.5	5.1	2.1
39	7.0	4.5	12	7.3
42	5.7	2.5	16	7.9
43	16	20	4.0	14
52–53	4.8	4.4	6.0	5.0
55	24	37	6.2	25
70	28	23	31	26
81	5.1	–	14	5.6
82	–	1.5	–	0.6

furfuryl alcohol (Table 29). The analysis of the latter peaks showed an eventual limitation of MS/MS.

The superposition of several 'pure' MIKE/CAD spectra on the same spectrum may indeed 'cloud the issue'. Nevertheless, we believe that by changing either collision gas pressure or type, it is possible to obtain sufficient information to resolve the ambiguities. Those which remain are due (relatively rarely) to the partial isomerization of molecular ions and fragment ions which are most often encountered. In the latter case, the examination of the MIKE/CAD spectra of other fragment ions belonging to the conventional mass spectrum may be useful for determining these structures.

Considerable experimental difficulties often arise when non-volatile and thermolabile molecules are analyzed. The exploitation of MIKE (and MIKE/CAD) spectra of these high-molecular-mass ions is rendered delicate by poor energy resolution.

Burlingame's group [254] adopted as an alternative, field desorption ionization combined with B/E linked scan mode, which led to a better resolution (± 1 amu). Biologically important molecules were studied with this technique.

For example, the studied metabolites are activated derivatives of benzo(*a*)pyrene, e.g., 7α,8β-dihydroxy-9β,10β-epoxy-7,8,9,10-tetrahydrobenzo(*a*)pyrene (Bap diol epoxide). This compound forms an adduct with DNA (Fig. 104).

After enzymatic hydrolysis and HPLC, a mixture of various products can be isolated. Subjected to FD (Table 30), this mixture generates a spectrum in which a certain number of molecular species are observed.

It is possible to show that the relative quantities of derivatives of deoxyguanosine, deoxyadenosine and deoxycytidine are present at 92, 7 and 1%. Collisional-induced decompositions of each of these peaks (B/E/CAD) are required, since spontaneous decompositions are relatively few.

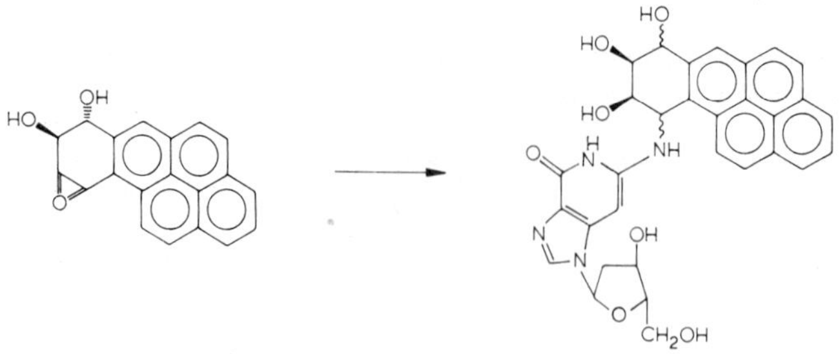

Figure 104. Structure of covalent DNA adduct.

TABLE 30
Compounds present in the mixture formed by enzymatic hydrolysis [254]

m/*z*	Intensity	Structure
592	100	$[569+Na]^+$
576	70	$[553+Na]^+$
569	80	deoxyguanosine adduct$^{+\cdot}$
553	20	deoxyadenosine adduct$^{+\cdot}$
552	5	$[529+Na]^+$
529	–	deoxycitidine adduct$^{+\cdot}$
320	5	Bap-tetraol$^{+\cdot}$

As an example, the cationized molecule $(M+Na)^+$ (*m*/*z* 592), under collision conditions, decomposed to form a large number of fragment ions.

The *B*/*E*/CAD spectra of the *m*/*z* 592 ion correspond to those of the deoxyguanosine adduct combined with Na. They are interpreted in the same way as those obtained from the *m*/*z* 569 ion itself (Fig. 105).

It must be stressed that it is useless to perform any chemical modification; it is thus possible to study the structures of metabolites produced in vivo directly, with all associated advantages.

The same group [254a,255] more recently used the same approach to demonstrate highly polar modified nucleosides (which were labile during pyrolysis) in samples on the order of 2 μg.

The FD spectra of these compounds are relatively complex. Although it is possible to attribute to the high-mass peaks at *m*/*z* 857 and 880 the respective structures of molecular ions ($M^{+\cdot}$, *m*/*z* 857) and adduct ions ($[M+Na]^+$, *m*/*z* 880), it is difficult to determine the origin of peaks corresponding to lower masses: *m*/*z* 554, 617 and 731. Collisional-induced metastable decomposition spectra (*B*/*E*/CAD) enable two of them to be attributed (Fig. 106).

McClusky et al. [256] identified the previously encountered polycyclic and aromatic hydrocarbons and analyzed the adducts.

It is difficult to localize and identify the metabolites of benzo(*a*)pyrene, since several isomeric structures are possible for these compounds.

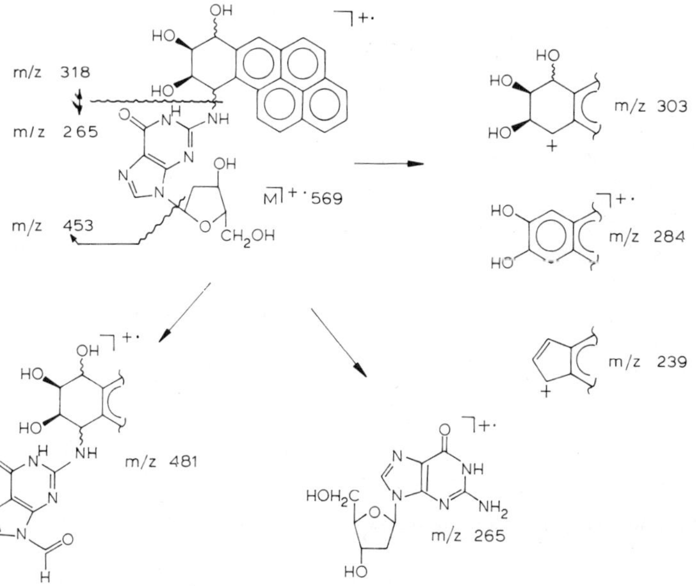

Figure 105. Mean ions observed in the B/E/CAD spectrum of m/z 569 ($M^{+\cdot}$) under FD conditions (the peaks at m/z 453 and 481 are shifted at m/z 476 and 504 from $[M + Na]^+$ ion) [254].

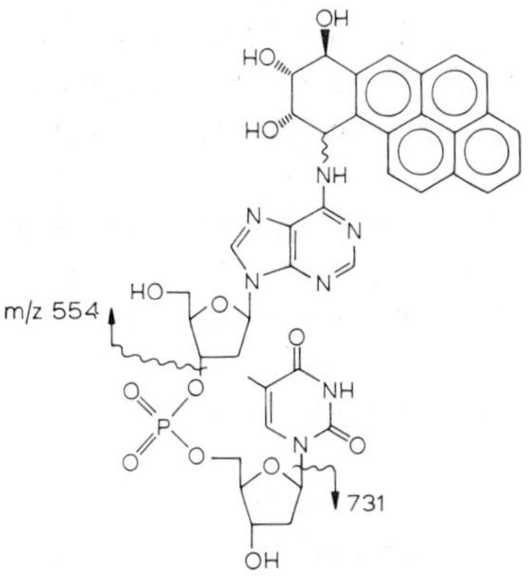

Figure 106. Interpretation of mean peaks of B/E/CAD spectrum of m/z 857 ion [255].

For example, a dozen isomers can be proposed for phenolic derivatives of Bap arising through various metabolic processes. Nevertheless, two isomers are retained (2-OH-Bap), since they are the only ones, with Bap itself, to be carcinogenic (Fig. 107).

In addition, 2-OH-Bap produces a common metabolite if incubated with certain cells. The small quantity of metabolites obtained was analyzed with high-resolution EI and the complex spectra obtained yield an ion at m/z 284, which appears to be a polyaromatic compound.

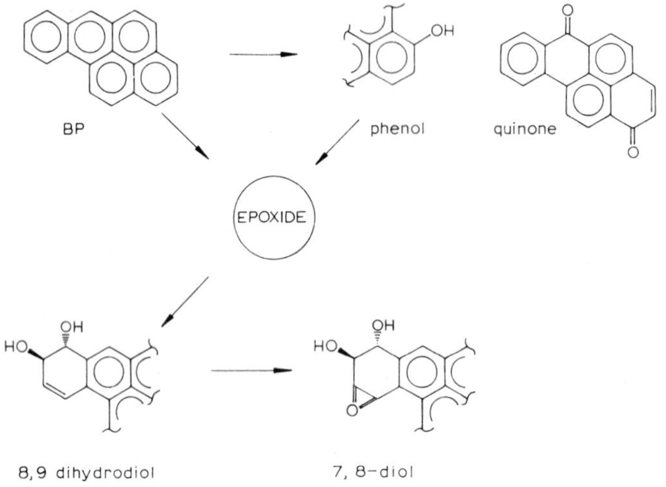

Figure 107. Possible isomers which might be carcinogenic diols.

The use of the magnetic field and ion separator makes it possible to study (MIKE mode) only the m/z 284 ion without interference from impurities or other compounds.

The MIKE/CAD spectrum of this $M^{+\cdot}$ ion is complex, but two groups of peaks are interesting: the first, of low intensity, is due to the loss of 17 and 34 u groups; the second, very intense group is related to the loss of 29- and 58-u fragments (Fig. 108). These processes correspond to the successive eliminations of one or two $OH^{\cdot}$ radicals and of $CHO^{\cdot}$ or two $CHO^{\cdot}$ fragments.

These fragmentations indicate the probable presence of two OH groups, enabling one to formulate the hypothesis according to which hydroxylated derivatives of 2-OH-Bap are present. Nevertheless, it is difficult to distinguish if this metabolite is 1,6- or 3,6-dihydroxybenzo(*a*)pyrene. Furthermore, the transformation of the –OTMS group totally shifts the m/z 284 peak to m/z 428, confirming the presence of OH groups (and not epoxide, which could have been proposed).

The MIKE and B/E linked scan modes were compared in this study (Fig. 108) and the authors noted that each technique has its advantages and drawbacks. B/E furnishes good resolution of secondary ions (as daughters) while MIKE leads to a better resolution of primary ions (as precursors).

These fragmentations are more or less fast, according to the isomer structure, enabling each of them to be distinguished.

Thus, the use of B/E/CAD combined with FD (which may lead to radical ions $M^{+\cdot}$ or to MH^{+} ions) will undergo increasing development as a result of the resolving qualities and the reproducibility furnished [257].

A series of nucleotides was systematically studied. The molecules analyzed were combinations of adenylyl (3′,5′) cytidine (or cytidylyl (3′,5′) adenosine) with the riboses ApC and CpA, and also the deoxyriboses ApC and CpA. This work is important,

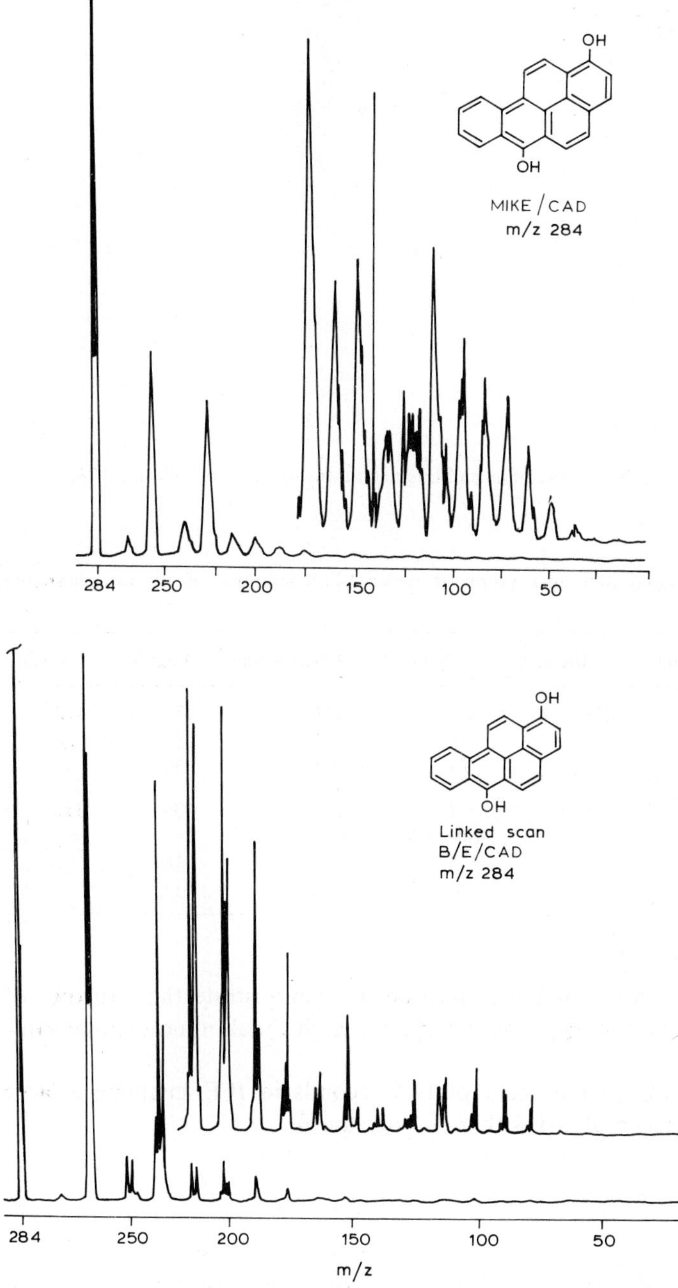

Figure 108. Comparison of MIKE/CAD and *B*/*E*/CAD spectra of *m*/*z* 284 ions [256].

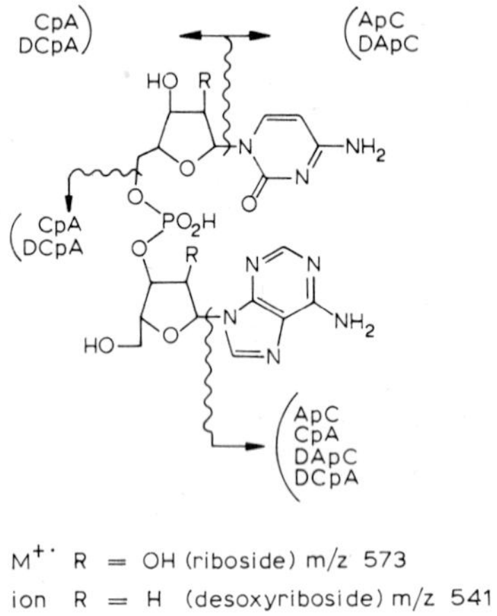

$M^{+\cdot}$ R = OH (riboside) m/z 573
ion R = H (desoxyriboside) m/z 541

Figure 109. Differences between principal ose units and fragmentations induced by collisions [257].

TABLE 31
B/*E*/CAD spectra of protonated molecules produced under FD conditions of various phosphate dinucleosides [257]

	MH^+	−Cytosine	−Adenine	−Cytidine	−Adenosine	Adenine	Cytosine
m/*z*	573	462	438	348	324	136	112
ApC	26	1	–	8	3	34	28
CpA	35	7	–	10	2	36	10
m/*z*	541	430	406	332 (-deoxy)	308 (-desoxy)	136	112
d(ApC)	40	2	1	5	2	18	32
d(CpA)	62	5	–	13	1	15	5

since it was previously shown that it is possible to demonstrate the structure of alkylated dinucleotide monophosphates by specific collisional-induced fragmentations (Table 31).

Measurement times are on the order of 16.5 seconds in *B*/*E*, enabling a large number of scans to be accumulated with a 3–5 μg sample.

4. Conclusion

This inexhaustive review shows the developments in mass spectrometry during these last years. Various aspects of analytical methods are improved in the bid to obtain an answer to quantitative determination and structure elucidations. The two orientations which interest the biochemistry field appear in a new methodology: MS/MS, which is complementary to other classical techniques. Thus, the metastable ions produced in the field free regions can be detected and studied at the same time as the ions formed in the source. The specificity of this technique is comparable to HR/GC/MS. Indeed, MS/MS can play a more valuable role than GC/MS or LC/MS for difficult cases.

The applications of this technique reported in the literature indicate various possibilities [258].

During the last 2 years, after this review had been written, a new soft ionization technique appeared: fast atom bombardment (FAB). First used by physicists [259] and, after improvement, by chemists [260], the FAB technique is comparable to SIMS, PDMS and laser desorption [261,262]. In addition, the probe for thermal desorption [263] is usually used to volatilize the thermolabile and nonvolatile compounds.

Recent reviews [264,265] show the possibilities of these different techniques of soft ionization, with various analyzers to separate the high-mass ions; in particular, with a time-of-flight, it has been possible to transmit ions of m/z 12 651 (± 10) [266]. Recently, resolution of TOF instruments has been improved [267], enabling the determination of metastable transitions. The FAB [268], SIMS [269] and laser desorption [270] techniques are consistent with MS/MS methodology.

On the other hand, the studies of mixtures necessitating the use of soft ionization techniques, such as SIMS [271], PDMS [272] and LD [273], are possible. Indeed, they can be 'coupled' with liquid chromatography. Thus, quantification should be one of the objectives of these new technologies in the future [274]. However, the 'search for zero' [275] is moving, and in particular the sensitivities obtained with the negative ions are already promising.

References

1a Grostic, M.F. and Rinehart, K.L. (1972) in Mass Spectrometry Techniques and Applications (Milne, G.W.A., ed.), Wiley Interscience, New York.
1b Millard, B.J. (1977) Quantitative Mass Spectrometry, Heyden and Son, Ltd., London.
1c Lehmann, W.D. and Schulten, H.R. (1978) Angew Chem. Int. Ed. Engl. 17, 221.
2 Schulten, H.R. (1977) in Methods of Biochemical Analysis, Vol. 24 (Glick, D., ed.), p. 313, Wiley Interscience, New York.
3a Frigerio, A. (ed.) (1978) Recent Developments in Mass Spectrometry in Biochemistry and Medicine, Vol. 1, Plenum Press, New York.
3b Frigerio, A. (ed.) (1979) Recent Developments in Mass Spectrometry in Biochemistry and Medicine, Vol. 2, Plenum Press, New York.
4a De Leenheer, A.P. and Roncucci, R.R. (1977) Quantitative Mass Spectrometry in Life Sciences I, Elsevier, Amsterdam.
4b De Leenheer, A.P., Roncucci, R.R. and Van Petegham, C. (1978) Quantitative Mass Spectrometry in Life Sciences II, Elsevier, Amsterdam.
5a Budzikiewicz, H., Djerassi, C. and Williams, D.H. (1964) Structure Elucidation of Natural Products by Mass Spectrometry, Vol. 1, Alkaloids; Vol. 2, Steroids, Terpenoids, Sugars and Miscellaneous Classes, Holden Day, San Francisco.
5b Budzikiewicz, H., Djerassi, C. and Williams, D.H. (1967) Interpretation of Mass Spectra of Organic Compounds, Holden Day, San Francisco.
6a Waller, G.R. (ed.) (1972) Biochemical Applications of Mass Spectrometry, Wiley Interscience, New York.

6b Waller, G.R. and Dermer, O.C. (eds.) (1980) Biomedical Applications of Mass Spectrometry, First Supplementary Volume, Wiley Interscience, New York.

7 Rosenstock, H.M., Wallenstein, M.B., Wahrhaftig, A.L. and Eyring, H. (1952) Proc. Natl. Acad. Sci. U.S.A. 38, 667.

8 Levsen, K. (1978) in Fundamental Aspects of Organic Mass Spectrometry, Progress in Mass Spectrometry, Vol. 4 (Budzikiewicz, H., ed.), Verlag Chemie, Berlin.

9a Viger, A., Marquet, A. and Tabet, J.C. (1976) Org. Mass Spectrom. 11, 1063.

9b Frappier, F., Guillerm, G. and Tabet, J.C. (1980) Biomed. Mass Spectrom. 7, 185.

10 Audier, H., Bottin, J., Diara, A., Fetizon, M., Foy, P., Golfier, M. and Vetter, W. (1964) Bull. Soc. Chim. France, 2292.

11 Das, B.C., Gero, S.D. and Lederer, E. (1967) Biochem. Biophys. Res. Commun. 29, 211.

12 Knapp, D.R. (1979) Handbook of Analytical Derivatization Reactions, Wiley Interscience, New York.

13 Lehmann, W.D. and Schulten, H.R. (1978) Angew. Chem. Int. Ed. Engl. 17, 221.

14 Facchetti, S., Fornari, A. and Montagna, M. (1980) Adv. Mass Spectrom. 8, 14a.

15a Millington, D., Jenner, D., Jones, T. and Griffiths, K. (1974) Biochemistry 139, 473.

15b Snedden, W. and Parker, R.B. (1976) Biomed. Mass Spectrom. 3, 295.

16 Jellum, E. (1977) Annali di Chimica 67, 571.

17 Krahmer, U.I. and McCloskey, J.A. (1978) Adv. Mass Spectrom. 7, 1483.

18a Klein, E.R. and Klein, P.D. (1978) Biomed. Mass Spectrom. 5, 91.

18b Klein, E.R. and Klein, P.D. (1978) Biomed. Mass Spectrom. 5, 521.

19a Klein, E.R. and Klein, P.D. (1978) Biomed. Mass Spectrom. 5, 373.

19b Klein, E.R. and Klein, P.D. (1978) Biomed. Mass Spectrom. 5, 425.

20 Gochmann, N., Bowie, L.J. and Bailey, D.N. (1979) Anal. Chem. 51, 525.

21 Arpino, P.J. (1978) in Instrumental Applications in Forensic Drug Chemistry, Proceedings of International Symposium (Klein, M., Kruegel, A.V. and Sobal, S.P., eds.), p. 151.

22 Arpino, P.J. and Guiochon, G. (1979) Anal. Chem. 51, 683A.

23 Brent, D.A., Bugge, C.J., Cuatrecasas, P., Hulbert, B.S., Nelson, D.J. and Sayhoun, N. (1978) in Chemical Applications of High Performance Mass Spectrometry (Gross, M.L., ed.), p. 429, American Chemical Society, Washington, DC.

24 Heller, S.R., McCormick, A. and Sargent, T. (1980) in Biochemical Applications of Mass Spectrometry, First Supplementary Volume (Waller, G.R. and Dermer, O.C., eds.), p. 103, J. Wiley and Sons, New York.

25 Kwiatkowski, J. and Riepe, W. (1979) Anal. Chim. Acta 112, 219. 51, 1236.

26 Dromey, R.G. (1976) Anal. Chem. 48, 1464.

27 Buchanan, B.G., Smith, D.H., White, W.C., Gritter, R.J., Feigenbaum, E.A., Lederberg, J. and Djerassi, C.J. (1976) J. Am. Chem. Soc. 98, 6168.

28 Dawson, P.W. (1976) Quadrupole Mass Spectrometry and Its Application, Elsevier, Amsterdam.

29 McFadden, W.H. (1979) J. Chromatogr. Sci. 17, 2.

30a Hunt, D.F., Stafford, G.C., Crow, F.W. and Russel, J.W. (1976) Anal. Chem. 48, 2098.

30b Hunt, D.F. and Crow, F.W. (1978) Anal. Chem. 50, 1781.

31a Field, F.H. (1968) Acc. Chem. Res. 1, 42.

31b Munson, M.S.B. (1977) Anal. Chem. 49, 772A.

32 Huntress, W.T. (1977) Chem. Soc. Rev. 6, 295.

33a Baldwin, M.A. and McLafferty, F.W. (1973) Org. Mass Spectrom. 7, 1353.

33b Arpino, P.J. and McLafferty, F.W. (1976) in Determination of Organic Structures by Physical Methods, Vol. 6 (Nachod, F.C., Zuckerman, J.J. and Randall, E.W., eds.), p. 1, Academic Press, New York.

33c Biemann, K. (1980) in Biochemical Applications of Mass Spectrometry, First Supplementary Volume (Waller, G.R. and Dermer, O.C., eds.), p. 469, J. Wiley and Sons, New York.

34 Radford, T. and Dejongh, D.C. (1980) in Biochemical Applications of Mass Spectrometry, First Supplementary Volume (Waller, G.R. and Dermer, O.C., eds.), p. 255, J. Wiley and Sons, New York.

35a Hunt, D.F., McEwen, C.N. and Upham, R.A. (1972) Anal. Chem. 44, 1292.
35b Lin, Y.Y. and Smith, L.L. (1978) 26th Annual Conference of Mass Spectrometry and Allied Topics, St. Louis, p. 148.
35c Hunt, D.F., Sethi, S.K. and Shabanowitz, J. (1978) 26th Annual Conference of Mass Spectrometry and Allied Topics, St. Louis, p. 146.
36 Hunt, D.F. and Ryan, J.F. (1972) J.C.S. Chem. Comm. 620.
37 Winkler, J.F. and Stahl, D. (1978) J. Am. Chem. Soc. 100, 6779.
38 Winkler, J.F. and Stahl, D. (1979) J. Am. Chem. Soc. 101, 3685.
39a Tabet, J.C. and Fraisse, D. (1981) Org. Mass Spectrom. 16, 45.
39b Maquestiau, A., Flammang, R. and Nielsen, L. (1980) Org. Mass Spectrom. 15, 376.
40a Lin, Y.Y. and Smith, L.L. (1978) Biomed. Mass Spectrom. 5, 604.
40b Bastard, J., Dokhac, M., Fetizon, M., Tabet, J.C. and Fraisse, D. (1981) J. Chem. Soc. I, 1591.
41 Tabet, J.C., unpublished results.
42 Harrison, A.G. and Li, Y.H. (1980) Adv. Mass Spectrom. 8A, 207.
43 Hunt, D.F. and Harvey, T.M. (1975) Anal. Chem. 47, 1965.
44 Jelus, B.L., Munson, B., Babiak, K.A. and Murray, R.K. (1974) J. Org. Chem. 39, 3250.
45a Arsenault, G.P. (1972) J. Am. Chem. Soc. 94, 8241.
45b Hunt, D.F. and Ryan, J.F. (1974) Anal. Chem. 46, 729.
46 Greathead, R.J. and Jennings, K.R. (1980) Org. Mass Spectrom. 15, 431.
47 Fales, H.M. (1971) in Mass Spectrometry: Techniques and Applications (Milne, G.W.A., ed.), p. 179, Wiley Interscience, New York.
48 Field, F.H. (1975) in Mass Spectrometry (Maccoll, A., ed.), M.T.P. International Review of Science, Series 1, Vol. 5, p. 133 Butterworths, London.
49 Williams, D.H. (1978) Adv. Mass Spectrom. 7, 1157.
50 Harrison, A.G. (1980) Environ. Sci. Res. 16, 265.
51 Hunt, D.F. and Sethi, S.K. (1978) in Chemical Applications of High Performance Mass Spectrometry (Gross, M.L., ed.), p. 150, American Chemical Society, Washington, DC.
52 Munson, B. (1977) Anal. Chem. 49, 772A.
53 Jennings, K.R. (1979) in Gas Phase Ion Chemistry, Vol. 2 (Bowers, M.T., ed.), p. 123, Academic Press, New York.
54 Dougherty, R.C., Dalton, J. and Biros, F.H. (1972) Org. Mass Spectrom. 6, 1171.
55a Dillard, J.G. (1973) Chem. Rev. 73, 589.
55b Dillard, J.G. (1980) in Biomedical Applications of Mass Spectrometry, First Supplementary Volume (Waller, G.R. and Dernier, O.C., eds.), p. 927, J. Wiley and Sons, New York.
56a Hunt, D.F. and Crow, F.W. (1978) in National Bureau of Standards Special Publication 519, Trace Organic Analysis: A New Frontier in Analytical Chemistry, Proceeding of the 9th Materials Research Symposium, p. 601.
56b Hunt, D.F. and Sethi, S.K. (1978) in Chemical Applications of High Performance Mass Spectrometry (Gross, M.L. ed.), p. 150, American Chemical Society, Washington, DC.
57a See Reference 48.
57b Field, F.H. (1980) 28th Annual Conferences on Mass Spectrometry and Allied Topics, American Society For Mass Spectrometry, New York, p. 2.
58 See Reference 50.
59a Jennings, K.R. (1978) in Chemical Applications of High Performance Mass Spectrometry (Gross, M.L., ed.), p. 1, American Chemical Society, Washington, DC.
59b Jennings, K.R. (1979) Phil. Trans. R. Soc. London 293A, 125.
59c Jennings, K.R. (1977) in Mass Spectrometry, Vol. 4 (Johnstone, R.A.W., ed.), p. 203, Specialist Periodical Reports, The Chemical Society, London.
60a Bowie, J.H. (1980) Acc. Chem. Res. 13, 76.
60b Bowie, J.H. and Williams, B.D. (1976) in Mass Spectrometry (Macoll, A., ed.), M.T.P. International Review of Science, Series 2, Vol. 5, p. 89, Butterworths, London.
61 Selva, A., Traldi, P., Camarda, L. and Nasani, G. (1980) Biomed. Mass Spectrom. 7, 148.

62 Roy, T.A., Field, F.H., Lin, Y.Y. and Smith, L.L. (1979) Anal. Chem. 51, 272.
63 Hass, J.R., Friesen, M.D., Harvan, D.J. and Parker, C.E. (1978) Anal. Chem. 50, 1474.
64 Markey, S.P., Lewy, A.J., Zavadil, A.P., Poppiti, J.A. and Hoveling, A.W. (1977) in 25th Annual Conference on Mass Spectrometry and Allied Topics, Washington, DC, p. 276.
65 Tabet, J.C., Kagan, B. and Poulain, J.C. (1985) Int. Spectrom. J., in press.
66 Horning, E.C., Horning, M.G., Caroll, D.L., Dzidic, L. and Stillwell, R.N. (1973) Anal. Chem. 45, 936.
67 Yinon, J. and Boettiger, H.G. (1977) in 25th Annual Conference of Mass Spectrometry and Allied Topics, Washington, DC, p. 711.
68 Anderson, W.R., Frick, W. and Daves, G.D. (1978) J. Am. Chem. Soc. 100, 1974.
69 Reed, R.I. and Reid, W.K. (1963) J. Chem. Soc. 5933.
70 Dell, A., Williams, D.H., Morris, H.R., Smith, G.A., Feeney, J. and Roberts, G.C.K. (1975) J. Am. Chem. Soc. 97, 2497.
71 Baldwin, M.A. and McLafferty, F.W. (1973) Org. Mass Spectrom. 7, 1141, 1353.
72a Constantin, E., Nakatani, Y. and Ourisson, G. (1980) Tetrahedron Lett. 4745.
72b Ohashi, M., Tsujimoto, K., Tamura, S., Nakayama, N., Okumura, Y. and Sakurai, A. (1980) Biomed. Mass Spectrom. 7, 153.
73 Blaskley, C.R., Carmody, J.J. and Vestal, M.L. (1980) J. Am. Chem. Soc. 102, 5931.
74 Blaskley, C.R., Carmody, J.J. and Vestal, M.L. (1980) Anal. Chem. 52, 1636.
75 Beckey, H.D. (1969) Angew. Chem. Int. 8, 623.
76 Brown, P., Bruschweiler, F. and Pettit, G.R. (1972) Helv. Chim. Acta 55, 531.
77 Promé, J.C. and Puzo, G. (1977) Org. Mass Spectrom. 12, 28.
78 Holland, J.F., Soltmann, B. and Sweeley, C.C. (1976) Biomed. Mass Spectrom. 3, 340.
79 Beckey, H.D. (1969) Int. J. Mass Spectrom. Ion Phys. 2, 500.
80 Beckey, H.D. and Schulten, H.R. (1975) Angew. Chem. Int. Ed. Engl. 14, 403.
81 Daves, G.D. (1979) Acc. Chem. Res. 12, 359.
82a Hunt, D.F. and Sethi, S.K. (1978) in Chemical Applications of High Performance Mass Spectrometry (Gross, M.L., ed.), p. 150, American Chemical Society, Washington, DC.
82b Schulten, H.R. (1978) Adv. Mass Spectrom. 7A, 83.
83 Schulten, H.R. (1979) Int. J. Mass Spectrom. Ion Phys. 32, 97.
84 Winkler, H.U., Beuhler, R.J. and Friedmann, L. (1976) Biomed. Mass Spectrom. 3, 201.
85 Deutsch, J. (1980) Org. Mass Spectrom. 15, 240.
86 Posthumus, M.A., Nibbering, N.M.M., Boerboom, A.J.H. and Schulten, H.R. (1974) Biomed. Mass Spectrom. 1, 352.
87 Hunt, D.F., Shabanowitz, J., Botz, F.K. and Brent, D.A. (1977) Anal. Chem. 49, 1160.
88a Hunt, D.F. and Sethi, S.K. (1978) in Chemical Applications of High Performance Mass Spectrometry (Gross, M.L., ed.), p. 150, American Chemical Society, Washington, DC.
88b Hostettman, K., Doumas, J. and Hardy, M. (1981) Helv. Chim. Acta 64, 297.
89 Arpino, P. and Devant, G. (1979) Analysis 7, 348.
90 Rapp, U., Dielmann, G., Games, D.E., Gower, J.L. and Lewis, E. (1980) Adv. Mass Spectrom. 8, 1660.
91 Rapp, U., Meyerhoff, G. and Dielmann, G. (1980) Osterreichische Chemie Zeitschrift, Heft 4, 120.
92 Torgerson, D.F., Skowronski, R.P. and MacFarlane, R.D. (1974) Biochem. Biophys. Res. Commun. 60, 615.
93a MacFarlane, R.D. (1978) in National Bureau of Standards Special Publication 519, Trace Organic Analysis: A New Frontier in Analytical Chemistry, Proceedings of the 9th Materials Research Symposium, p. 673.
93b MacFarlane, R.D. (1980) in Biochemical Applications of Mass Spectrometry, First Supplementary Volume (Waller, R.G., ed.), p. 1209, Wiley Interscience, New York.
94a MacFarlane, R.D. and Forgerson, D.F. (1976) Int. J. Mass Spectrom. Ion Phys. 21, 81.
94b See Reference 93a.
94c Chait, B.T., Field, F.H. and Agosta, W.C. (1980) in 28th Annual Conference on Mass Spectrometry and Allied Topics, New York, FAMOB 11, p. 657.

94d Hunt, J.E. and MacFarlane, R.D. (1980) in 28th Annual Conference on Mass Spectrometry and Allied Topics, New York, FAMOB 11, p. 659.
95 McNeal, C.J., Narang, S.A., MacFarlane, R.D., Hsiung, H.M. and Brousseau, R. (1980) Proc. Natl. Acad. Sci. U.S.A. 77, 735.
96 Posthumus, M.A., Kistemaker, P.G., Meuzelaar, H.L.C. and Ten Noever de Brauw, M.C. (1978) Anal. Chem. 50, 985.
97 Hardin, E.D. and Vestal, M.L. (1980) in 28th Annual Conference on Mass Spectrometry and Allied Topics, New York, RPMP 16, p. 616.
98a Cotter, R.J. (1980) Anal. Chem. 52, 1767.
98b Cotter, R.J. and Fenselau, C. (1980) in 28th Annual Conference in Mass Spectrometry and Allied Topics, New York, RPMP 17, p. 618.
99 Hunt, D.F., Bone, W. and Shabanowicz, J. (1980) in 28th Annual Conference in Mass Spectrometry and Allied Topics, New York, RPMP 18, p. 620.
100 Schueler, B. and Krueger, F.R. (1980) Org. Mass Spectrom. 15, 295.
101 Kistemaker, P.G., Van der Peyl, G.J.Q., Boerboom, A.J.H. and Haverkamp, J. (1980) in 28th Annual Conference on Mass Spectrometry and Allied Topics, New York, FAMOB 7, p. 652.
102a Benninghoven, A. (1978) in National Bureau of Standards Special Publication 519, Trace Organic Analysis: A New Frontier in Analytical Chemistry, Proceedings of the 9th Materials Research Symposium, p. 627, Washington, DC.
102b Benninghoven, A. and Sichtemann, W.K. (1978) Anal. Chem. 50, 1180.
102c Benninghoven, A., Powell, R.A., Shimizu, R. and Storme, H.A. (1979) in Secondary Ion Mass Spectrometry, SIMS II, Springer in Chemical Physics 9, Springer Verlag, Berlin, pp. 116, 127.
103 Eicke, A., Sichtermann, N. and Benninghoven, A. (1980) Org. Mass Spectrom. 15, 289.
104a Jennings, K.R. (1971) in Mass Spectrometry, Techniques and Applications (Milne, G.W.A., ed.), p. 419, Wiley Interscience, New York.
104b Weston, A.F., Jennings, K.R., Evans, S. and Elliott, R.M. (1976) Int. J. Mass Spectrom. Ion Phys. 20, 317.
104c Jennings, K.R. (1978) in High Performance Mass Spectrometry: Chemical Applications (Gross, M.L., ed.), p. 1, American Chemical Society, Washington, DC.
105a Beynon, J.H., Cooks, R.G., Amy, J.W., Battinger, W.E. and Ridley, T.Y. (1973) Anal. Chem. 45, 1023A.
105b Cooks, R.G. and Beynon, J.H. (1974) J. Chem. Ed. 51, 437.
105c Cooks, R.G., Beynon, J.H., Caprioli, R.M. and Lester, G.R. (1973) Metastable Ions, Elsevier, Amsterdam.
106a McLafferty, F.W. and Bockhoff, F.W. (1978) Anal. Chem. 50, 69.
106b McLafferty, F.W., Todd, P.J., McGilvery, D.C. and Baldwin, M.A. (1980) J. Am. Chem. Soc. 102, 3360.
107a Schlunegger, P. (1975) Angew. Chem. Int. Ed. Engl. 14, 679.
107b Schlunegger, P. (1980) Advanced Mass Spectrometry, Applications in Organic and Analytical Chemistry, Pergamon Press, London.
108a Bruins, A.P., Jennings, K.R. and Evans, S. (1978) Int. J. Mass Spectrom. Ion Phys. 26, 395.
108b Jennings, K.R. (1968) Int. J. Mass Spectrom. Ion Phys. 1, 227.
108c Cooks, R.G. and Beynon, J.M. (1975) in Mass Spectrometry (Maccoll, A., ed.), M.T.P. International Review of Science, Series 2, Vol. 5, p. 159, Butterworths, London.
109a McLafferty, F.W., Bente, P.F., Kornfeld, R., Tsai, S.C. and Howe, I. (1973) J. Am. Chem. Soc. 95, 2120.
109b McLafferty, F.W. (1978) High Performance Mass Spectrometry: Chemical Applications (Gross, M.L., ed.), p. 45, American Chemical Society, Washington, DC.
110a Beynon, J.H., Cooks, R.G., Amy, J.W., Baitinger, W.E. and Ridley, T.Y. (1973) Anal. Chem. 45, 1023A.
110b Keough, T., Beynon, J.H. and Cooks, R.G. (1973) J. Am. Chem. Soc. 95, 1695.
110c Cooks, R.G. (1978) Collision Spectroscopy, Plenum Press, New York.
111 Maquestiau, A., Van Haverbeke, Y., De Meyer, C., Flammang, R. and Perlaux, J. (1976) Bull. Soc. Chim. Belge 85, 69.

112 Levsen, K. and Schwarz, H. (1976) Angew. Chem. Int. Ed. Engl. 15, 509.
113 Ast, T., Bozorgzadeh, M.H., Wiebers, J.L., Beynon, J.H. and Brenton, A.G. (1979) Org. Mass Spectrom. 14, 313.
114 Jennings, K.R. (1965) J. Chem. Phys. 43, 4176.
115 Barber, M. and Elliott, R.M. (1964) in 12th Annual Conference on Mass Spectrometry and Allied Topics, Montreal, Canada.
116 Gallegos, E.J. (1978) in High Performance Mass Spectrometry: Chemical Applications (Gross, M.L., ed.), p. 274, American Chemical Society, Washington, DC.
117 Ferguson, A.M., Gwyn, S.A., Pannell, L.K. and Wright, G.J. (1977) Anal. Chem. 49, 174.
118 Beynon, J.H. (1970) Anal. Chem. 42, 97A.
119a Beynon, J.H., Caprioli, R.M. and Ast, T. (1972) Org. Mass Spectrom 6, 273.
119b Caprioli, R.M., Beynon, J.H. and Ast, T. (1971) Org. Mass Spectrom. 5, 417.
120 Safe, A., Jamieson, W. and Hutzinger, O. (1972) Org. Mass Spectrom. 6, 33.
121 Safe, S., Hutzinger, Q. and Cook, M. (1972) J.C.S. Chem. Comm., 260.
122 Beynon, J.H. and Cooks, R.G. (1971) Res. Dec. 22, 26.
123 Morgan, R.P., Beynon, J.H., Bateman, R.H. and Green, B.N. (1978) Int. J. Mass Spectrom. Ion Phys. 28, 171.
124 Tabet, J.C., Stahl, D. and Gaumann, T., unpublished results.
125 Jaudon, P. and Tabet, J.C. (1980) Org. Mass Spectrom. 15, 65.
126 Bozorgzadch, K.H., Morgan, R.P. and Beynon, J.H. (1978) Analyst. 103, 613.
127a Brentoa, A.G. and Beynon, J.H. (1980) Eur. Spectros. News 29, 39.
127b Beynon, J.H. and Caprioli, R.M. (1980) in Biochemical Applications of Mass Spectrometry, First Supplementary Volume (Waller, R.G., ed.), p. 89, Wiley Interscience, New York.
128 Frappier, F., Jouany, M., Fraisse, D., Marquet, A. and Tabet, J.C. (1980) in 28th Annual Conference on Mass Spectrometry and Allied Topics, New York, p. 686.
129 Maquestiau, A., Van Haverbeke, Y., De Meyer, C., Duthoit, C., Meyrant, P. and Flammang, R. (1979) Nouv. J. Chimie 3, 517.
130 Boyd, R.K. and Beynon, J.H. (1977) Org. Mass Spectrom. 12, 163.
131 Lacey, M.J. and MacDonald, C.G. (1980) Org. Mass Spectrom. 15, 135.
132 Evans, S. and Graham, R. (1974) Adv. Mass Spectrom. 6, 429.
133 Weston, A.F., Jennings, K.R., Evans, S. and Elliott, R.M. (1976) Int. J. Mass Spectrom. Ion Phys. 20, 317.
134 Lacey, J. and MacDonald, C.G. (1975) J.C.S. Chem. Comm., 421.
135a Jennings, K.R. (1978) in High Performance Mass Spectrometry: Chemical Applications (Gross, M.L., ed.), p. 1, American Chemical Society, Washington, DC.
135b Bruins, A.P., Jennings, K.R. and Evans, S. (1978) Int. J. Mass Spectrom. Ion Phys. 26, 395.
136 Haddon, W.F. and Young, R. (1978) in 26th Annual Conference on Mass Spectrometry and Allied Topics, St. Louis, FA 7, p. 472.
137 Shushan, B. and Boyd, R.K. (1980) Int. J. Mass Spectrom. Ion Phys. 34, 37.
138 Tulp, M.T.M., Safe, S.M. and Boyd, R.K. (1980) Biom. Mass Spectrom. 7, 109.
139 Jennings, K.R., Stradling, R.S. and Warburton, G.A. (1979) 27th Annual Conference on Mass Spectrometry and Allied Topics, Seattle, FD 13, p. 542.
140a Warburton, G.A., McDowell, R.A., Hazelby, D., Evans, S. and Compton, K.R. (1979) 27th Annual Conference on Mass Spectrometry and Allied Topics, Seattle, p. 133.
140b Warburton, G.A., McDowell, R.A. and Compton, K.R. (1979) in 27th Annual Conference on Mass Spectrometry and Allied Topics, Seattle, p. 469.
140c Warburton, G.A., McDowell, R.A., Taylor, K.T. and Chapman, J.R. (1980) Adv. Mass Spectrom. 8B, 1953.
141 Morgan, R.P., Forter, C.J. and Beynon, J.H. (1978) 26th Annual Conference on Mass Spectrometry and Allied Topics, St. Louis, p. 539.
142 Lacey, M.J. and McDonald, C.G. (1979) Org. Mass Spectrom. 14, 465.
143a Shushan, B., Safe, S.H. and Boyd, R.K. (1979) Anal. Chem. 51, 156.

143b Shushan, B. and Boyd, R.K. (1980) Org. Mass Spectrom. 15, 445.
144 Porter, C.J., Brenton, A.G. and Beynon, J.H. (1980) Int. J. Mass Spectrom. Ion Phys. 36, 69.
145 Haddon, W.F. (1979) Anal. Chem. 51, 983.
146a Lacey, M.L. and MacDonald, C.G. (1979) Anal. Chem. 51, 691.
146b Lacey, M.L. and MacDonald, C.G. (1979) Org. Mass Spectrom. 14, 465.
147 Haddon, W.F. (1980) Org. Mass Spectrom. 15, 539.
148 Shushan, B. and Boyd, R.K. (1979) in 27th Annual Conference on Mass Spectrometry and Allied Topics, Seattle, FAMP 19, p. 747.
149 Zakett, D., Schoen, A.E., Kondrat, R.W. and Cooks, R.G. (1979) J. Am. Chem. Soc. 101, 6781.
150 Haddon, W.F. and McLafferty, F.W. (1965) J. Am. Chem. Soc. 90, 4745.
151 McLafferty, F.W. and Schuddemage, H.D.R. (1969) J. Am. Chem. Soc. 91, 1866.
152 Tabet, J.C., Stahl, D. and Schwarz, H., unpublished results.
153 Audier, H., Milliet, A., Mruzek, M. and Tabet, J.C. (1980) Bull. Soc. Chim. Belge 89, 157.
154a Stahl, D. and Tabet, J.C. (1979) Chimia 33, 287.
154b McLafferty, F.W., Todd, P.J., McGilvery, D.C. and Baldwin, M.A. (1980) J. Am. Chem. Soc. 102, 3360.
155 Stradling, R.S., Jennings, K.R. and Evans, S. (1978) Org. Mass Spectrom. 13, 429.
156 Kambara, H., Walls, F.C., McPherron, R., Straub, K. and Burlingame, A.L. (1979) in 27th Annual Conference on Mass Spectrometry and Allied Topics, Seattle, MPMP 13, p. 184.
157 Harris, D., McKinnan, S. and Boyd, R.K. (1979) Org. Mass Spectrom. 14, 265.
158 Zakett, O., Shaddock, V.M. and Cooks, R.G. (1979) Anal. Chem. 51, 1849.
159 Bowie, J.H. and Hart, S.G. (1974) Int. J. Mass Spectrom. Ion Phys. 13, 319.
160 Bowie, J.H. and Blumenthal, T. (1975) J. Am. Chem. Soc. 97, 2959.
161 Bowie, J.H. and Ho, A.C. (1977) Aust. J. Chem. 30, 657.
162 Bowie, J.H. and White, P.Y. (1978) Aust. J. Chem. 31, 1511.
163 Bowie, J.H., White, P.Y., Wilson, J.C., Larsson, F.C.V., Lawesson, S.O., Madsen, J.O., Nolde, C. and Schroll, G. (1977) Org. Mass Spectrom. 12, 191.
164 Cooks, R.G. (1978) in National Bureau of Standards Special Publication 519, Trace Organic Analysis: A New Frontier in Analytical Chemistry, Proceeding of the 9th Materials Research Symposium, p. 609.
165 McCluskey, G.A., Kondrat, R.W. and Cooks, R.G. (1978) J. Am. Chem. Soc. 100, 6045.
166a Winkler, F.J., Stahl, D. and Guenat, C. (1980) Euchem Conference, Ion Beam, Utrecht.
166b Beloeil, J.C., Bertranne, M., Stahl, D. and Tabet, J.C. (1983) J. Am. Chem. Soc. 105, 1335.
167 Weiss, M., Karnofsky, J., Hass, R. and Harvan, D. (1979) in 27th Annual Conference on Mass Spectrometry and Allied Topics, Seattle, p. 270.
168 Gaumann, T., Guenat, C. and Stahl, D., unpublished results.
169 Bolton, P.D., Trott, G.W., Morgan, R.P., Brenton, A.G. and Beynon, J.H. (1979) Int. J. Mass Spectrom. Ion Phys. 29, 179.
170 Mandli, H., Robbiani, R., Kuster, T. and Seibl, J. (1979) Int. Mass Spectrom. Ion Phys. 31, 57.
171 Haddon, W.F. (1979) Anal. Chem. 51, 983.
172 Bill, J.C., Gilbert, A.J., Wallington, M.J. and Powers, P. (1980) in 28th Annual Conference on Mass Spectrometry and Allied Topics, New York, MPMP 23, p. 193.
173a McLafferty, F.W., Todd, P.J., McGilvery, D.C. and Baldwin, M.A. (1980) J. Am. Chem. Soc. 102, 3360.
173b McLafferty, F.W. (1980) Acc. Chem. Res. 13, 33.
173c McLafferty, F.W., Todd, P.J. (1980) Org. Mass Spectrom. 15, 272.
173d Barbalas, M., Todd, P.J., Pagano, P.F., Pegues, R.F. and McLafferty, F.W. (1980) in 28th Annual Conference on Mass Spectrometry and Allied Topics, New York, TPMP 19, p. 351.
174 Maquestiau, A., Van Haverbeke, Y., Flammang, R., Abrassant, M. and Finet, D. (1978) Bull. Soc. Chim. Belge 87, 765.
175a Russell, D.H., Smith, D.H. and Warmack, R.J. (1979) 27th Annual Conference on Mass Spectrometry and Allied Topics, Seattle, p. 703.

175b Russell, D.H., Smith, D.H., Warmack, R.J. and Bertram, L.K. (1980) Int. J. Mass Spectrom. Ion Phys. 33, 381.

175c Russell, D.H., McBay, E.H. and Mueller, T.R. (1980) Int. Lab. 49.

175d Russell, D.H., McBay, E.H. and Mueller, T.R. (1980) in 28th Annual Conference on Mass Spectrometry and Allied Topics, New York, TPMP 2, p. 323.

176 Vrscaj, V., Kramer, V., Medved, M., Kralj, B., Marsel, J., Beynon, J.H. and Ast, T. (1980) Int. Mass Spectrom. Ion Phys. 33, 409.

177 Gross, M.L., Lyon, P.A., Crow, F.W. and Evans, S. (1980) in 28th Annual Conference on Mass Spectrometry and Allied Topics, New York, MPMOA 3, p. 107.

178 Todd, P.J., McGilvery, D.C., Baldwin, M.A. and McLafferty, F.W. (1980) in 28th Annual Conference on Mass Spectrometry and Allied Topics, New York, MPMOA 3, p. 48.

179a Louter, G.J., Boerboom, A.J.H., Stalmeirer, P.F.M., Tuithof, H.M. and Kistemaker, J. (1980) Int. J. Mass Spectrom. Ion Phys. 33, 335.

179b Boerboom, A.J.H., Louter, G.J. and Haverkamp, J. (1980) in 28th Annual Conference on Mass Spectrometry and Allied Topics, New York, MPMOA 1, p. 103.

180a Yost, R.A. and Enke, C.G. (1978) J. Am. Chem. Soc. 100, 2274.

180b Latven, R.K., Yost, R.A. and Enke, C.G. (1979) in 27th Annual Conference on Mass Spectrometry and Allied Topics, Seattle, MPMOB 6, p. 131.

180c Enke, C.G., Chakel, J.A. and Darland, E.J. (1979) in 27th Annual Conference on Mass Spectrometry and Allied Topics, Seattle, MPMOC 4, p. 151.

180d Yost, R.A. and Enke, C.G. (1979) in 27th Annual Conference on Mass Spectrometry and Allied Topics, Seattle, WAMP 12, p. 461.

180e Yost, R.A. and Enke, G.C. (1979) Anal. Chem. 51, 1251A.

181 Boitnott, C.A., Steiner, U. and Story, M.S. (1980) in 28th Annual Conference on Mass Spectrometry and Allied Topics, New York, MPMOA 2, p. 105.

182 Hunt, D.F., Shabanowitz, J. and Giordani, A.B. (1980) Anal. Chem. 52, 386.

183 Hunt, D.F., Shabanowitz, J. and Giordani, A. (1979) in 27th Annual Conference on Mass Spectrometry and Allied Topics, Seattle, FAMOC 2, p. 701.

184 Hunt, D.F., Shabanowitz, J. and Giordani, A.B. (1980) in 28th Annual Conference on Mass Spectrometry and Allied Topics, New York, RAMOA 8, p. 476.

185a Siegel, M.W. (1980) in 28th Annual Conference on Mass Spectrometry and Allied Topics, New York, RAMOA 9, p. 478.

185b Siegel, M.W. (1980) Anal. Chem. 52, 1790.

186 Zakett, D. and Cooks, R.G. (1980) in 28th Annual Conference on Mass Spectrometry and Allied Topics, New York, MAMOA 3, p. 18.

187a Zakett, D., Cooks, R.G. and Fites, W.J. (1980) Anal. Chim. Acta 119, 129.

187b Glish, G.L., Hemberger, P.H. and Cooks, R.G. (1980) Anal. Chim. Acta 119, 137.

187c Glish, G.L. and Cooks, R.G. (1980) Anal. Chim. Acta 119, 145.

187d Zakett, D., Hemberger, P.H. and Cooks, R.G. (1980) Anal. Chim. Acta 119, 149.

187e Busch, K.L., Kruger, T.L. and Cooks, R.G. (1980) Anal. Chim. Acta 119, 153.

188 Glish, G.L., Zakett, D., Hemberger, P.H. and Cooks, R.G. (1980) in 28th Annual Conference on Mass Spectrometry and Allied Topics, New York, TPMP 10, p. 335.

189a Kondrat, R.W. and Cooks, R.G. (1978) Anal. Chem. 50, 81A.

189b Kruger, T.L., Litton, J.F., Kondrat, R.W. and Cooks, R.G. (1976) Anal. Chem. 48, 2113.

189c Kondrat, R.W., McClusky, G.A. and Cooks, R.G. (1978) Anal. Chem. 50, 2017.

190 Harvan, D.J., Hass, J.R. and Busch, K.L. (1979) in 27th Annual Conference on Mass Spectrometry and Allied Topics, Seattle, RAMOC 10, p. 529.

191 Fraisse, D. and Maquin, F. (1980) in Congres de Spectrometrie de Masse Fondamentale et Appliquées (G.A.M.S.), Dijon.

192a Bobenrieth, M.J., Hass, J.R., Liao, W.T., Christman, R.F. and Pfaender, F.K. (1979) in 27th Annual Conference on Mass Spectrometry and Allied Topics, Seattle, RAMP 12, p. 550.

192b Harvan, D.J., Hass, J.R. and Busch, K.L. (1979) in 27th Annual Conference on Mass Spectrometry and Allied Topics, Seattle, RAMOC 10, p. 529.

193a Chess, E.K., Crow, F.W., Lyon, P.A. and Gross, M.L. (1980) in 28th Annual Conference on Mass Spectrometry and Allied Topics, New York, p. 596.
193b Chess, E.K. and Gross, M.L. (1980) Anal. Chem. 52, 1880.
194 Tabet, J.C. and Gaudemer, A., unpublished results.
195 McLafferty, F.W., Kornfeld, R., Haddon, W.F., Levsen, K., Sakai, I., Bente, P.F. III, Tsai, S.C. and Schuddenage, H.D.R. (1973) J. Am. Chem. Soc. 95, 3886.
196a Smith, D.F., Djerassi, C., Maurer, K.H. and Rapp, U. (1974) J. Am. Chem. Soc. 96, 3482.
196b Djerassi, C. (1978) Pure Appl. Chem. 50, 171.
197a Levsen, K. and Beckey, H.D. (1974) Org. Mass Spectrom. 9, 570.
197b Weber, R. and Levsen, K. (1980) Biomed. Mass Spectrom. 7, 314.
198 Baczynsky, J.L. and Duchamp, D.J. (1978) Adv. Mass Spectrom. 7, 1243.
199 Horvath, G.Y. and Ambrus, G. (1978) Adv. Mass Spectrom. 7, 1280.
200 Schlunegger, U.P. (1978) Chimia 32, 9.
201 Brown, F.J. and Djerassi, C. (1980) J. Am. Chem. Soc. 102, 807.
202 Bozorgzadeh, M.H., Brenton, A.G., Wieber, J.L. and Beynon, J.H. (1979) Biomed. Mass Spectrom. 6, 340.
203 Zaretskii, Z.V. (1980) Adv. Mass Spectrom. 8, 1239.
204 Zaretskii, Z.V., Larka, E.A., Howe, I. and Beynon, J.H. (1980) in 28th Annual Conference on Mass Spectrometry and Allied Topics, New York, TPMP 23, p. 358.
205 Patterson, D.G., Lavanchy, A. and Djerassi, C. (1980) Org. Mass Spectrom. 15, 41.
206a Gaskell, S.J., Pike, A.W. and Millington, D.S. (1979) Biomed. Mass Spectrom. 6, 78.
206b Gaskell, S.J. and Millington, D.S. (1978) in 26th Annual Conference on Mass Spectrometry and Allied Topics, St. Louis, RA 10, p. 350.
207a Millington, D.S., Smith, J.A. and Gaskell, S.J. (1979) in 27th Annual Conference on Mass Spectrometry and Allied Topics, Seattle, RPMP 8, p. 638.
207b Gaskell, S.J. and Millington, D.S. (1978) Biomed. Mass Spectrom. 5, 557.
208a Gaskell, S.J., Finlay, E.M.H. and Millington, D.S. (1980) Adv. Mass Spectrom. 8, 1908.
208b Gaskell, S.J. and Pike, A.W. (1980) Adv. Mass Spectrom. 8, 279.
209 Maquestiau, A., Van Haverbeke, Y., Flammang, R., Misprevue, H., Kaisin, M., Braekman, J.C., Daloze, D. and Tursch, B. (1978) Steroids 31, 31.
210 Krugger, T.L., Kondrat, R.W., Joseph, K.T. and Cooks, R.G. (1979) Anal. Biochem. 96, 014.
211 Duholke, W.K. and Fox, E. (1980) in 28th Annual Conference Mass Spectrometry and Allied Topics, New York, p. 353.
212 Sun, T. and Lovins, R.E. (1972) Org. Mass Spectrom. 6, 39.
213a Russell, D.W., Jamiesm, W.D., Taylor, A. and Das, R.C. (1976) Can. J. Chem. 54, 1355.
213b Rahman, R., Taylor, A., Das, R.C. and Verpoorke, J.A. (1976) Can. J. Chem. 54, 1360.
214a Schlunegger, U.P., Hirter, P. and Von Feton, H. (1976) Helv. Chim. Acta 59, 406.
214b Schlunegger, U.P. and Hirter, P. (1978) Isr. J. Chem. 17, 168.
214c Steinaver, R., Walther, H. and Schlunegger, U.P. (1980) Helv. Chem. Acta 63, 610.
215 Schlunegger, U.P., Steinaver, R. and Walther, H.J. (1979) in 8th International Mass Spectrometry Conference, Oslo.
216 Ronje, H.W. and Grutzmatcher, H.F. (1980) in 7th International Symposium on Mass Spectrometry in Biochemistry, Medicine and Environmental Research, Milan.
217 Millard, B.J. (1965) Tetrahedron Lett. 3041.
218 Levsen, K., Wipf, H.K. and McLafferty, M.W. (1974) Org. Mass Spectrom. 8, 117.
219 Hunt, D.F., Buko, A.M., Baillard, J. and Shabanowitz, J. (1979) in 27th Annual Conference on Mass Spectrometry and Allied Topics, Seattle, FAMOB 5, p. 680.
220 Baillard, J.M., Hunt, D.F. and Buko, A.M. (1980) in 28th Annual Conference on Mass Spectrometry and Allied Topics, New York, WAMOP 19, p. 150.
221 Hunt, D.F., Sisak, M., Buko, A., Baillard, J., Giordani, A. and Shabanowitz, J. (1980) in 28th Annual Conference on Mass Spectrometry and Allied Topics, New York, WAMOB 4, p. 390.
222 Bradley, C.V., Howe, I. and Beynon, J.H. (1980) J.C.S. Chem. Comm., 562.

223 Burlingame, A.L., Kambara, H. and Walls, F.C. (1978) in 26th Annual Conference on Mass Spectrometry and Allied Topics, St. Louis, p. 184.
224 Baillie, T.A., Kambara, H., Nelson, S.D. and Vaishna, Y. (1979) in 27th Annual Conference on Mass Spectrometry and Allied Topics, Seattle, WAMOA 1, p. 370.
225 McReynolds, J.H. and Anbar, M. (1977) Anal. Chem. 49, 1832.
226 Stapleton, B.J. and Bowie, J.H. (1976) Org. Mass Spectrom. 11, 429.
227 Kondrat, R.W., McClusky, G.A. and Cooks, R.G. (1978) Anal. Chem. 50, 1222.
228 Addeo, F., Malorni, A. and Marino, G. (1975) Anal. Biochem. 64, 98.
229 Klein, R.A. (1972) J. Lipid. Res. 13, 672.
230 Das, K.G. and Thayumanavan, B. (1975) Org. Mass Spectrom. 10, 455.
231 Daniels, P.J.L., Mallams, A.K., Wainstein, J., Wright, J.J., Milne, G.W.A. and Perkin, ?. (1976) J.C.S. I, 1078.
232 Ando, S., Ariga, T. and Yamakawa, T. (1976) Bull. Chem. Soc. Japan 49, 1335.
233 Warburton, G., McDowell, R., Hazelby, D., Evans, S and Compson, K.R. (1979) in 27th Annual Conference on Mass Spectrometry and Allied Topics, Seattle, p. 133.
234 David, L., Scanzi, E., Fraisse, D. and Tabet, J.C. (1982) Tetrahedron, 38, 1609.
235 Kambara, H. and Burlingame, A.L. (1979) in 27th Annual Conference on Mass Spectrometry and Allied Topics, Seattle, FAMOB 15, p. 697.
236 De Jong, E.G., Heerma, W. and Dijkstra, G. (1980) Biomed. Mass Spectrom. 7, 127.
237 VG Instruments (1979) technical publication.
238 D'Angona, J., Linscheid, M., Burlingame, A.L., Yamada, H., Ballou, C.E. and Dell, A. (1980) in 28th Annual Conference on Mass Spectrometry and Allied Topics, New York, p. 495.
239 Puzo, G., Promé, J.C. and Maxime, B. (1980) Adv. Mass Spectrom. 8, 1003.
240 McClusky, G.A., Kondrat, R.W. and Cooks, R.G. (1978) J. Am. Chem. Soc. 100, 6045.
241a Krugger, T.L., Litton, J.F. and Cooks, R.G. (1976) in 24th Annual Conference on Mass Spectrometry and Allied Topics, San Diego, R-11, p. 404.
241b Krugger, T.L., Cooks, R.G., McLaughlin, J.L. and Ranieri, R.L. (1977) J. Org. Chem. 42, 4161.
242 Tamas, J., Czira, G. and Bujtas, G. (1980) Adv. Mass Spectrom. 8, 733.
243 Youssefi, M., Cooks, R.G. and McLaughlin, J.L. (1979) J. Am. Chem. Soc. 101, 3400.
244a McReynolds, J.H. and Anbar, W. (1977) in 25th Annual Conference on Mass Spectrometry and Allied Topics, Washington, DC, p. 570.
244b McReynolds, J.H. and Anbar, W. (1977) Int. J. Mass Spectrom. Phys. 24, 37.
245 Maquestiau, A., Van Haverbeke, Y., De Meyer, C., Duthoit, C., Meyrant, P. and Flammang, R. (1979) Now. J. Chimie 3, 517.
246 Zakett, D. and Cooks, R.G. (1980) in 28th Annual Conference on Mass Spectrometry and Allied Topics, New York, p. 18.
247 Safe, S., Jamieson, W.D. and Hutzinger, O. (1972) Org. Mass Spectrom. 6, 33.
248 Occolowitz, J.C., Barbalas, M.P. and McLafferty, F.W. (1980) in 28th Annual Conference on Mass Spectrometry and Allied Topics, New York, p. 474.
249 Haddon, W.F. and Molyneux, R.J. (1980) in 28th Annual Conference on Mass Spectrometry and Allied Topics, New York, p. 472.
250 Puzo, G., Schram, K.H., Liehr, J.G. and McCloskey, J.A. (1978) J. Org. Chem. 43, 767.
251a Schoen, A.E., Wiebers, J.L. and Cooks, R.G. (1978) in 26th Annual Conference on Mass Spectrometry and Allied Topics, St. Louis, FB 5, p. 491.
251b Schoen, A.E., Cooks, R.G. and Wiebers, J.L. (1979) Science 203, 1249.
252 Bozorgzadeh, M.N., Beynon, J.H. and Wiebers, J.L. (1980) Adv. Mass Spectrom. 8, 97.
253 Levsen, K. and Schulten, H.R. (1976) Biomed. Mass Spectrom. 3, 137.
254a Straub, K.M. and Burlingame, A.L. (1979) in 27th Annual Conference on Mass Spectrometry and Allied Topics, Seattle, WAMOA 16, p. 394.
254b Straub, K.M., Meehan, H., Kambara, H., Burlingame, A.L. and Evans, S. (1978) in 26th Annual Conference on Mass Spectrometry and Allied Topics, St. Louis, p. 499.
255 Straub, K.M. and Burlingame, A.L. (1980) Adv. Mass Spectrom. 8, 1127.

256 McClusky, G.C., Huang, S.K.S., Moore, C.J. and Selkirk, J.K. (1980) in 28th Annual Conference on Mass Spectrometry and Allied Topics, New York, RAMOA 5, p. 469.
257 Linscheid, M. and Burlingame, A.L. (1980) 28th Annual Conference on Mass Spectrometry and Allied Topics, New York, p. 467.
258 Cooks, R.G. and Glish, G.L. (1981) Chem. Eng. 40, Nov. 30.
259 Devienne, F.M. (1967) Entropie 18, 61.
260 Barber, M., Borodoli, R.S., Sedgwick, R.D. and Tyler, A.N. (1981) J. Chem. S. Chem. Comm. 325.
261 Busch, K.I., Unger, S.E., Vineze, A., Cooks, R.G. and Keough, T. (1982) J. Am. Chem. Soc. 104, 1507.
262 Krueger, F.R. and Schulten, B. (1980) Adv. Mass Spectrom. 8B, 918.
263a Cotter, R.J. (1979) Anal. Chem. 51, 317.
263b Cotter, R.J. and Yergey, A.L. (1981) J. Am. Chem. Soc. 103, 1596.
263c Yergey, A.L. and Cotter, R.J. (1982) Biom. Mass Spectrom. 9, 286.
264 McNeal, C.J. (1982) Anal. Chem. 54, 43A.
265 Fenselau, C. (1982) Anal. Chem. 54, 105A.
266 McNeal, C.J. and MacFarlane, R.D. (1981) J. Am. Chem. Soc. 103, 1609.
267 Chait, B.T. and Field, F.H. (1981) Int. J. Mass Spectrom. Ion Phys. 41, 17.
268 Sindona, G., Uccella, N. and Weclawek, K. (1982) J. Chem. Res. 5, 184.
269 Liu, L.K., Unger, S.E. and Cooks, R.G. (1981) Tetrahedron 37, 1067.
270 Zakett, T., Schoem, A.E. and Cooks, R.G. (1981) J. Am. Chem. Soc. 103, 1295.
271 Smith, R.D., Burger, J.A. and Johnson, A.L. (1981) Anal. Chem. 53, 1603.
272 Jungclas, H., Danigen, H. and Smith, L. (1982) Org. Mass Spectrom. 17, 86.
273 Hardlin, E.D. and Vestal, M.L. (1981) Anal. Chem. 53, 1492.
274 Heniom, J.D., Thomson, B.A. and Dawson, P.H. (1982) Anal. Chem. 54, 451.
275a Miller, S.A. (1981) Biomed. Mass Spectrom. 8, 376.
275b Beynon, J.H. (1981) Biomed. Mass Spectrom. 8, 380.

Neuberger/Van Deenen (eds.) Modern Physical Methods in Biochemistry, Part A

CHAPTER 4

Absorption, circular dichroism and optical rotatory dispersion of polypeptides, proteins, prosthetic groups and biomembranes

DAN W. URRY

Laboratory of Molecular Biophysics, University of Alabama in Birmingham, School of Medicine, Birmingham, AL 35294, U.S.A.

1. Introduction

The physical methods of absorption, circular dichroism (CD) and optical rotatory dispersion (ORD) are invaluable laboratory tools which are useful at every degree of familiarity with the method. With no knowledge of the physical processes being necessary, these tools are used routinely to demonstrate the satisfactory completion of a preparation or to indicate the purity of a product, as the sodium D-line rotation, $[\alpha]_D^t$, and refractive index, n_D, have been used routinely by the synthetic chemist. Again, no understanding of the mechanism of absorption or rotation is required to detect and characterize interactions and reactions in terms of binding constants and rate constants. The use of these methods to obtain structural information requires knowledge of characteristic spectra of a given absorbing unit (chromophore) or array of chromophores. This can often be achieved by pattern recognition with only limited knowledge of the method. Even here, however, it becomes important to be aware of the problems of instrumental artifact and of distortions of the spectra. Bringing the full power of these optical methods to a structure-function problem in biochemistry requires an understanding of the physical processes and of the causes of spectral band increases, decreases and shifts. In this regard it is important to understand the processes and to be familiar with the relevant mathematical expressions and with their graphical representations.

The physical processes of absorption, refraction, circular dichroism and optical rotatory dispersion are closely inter-related, such that mechanistic discussions of the latter two require adequate consideration of the former two. Accordingly, this presentation begins with consideration of basic elements and terminologies of absorption and refraction, and then treats circular dichroism and optical rotatory dispersion, and in so doing provides prosthetic group examples of excitation resonance interaction and of the dispersion force interactions giving rise to hyper- and hypochromism in absorption and to reciprocal relations in optical rotation. In the subsequent section, characteristic absorption and circular dichroism spectra of some five or more regular protein and polypeptide conformations are presented, as well as of heme peptides, heme protein and adenosine dinucleotides. Finally, in the last section, the particular problem of applying these methods to the timely study of biomembranes is presented by considering the model poly-L-glutamic acid particulate system and by applying the analysis to the purple membrane of *Halobacterium halobium*.

2. Fundamental aspects of absorption and optical rotation

(a) Absorption of ultraviolet and visible light

Light may be described as an oscillating electromagnetic vector with electric and magnetic moments at right angles. This is shown graphically in Figure 1a for light traveling in a vacuum in the z direction with the electric vector oscillating in the xz plane (i.e., plane polarized). The magnitude of the electric vector, E_x, oscillates as a transverse wave according to the expression

$$E_x = E_x^0 \cos 2\pi \nu (t - z/v) \tag{1}$$

where E_x^0 is the amplitude; ν, the frequency; v, the velocity of light in the medium; and t, the time of propagation. The intensity of the light, I, equals Uv where U, the energy density, is proportional to the square of the amplitude, $(E_x^0)^2$. The loss of intensity, $-dI$, as light passes through a length, dl, of an absorbing medium is proportional to the intensity entering the increment of length and to the concentration, C, of absorbing units, i.e., $-dI = kCIdl$ where k is the proportionality constant. Integrating the loss of intensity over the entire path length for a solution of non-interacting absorbing units gives the familiar Beer's Law for absorbance, A,

$$A = -\log I/I_0 = \varepsilon Cl \tag{2}$$

where I_0 is the initial beam intensity; I, the intensity of the emergent beam; C, the concentration in moles per liter; l, the path length in cm; and ε, the molar extinction coefficient. A is also commonly referred to as optical density (OD).

Chromophores of interest in biochemistry are often heteronuclear groupings of atoms, such as the peptide chromophore in Figure 2. Such a grouping of atoms with different electron-negativities and bonding orbitals results in an unequal sharing of electrons. As shown in Figure 2, the oxygen and nitrogen atoms have an excess of electron distribution and have net negative charges whereas the carbon and hydrogen atoms have a deficit of electron distribution and carry partial positive charges. This charge separation in the molecule is represented as a permanent dipole moment, $\vec{\mu}$, shown as an electric vector pointing, by convention, toward the positive charge.

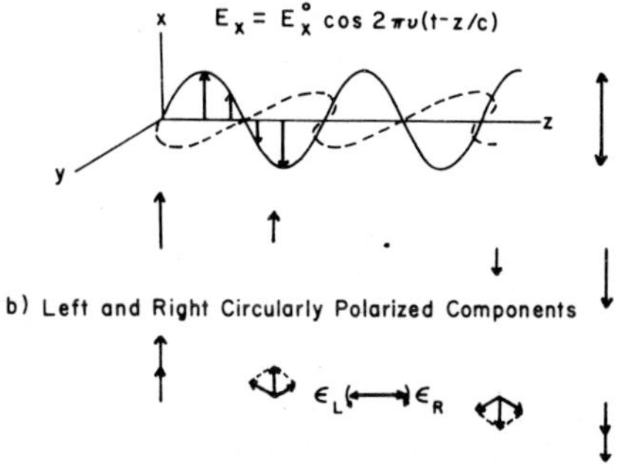

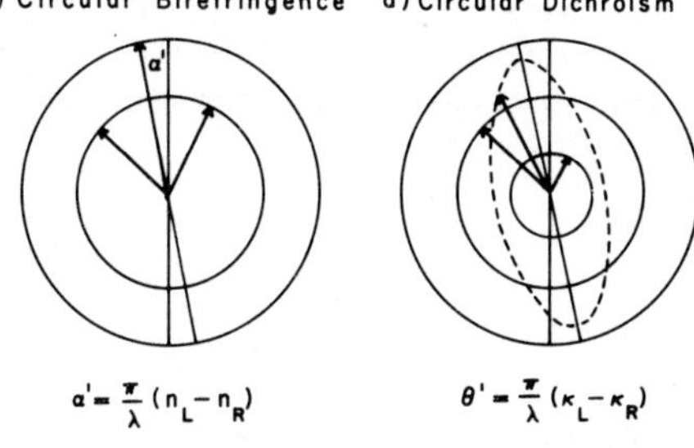

Figure 1. a, Plane polarized electromagnetic radiation traveling in the z direction with the electric vector polarized in the xz plane and the magnetic component in the yz plane. b, Showing plane polarized electric vector as describable by the sum of left and right circularly polarized components traveling in phase. On passing through a medium containing optically active (dysymmetric) molecules, the two components will travel with different velocities and, if in an absorption band, will be differentially absorbed within the medium due to differential interaction of the circularly polarized components with the optically active molecule. c, Circular birefringence (circular double refraction). Since $n_L = c/v_L$ and $n_R = c/v_R$, when the right circularly polarized component is slowed more than the left circularly polarized component, n_R will be greater than n_L and the angle, α', will be negative in accordance with the Fresnel equation given below the drawing. On leaving the optically active medium, the two circularly polarized components combined to give plane polarized light, but the plane of polarization is seen to be rotated by the angle α'. d, Circular dichroism occurs in absorption regions when there is greater absorption of one circularly polarized component. Due to the differential absorption, the circularly polarized component that was preferentially absorbed is of a lesser magnitude such that when the two components are summed on leaving the medium the resultant vector scribes an ellipse, the major axis of which coincides with the angle of rotation. The magnitude of a circular dichroism signal, therefore, depends on the magnitude of the ellipticity. This is the reason for using the units of ellipticity. In current instruments, what is directly measured is the difference absorbance of the left and right circularly polarized components by a phototube that is alternately receiving left and then right circularly polarized light, produced as shown in Figure 13. Reproduced, with permission, from [3].

(i) Electric transition dipole moment and experimental determination of dipole strength

Since light is describable as an oscillating electric vector, it is not surprising that the absorption of light by a chromophore results in a change in the electric vector or electric dipole moment of the absorbing unit. The change in dipole moment of a

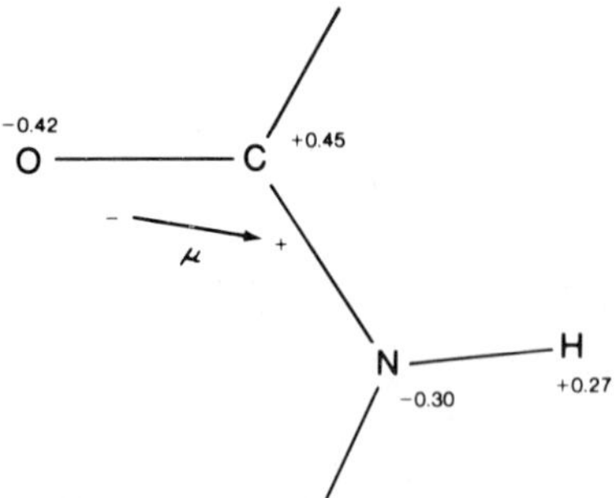

Figure 2. The peptide moiety showing charge distributions on each of four atoms. The result is a permanent (ground state) dipole moment indicated by the arrow with the head of the arrow pointing in the positive direction. On absorption of a photon, the electron distribution changes. This results in the atoms having a new distribution of charge and the excited state will have a different dipole moment. The difference dipole moment (between the ground and excited states) is called the electric transition dipole moment, μ_i, and its magnitude can be calculated from the area of the absorption curve as shown in Figure 3.

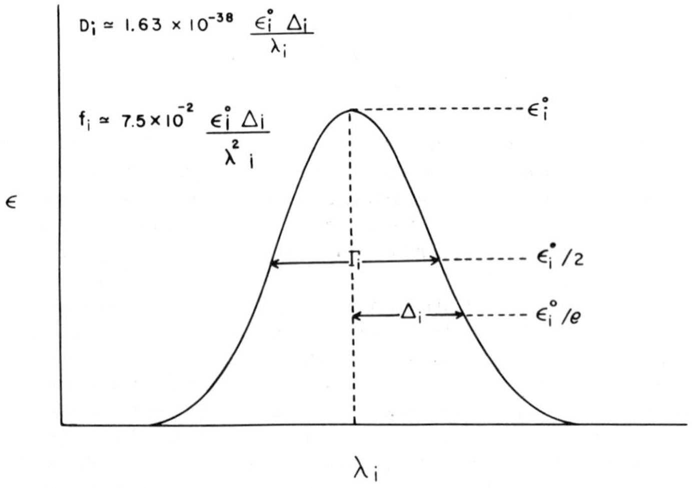

Figure 3. Absorption curve defining the maximum molar extinction coefficient as ε_i^0 and the half band width at ε_i^0/e as Δ_i. The dipole strength, D_i, which is the square of the magnitude of the electric transition dipole moment is calculated by the equation shown, as is the oscillator strength, f_i. Reproduced, with permission, from [3].

chromophore is called an electric transition dipole moment, $\vec{\mu}_i$ [1]. This transition occurs when the energy of light, hv, is the same as the energy required to move the electron to the excited state, i.e., to a higher unoccupied orbital with a different electron distribution. Because the time for the electronic transition is so fast, the absorption process occurs with no change in the position of the nuclei and the electric transition dipole moment is the difference between the permanent dipole moment of the ground state and the dipole moment of the excited state. The magnitude of μ_i is obtainable from the absorption spectrum as shown in Figure 3, where [2]

$$|\vec{\mu}_i^2| = D_i = 1.63 \times 10^{-38} \frac{\varepsilon_i^0 \Delta_i}{\lambda_i} \tag{3}$$

with D_i being the dipole strength, ε_i^0 the molar extinction coefficient at the absorption maximum, λ_i the wavelength in nm of the maximum, and Δ_i the half band width in nm at ε_i^0/e. The interest in determining $\vec{\mu}_i$ is that this quantity, and changes in its magnitude on association of chromophores, provides information on the relative orientation of the chromophores. In a polymer such as a polypeptide this relative orientation is the conformation of interest in structure studies.

(ii) Magnetic transition dipole moment

Some absorptions have only a small electric component, i.e., a small ε_i^0, and a relatively large magnetic transition dipole moment, because of the particular change in electron distribution that is dictated by the ground state and excited state orbitals. A transition of this type is important in the peptide chromophore and is represented in prototype form in Figure 4. It is referred to as an n–π^* transition [5] and involves a transition from a nonbonding $2p_y$ orbital on the oxygen and symmetrically distributed about the xz plane to an antibonding orbital distributed primarily about the yz plane but also delocalized over the 4-atom peptide moiety. Looking along the O–C bond, the change in electron distribution has a circular motion. And, as with any circular motion of electrons, there is a resultant magnetic moment. This magnetic

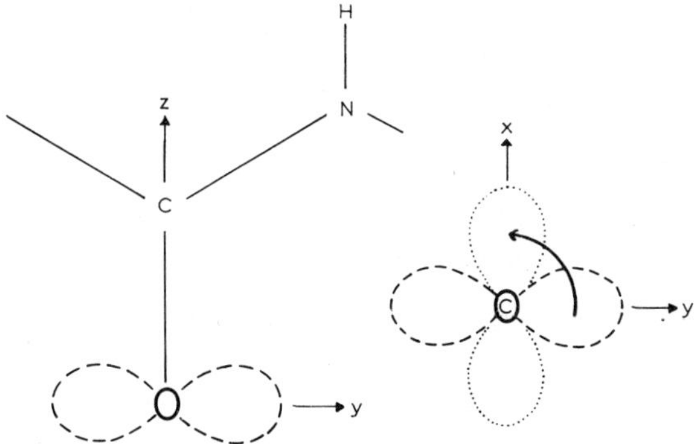

Figure 4. Prototype of an n–π^* transition where an electron in a non-bonding orbital localized on the oxygen atom (in this case a $2p_y$ orbital) on transition redistributes in a $2p_x$ orbital. This results in a circular motion of charge, giving rise to a magnetic moment directed along the O–C bond. Actually, the non-bonding orbital may be hybridized and the π^* orbital is delocalized over the chromophoric moiety, but the circular motion of charge for a carbonyl moiety would arise from a change in distribution from the yz plane to the xz plane. Reproduced, with permission, from [3].

transition dipole moment, $\vec{m}_i$, is directed approximately along the O–C bond. While the $\vec{\mu}_i$ for this transition is small, the m_i is large and, as will be seen below, in optical rotation it is the dot product $\vec{\mu}_i \cdot \vec{m}_i$ that is important such that this can give rise to a large band in optical rotation spectra.

(iii) Effects of polymeric arrays of interacting chromophores

The absorbance of a solution is linearly proportional to the concentration of absorbing solute, as indicated in equation 2, only for solutions in which the chromophores are not interacting or associating in a regular manner. When chromophores aggregate or when they are chemically tied together, as in a polymer with a nonrandom relation between the chromophores, Beer's Law is no longer valid. The absorption band can shift to shorter wavelengths (blue shift), to longer wavelengths (red shift) or split into two bands, and it can exhibit an increase in intensity (hyperchromism) or a decrease in intensity (hypochromism). As is so often the case, such special cases become a source of additional information. Our understandings of these effects are due to the important contributions of Davydov [6], Kasha and co-workers [7,8], Rhodes [9] and Tinoco [10].

The absorption spectrum of a solution of dimers, with a relatively fixed geometrical arrangement or configuration for the dimer, can differ from that of a solution of monomers. The difference depends on the configuration of the dimer and the difference, therefore, can be used to provide information on the configuration. Considering an isolated absorption band, schematic representation of three classes of difference absorbance, ΔOD, curves are seen in Figure 5 for an experimental situation using a split-beam spectrophotometer with a solution of monomers in the reference beam and with the same number of chromophores in the sample beam but with the chromophores associated as dimers in solution. Concentration-dependent dimerization is a simple way to achieve these conditions, in which case the cuvette in the

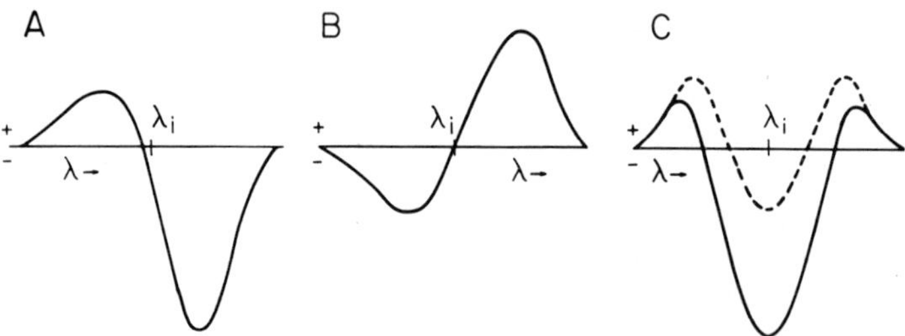

Figure 5. Schematic difference absorbance curves to exemplify absorption curve intensity and energy shifts. The monomeric state is in the reference beam and the associated state is in the sample beam, with the same number of monomers as in the reference beam. The difference curves are a sensitive measure of the changes in absorption that can occur when identical chromophores associate in a regular way. A, Hypochromism with a shift in the band to shorter wavelengths. B, Hyperchromism with a shift in the band to longer wavelengths. C, Band splitting with either a hypochromism (solid curve) or a hyperchromism (dashed curve). λ_i is the monomer band center.

reference beam would have a very long path length, e.g., 10 cm, whereas in the sample beam would be a cuvette with a short path length, e.g., 0.01 cm, but containing a solution with a 10^3-fold greater concentration. In Figure 5A the absorption band is seen to have shifted to shorter wavelengths, but since the area of the negative part of the difference curve is greater than that of the positive part there has been a net decrease in the area of the absorption band, i.e., a hypochromism. In Figure 5B the absorption band is seen to have shifted to longer wavelengths. And, since the positive lobe of the difference absorbance is greater than the negative lobe, the overall area of the absorption curve has increased, i.e., the dimerization has resulted in hyperchromism. In Figure 5C, the two positive lobes, which straddle the wavelength, λ_i, of the monomer absorption maximum, demonstrate the band to have split. This can occur either with a net hypochromism, the solid curve, or a net hyperchromism, the dashed curve. These effects can be interpreted in terms of the configuration of the dimer or of a polymer, but first it is useful to develop a shorthand for representing the relative orientation of the chromophores.

It was noted above (see Eqn. 3 and Fig. 3) that the area of an absorption band divided by the wavelength is the measure of the magnitude of electron redistribution and of the dipole moment change attending absorption of the photon and that the process can be represented as a vector quantity. For strong absorbances by unsaturated chromophores, e.g., π–π^* transitions from a delocalized bonding to a delocalized antibonding orbital [5], the vector is in the plane of the chromophore. Once the orientation of the electric transition dipole moment is fixed with respect to the chromophore, the relative orientation of vectors of different identical chromophores is entirely equivalent to the relative orientation of chromophores. With this in mind, the effects of shifting and splitting of an absorption band and of its hypo- and hyperchromism can be analyzed in terms of a vector representation for the interaction of electric transition dipole moments.

(iii-a) The shifting and splitting of absorption bands and excitation resonance interactions [6–8]. Because the chromophores of a dimer absorb light at the same frequency, on excitation the two chromophores can exhibit a resonance interaction. A common analogy to explain this resonance effect is made of two tuning forks on a resonating board. Giving energy to one tuning fork, by striking it, causes it to vibrate. Depending on the properties of the resonating board, however, the energy can transfer from one tuning fork to the other, with first one vibrating and then the other. The transfer of energy back and forth occurs at a rate which depends on the distance between the two vibrators and the mechanical properties of the resonating board. Similarly, a resonance can occur between two chromophores and the interaction between the two electric transition dipole moments can be approximated by a dipole interaction potential, V_{12}, which is written in terms of the transition dipole moments, $\vec{\mu}_1$ and $\vec{\mu}_2$, of the two identical interacting chromophores, i.e.,

$$V_{12} = \frac{1}{|\vec{r}_{12}|^3}\left[\vec{\mu}_1 \cdot \vec{\mu}_2 - \frac{3(\vec{r}_{12} \cdot \vec{\mu}_1)(\vec{r}_{12} \cdot \vec{\mu}_2)}{|\vec{r}_{12}|^2}\right] \tag{4}$$

where

$$\vec{\mu}_1 \cdot \vec{\mu}_2 = |\vec{\mu}_1||\vec{\mu}_2| \cos \theta_{12}$$

$$\vec{r}_{12} \cdot \vec{\mu}_1 = |\vec{r}_{12}||\vec{\mu}_1| \cos \theta_{r1} \tag{5}$$

$$\vec{r}_{12} \cdot \vec{\mu}_2 = |\vec{r}_{12}||\vec{\mu}_2| \cos \theta_{r2}$$

with the scheme in Figure 6 defining the angles and the vector distance, $\vec{r}_{12}$.

Because of the interaction potential, the energy for the electronic transition of the dimer is different from that of a monomer in the manner shown in Figure 7, where M, D, G and E stand for monomer, dimer, ground state and excited state, respectively, and where $V_{12} = (E'' - E')/2$. Depicted in Figure 7 are three classes of dimeric configurations for the chromophores which are represented by their electric transition dipole moments. In the parallel stacked configuration, since $\vec{r}_{12}$ is perpendicular to $\vec{\mu}_1$ and $\vec{\mu}_2$, the second term in equation 4 is zero, i.e., $\theta_{r1} = \theta_{r2} = 90°$ and $\cos 90° = 0$, and with $\theta_{12} = 0°$, V_{12} becomes $|\vec{\mu}_1||\vec{\mu}_2|/|\vec{r}^{\,3}_{12}|$, i.e., the transition occurs at higher energy. In the antiparallel stacked configuration, the second term in equation 4 is

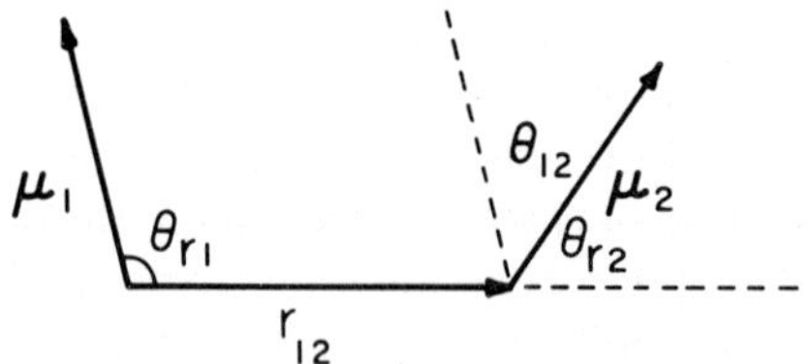

Figure 6. Definition of the vectors and angles of equations 4 and 5.

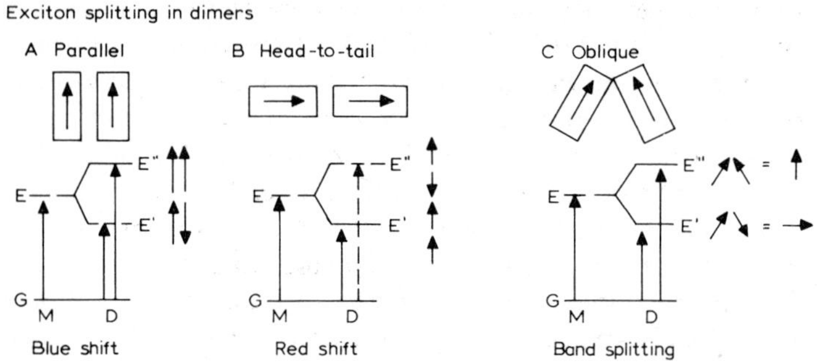

Figure 7. Vector model of exciton splitting in dimers (redrawn following Kasha et al. [8]). A, Parallel stacking of chromophores resulting in a shift to higher energies (blue shift). B, Head-to-tail coplanar alignment of chromophores, resulting in a shift to lower energies. C, Oblique arrangement of chromophores, leading to band splitting. These three states of association correspond to the three examples of difference absorbance curves given in Figure 5. Reproduced, with permission, from [128].

again zero, but $\theta_{12}=180°$ and $\cos\theta_{12}=-1$ such that $V_{12}=-|\vec{\mu}_1||\vec{\mu}_2|/|\vec{r}_{12}|^3$. However, for this configuration, the transition dipole moments sum to zero and there would be no absorption band centered at lower energies, E'. By these considerations alone, the absorption band becomes blue-shifted and the intensity of the dimer transition, $2|\vec{\mu}_1|$, is the same as that of two monomers, i.e., the area of the absorption curve would be unchanged. As represented in Figure 5A, a blue shift can be associated with a decrease in absorption. This is due to additional interactions of the transition dipole moment occurring at one energy as measured by an absorption curve centered at one wavelength with other transitions that can occur at other wavelengths. These effects will be discussed in the next section, on dispersion force interactions.

In the end-to-end antiparallel alignment of chromophoric transition dipole moments not seen in Figure 7B, $\vec{\mu}_1+\vec{\mu}_2=0$ for the E'' excited state so that the high-energy absorption does not occur and only the lower energy head-to-tail alignment as shown, where $\vec{\mu}_1+\vec{\mu}_2=2|\vec{\mu}_1|$, is to be considered. Since θ_{12}, θ_{r1} and θ_{r2} are all 180°, the cosine terms in equation 5 become -1 and V_{12} becomes $-2|\vec{\mu}_1||\vec{\mu}_2|/|\vec{r}_{12}|^3$. Accordingly, $E'=E-2|\vec{\mu}_1||\vec{\mu}_2|/|\vec{r}_{12}|^3$. This decrease in energy results in a red-shifted absorption curve, as shown in Figure 5B in terms of a difference absorbance curve.

In the oblique orientation shown in Figure 7C, both orientations, ↗ ↖ and ↗ ↘, lead to net dipole moments, so that both transitions to E' and E'' can occur. This causes the absorption band to split. Interestingly, the directions of the two transition dipole moments are at right angles. Because of this, orientable polypeptides with highly ordered chromophores, such as in an α-helix, exhibit absorption bands polarized parallel and perpendicular to the helix axis.

With an experimental value for V_{12} and by means of equation 4, it becomes possible to approximate the maximal distance between chromophores. Taking λ_i as the wavelength of the absorption maximum of the monomer and λ_+ or λ_- as wavelengths of the shifted absorption maxima, the experimental value for V_{12} is

$$V_{12}=(\lambda_{\pm}^{-1}-\lambda_i^{-1})ch \tag{6}$$

where h is Planck's constant and c is the velocity of light in a vacuum. For a blue shift the maximal distance between chromophores is

$$|\vec{r}_{12}|=(|\vec{\mu}_i|^2/V_{12})^{1/3} \tag{7}$$

and for a red shift

$$|\vec{r}_{12}|=(2|\vec{\mu}_i|^2/V_{12})^{1/3} \tag{8}$$

where μ_i has been used in place of μ_1 and μ_2, since they are the same, i.e., they contribute to the same absorption band. Therefore, once the transition dipole moment can be fixed within the chromophore, the configuration of a pair of chromophores can be obtained from observing the change in an absorption band on association, the relative orientation from the direction of the shift and the distance between chromophores from the magnitude of the shift. Often it is sufficient to know that the transition dipole moment is within the plane of the chromophore, as it usually is for aromatic chromophores, to obtain the desired structural information.

(iii-b) Hypochromism and hyperchromism and dispersion force interactions [9–10]. When nucleic acids are denatured, there is a large increase in the intensity of the 260 nm absorption band. Similarly, when α-helical proteins, such as myoglobin, hemoglobin and tropomyosin, are denatured there is a large increase in the intensity of the 190 nm absorption band. Denaturation can also lead to hypochromism, as occurs with proteins which have the β-pleated sheet conformation, such as the natural silks, e.g., silk fibroin and chrysopa silk. Generally reference is made to the change which occurs on ordering. For example, the 190 nm band is hypochromic on forming the α-helical conformation and the 260 nm band is hypochromic on forming the double-stranded helix of DNA.

Since the dipole strength of an absorption band, D_i, is proportional to the product of the molar extinction coefficient and the half band width (see Eqn. 3 and Fig. 3), it is proportional to the area of an absorption band. In these terms

$$(D_i^R - D_i^o) = + \text{ for hypochromism} \tag{9}$$

and

$$(D_i^R - D_i^o) = - \text{ for hyperchromism} \tag{10}$$

where the superscripts R and o stand for random and ordered, respectively. Monomers molecularly dispersed in solution would represent a random state for the chromophores. Aggregation to form specific dimers, trimers, etc., would result in an ordered state. Also for covalent polymers, where there is no regular relation between repeating chromophores, the polymer is considered to be in a random state.

The Rhodes equation [9] for expressing hypo- and hyperchromism may be written in terms of the above difference in dipole strength [4]

$$(D_i^R - D_i^o) = K \sum_{j \neq i} \frac{\lambda_j \lambda_i}{(\lambda_i^2 - \lambda_j^2)} V_{ij}\, \vec{\mu}_i \cdot \vec{\mu}_j \tag{11}$$

where K is a numerical constant, λ_i and λ_j the wavelengths of the absorption maximum of the ith and jth absorption bands, respectively; V_{ij} is the dipole-dipole interaction potential for the interaction of the $\vec{\mu}_i$ and $\vec{\mu}_j$ transition dipole moments. The fundamental element to keep in mind with equation 11 is that, while the band of interest is due to the ith transition, the summation $\sum_{j \neq i}$ is over all of the other transitions which occur at other wavelengths. With respect to chromophores, this means that it is the interaction of the ith transition in one chromophore with all of the j transitions of neighboring chromophores, with the j transitions being all transitions except the ith transition. This $j \neq i$ summation and the difference quantity $(\lambda_i^2 - \lambda_j^2)$ define these as dispersion force interactions.

The interaction potential, V_{ij}, is again written

$$V_{ij} = \frac{1}{|\vec{r}_{ij}|^3} \vec{\mu}_i \cdot \vec{\mu}_j - 3\frac{(\vec{r}_{ij} \cdot \vec{\mu}_i)(\vec{r}_{ij} \cdot \vec{\mu}_j)}{|\vec{r}_{ij}|^2} \tag{12}$$

Since the product $V_{ij}\vec{\mu}_i \cdot \vec{\mu}_j$ of equation 11 does not change sign or magnitude on going from a parallel to an antiparallel orientation of transition dipole moments, only parallel alignments need be considered. Also, since experimentally the longest wavelength band is always observable, this will be called the ith transition such that $\lambda_i - \lambda_j$ will always be positive for this band. Accordingly, the sign of $D_i^R - D_i^o$ for the long wavelength absorption band follows the sign of V_{ij}. As before, when the interest was in the interaction of the ith transition in chromophore 1 with the ith transition in chromophore 2 for the purpose of analyzing the splitting and shifting of bands, the sign of V_{ij} depends on the relative orientation of the transition dipole moments but in this case it is the ith transition dipole moment in chromophore 1 that is interacting with the jth transition dipole moment in chromophore 2. The geometrical considerations remain the same. Thus,

$$V_{ij} = + \text{for } \vec{\mu}_i \uparrow \uparrow \vec{\mu}_j \tag{13}$$

i.e., for hypochromism and

$$V_{ij} = - \text{for } \underset{\mu_i}{\longrightarrow} \underset{\mu_j}{\longrightarrow} \tag{14}$$

i.e., for hyperchromism. Since the strong electric transitions are in the plane of an unsaturated or aromatic chromophore, hypochromism occurs when there is a stacking of chromophores and hyperchromism occurs when there is a head-to-tail alignment or coplanar arrangement of chromophores.

(iii-c) The heme chromophore and heme-heme association. The heme chromophores of myoglobin, hemoglobin and the cytochromes exhibit one of the most intense absorption bands known. It is the Soret, or γ, band which is found near 400 nm with a molar extinction coefficient of greater than 100 000. In Figure 8 is shown the hemochromogen spectrum of the heme undecapeptide of cytochrome *c* [11] where the Soret band is seen to have a molar extinction coefficient of 130 000. The molar extinction coefficient of the most intense solution accessible band of the peptide chromophore, on the other hand, can vary between 4000 for the α-helical conformation to about 8000 for the β-pleated sheet conformations. Because of this, analysis of the Soret band intensity changes as a function of heme peptide association is reduced to an analysis of relative heme orientations.

As the protein sequence is covalently attached to the heme in cytochrome *c*, digestion by pepsin gives rise to a heme undecapeptide, and following by tryptic digestion decreases the length of the attached peptide from eleven residues to eight residues, i.e., to a heme octapeptide [12]. These two peptides aggregate in aqueous

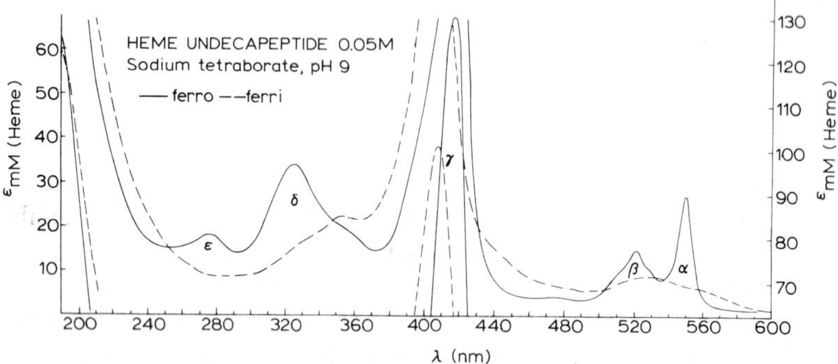

Figure 8. Absorption spectrum of ferriheme undecapeptide (---) and of ferroheme undecapeptide (hemochromogen) (—) in 0.05 M sodium tetraborate, pH 9. The right-hand ordinate is for the Soret (γ) band and for the absorbances at wavelengths shorter than 210 nm. The hemochromogen bands are labelled α–ε. As there are no residues in the undecapeptide with chromophores absorbing at wavelengths greater than 240 nm, the 280 nm, ε and δ bands are clearly due to the heme chromophore. Reproduced, with permission, from [11].

solutions at mM concentrations and largely dissociate at μM concentrations, and they do so with quite different effects on the Soret band. The heme octapeptide exhibits a very large hypochromism and splitting [13], as shown in Figure 9, whereas the heme undecapeptide exhibits hyperchromism and a red shift [14], as seen in the difference spectra in Figure 10. Because the hypochromism of the heme octapeptide is so extensive, the positive lobes of the ΔOD curves in the lower part of Figure 9 are small but the ΔOD pattern of Figure 5C is still apparent. On the basis of the above discussion of excitation resonance interactions (excitons) and of dispersion force interactions, it is concluded that there is an oblique stacking of the heme moieties of the heme octapeptide, as shown schematically in Figure 7C. Using equation 7 with inclusion of the quadrupolar term in the dipole-dipole interaction potential because of the large transition dipole moment length with respect to the interheme distance, $\vec{r}_{12}$, the maximum distance between hemes is less than 7 Å [13]. In the case of the heme undecapeptide aggregation, the red shift and hyperchromism, apparent in Figure 10, are very close to the schematic example of Figure 5B and a more nearly coplanar alignment of hemes is indicated, as shown schematically in Figure 7B (see below for use of circular dichroism data).

In addition to being useful examples of these interesting association-dependent absorption effects, the study of the properties of interacting hemes is relevant to understanding structure and function in hemoglobin and in multiheme cytochromes. With respect to the latter, the circular dichroism and optical rotatory dispersion counterparts to these effects (considered in general in sections 2(c) and (d) below) provided early evidence for the proximity of hemes in multiheme cytochromes [15–17], a suggestion which has since been supported by nuclear magnetic resonance studies [18]. The proximity of hemes in multiheme cytochromes is fundamental to the mechanism of electron transport. These effects also have relevance to the sigmoid

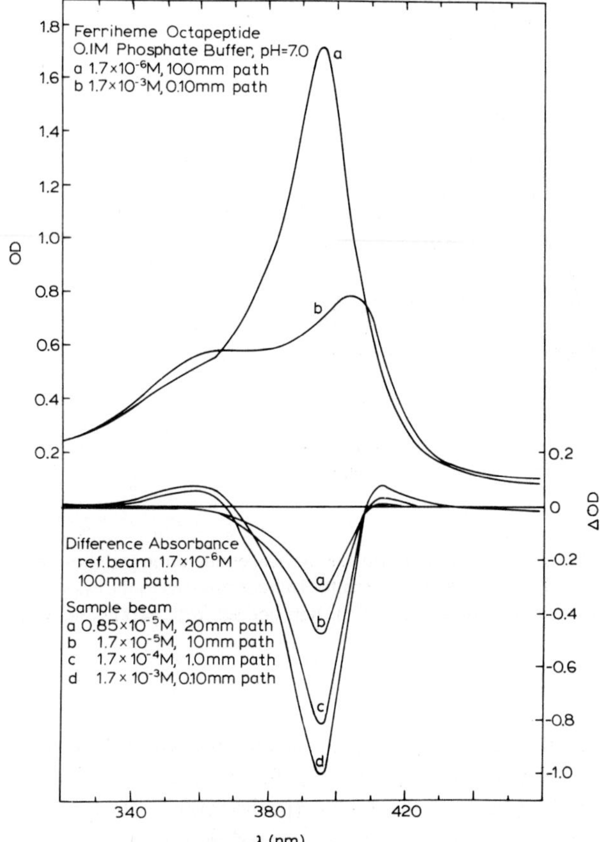

Figure 9. Dramatic hypochromism of the heme Soret band on association of ferriheme octapeptide molecules. Note that the difference absorbance curves in the lower part of the figure resemble that of Figure 5C (solid curve), and therefore would result from an oblique orientation as shown in Figure 7C. Reproduced, with permission, from [13].

oxygen-binding curve of hemoglobin, which is central to its O_2 transport function. Using the oscillator strength, f_i, of an absorption band as defined in Figure 3 in combination with the Kuhn-Thomas sum rule [19,20],

$$\sum_i f_i = n \tag{15}$$

where n is the number of electrons in the group, using the expression for the total polarizability, α, of a moiety i.e.,

$$\alpha = \sum_i \alpha_i = \frac{e^2}{4\pi^2 mc^2} \sum_i f_i \lambda_i^2 = K \sum_i D_i \lambda_i \tag{16}$$

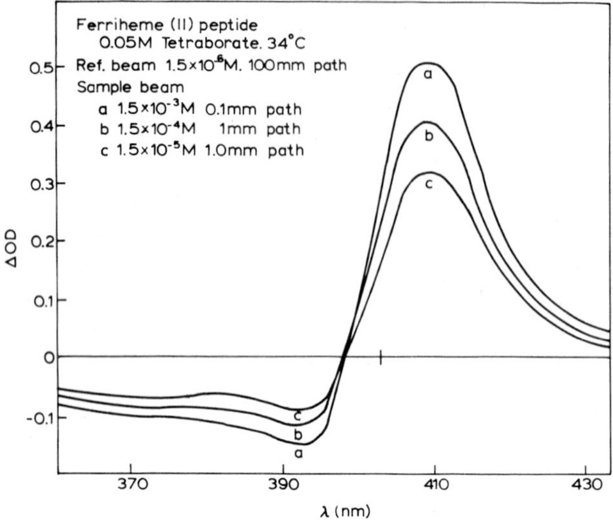

Figure 10. Hyperchromism of the heme Soret band of the ferriheme undecapeptide on association. Also observed is a shift of the band intensity to longer wavelengths (red shift). This corresponds to the example in Figure 5B and the head-to-tail (coplanar) orientation of Figure 7B. Reproduced, with permission, from [14].

where e, m, c and K are the electronic charge, the mass of the electron and the velocity of light in a vacuum and a constant, respectively, in combination with the hemochromogen spectrum in Figure 8, and utilizing the experimental enthalpies for aggregation of the heme peptides, an estimate of the energy of interaction between hemes is possible. This estimate and the shape of the circular dichroism band exhibited by the Soret band of oxyhemoglobin allowed the conclusion that the direct heme-heme interaction energy, with heme centers separated by 25 Å in hemoglobin, is too small (<0.1 kcal/mole) to account for the cooperative (sigmoidal) oxygenation curve [13], with required energies being an order of magnitude greater [21,22]. By elimination, it could be concluded that the increased affinity for oxygen at a second heme after binding O_2 at one of the four hemes is an effect that must be transmitted by the polypeptide chains of the protein [13].

(b) Refractive index (ordinary dispersion)

The velocity of light is decreased when it passes from a vacuum into a medium, and the amount of slowing depends on the properties of the medium and specifically on the interaction of light with the transition dipole moments of the medium. This effect is expressed in terms of the refractive index, n,

$$n = c/v \tag{17}$$

where, as before, c and v are the velocity of light in a vacuum and in the medium, respectively. The refractive index at a given wavelength, n^{λ}, becomes

$$n^{\lambda} = 1 + \sum_{i} n_i^{\lambda} \tag{18}$$

where the summation is over each of the electronic transitions. Appropriately, when there are no electronic transitions, as in a vacuum, the summation is zero and n^{λ} becomes 1, consistent with equation 17 when $v = c$. As the density of the electronic transitions increases, n becomes greater than 1. The n_i^{λ} can be expressed as

$$n_i^{\lambda} = 2.02 \times 10^{-5} \frac{\rho \varepsilon_i^0 \Delta_i}{M} \frac{\lambda^2(\lambda^2 - \lambda_i^2)}{(\lambda^2 - \lambda_i^2)^2 + 4\Delta_i^2 \lambda_i^2} \tag{19}$$

where ρ is the density in grams per cm^3, Δ_i is the half band width as defined in Figure 3 in nm, ε_i^0 is the molar extinction coefficient of the absorption maximum, M is the grams per mole of chromophores, λ_i is the wavelength of the absorption maximum, and λ is the wavelength of the light. The wavelength dependence of the magnitude of n_i, the refractive index contribution of a single isolated absorption band, is given in Figure 11. The reversal of sign is referred to as anomalous dispersion. When not in the

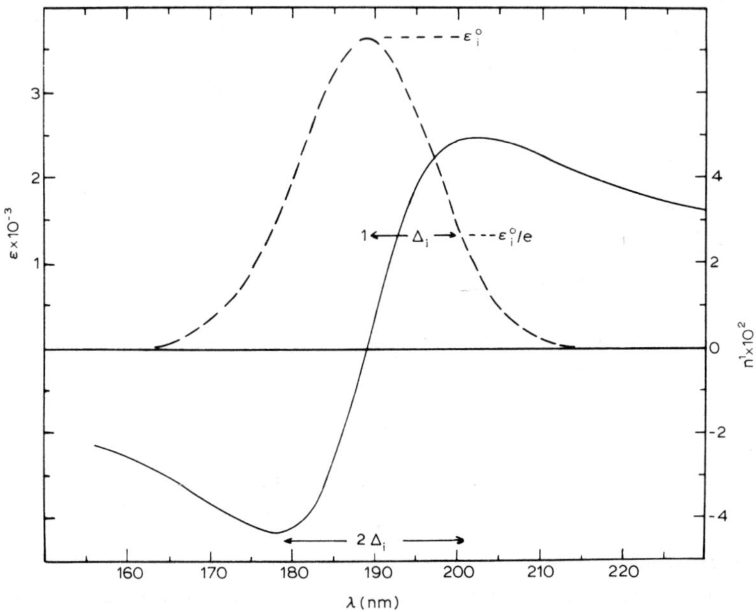

Figure 11. Resolved perpendicularly polarized 189 nm absorption band of α-helical poly-L-alanine (see Table 2A and Figure 21B), left-hand ordinate, dashed curve. Plotted on the right-hand ordinate as the solid curve is the contribution to the refractive index of the single band, calculated using equation 19. Note that the dispersion curve is anomalous in an absorption region rather than monotonic, as is the water curve of Figure 12.

region of the absorption band the quantity $4\Delta_i^2\lambda_i^2$ is negligible and the wavelength-dependent portion of equation 19 reduces to $\lambda^2/(\lambda^2-\lambda_i^2)$. This difference term in the denominator is the result of dispersion force interactions and is analogous to the similar denominator in equation 11. In equation 11, however, it was the transition dipole moment for the absorption band of interest in one chromophore interacting with the transition dipole moments responsible for other absorption bands in a second chromophore. In this situation, it is the oscillating electric vector of the light that is directly interacting with the transition dipole moments of absorptions that occur at other wavelengths. This is the mechanism for the dispersion of ordinary light.

For a solvent such as water which contains no absorption bands, the refractive index as a function of wavelength increases monotonically [23], as shown in Figure 12. If the solvent contains a peptide chromophore, as in the model *N*-methyl acetamide, the refractive index is larger and it does not continue monotonically through the region of the absorption band. The dispersion is said to be anomalous in the absorption region. In the case of optical rotatory dispersion considered below, the dispersion of interest is that of circularly polarized light and the anomalous dispersions in the absorption regions are called Cotton effects.

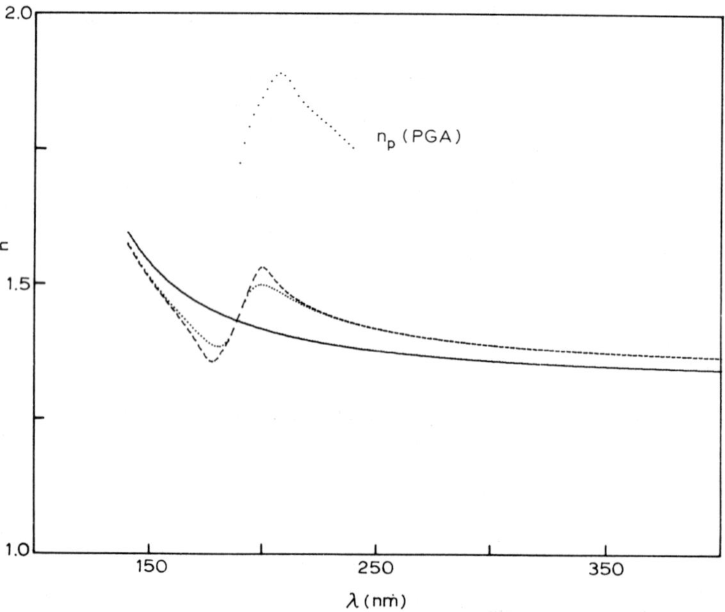

Figure 12. Solid curve: the refractive index of water showing a simple, monotonic dispersion curve. Dotted curve: the contribution of the 189 nm band of *N*-methylacetamide at a 1 M concentration added on to the water curve using equation 19. The point to be made is that there is no way to match the background curve of a good solvent to the anomalous dispersion of a chromophoric system under study. Such matching is commonly attempted to remove light scattering problems which depend on the difference in refractive index of the particle, n_p, with that of the solvent, n_s, i.e., $(n_p^2-n_s^2)$. The dashed curve adds the second dispersion term in equation 25. Also included is the calculated refractive index of particulate poly-L-glutamic acid (PGA) (see section 4(c)(ii)).

(c) Optical rotation

Optical rotation is a general term which is used to include both circular dichroism and optical rotatory dispersion. These are closely related phenomena in the same way that absorption and ordinary dispersion (refractive index) are related. They can be interconverted by mathematical transforms. Whereas the common element in absorption and ordinary dispersion is the dipole strength of the transition, the common quantity in circular dichroism and optical rotatory dispersion is rotational strength. And, as will be seen below, the rotational strength of a transition may be obtained from either measurement.

Both circular dichroism (CD) and optical rotatory dispersion (ORD) are difference measurements; CD is a measurement of the difference absorbance of left and right circularly polarized light, and ORD is a difference measurement between the refractive indices for left and right circularly polarized light. Both involve the interaction of circularly polarized light with chromophores. However, the chromophore must be part of a special molecular structure. If the molecule has a plane of symmetry, then both the left and right circularly polarized light interact equally and there is no difference to measure. But if the molecular system of which the chromophore is part is dissymmetric and non-superimposable mirror images occur, then there will be a differential interaction. The molecule is said to be optically active and it will exhibit a difference in its interaction with left and with right circularly polarized light.

To be dissymmetric, a molecule must have a non-superimposable mirror image. This includes asymmetry, as occurs with the tetrahedrally substituted carbon atom in which each of the four substituents is different, and it also includes structures with a resolved chirality, i.e., a left-handed or a right-handed screw sense. An example of resolved chirality or screw sense is that of the disulfide bridge in proteins where, for example, on looking along the S–S bond the S–C′ and S–C″ bonds are at an angle of 90° and the direction of rotation by 90° on going from the near S–C′ bond to the far S–C″ bond may be clockwise or counterclockwise. A right-handed screw sense would be clockwise rotation.

(i) Plane polarization and the physical optics of rotatory polarization

When a beam of ordinary light is directed into a crystal, as long as it is not directed along the optic axis and the crystal is not of the cubic space group, the beam splits, with the two components traveling with different velocities. Each component is plane polarized but the polarizations are in perpendicular directions. This is birefringence or double refraction. (Recall that when the absorption band splits into two components for a particular polymeric array of chromophores, as shown in Figures 5C and 7C, the two components are polarized at right angles. The above is a dispersion counterpart.) For optical rotation studies one of the components is masked and the other can be directed into a solution containing optically active molecules for ORD studies or, as will be discussed below, the plane polarized light can be converted to circularly polarized light by using a special birefringent plate in preparation for CD studies.

As demonstrated in Figure 1b, plane polarized light with its oscillating electric vector can be represented as left and right circularly polarized components traveling in phase. If one of the components is slowed due to its greater dispersion force interactions with dissymmetric molecules having the same chirality, there will be a larger refractive index for this component and $n_L - n_R$ will be nonzero. This is circular birefringence, as seen in Figure 1c. If the wavelength of light is in the absorption region, there is a differential absorbance of the left and right circularly polarized components, i.e., a circular dichroism, as shown in Figure 1d, where the sum of the two vector components of different magnitudes is seen to scribe an ellipse.

(ii) Circular dichroism

For the circular dichroism experiment, it is necessary to obtain left and right circularly polarized beams of light. This can be achieved in a number of ways [24]. One approach is to use a quarter wave retarder, which can also be achieved in several ways [24]. When plane polarized light is directed at a birefringent plate with the plane of polarization exactly bisecting the fast and slow axes, it splits into plane polarized components of equal intensity, one oscillating with its plane of polarization along the fast axis and the other polarized along the slow axis. If the birefringent plate is a perfect quarter wave retarder, then the two beams will emerge from the plate 90° out of phase. This is depicted in Figure 13A. With the two plane polarized beams of equal intensity and traveling 90° out of phase, as shown in Figure 13B, combination of the two components results in perfectly circularly polarized light. In a single beam instrument, the trick is to cause the slow and fast axes to alternate at a modulation frequency so that a difference can be measured after passing through the sample and to maintain perfect quarter wavelength retardation as the wavelength is scanned.

(ii-a) Ellipticity and experimental determination of rotational strength. The difference absorbance measurement for left and right circularly polarized light is

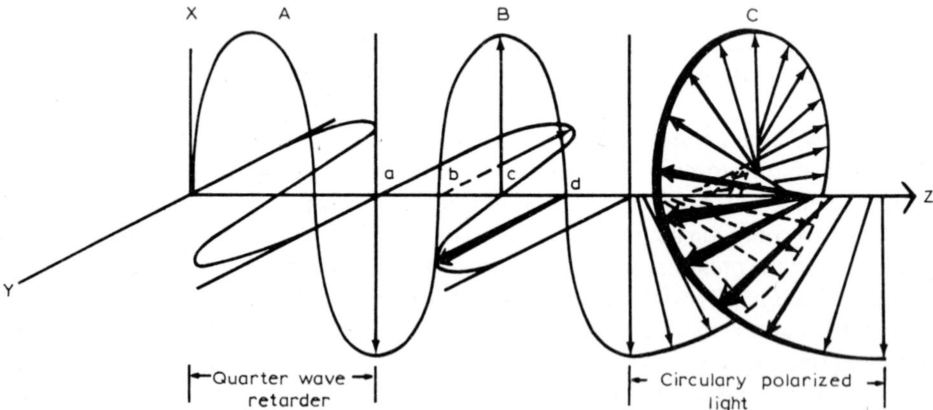

Figure 13. Producing circularly polarized light by means of a quarter wavelength retarder. A, Two beams, plane polarized at right angles, traverse a birefringent plate. B, On emerging the two beams are 90° (one quarter of a wavelength) out of phase. C, Shows the two beams recombined. The result is perfectly circularly polarized light. See text for discussion. Reproduced, with permission, from [4].

reported in terms of molar ellipticity, $[\theta]$, which at a given wavelength is defined as

$$[\theta] = 3300\,(\varepsilon_L - \varepsilon_R) = 3300\,\frac{(A_L - A_R)}{Cl} \tag{20}$$

where ε_L and ε_R are the molar extinction coefficients and A_L and A_R are the absorbances for left and right circularly polarized light, respectively; C is the concentration in moles per liter and l is the path length in cm. The term ellipticity comes from the nature of the emergent beam. When there has been a greater loss in intensity of one circularly polarized component of a plane polarized beam, the recombined emergent beam scribes an ellipse, as seen in Figure 1d, and the molar ellipticity is a measure of the ellipticity that would occur for a molar concentration.

The rotational strength, in direct analogy to the dipole strength, is measured from the area of the circular dichroism band; however, it can be either positive or negative since the CD band may be either positive for $\varepsilon_L > \varepsilon_R$ or negative when $\varepsilon_L < \varepsilon_R$. The rotational strength, R_i, is calculated from the CD band by the expression [2]

$$R_i = 1.23 \times 10^{-42}\,\frac{[\theta_i^0]\Delta_i}{\lambda_i} \tag{21}$$

where the quantities are defined in Figure 14. The magnitude and sign of R_i are a result of the conformation or absolute configuration of the dissymmetric absorbing molecule.

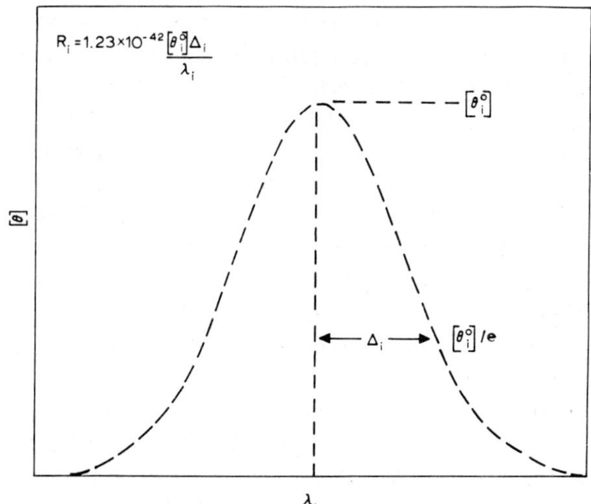

Figure 14. Gaussian shaped circular dichroism band, defining quantities required to calculate rotational strength, R_i, of the band. $[\theta_i^0]$ is the molar ellipticity at the band maximum and Δ_i is the half-band width at $[\theta_i^0]/e$. Reproduced, with permission, from [3].

(iii) Optical rotatory dispersion

(iii-a) Molar rotation. Optical rotatory dispersion data are reported in terms of molar rotation, $[M]$, i.e.

$$[M]=\frac{100\alpha}{Cl} \tag{22}$$

where α is the angle of rotation of the plane of polarized light in degrees, C is the concentration in moles per liter and l is the path length in cm. As depicted in Figure 1c, α is proportional to $n_L - n_R$, the difference in refractive indices for left and right circularly polarized light.

The utilization of optical rotatory dispersion data is complicated by its dispersion mechanism, in which the contributions of all of the electronic transitions in the molecule occur at all wavelengths, as shown in what is called the Drude equation

$$[M]_\lambda=\sum_i [M_i]_\lambda=\sum_i \frac{a\lambda_i^2}{(\lambda^2-\lambda_i^2)} \tag{23}$$

for wavelengths well removed from absorption regions. Utilization is also complicated because the refractive index of the solvent, n_λ is required over the wavelength range of interest, i.e.

$$[M]_\lambda=\frac{n_\lambda+2}{3}[M']_\lambda \tag{24}$$

in order to obtain conformationally comparable quantities obtained from different solvents. This means that following a process, for example, such as the denaturation of a protein by successive additions of urea to the solution, requires the refractive index of the urea containing solvent over the entire wavelength of interest. For this reason refractive index data has been reported for numerous solvents of interest [23,25,26].

ORD data, however, can be obtained for a number of systems of interest where little or no data are possible by circular dichroism, or when the absorption bands cannot be reached due to the high absorbance of a solvent or because the absorption bands of interest are beyond the usual 180–190 nm limit of solution studies. Also, most of the optical rotation data in the literature prior to the mid-1960s are reported in terms of optical rotatory dispersion. Additionally, when correcting the distortions in the CD spectra of biomembranes (see below), it is convenient to have the ORD curve of what is referred to as the pseudoreference state.

(iii-b) Rotational strengths from ORD data. For comparison both the ORD curve and the CD curve for the isolated n–π^* transition of d-camphor-10-sulfonate in water are plotted in Figure 15. It is the $[M_{n\text{-}\pi^*}]$ that is plotted. This curve is calculated from the expression [28]

$$[M_i]_\lambda=\frac{96\pi N}{hc}\frac{n_\lambda^2+2}{3}R_i\sum_{x=1,3}^{\text{odd}}\frac{2^x}{(2x)!}\frac{\lambda^{2x}(\lambda^2-\lambda_i^2)^x}{(\lambda^2-\lambda_i^2)^{2x}+(2\Delta_i)^{2x}\lambda_i^{2x}} \tag{25}$$

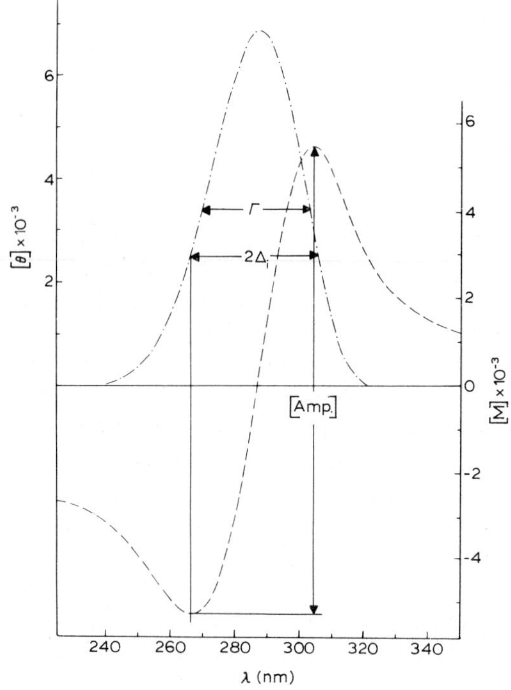

Figure 15. Circular dichroism curve (-·-·-) plotted on the left-hand ordinate and optical rotatory dispersion (ORD) curve (---) plotted on the right-hand ordinate of d-camphor-10-sulfonate. It is seen that the band width at $[\theta_i^0]/e$ and the distance between the ORD extrema coincide better than does the band width at $[\theta_i^0]/2$, i.e., Γ. The amplitude of the ORD anomalous dispersion curve (Cotton effect) is indicated as [Amp]. Reproduced, with permission, from [3].

where N, h and c are Avogadros number, Planck's constant and the velocity of light, respectively, and where the summation is over the set of odd integer values of x. Using only the first term, $x = 1$, gives the familiar Condon expression [28] which was written in analogy to ordinary dispersion but this term alone does not give an adequate steepness to the anomalous dispersion [2,27] (i.e., the Cotton effect). With equation 25 the rotational strength may be approximated using two terms in the summation [27] as

$$R_i \simeq 2.2 \times 10^{-42}\,[\mathrm{Amp}']\,\frac{\Delta_i^3}{\Delta_i^2 \lambda_i + 0.003\lambda_i^3} \tag{26}$$

where [Amp′], the amplitude of the Cotton effect is

$$[\mathrm{Amp}'] = [M_i']_{\lambda+} - [M_i']_{\lambda-} \tag{27}$$

where the terms are defined in Figure 15 and the prime is on the amplitude to indicate the solvent refractive index correction (the Lorentz correction factor) has been made (see Eqn. 24). Actually, if the Cotton effect is on a steep background dispersion curve, it will be necessary to measure $[M_i']_{\lambda+}$ and $[M_i']_{\lambda-}$ as excursions from the background curve.

(iv) Analysis of optical rotation data in terms of rotational strengths

A molecule understudy may be divided operationally into groupings of atoms or moieties, which may in turn be termed chromophoric or non-chromophoric. Whether or not a moiety is a chromophore depends to some extent on the wavelength accessible in the study. For example, a methyl group is not generally considered to be a chromophore, yet at short enough wavelengths it does become one. Spectrophotometers, spectropolarimeters and dichrographs commonly have a low wavelength limit for solution studies of 180–190 nm. In considering chromophores with accessible absorption bands the first concern is whether the chromophore is symmetric or non-symmetric (inherently dissymmetric). Most commonly chromophores are symmetric due to the presence of a plane of symmetry. The next concern is the nature of the electronic transition. Is it a strong electric transition with a large molar extinction coefficient (e.g., >1000) or is it a weak electric transition with a small molar extinction coefficient (e.g., <1000) but with a significant molar ellipticity due to a strong magnetic transition dipole moment? The analysis of each of these transitions in a symmetric chromophore as well as transitions in an inherently dissymmetric chromophore proceed quite differently when relating optical rotation to structure. This approach is outlined in Figure 16.

(iv-a) Strong absorption bands: Large electric transition dipole moments. In the Kirkwood coupled oscillator mechanism [1,29], the rotational strength, R_{ij}, of one strong electric transition dipole moment, $\vec{\mu}_i$, at the wavelength, λ_i, derives from coupling with a second large electric transition dipole moment, $\vec{\mu}_j$, in a second

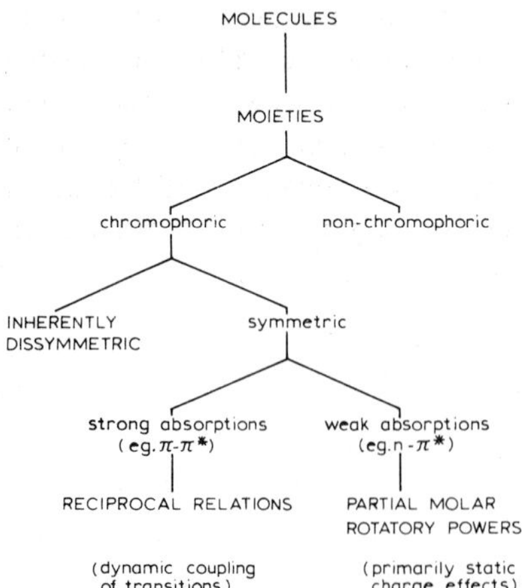

Figure 16. Flow chart for approach to analyzing optical rotation data of molecules. Reproduced, with permission, from [3].

chromophore and occurring at λ_j. The expression is

$$R_{ij} = \frac{2\pi}{hc} \frac{V_{ij}\lambda_i\lambda_j\vec{r}_{ij} \cdot \vec{\mu}_i \times \vec{\mu}_j}{(\lambda_i^2 - \lambda_j^2)} \tag{28}$$

The interaction potential, V_{ij}, is written as in equation 12. The quantity, $\vec{r}_{ij} \cdot \vec{\mu}_i \times \vec{\mu}_j$, is the triple scalar product. The cross-product, $\vec{\mu}_i \times \vec{\mu}_j$, is taken as shown in Figure 17A and then the dot product is then taken with the vector resulting from the cross product, as demonstrated in Figure 17B. The result is the signed quantity $|\vec{r}_{ij}||\vec{\mu}_i||\vec{\mu}_j| \sin \theta_{ij} \cos \theta$. Thus, with the values of $|\vec{\mu}_i|$ and $|\vec{\mu}_j|$ determined from their absorption bands and with their orientation within the chromophores known, the geometry responsible for R_{ij} can be determined.

(1) Reciprocal relations. There is an interesting and useful relationship between R_{ij}, the rotational strength of the ith transition due to coupling with the jth transition in an adjacent second chromophore, and R_{ji}, the rotational strength of the jth transition in the second chromophore due to coupling with the ith transition in the first chromophore. R_{ji} is simply written by interchanging the indices of equation 28, i.e.,

$$R_{ji} = \frac{2\pi}{hc} \frac{V_{ji}\lambda_j\lambda_i\vec{r}_{ji} \cdot \vec{\mu}_j \times \vec{\mu}_i}{(\lambda_j^2 - \lambda_i^2)} \tag{29}$$

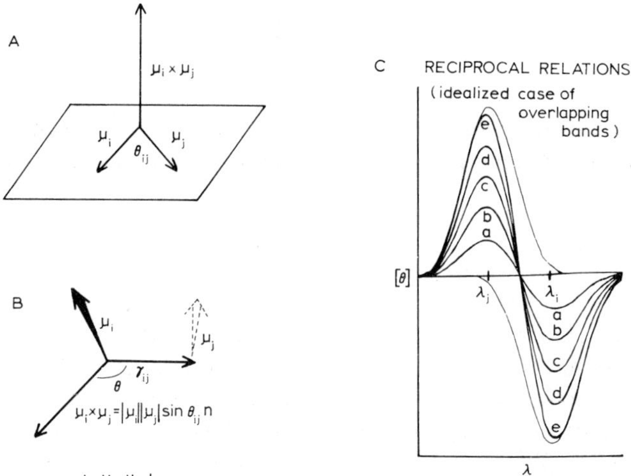

Figure 17. A, Definition of the cross product of two vectors, $\mu_i \times \mu_j$. B, Description of the triple scalar product, $r_{ij} \cdot \mu_i \times \mu_j$. C, Schematic representation of reciprocal relations wherein the rotational strength of the band at λ_i is derived entirely from interaction of its electric transition dipole moment, $\vec{\mu}_i$, with the electric transition dipole moment, $\vec{\mu}_j$, of the band at λ_j and similarly for the band at λ_j in accord with equations 28 and 29. The transitions at λ_i and λ_j are in different chromophores but the chromophores must be juxtaposed and have a favorable orientation to give the triple scalar product of part B a significant magnitude.

Since

$$(\lambda_i^2 - \lambda_j^2) = -(\lambda_j^2 - \lambda_i^2), \tag{30}$$

$$\vec{r}_{ij} = -\vec{r}_{ji} \tag{31}$$

that is, a vector changes sign on reversing direction,

$$\vec{\mu}_i \times \vec{\mu}_j = -\vec{\mu}_j \times \vec{\mu}_i, \tag{32}$$

that is, the cross product is anticommutative, and

$$V_{ij} = V_{ji}, \tag{33}$$

it follows that

$$R_{ij} = -R_{ji} \tag{34}$$

Plotted in terms of ellipticities in Figure 17, there is a reciprocal behavior of the bands; as the band at λ_i gets more negative, the band at λ_j gets more positive. This has been termed reciprocal relations in optical rotation [30,31] and represents the specific way in which the electric transitions adhere to the sum rule in optical rotation [28,32], i.e.

$$\sum_i R_i = 0 \tag{35}$$

Of course, the situation is somewhat more complex. The rotational strength of the ith transition, R_i, is the result of $\vec{\mu}_i$ coupling with all of the other transitions in all of the adjacent chromophores, i.e.,

$$R_i = \sum_i R_{ij} \tag{36}$$

If the two wavelengths, λ_i and λ_j, and the two chromophores are very close, however, the denominator of these two terms, R_{ij} and R_{ji}, becomes quite small and these terms can dominate in the summations for R_i and R_j. This has been demonstrated in a number of dinucleotide prosthetic groups of enzymes, such as flavin-adenine dinucleotide and nicotinamide-adenine dinucleotide. An example is given below for the adenine and nicotinate chromophores.

The long wavelength absorption band of adenine is at 259 nm and that of nicotinate is near 265 nm. When these two chromophores are attached to the same side of a ribose ring as shown in Figure 18A, the result is adenosine-5′-mononicotinate. This structure allows the two chromophoric moieties to stack, as shown, and to separate. A number of experimental parameters can be used to effect stacking or destacking. These are solvent polarity, pH and temperature, all of which show

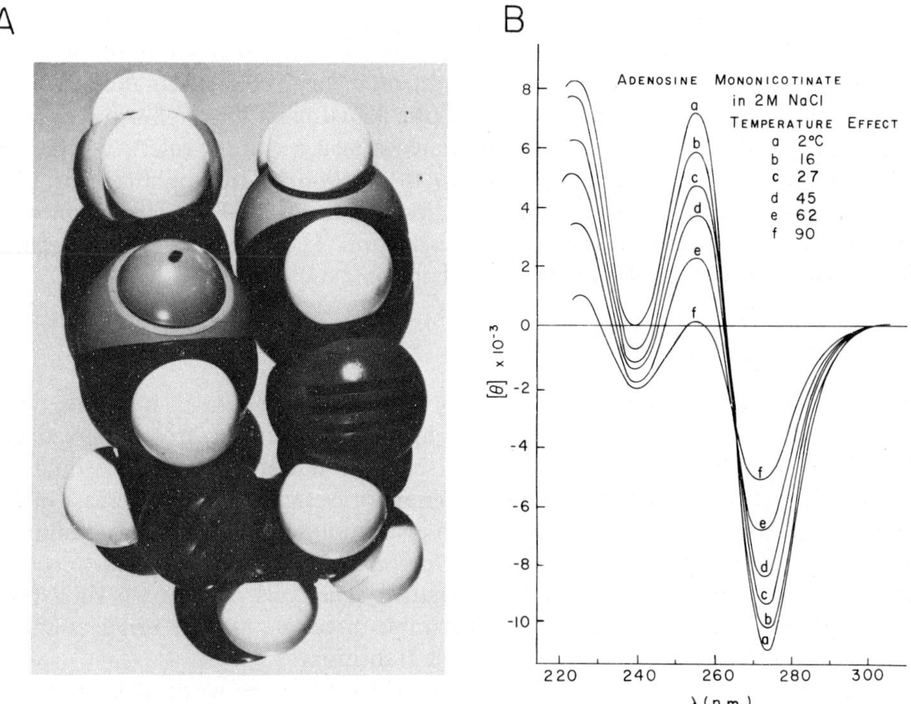

Figure 18. A, Molecular structure of adenosine-5′-mononicotinate showing the facility with which the two bases can stack. B, Temperature study of the circular dichroism of adenosine-5′-mononicotinate. A reciprocal relation is seen between the 260 nm band of adenine and the band near 270 nm of the nicotinate chromophore as the two aromatic chromophores stack at lower temperatures. Reproduced, with permission, from [31].

reciprocity in the circular dichroism spectra [31]. Perhaps the best variable is temperature as this perturbs the chromophores the least. The temperature dependence of the adenosine-5′-mononicotinate circular dichroism spectrum is given in Figure 18B, where the reciprocal relations are very striking.

Clearly, optical rotation can be utilized in favorable cases to identify proximity of a pair of chromophores, whether the pair of chromophores are in a prosthetic group or whether one is in a prosthetic group or other binding molecule and the second is in a protein, e.g., the Trp or Tyr side chains. Also, if an intense absorbance due, for example, to the Soret band of a heme, which is inherently symmetric, does exhibit a large rotational strength, then concern should be present that the protein transitions to which the Soret transition is coupling contain reciprocal contributions due to that pair-wise coupling. This has been described as protein-heme interactions in heme proteins [30]. Such interactions, while complicating interpretations of optical rotation data in terms of the structure of the protein, do provide information on the stereochemistry of the interaction.

(iv–b) Weak absorption bands with large magnetic transition dipole moments. The origin of our understanding about the source of rotational strength of weak absorption bands is due to the Eyring one-electron theory of optical rotation [33]. As originally formulated, it was an octant rule (the actual basis for the much discussed octant rule [34]), though it is now appreciated that a quadrant rule is correct for chromophores with C_{2v} symmetry, such as the carbonyl chromophore [35]. The useful simplicity of the original Condon, Altar and Eyring [33] statement on one-electron rotatory power warrants a short description. The potential for the chromophoric electron was that of an anisotropic harmonic oscillator

$$V=\tfrac{1}{2}(k_1 x^2 + k_2 y^2 + k_3 z^2) + Axyz \tag{37}$$

with unequal force constants, i.e., $k_1 \neq k_2 \neq k_3$. The term $Axyz$ introduces the dissymmetry due to a perturbing vicinal group. If the perturbing group is positively charged or uncharged with incompletely screened nuclei [36], $Axyz$ would be negative, i.e., the potential energy of the chromophoric electron would decrease in the vicinity of this vicinal group and the electron distribution would distort, extending in the direction of the perturbing group. If the perturbing group is in a positive octant then the rotational strength would be negative. The axes are chosen such that $k_1 > k_2 > k_3$, i.e., the z axis is the most polarizable direction, and the origin is at the atom representing the origin of the electronic transition.
(1) Partial molar rotatory powers [37]. Due primarily to the work of Litman and Schellman [38], it now appears that a quadrant rule is most appropriate for a peptide chromophore, in which case the sign does not change with the $+z$ and $-z$ directions. As for distance dependence, there are two dominant terms, r^{-5} and r^{-3}, which are of opposite sign with the r^{-3} term relevant to the above discussion and dominating at distances greater than 3 Å. A statement of $R_{n-\pi^*}$ has been written for the methyl moiety perturbing a peptide n–π^* transition [38]

$$R_{n-\pi^*} = \frac{\gamma_x \gamma_y}{r^5}[P^1(k)] + \frac{\gamma_x \gamma_y}{r^3}[P^2(k)] \tag{38}$$

where in analogy to partial molar refraction, the P^i are referred to as partial molar rotatory powers of moiety, k. When these quantities can be determined for a given moiety, it becomes possible to compute the rotational strength due to the moiety wherever it may be placed. This has been achieved for the methyl moiety [38] and the reported values are

$$[P^1(CH_3)] = 9.62 \times 10^{-76} \quad \text{and} \quad [P^2(CH_3)] = -1.25 \times 10^{-60}. \tag{39}$$

The partial molar rotatory power approach requires simple compounds from which to solve for the values of the $[P^i(k)]$. In this regard a particularly useful model of the peptide chromophore is contained in pyrrolid-2-one, the structure of which is given in Figure 19. The reason this structure is so useful is that it is essentially planar, as has

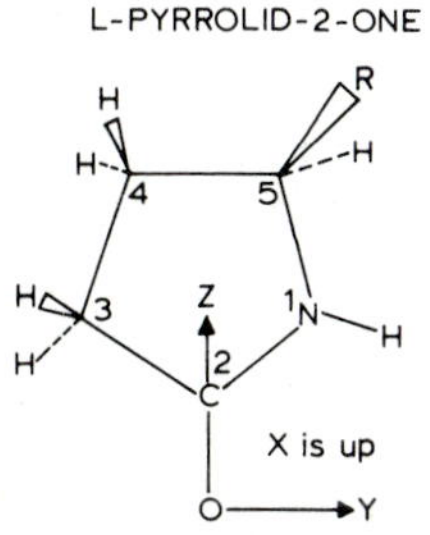

L-PYRROLID-2-ONE

X is up

R	NAME
CH_3	L-5-METHYL PYRROLID-2-ONE
CH_2OH	L-5-HYDROXYMETHYL PYRROLID-2-ONE
COO^-	L-PYROGLUTAMATE
COOH	L-PYROGLUTAMIC ACID
$COOCH_3$	METHYL-L-PYROGLUTAMATE
CH_2I	L-5-IODOMETHYL PYRROLID-2-ONE

Figure 19. Structure of L-pyrrolid-2-ones with various substituents indicated for the R group on position 5. The ring is essentially planar so that the character of the R group is responsible for the rotational strength of the peptide n–π^* transition. The CD spectra resulting from these molecules are given in Figure 20. Reproduced, with permission, from [37].

been shown in crystal structures of pyrrolid-2-one-5-carboxamide and L-5-iodomethyl-pyrrolid-2-one [39]. This means that the source of optical activity and rotational strength derives from the substituents on the ring. A series of compounds with different substituents at position 5 provides a unique opportunity to demonstrate the dependence of rotational strength on the nature of the substituent.

The circular dichroism spectra of a series of L-5-substituted pyrrolid-2-ones are given in Figure 20. The band due to the n–π^* transition is the long wavelength band near 210 nm. Experimentally derived rotational strengths are given in Table 1. Even though all R groups are in the same quadrant and at approximately the same distance from the peptide oxygen, both the sign and magnitude change, and they change in a systematic way with the chemical nature of the R group, the more electron-negative the R group the more positive the rotational strength. This is in accord with equation 37. The carboxylate or carboxyl being an electron-negative grouping acts as a positive potential in the positive quadrant, increasing the potential energy of the chromophoric electron in this positive quadrant. The result is a positive rotational strength. When the R group is a neutral CH_3 moiety, the rotational strength has changed sign and become negative. Following Kauzmann et al. [36], with the nuclei of the methyl moiety being incompletely screened by their surrounding electrons, this vicinal group acts as a partial positive charge, giving a negative potential in the positive quadrant for the chromophoric electron. The result is a negative rotational strength. The large negative rotational strength due to the iodomethyl substituent would derive from incomplete screening and the polarizability of this moiety. The rotational strength due to polarizability originates from a magnetic-electric coupled

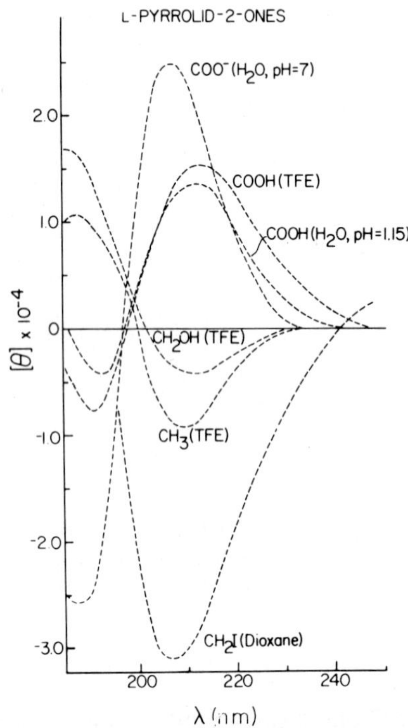

Figure 20. Circular dichroism spectra of a series of L-pyrrolid-2-ones, the structures of which were indicated in Figure 19. The magnitude and sign of the n–π^* band near 210 nm is due essentially entirely to the properties of the substituent in position 5, i.e., the moiety noted beside each curve. Note that as the group becomes less negative the positive CD band decreases in magnitude and finally for the neutral methyl moiety a negative band occurs. See text for discussion. TFE, trifluoroethanol. Reproduced, with permission, from [37].

oscillator term [1,40], i.e.,

$$\mathrm{R}'(n\text{–}\pi^*) = -\frac{2}{hc}\sum_{j\neq i}\frac{V_{ij}\lambda_j^2\,\vec{\mu}_j\cdot\vec{m}_i}{(\lambda_i^2-\lambda_j^2)} \tag{40}$$

where the $\vec{\mu}_j$ are the electric transition dipole moments of the vicinal group and $\vec{m}_i$ is the magnetic transition dipole moment of the n–π^* transition. All three effects, partial charge, polarizability and incomplete screening, have been explicitly treated by Caldwell and Eyring [41].

Considering these several different contributions, the partial molar rotatory powers may be written in general for a polymer as

$$\mathrm{R}_{n\text{–}\pi^*} = \frac{1}{N}\sum_{n=1}^{N}\sum_{k}\sum_{l}\sum_{m} G_n^m(k)l[P_n^m(k)l] \tag{41}$$

TABLE 1
Rotational strengths of n–π^* transitions in pyrrolidones[a]

Compound	Wavelength of CD extremum	Molar ellipticity	Rotational strength
L-Pyroglutamate (H_2O, pH = 7)	206	25×10^3	20×10^{-40}
Methyl-L-pyroglutamate (H_2O)	209	18×10^3	17×10^{-40}
L-Pyroglutamic acid (trifluoroethanol)	212	15×10^3	14×10^{-40}
L-Pyroglutamic acid (H_2O, pH = 1.1)	212	14×10^3	12×10^{-40}
L-Pyroglutamic acid amide (trimethylphosphate)	222	11.3×10^3	9.4×10^{-40}
L-5-Hydroxymethyl pyrrolid-2-one (trifluoroethanol)	211	-3.2×10^3	-2.3×10^{-40}
L-5-Methyl pyrrolid-2-one (trifluoroethanol)	208	-8.5×10^3	-6.5×10^{-40}
L-5-Iodomethyl pyrrolid-2-one (dioxane)	212	-31×10^3	-35×10^{-40}

[a]From [37].

The first summation is over the peptide moeities in the polymer and the division by N, the number of peptide moieties, gives a mean residue rotational strength. Here k is used for the particular type of vicinal moiety. The summation of k is over all the different vicinal groups in the polymer. The summation over l is the sum over all of the groups of the same type, e.g., if $k = CH_3$, then a summation over all of the methyl groups in the polymer. Finally, the summation over m is the sum over the relevant physical property each with its own appropriate geometric factor, G, and corresponding partial molar rotatory power, P. Even considering the n–π^* transition, the sources of rotational strength are several, but it should be remembered that this becomes a relatively simple expression for a group like a methyl moiety (see Eqn. 38). The recognition of the several relevant physical properties demonstrates further the utility of the series of compounds such as the L-5-pyrrolid-2-one molecules shown in Figure 19, with CD data in Figure 20 and values of rotational strength in Table 1.

(iv-c) The inherently dissymmetric chromophore. When the chromophore is inherently dissymmetric, i.e., when the chromophore has a non-superimposable mirror image, the electronic transition itself has substantial electric and magnetic transition dipole moments with parallel components without considering perturbations of nearby neighboring groups. The rotational strength of such a chromophore is given directly by the dot product of the electric and magnetic transition dipole moments. Because there is no steep distance dependence of an interaction potential term, in general, the rotational strength derived from inherent dissymmetry can be expected to dominate. Molecules comprised of or containing chromophores with inherent dissymmetry are ergosterol and hemisterol, containing skewed dienes, hexahelicene, the urobilins and, most relevant to proteins, the disulfide bridge. The contributions to the rotational strengths of electronic transitions in a disulfide depend on the dihedral angle made by the S–C′ and S–C″ bands when looking along the S–S bond axis. Due to repulsion between unshared electron pairs on each sulfur atom the preferred dihedral angle is $\pm 90°$ [42]. At this angle, however, the transitions originating in these

non-binding orbitals are nearly degenerate and have rotational strengths of opposite sign [43], such that they tend to cancel. Under these circumstances the rotational strength is again derived from vicinal perturbations. When the dihedral angle is made larger than $\pm 90°$, the longest wavelength band would be positive for a left-handed helix sense [43]. When the dihedral angle is less than $+90°$ the sign is reversed. And the reverse is true for the negative dihedral angles [44,45], such that a quadrant rule can be said to be relevant for the inherent dissymmetry term. These effects are reviewed by Boyd [44], where the changes in wavelength of the band with dihedral angle are also considered.

3. Circular dichroism and absorption spectra of polypeptide conformations and prosthetic groups

(a) Polypeptide conformations

The classical polypeptide conformations are the α-helix and the parallel and antiparallel β-pleated sheets due to Pauling and colleagues [46,47]. These conformations can occur separately in fibrous proteins or they can often be beautifully combined, as in certain globular proteins such as triose phosphate isomerase [48] and carboxypeptidase A [49]. The interesting motifs constituting the anatomy of some globular proteins have been emphasized by the work of Richardson [50], and a combination of pleated sheet and α-helix was early proposed as a structural motif for dynamic voltage dependent channel formation [51,52]. Our concern here, however, is to characterize separately these conformations in order that a view of more complex combinations of these structures can be correctly obtained.

The situation of polypeptide conformation is yet more complex than the consideration of these classical structures. Ramachandran and Kartha [53,54] and Rich and Crick [55] demonstrated that the special polypeptide sequences in which glycine repeated rigorously every third residue resulted in a triple-stranded helix which was the structure of collagen, a connective tissue protein important among other things in providing requisite tensile strength. This structure also occurs in complement C1q, a component of the immune response system [56]. Furthermore, repeating sequences have been found in elastin, the most prominent of which is the polypentapeptide, (L·Val-L·Pro-Gly-L·Val-Gly)$_n$, where n has been found to be greater than 11 [57]. This structure has been shown to have a regularly repeating β-turn, a 10-atom hydrogen-bonded ring [58,59], which is thought to form a new class of helices, called β-spirals [58,60,61]. Finally, there is yet another class of helices that utilizes a repeating sequence. These are polydipeptides that contain an L-D dipeptide repeat. These have been proposed to form single [62–65] and double [66,67] stranded β-helices. The term β-helix has been used to emphasize that the hydrogen-bonding patterns between turns of the single-stranded helix and between chains in the double-stranded helix are those of the β-pleated sheet structures with both parallel and antiparallel hydrogen-bonding patterns being possible [68]. A single-stranded β-helix forms a monovalent cation selective transmembrane channel across biological membranes.

(i) The α-helix

From the standpoint of optical properties of interest here, the α-helix is the most striking polypeptide conformation. On ordering, α-helix-forming polypeptides exhibit the greatest amount of hypochromism with the mean residue molar extinction coefficient of the 190 nm band going from about 7000 to 4000 [69]. Also, the magnitude of mean residue molar ellipticities of the α-helix is generally larger, in the 205–230 nm range with values of the order of $(-3 \text{ to } -4) \times 10^4$. In order to obtain the pure optical properties for an α-helix, a model is required that contains only the peptide chromophore, i.e., a model without complicating absorbances in the amino acid side chains. This is achieved with poly-L-alanine. The CD spectrum of poly-L-alanine in the right-handed α-helical conformation is given in Figure 21A in the solvent trifluoroethanol [70]. The corresponding absorption curve is given in Figure 21B. In both cases, the bold-faced curve is the experimental curve. As a means of achieving a more unique characterization of the electronic transitions giving rise to the absorption and CD spectra, the two sets of data can be simultaneously resolved into bands of a Gaussian shape. It should be noted, here, that Gaussian in wavelength has been shown to fit such bands better than Gaussian in wave number [3]. The set of Gaussian bands are shown in Figure 21 and listed in Table 2A.

The long wavelength band near 220 nm is readily assignable to the n–π* transition because of its energy but also because of the low absorbance yet high rotational

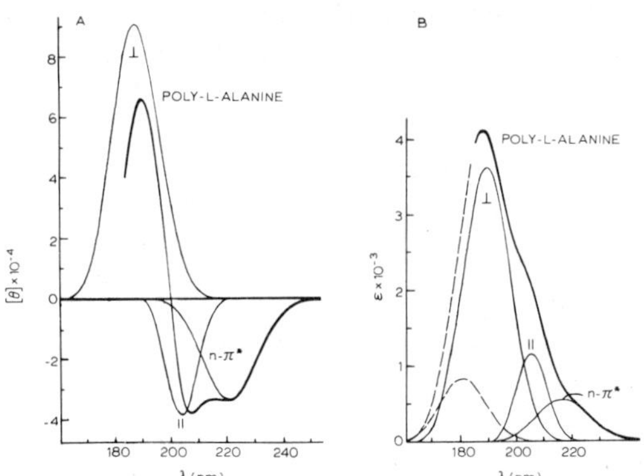

Figure 21. Simultaneous resolution of circular dichroism and absorption curves of α-helical poly-L-alanine. The constraint of correspondence of band maxima and widths was relaxed for the long wavelength (n–π*) band. A, Circular dichroism curve resolved into three bands: the positive perpendicularly polarized band near 190 nm, a negative parallel polarized band resolved near 204 nm, and a negative n–π* band near 220 nm. B, Absorption curve also resolved into the same three bands. The critical values for these bands are given in Table 2A. Reproduced, with permission, from [70].

TABLE 2
Optical values for α-helix, antiparallel β-pleated sheet and type II β-turn

Wavelength of extremum	Molar extinction coefficient ($\times 10^{-3}$)	Oscillator strength	Dipole strength ($\times 10^{36}$)	Molar ellipticity ($\times 10^{-3}$)	Rotational strength ($\times 10^{40}$)	Anisotropy $\|R_i/D_i\|$ ($\times 10^4$)
A. α-Helix (poly-L-alanine)[a]						
216 (221)	0.55	0.013	0.58	−33	−24	40
204	1.16	0.017	0.71	−38	−18	24
189	3.62	0.088	3.62	91	71	20
B. Antiparallel β-pleated sheet (poly-L-serine)[b]						
214 (222)	0.6	0.019	0.9	−9	−5.6	6.2
197	3.5	0.065	2.85	52	32	11.2
C. Type II β-turn $(Val_1\text{-}Pro_2\text{-}Gly_3\text{-}Gly_4)_n$[c]						
216	0.78	0.017	0.82	−2.1	−1.6	2
203	1.18	0.024	1.0	5.2	3.5	3.5
190	5.2	0.15	6.2	$>\|-4.2\|$	$>\|-3.8\|$	0.6

[a]From [72].
[b]From [73].
[c]From [87].

strength. This is seen in the ratio R_i/D_i, referred to as the anisotropy of the band [36]. Another feature of a largely forbidden electric transition is that the absorption band can be expected to occur shifted by several nm to higher energies (lower wavelength) than the corresponding CD band, as shown by Moffitt and Moscowitz [71]. The other two bands, the negative band near 205 nm and the positive band near 190 nm are the result of excitation resonance interaction of the sort depicted in Figure 7C, where the two bands are polarized but at right angles. In the case of the α-helix, the negative band near 205 nm is polarized parallel to the helix axis and the positive band near

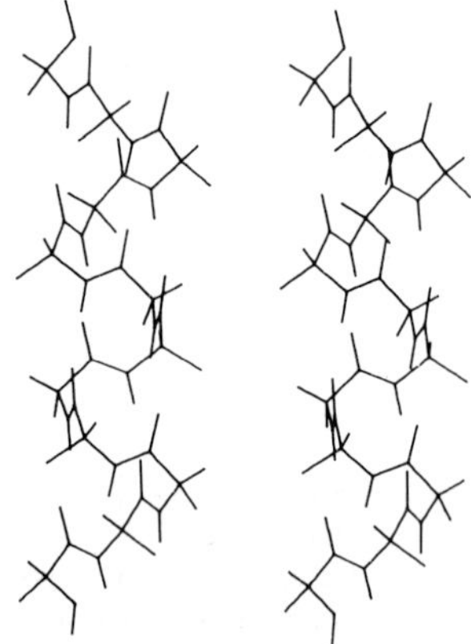

Figure 22. Stereo pair drawing of poly-L-alanine in the right-handed α-helical conformation of Pauling et al. [46]. These stereo pairs are arranged for cross-eye viewing.

190 nm is polarized perpendicular to the helix axis [72]. A stereo pair drawing of poly-L-alanine in the right-handed α-helical conformation is given in Figure 22.

(ii) The β-pleated sheet conformations

Again, for the β-pleated sheet conformations, a model system is sought for which there are no interfering electronic transitions in the side chains and for which simultaneous resolution of CD and absorbance data has been carried out. This has been achieved with poly-L-serine with a degree of polymerization of 20 in 80% trifluoroethanol/20% water. As will be noted below, under these conditions this structure is dominantly antiparallel-β-pleated sheet. The resolved CD and absorption spectra are given in Figure 23 and the resolved values are listed in Table 2B [63]. The band near 220 nm is again the n–π^* transition and the resolved CD band near 197 nm is considered to result from the splitting of the π–π^* (190 nm) band due to excitation resonance interactions. An important complication of the β-pleated sheet conformations with respect to properties being considered here is the significant dependence on dimension and twist of the sheet [74–77]. Of course, there is also the problem of differentiating between parallel and antiparallel β-pleated sheet conformations. These conformations are depicted in stereo pairs in Figure 24.

Efforts to obtain characteristic CD spectra of antiparallel and parallel β-pleated sheet conformations have utilized the vacuum ultraviolet and drawn on infrared spectra to substantiate the two different states. It was found by Balcerski et al. [78] that films of Boc-(L · Ala)$_7$-OMe formed antiparallel β-pleated sheets and films of Boc-

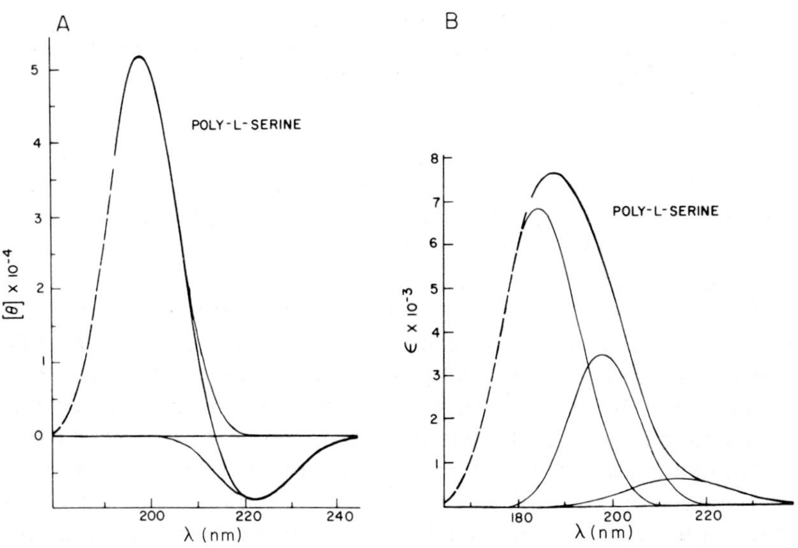

Figure 23. Simultaneous resolution of circular dichroism (A) and absorption (B) curves of poly-L-serine in the antiparallel β-pleated sheet conformation. The critical values for the resolved bands are included in Table 2B. Reproduced, with permission, from [73].

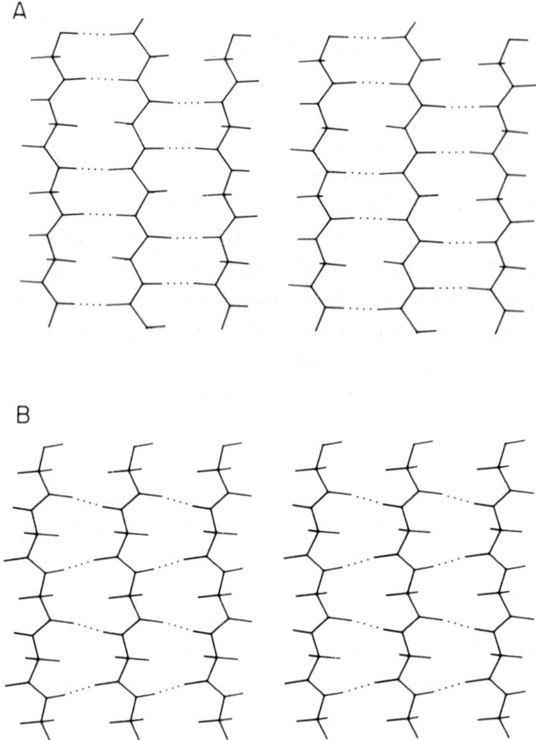

Figure 24. A, Stereo pair plot of poly-L-alanine in the antiparallel β-pleated sheet conformation of Pauling and Corey [47]. Sheet is tilted 10°. Stereo pairs arranged for cross-eye viewing. B, Stereo pair plots of poly-L-alanine in the parallel β-pleated sheet conformation of Pauling and Corey [47]. Stereo pairs are arranged for cross-eye viewing.

$(\text{L} \cdot \text{Val})_7$-OMe formed parallel β-pleated sheets. The vacuum ultraviolet CD spectra are shown in Figure 25 [79]. Significant differences are seen in the magnitude and wavelength of the positive band near 200 nm. The parallel β-pleated sheet has a more intense positive band at wavelengths longer than 200 nm and the antiparallel structure has a less intense positive band at wavelengths of 200 nm or shorter. Recall that in Figure 23A the peak is at 197 nm and is of a magnitude close to that of the antiparallel β-pleated sheet. The most striking difference between the CD patterns of the two structures appears to be at shorter wavelengths, where the parallel structure has a negative band near 180 nm and the antiparallel structure has a positive shoulder in this range. Since solution studies usually cannot get to this wavelength, reliance will generally have to depend on the positive bands for indicating the type of β-structure. Two notes of caution, however, are that film data can contain light scattering distortions and absorption flattening effects, which can cause red shifting and dampening of bands (see below). Also there are the problems of sheet dimension and twist.

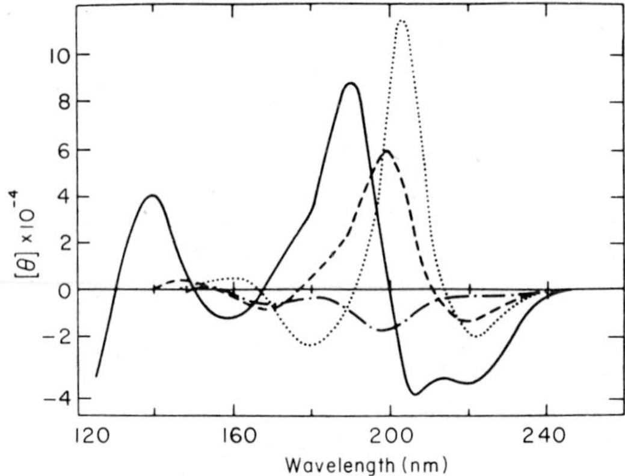

Figure 25. Circular dichroism spectra of the classical polypeptide conformations extended into the vacuum ultraviolet region. Solid curve, α-helical pattern averaged from poly-L-alanine and poly(γ-methyl-L-glutamate) data. Dashed curve, antiparallel β-pleated sheet CD pattern due to films of Boc-(L · Ala)$_7$-OMe [78]. Dotted curve, parallel β-pleated sheet patterns were calibrated by solution spectra. Dash-dot curve, disordered collagen to provide a measure of a random structure. Reproduced, with permission, from [79].

(iii) The collagen triple-stranded helix

The CD and absorption spectra of collagen are given in Figure 26, where the CD spectrum is seen to be distinct from those of the α-helix and β-pleated sheet conformations. This form of CD pattern for the peptide fragment of C1q and the effect of digestion by collagenase after heat denaturation was the basis for concluding that complement contained a triple-stranded collagen-like helix [56]. The importance of prolyl hydroxylation in the stability of this structure and the requirement of vitamin C for hydroxylation has been presented as a basis for the much discussed role of vitamin C in enhancing proper immune response [80].

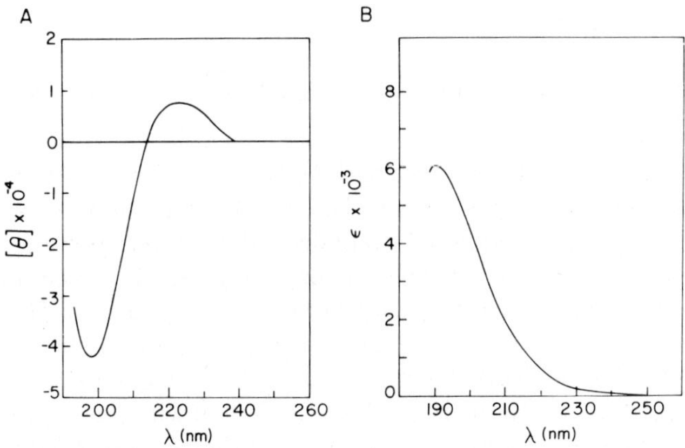

Figure 26. Circular dichroism (A) and absorption (B) spectra of calf skin type III collagen.

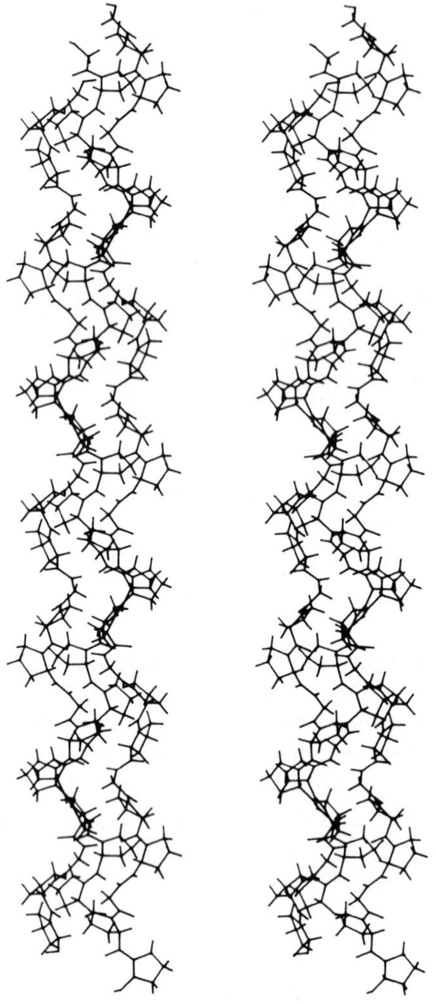

Figure 27. Stereo pair plots of the collagen triple-stranded helix using the polytripeptide (L · Pro-L · Pro-Gly)$_n$ coordinates of Miller and Scheraga [129].

As with the previous CD patterns, the long wavelength positive CD band has been assigned to the n–π^* transition on the basis that theoretical calculations show electrostatic effects giving rise to positive rotational strengths in this wavelength range for the poly-L-proline II model of the collagen structure [81]. As discussed in section 2(c)(iv-b), it is the n–π^* transition that is sensitive to static charge effects. This was the original basis of the Eyring one-electron theory of optical rotation [33]. On the basis of the absorption of plane polarized light by oriented films of collagen, the negative CD band is polarized parallel to the helix axis [82], consistent with theoretical treatments [81,82]. A band of positive rotational strength is at shorter wavelengths and is polarized perpendicular to the helix axis.

The conformation of the triple-stranded helix is shown by means of stereo pairs in Figure 27 for the polytripeptide, (L · Pro-L · Pro-Gly)$_n$. On careful examination it is possible to see why every third residue must be a glycine. These stereo pairs are given for cross-eye viewing rather than the usual wall-eye (distance) viewing.

(iv) β-turns and β-spirals

(iv-a) The type II β-turn. The β-turn conformational feature is shown in Figure 28 for the Pro_{i+1}-Gly_{i+2} sequence. The hydrogen bond is between the C–O of the *i*th residue and the N–H of residue $i+3$. Two general types of β-turns are immediately apparent. One has the C–O of the residue $i+1$-$i+2$ peptide moiety (the end peptide moiety) pointing out on the same side of the structure as the α-hydrogen of residue $i+1$. This is called a type I β-turn (see Figure 28A). The other has the C–O of the end peptide moiety on the opposite side of the mean plane and is called a type II β-turn (see Figure 28B). This general conformational feature was first appreciated by Venkatachalam [83]. The specific type II Pro-Gly β-turn has been demonstrated in recurring peptide sequences of elastin [58,84]. In a survey of the crystal structures of 29 globular proteins, Chou and Fasman [85] have found that the β-turn as a conformational feature is more common than the β-pleated sheet and almost as common as the α-helix. It was also found that the Pro-Gly sequence was the most probable β-turn. That the Pro-Gly β-turn is type II was demonstrated by means of the nuclear Overhauser effect [86], and this was subsequently verified by crystal structure [59].

Four residues are required for a recurring Pro-Gly β-turn. This is provided by the repeating tetrapeptide of elastin, (L · Val_1-L · Pro_2-Gly_3-Gly_4)$_n$. The CD and absorption spectra of the polytetrapeptide of elastin are given in Figure 29 [87] and the resolved bands are characterized in Table 2C. Interestingly, this is similar to the pattern calculated by Woody [88] for β-turns. As usual, the long wavelength band near 220 nm is the n–π^* transition and the other bands derive from the π–π^* transition. It should be noted that the CD pattern is quite distinct from the CD spectra of the previously considered conformations. Significant differences with the

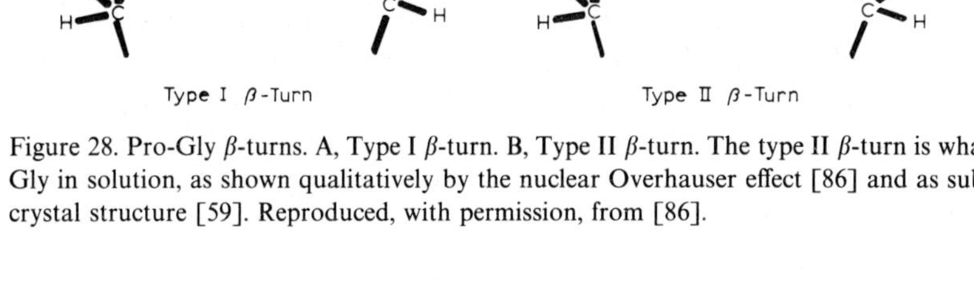

Type I β-Turn　　Type II β-Turn

Figure 28. Pro-Gly β-turns. A, Type I β-turn. B, Type II β-turn. The type II β-turn is what occurs for Pro-Gly in solution, as shown qualitatively by the nuclear Overhauser effect [86] and as substantiated in the crystal structure [59]. Reproduced, with permission, from [86].

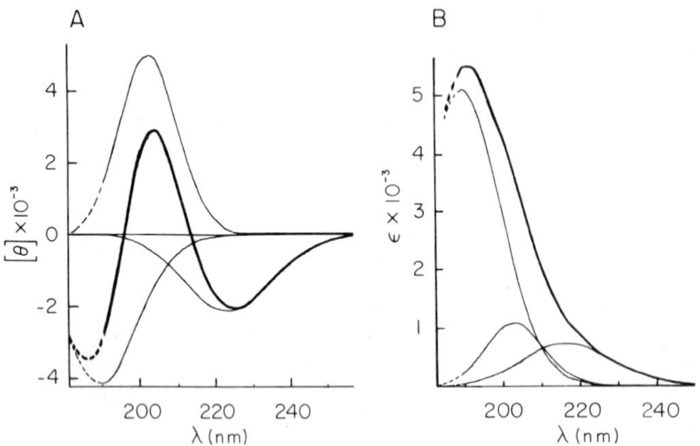

Figure 29. Circular dichroism (A) and absorption (B) data on the polytetrapeptide, (L · Val-L · Pro-Gly-Gly)$_n$, which is a model for a poly β-turn structure. The critical values for the resolved bands are given in Table 2C. Resolution due to simultaneous fitting of both curves. Reproduced, with permission, from [87].

parallel β-pleated sheet, which is most similar, are the low molar ellipticity of the β-turn, more than an order of magnitude lower, and the energies and relative magnitudes of the two extrema.

(iv-b) The β-spiral of the polypentapeptide of elastin. A β-spiral has been defined as the helical generation of repeating units containing the β-turn as the dominant conformational feature [84]. A β-spiral conformation has been developed for the polypentapeptide of elastin, (L-Val_1-L · Pro_2-Gly_3-L · Val_4-Gly_5)$_n$. This peptide sequence differs from the foregoing polytetrapeptide by the insertion of a valyl residue between the two glycine residues. The Pro-Gly sequence is kept intact and, therefore, it is not surprising that this polypentapeptide with $n \simeq 12$ can exhibit a type II β-turn CD pattern, as shown in Figure 30A in trifluoroethanol [89]. In water the CD pattern shows components of the β-turn pattern superimposed on a large negative band at 198 nm. As will be seen below, the magnitude of this negative band is less for higher molecular weight polymers. On raising the temperature of very dilute aqueous solutions, this negative band becomes less intense. When concentrated aqueous solutions are warmed above 30°C, they become cloudy, and on standing a viscoelastic layer is formed, which is called a coacervate. This is considered to be the state under physiological conditions and to be a state comprised of filaments [90]. The CD pattern of a coacervate film, while red-shifted, is that of the type II β-turn. Accordingly, the polypentapeptide (PPP) in the coacervate state can be considered to be a β-spiral.

(1) Concept of cyclic conformations with linear conformational correlates. More information on the details of the PPP β-spiral can be obtained by considering a series of cyclic analogs. The concept being utilized here is that of cyclic conformations with linear conformational correlates. The idea is that a helix with a large number of residues per turn can have a low pitch and only small changes in torsion angles are sufficient to interconvert between cyclic and linear helical conformations. With this

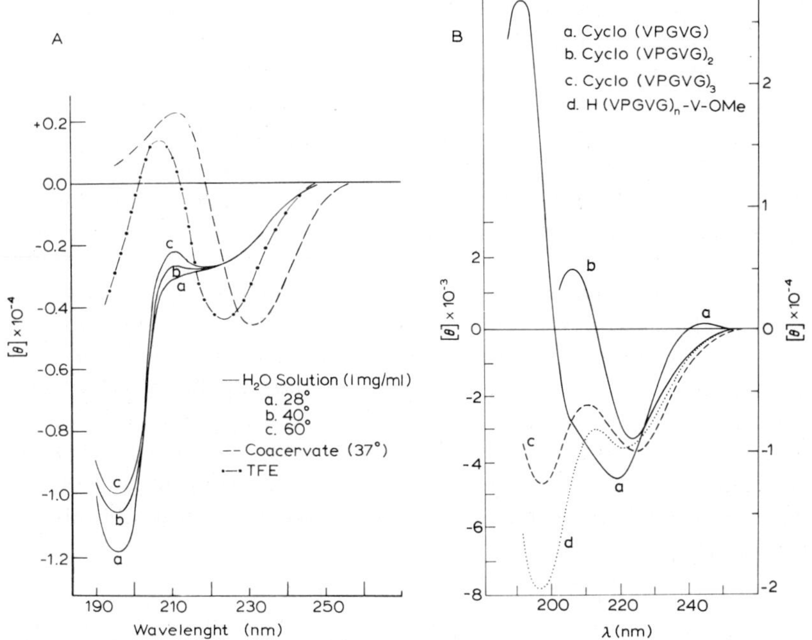

Figure 30. A. Circular dichroism spectra of the polypentapeptide of elastin, (L · Val-L · Pro-Gly-L · Val-Gly)$_n$, with $n \simeq 12$. Solid curve, a dilute solution in water at (a) 28°C, (b) 40°C and (c) 60°C. Dash-dot curve, in trifluoroethanol where the type II β-turn CD pattern is apparent. Dashed curve, the coacervate film of the polypentapeptide, again showing the type II β-turn CD pattern. The coacervate is a heat-aggregated state that occurs at physiological temperatures and is about 50% peptide and 50% water by weight. Reproduced, with permission, from [89]. B, Circular dichroism spectra of the polypentapeptide of elastin with $n \simeq 40$ and of cyclic analogs with $n = 1$, 2 and 3. Note that the CD spectrum of the cyclopentapeptide, the values for which are plotted on the right-hand ordinate, is strikingly similar to that of the α-helical conformation. The spectrum for the cyclodecapeptide is representative of a type II β-turn structure and that of the cyclopentadecapeptide is representative of a type II β-turn spectrum which is displaced by a more intense negative band at shorter wavelengths, much like that of the linear polypentapeptide.

concept the experimental approach is to synthesize a set of cyclic oligomers and to compare their physical properties to those of the linear polymer in the hope that one of the cyclics will have properties nearly identical to that of the linear high polymer. In Figure 30B are the CD patterns in water at 25°C of the cyclopentapeptide, cyclo(L-Val$_1$-L · Pro$_2$-Gly$_3$-L · Val$_4$-Gly$_5$); of the cyclodecapeptide, cyclo(L · Val$_1$-L · Pro$_2$-Gly$_3$-L · Val$_4$-Gly$_5$)$_2$; of the cyclopentadecapeptide, cyclo(L · Val$_1$-L · Pro$_2$-Gly$_3$-L · Val$_4$-Gly$_5$)$_3$; and of the linear polypentapeptide with $n \simeq 40$. The CD patterns of the three cyclic molecules are very different, but that of the cyclopentadecapeptide is similar to that of the linear polypentapeptide. Accordingly, the cyclopentadecapeptide is a candidate for the cyclic conformational correlate of the linear polypentapeptide. Detailed proton and carbon-13 nuclear magnetic resonance studies have verified this relationship [60]. Fortunately, a crystal structure has been obtained for the cyclopentadecapeptide [59]; this is shown in Figure 31A. In

Figure 31B is the solution-derived conformation using nuclear magnetic resonance [91]. Figure 31C is the β-turn perspective, showing the type II β-turn, which had previously been derived in solution from nuclear magnetic resonance studies [86]. Also included, in Figure 31D, is the conformation of the cyclopentapeptide [92]. An important point to be made with this structure and its corresponding CD pattern (curve a of Figure 30B) is that the CD pattern looks very much like that of an α-helix, particularly, for example, the α-helix CD spectrum of myoglobin or hemoglobin where the second negative band is less intense than in Figure 21A (see Figures 35C and 46B below).

The purpose of this approach of studying cyclic conformations is to derive the conformation of the linear conformational correlate. This has been done for the polypentapeptide and one of a closely related class of β-spirals is shown in Figure 32A [61]. The upper stereo pair is a view along the spiral axis showing the space for water within the β-spiral, as in the crystal structure of the cyclic analog in Figure 31A [59]. The side view shows the β-turns functioning as spacers between the turns of the spiral. This is schematically depicted in Figure 32B [93]. An interesting feature of this structure is that it provides a conformational basis for elasticity, called a librational entropy mechanism [94].

(v) β-helices

The circular dichroism pattern of yet another polypeptide conformation has been established [95]. This is that of the Gramicidin A transmembrane channel, the conformation for which is shown in stereo pairs in Figure 33 [96]. The primary structure of Gramicidin A is HCO-L · Val_1-Gly_2-L · Ala_3-D · Leu_4-L · Ala_5-D · Val_6-L · Val_7-D · Val_8-L · Trp_9-D · Leu_{10}-L · Trp_{11}-D · Leu_{12}-L · Trp_{13}-D · Leu_{14}-L · Trp_{15}–$NHCH_2CH_2OH$ [97]. The channel is formed by association of two monomers, each in a left-handed, single-stranded $\beta^{6.3}$-helical conformation and hydrogen bonding head-to-head (formyl end to formyl end) by means of six intermolecular hydrogen bonds, as first described over a decade ago [62,63]. The basis for this structure of the channel has recently been reviewed in detail [95]. The function of the channel is one of spanning the lipid layer of a lipid bilayer membrane, such as a cell membrane, and allowing selective passage of monovalent cations while excluding anions and divalent cations. The unique requirements of such a structure are a polar channel, polar ends and hydrophobic sides. The structure (originally termed a $\pi_{L,D}$-helix) is named a β-helix because the hydrogen bonding pattern between turns of the helix is that of the parallel β-pleated sheet (compare Figure 24B with Figure 33) and the hydrogen bonding pattern at the head to head junction, i.e., between monomers, is that of the antiparallel β-pleated sheet. Actually, the helical conformation is readily derived from a single chain in the β-pleated sheet conformation by interchanging backbone and side chains at alternating residues, i.e., at the D-residues. This generation of a helical structure can also be achieved with two chains associated either in the parallel or antiparallel configuration [68]. The result is a series of double-stranded β-helices first described by Veatch et al. [66].

The CD spectrum has been unequivocally established for the left-handed, single-

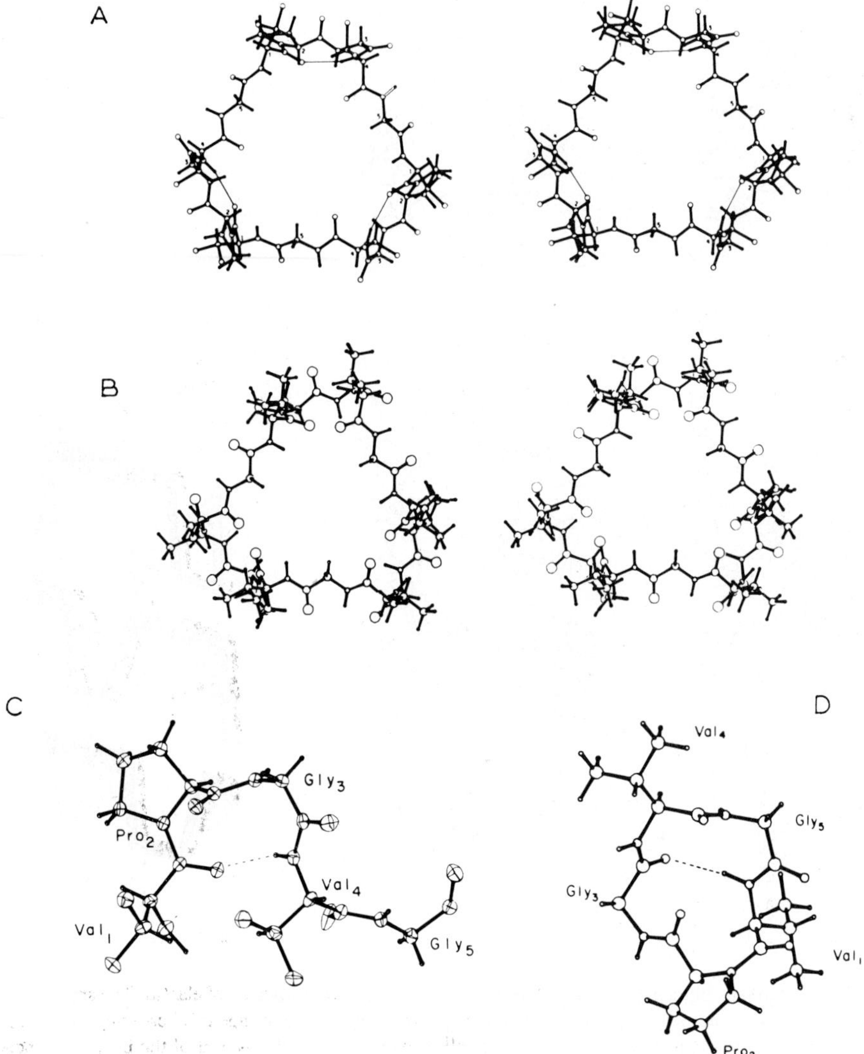

Figure 31. A, Stereo pair plot of the molecular structure of crystalline cyclopentadecapeptide. Reproduced, with permission, from [59]. B, Stereo plots of the molecular structure of the cyclopentadecapeptide in solution. Reproduced, with permission, from [91]. C, β-Turn perspective of crystalline cyclopentadecapeptide. Reproduced, with permission, from [59]. D, Solution derived conformation of the cyclopentapeptide which gave the α-helix type of CD spectrum in Figure 30B. Reproduced, with permission, from [92]. In parts A and B, the stereo pairs are plotted for cross-eye viewing.

stranded $\beta^{6.3}$-helix in a lipid bilayer structure [95] to be that given in Figure 34A, curve a, and the absorption curve for the system is given in Figure 34B. Again, this is a unique CD pattern that is quite distinct from the CD patterns of the above described

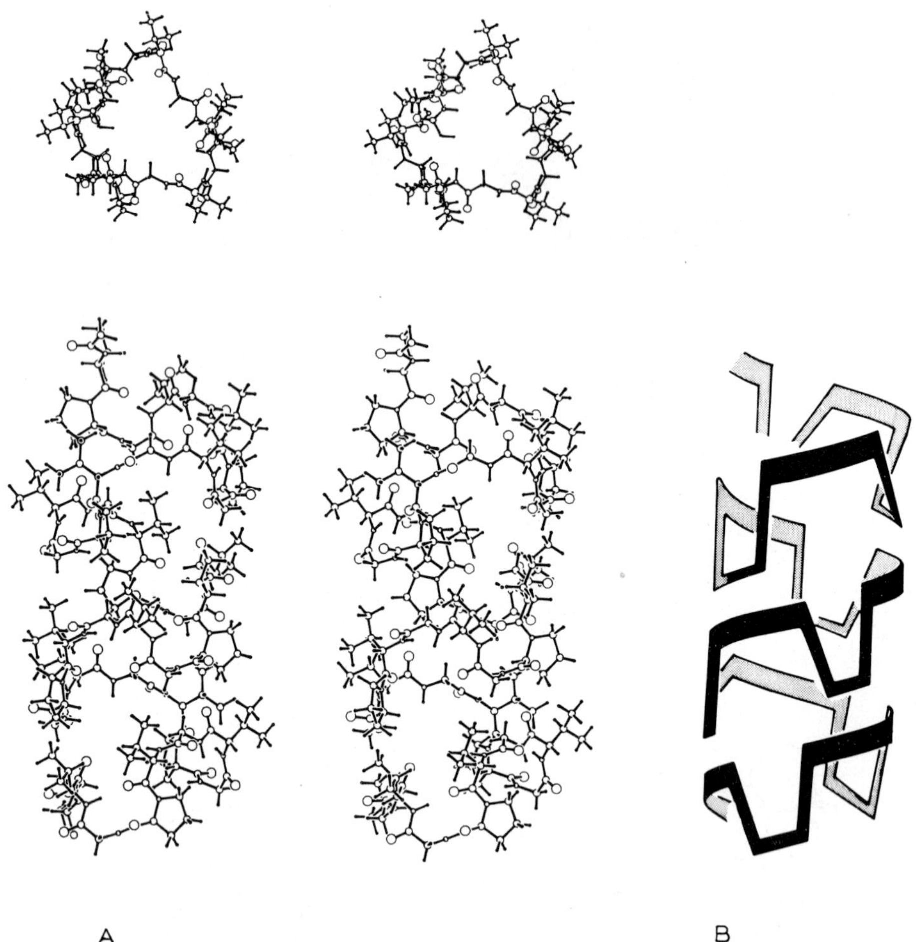

Figure 32. A, Stereo pair plots of a β-spiral conformation of the polypentapeptide of elastin. The spiral axis view is above the view perpendicular to the spiral axis. Stereo pairs are arranged for cross-eye viewing. Reproduced, with permission, from [61]. B, Schematic representation of the β-spiral of the polypentapeptide shown in part A. The conformation is one of a helix in which the β-turns function as spacers between the turns of the spiral. Reproduced, with permission, from [93].

polypeptide conformations. The negative deflection near 230 nm is due to the tryptophan residues as well as some deflection below 210 nm. The pattern of a positive CD band with an extremum occurring at 220 nm and the negative band with an extremum below 200 nm can be considered as due to the polypeptide backbone. The absorption curve in Figure 34B, of course, is dominated by the tryptophan residues. While other methods were used to determine the conformation of the Gramicidin A transmembrane channel, the CD pattern of curve a of Figure 34A has become the simple means of establishing the presence of the channel conformation for a given study. Curve b of Figure 34A is the conformation of hydrogenated Gramicidin A at

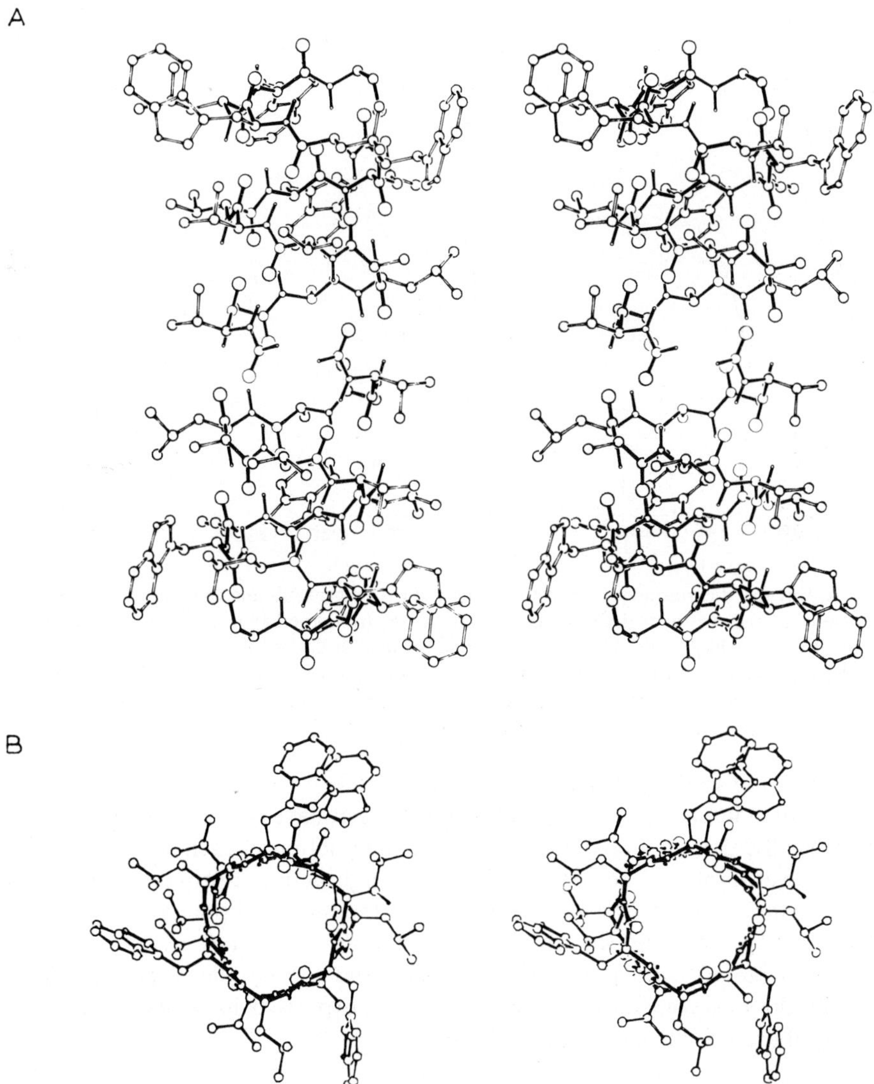

Figure 33. Stereo pair plots of the Gramicidin A transmembrane channel. It is a single-stranded, left-handed β-helix with approximately six residues per turn. A, Side view. Two molecules are hydrogen-bonded head to head (amino end to amino end) by means of six hydrogen bonds. The intermolecular hydrogen bonds have the pattern of an antiparallel β-pleated sheet (see Figure 24A), whereas the intramolecular hydrogen bonding pattern is a parallel β-pleated sheet (see Figure 24B). B, Channel view of a monomer. Reproduced, with permission from [96].

high concentrations and elevated temperatures (hydrogenation is to remove side chain chromophores). These are the conditions which favor the double-stranded β-helices [67,68].

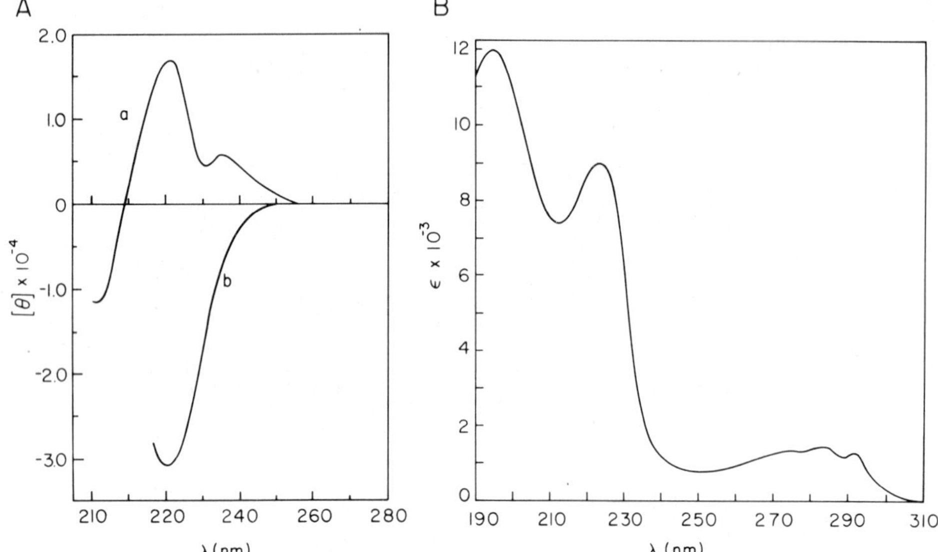

Figure 34. A, Circular dichroism spectra of the Gramicidin A transmembrane channel in phospholipid bilayers (curve a) and of hydrogenated Gramicidin A in trifluoroethanol at high concentration (~100 mg/ml), which is likely a double-stranded β-helix. See text for discussion. Mean residue ellipticities are given. B, Absorption spectrum of the Gramicidin A transmembrane channel in phospholipid bilayers. The absorption spectrum is dominated by the four tryptophan residues per pentadecapeptide. The extinction coefficient is given on a per residue basis.

(vi) Estimations of conformational fractions in a protein

When a system of interest is a mixture of two states that have each been characterized spectroscopically, then it is possible to calculate the fraction of each state within the mixture. As an example, let the system of interest be a protein comprised of α-helix and parallel β-pleated sheet and let the observable be the molar ellipticity at a specified wavelength, $[\theta]^{\lambda}_{\text{obs}}$. Taking the characteristic molar ellipticities at the specified wavelength to be $[\theta]^{\lambda}_{\alpha}$ and $[\theta]^{\lambda}_{\beta\text{p}}$ for the α-helix and the parallel β-pleated sheet, respectively, the observed molar ellipticity may be expressed in terms of the mole fractions χ_i of each of the conformations, i.e.,

$$[\theta]^{\lambda}_{\text{obs}} = \chi_{\alpha}[\theta]^{\lambda}_{\alpha} + \chi_{\beta\text{p}}[\theta]^{\lambda}_{\beta\text{p}} \tag{42}$$

Since

$$1 = \chi_{\alpha} + \chi_{\beta\text{p}} \quad \text{and} \quad \chi_{\alpha} = 1 - \chi_{\beta\text{p}},$$

then

$$\chi_{\alpha} = \frac{[\theta]^{\lambda}_{\text{obs}} - [\theta]^{\lambda}_{\beta\text{p}}}{[\theta]^{\lambda}_{\alpha} - [\theta]^{\lambda}_{\beta\text{p}}} \tag{43}$$

This may be written in general for any number of states and for the general observable, a, as

$$\chi_i = \frac{a_{\text{obs}} - \sum_{j \neq i} \left(1 - \sum_{k \neq i, j} \chi_k\right) a_j}{a_i - \sum_{j \neq i} a_j} \tag{44}$$

The complication with utilizing such an approach to the circular dichroism and absorption spectra of proteins is the number of different conformations and the variable extent of the structure, the immediate environment, etc. It is correct that the complete spectra are rich with differences between conformations and a selected set of wavelengths could be chosen which would best delineate the different structures and that simultaneous equations could be used to calculate the mole fractions. This approach has been usefully applied by a number of researchers. One significant problem, however, is that in many proteins there are residues with torsion angles which do not conform to any of the conformations considered here. Also, there can be cases where single residues do have the torsion angles of the characterized conformations but they do not have the same optical rotation and absorbance properties of the parent conformation. This is because those properties for the α-helix or β-pleated sheets, for example, derive from the excitation resonance, dispersion force interactions and static effects due to the presence of neighboring residues in the same conformational state. Thus, a general approach seems quite unlikely. Each protein system should be considered separately. In some cases the application of equations of the form of equation 44 can be quite informative. In other cases it can be an exercise of little consequence. In all cases, the more that is understood about the basic aspects of absorption and optical rotation the more correct the information that can be learned about the protein or polypeptide system of interest. What is often of interest, of course, is the capacity to monitor changes, and in this regard the optical spectroscopic methods are of great value.

(b) Prosthetic groups

There is an enormous amount of literature on the optical rotation and absorption of prosthetic groups of proteins. Only a few examples will be given here, which derive from the past interests of the reviewer. The points to be made, hopefully, will be of general utility and are chosen to relate to the previous discussions.

(i) Heme moieties

In Figure 8 the ferro- (hemochromagen) and ferri-heme spectra were given in the 600–200 nm range utilizing the heme undecapeptide of cyctochrome c. The undecapeptide contains no amino acids with chromophores in the wavelength range down to 220 nm, and therefore the bands in this range are due to the heme moiety. Changes in ellipticity, even in the 220 nm band which in proteins is dominated by the peptide

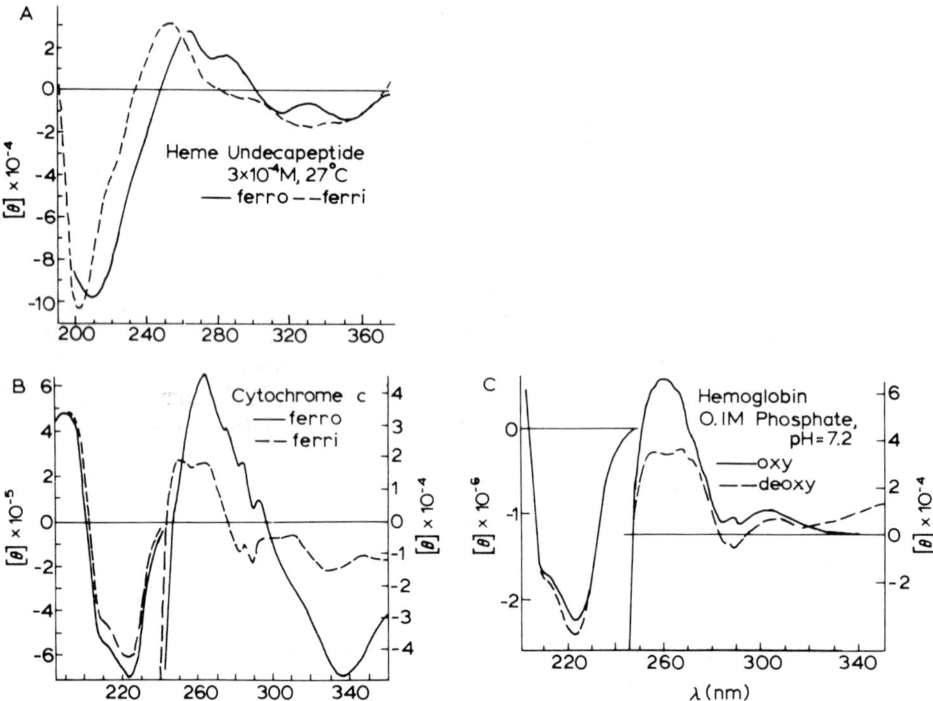

Figure 35. Circular dichroism spectra of the ferro- and ferriheme undecapeptide (A), of ferro- and ferricytochrome *c* (B) and of deoxy- and oxyhemoglobin (C). The molar ellipticities are plotted with respect to the heme moiety concentration. For the proteins, the greater than 240 nm region is plotted on the right-hand ordinate. On reduction of the heme undecapeptide there is a negative shift of ellipticity in the 205–250 nm wavelength range which is due to the heme moiety. A negative shift of similar magnitude is seen on reduction of cytochrome *c*. Even though the change nicely encompasses the negative n–π^* and parallel polarized bands attributed to peptide backbone in an α-helical conformation, the change is most reasonably taken as being due to the heme chromophore. A similar negative shift is seen on deoxygenation of oxyhemoglobin. Superimposed on the heme bands in the 240 nm and longer wavelengths are the local bands due to Phe, Tyr and Trp side chains. Reproduced, with permission, from [11].

n–π^* transition, cannot be utilized to conclude changes in protein conformation when they are derived by changes in the state of the heme. On reduction, the heme undecapeptide shows marked ellipticity changes in the 210–250 nm range which are due to heme bands (see Fig. 35A). Similar changes take place on forming the hemochromogen spectrum in cytochrome *c* (see Fig. 35B) and in hemoglobin (see Fig. 35C).

(i-a) Aggregation of heme peptides (heme-heme interactions). What is demonstrated in Figure 36 is the relative richness of information available from circular dichroism spectra of the Soret band. At a concentration of 3×10^{-4} M and at 10°C, the Soret (γ) absorption band of the heme undecapeptide appears quite simple, yet the circular dichroism curve shows multiple extrema. The CD spectrum with alternating positive and negative extrema requires that there be at least three electronic transitions within

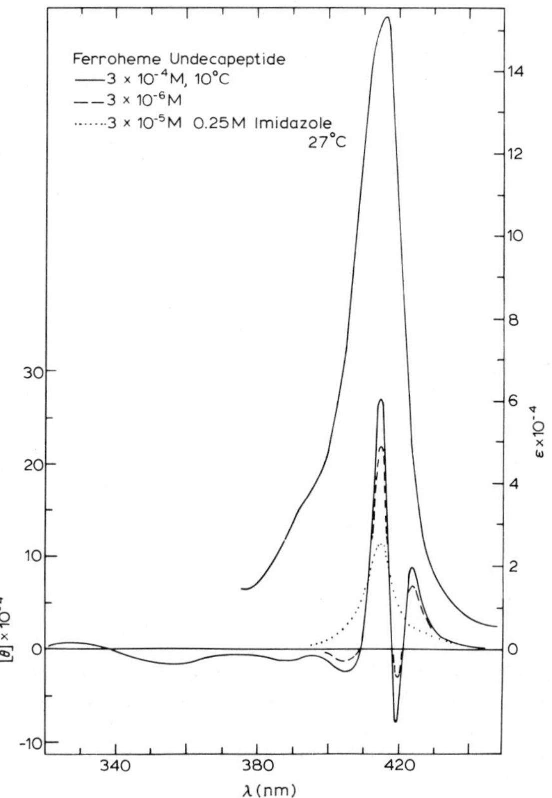

Figure 36. Circular dichroism of the Soret (γ) band of the ferroheme undecapeptide. Dotted curve, in the presence of imidazole which causes hemes to dissociate with a resulting monomer spectrum that is a simple near-Gaussian curve. The dashed and solid curves are at different stages of association and show at least three bands within the envelope of the simple appearing absorption spectrum (upper curve), which is plotted with respect to the right ordinate. Reproduced, with permission, from [16].

the width that could be considered a single band in the absorption curve. These are due in major part to the excitation resonance splitting of the monomer band. The simple Gaussian shape of a monomer band is demonstrated (see the dotted curve of Figure 36) by the addition of imidazole, which causes disaggregation. The heme undecapeptide was demonstrated in Figure 10 to exhibit a significant hyperchromism on association and, as discussed in section 2(a)(iii-b), this indicates a more nearly coplanar arrangement of the heme moieties in the aggregate. The narrow splitting energies seen in the CD curve would be consistent with larger distances between heme centers. If stacking could occur, the heme centers could be at shorter distances; splitting energies would be greater, and hypochromism would be expected. A dramatic hypochromism was observed for the heme octapeptide (see Fig. 9) and in Figure 37 the splitting energies are seen to be very large. Comparing parts A and B of

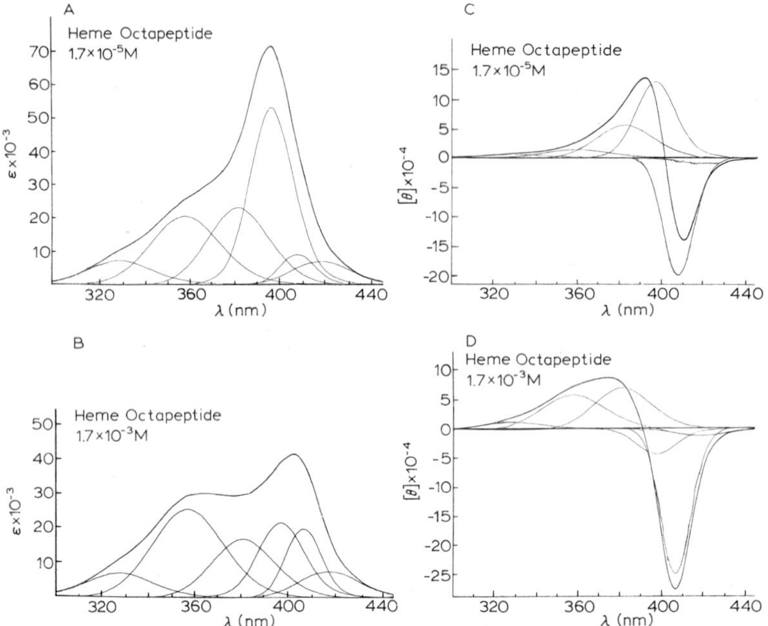

Figure 37. Absorption (A and B) and circular dichroism (C and D) spectra of the heme octapeptide at two different concentrations, 1.7×10^{-5} M (A and C) and 1.7×10^{-3} M (B and D) with simultaneous resolution of curves A and C and of curves B and D. On going to higher concentration, the positive 358 nm band is seen to increase markedly; the 397 nm band changes sign and the negative 408 nm band becomes more intense. Reproduced, with permission, from [13].

Figure 37, the most dramatic change is the increase in intensity of the band at 358 nm as aggregation proceeds on increasing the concentration. This is seen in parts C and D of Figure 37 to correspond with an increase in negative ellipticity at 408 nm arising in the resolved curves in large part from the 397 nm band. The two wavelengths, 358 and 408 nm, can be utilized to obtain the splitting energies. Following equation 6 using λ_- of 358 nm and λ_0 of 383 nm gives a splitting energy (V_{12}) of 3.6×10^{-13} ergs. Using the dipole strengths of the 358 nm band (25×10^{-36} esu^2 cm^2) and the combined 397 nm and 408 nm bands (20×10^{-36} esu^2 cm^2), taking the square root of these values and multiplying them, and then multiplying by $2/V_{12}$ and taking the cube root gives a distance of the order of 5 Å. Addition of the quadrupolar term is not expected to increase this distance by more than about 25% [13]. Accordingly, the hemes of the heme octapeptide are in a nearly stacked configuration. An oblique orientation as in Figure 7C to correspond with the data on Figure 9 would be the best description.

(i-b) Applications to multiheme proteins. The above heme peptide models for heme-heme interactions can be used to provide information on the nature of heme-heme interaction in multi-heme proteins. One classical problem is the heme-heme interaction responsible for the important sigmoid oxygen-binding curve of hemoglobin. The nature of the binding curve indicates certain energies of interaction between hemes

[21,22]. Knowing the approximate geometries for the different heme peptide associations, by means of temperature studies on those aggregated states it becomes possible to estimate energies of interactions between hemes in hemoglobin. These direct, through-space interactions arise due to the polarizability, α, of the heme moiety. Using the hemochromogen spectrum of Figure 8 and equations 15 and 16, it is possible to place a maximum polarizability on the heme moiety, i.e.,

$$\alpha \leqslant \frac{e^2}{4\pi^2 mc^2}\left[f_{550}\,\lambda_{550}^2 + f_{510}\,\lambda_{510}^2 + f_{418}\,\lambda_{418}^2 + f_{325}\,\lambda_{325} + f_{277}\,\lambda_{277} + \left(n - \sum_{i=500}^{220} f_i\right)\lambda_{190}^2\right] \tag{45}$$

With a maximum polarizability, maximal energies of interaction can be calculated

$$E = -\frac{9I}{8r_{ij}^6}\alpha^2 \tag{46}$$

where I is the ionization potential. Relating the experimental and spectrally derived energies, it becomes possible to conclude that the interactions responsible for the sigmoid binding curve of hemoglobin are not due to through-space interactions but rather must be transmitted through the protein structure [13].

The question of proximal hemes in multiheme proteins may be approached directly by means of the Soret CD spectrum. In general, a complex Soret CD spectrum can be considered a necessary condition for heme-heme distances of less than 10 Å. In oxyhemoglobin, the Soret CD band is a simple near Gaussian curve [13], as was observed in Figure 35 when imidazole was used to produce monomeric heme undecapeptide. Cytochrome oxidase, on the other hand, has a Soret CD band which is complex when reduced and when reduced and liganded with carbon monoxide, but it is a simple curve when oxidized. A multiple extrema Soret CD band cannot be used a priori to conclude that hemes are within a given distance because binding to the protein itself can remove degeneracies in the heme transitions and because the hemes can be chemically different moieties, but when properly combined with other information about the multiheme protein, such as the identity of the hemes, it can be a cogent argument.

(ii) Dinucleotides

There are many heteronuclear aromatic moieties in biomolecules with characteristic absorption bands in the 250–290 nm range. Because of this, a method for demonstrating the pairwise proximity of these groups can be of great use, particularly a relatively fast, sensitive method that can be carried out over wide ranges of concentration and other conditions. Circular dichroism is a method well-suited to identifying such interactions because of the nature of the coupling of strong electric transition dipole moments and the resulting possibility of observing reciprocal relations in the rotational strengths of electronic transitions in a pair of moieties. This was demonstrated above (see section 2(c)(iv-a)) for adenosine mononicotinate in Figure 18. Reciprocal relations have been observed which demonstrate the presence of and conditions for stacking of the two nucleotides in both the reduced and oxidized states of nicotinamide-adenine dinucleotides and for both the α- and β-epimers [98]. One example is given in Figure 38A for the oxidized state of β-nicotinamide-adenine dinucleotide. In this demonstration, temperature is the variable. As the temperature is decreased a band near 270 nm in the nicotinamide is seen coupling with the band near 260 nm in adenine. Another example is provided by flavin-adenine dinucleotide, as shown in Figure 38B [99]. In this case the variable used is solvent. As the volume fraction of dioxane is increased, there is an increasingly apparent reciprocity in the rotational strengths of the isoalloxazine band near 270 nm and again of the adenine band near 260 nm. FAD reciprocal relations may also be observed by varying the temperature. The original papers can be examined for the efforts to arrive at a more detailed description of the stacking. In both molecules, additional methods have been used to see the stacking [100–102], including hypochromism [103,104]. Similarly, proximity of a nucleotide prosthetic group to a tryptophan or tyrosine side chain can give reciprocal relations as binding occurs.

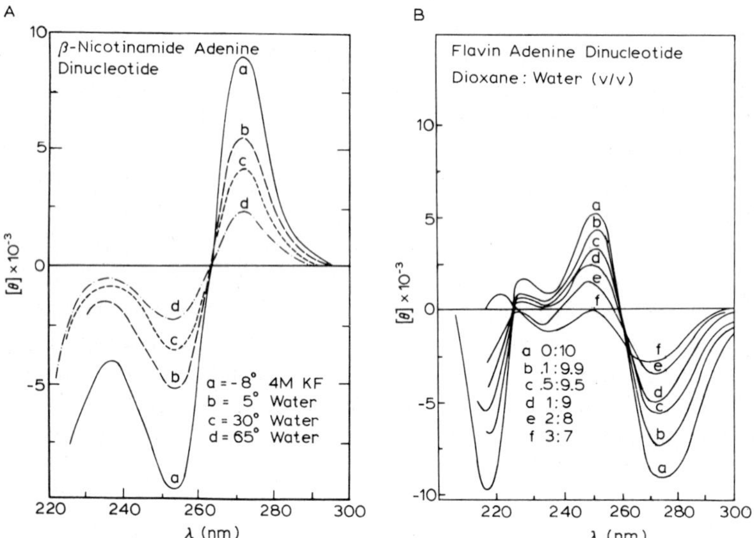

Figure 38. A, Circular dichroism spectra of β-nicotinamide-adenine dinucleotide (β-NAD) as a function of temperature showing the reciprocity between the adenine band just below 260 nm and the nicotinamide band near 270 nm. These reciprocal relations in optical rotation qualitatively demonstrate a close interaction of the two aromatic rings. Reproduced, with permission, from [98]. B, Circular dichroism curves of flavin-adenine dinucleotide (FAD) as a function of water/dioxane mixtures. As the solvent becomes more polar, there is increased proximity of the two aromatic rings, as shown by the reciprocal changes in the intensity of the adenine band below 260 nm and the isoalloxazine band of flavin near 270 nm. Reproduced, with permission, from [99].

4. Circular dichroism, absorption and optical rotatory dispersion of biomembranes

The problem of utilizing the optical rotation properties of membranes to obtain conformational information is demonstrated in Figure 39. The circular dichroism spectra of mitochondrial membranes is found to vary in a regular manner as a function of the extent of sonication [105]. In this case sonication causes a decrease in particle size. Starting from the sonicated membrane and approaching the intact unsonicated mitochondria, the magnitudes of all extrema become dramatically dampened and the 222 nm extremum becomes progressively red-shifted. In order to utilize CD data on biomembranes, it is essential to understand the effect of particulate systems on CD spectra. In this regard, it was early proposed that these effects arise due to the particulate nature of membrane systems and that the two basic effects leading to the distortion of the spectra were an absorption flattening (similar to the effect of Duysens [106]) and light scattering and that 'differential treatment of left and right circularly polarized beams' could be anticipated [107]. As outlined below, these effects were worked out on the poly-L-glutamic acid model system.

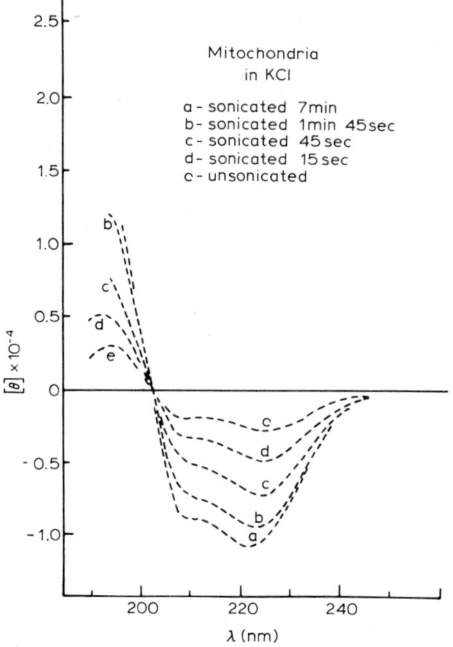

Figure 39. Circular dichroism spectra of heavy beef heart mitochondria in KCl as a function of sonication, which causes the membrane aggregates to decrease in size. As the size of the aggregate decreases, there is a progressive increase in the magnitude of the ellipticity and a shifting of the location of the negative extremum to 222 nm. These changes are due to particle size-dependent distortions in the CD spectra. Most pronounced is the effect on the magnitude of the bands, including the 222 nm band. Reproduced, with permission, from [105].

(a) Poly-L-glutamic acid as a model particulate system

Since poly-L-glutamic acid (PGA) has an ionizable side chain, in water its conformation is pH dependent [108]. Near neutral pH, when all of the carboxyl moieties are ionized, PGA is disordered and has even been used as a model for the random state, though repulsion between negatively charged side chains can reasonably be expected to cause some extension of the chain [109]. At pH 5, where about 30% of the side chains are ionized, the conformation is almost entirely α-helical. Lowering the pH to 4, however, causes the CD spectrum to continue to change; the magnitude of the ellipticity continues to increase, with the n–π^* band showing a greater effect than the parallel polarized band, as shown in Figure 40 (Quadrifoglio and Urry, reported in Ref. 37). This has been attributed to the effect of charge on rotational strength [37], as discussed in section 2(c)(iv). Changing the pH from 5 to 4 reduced the percent of

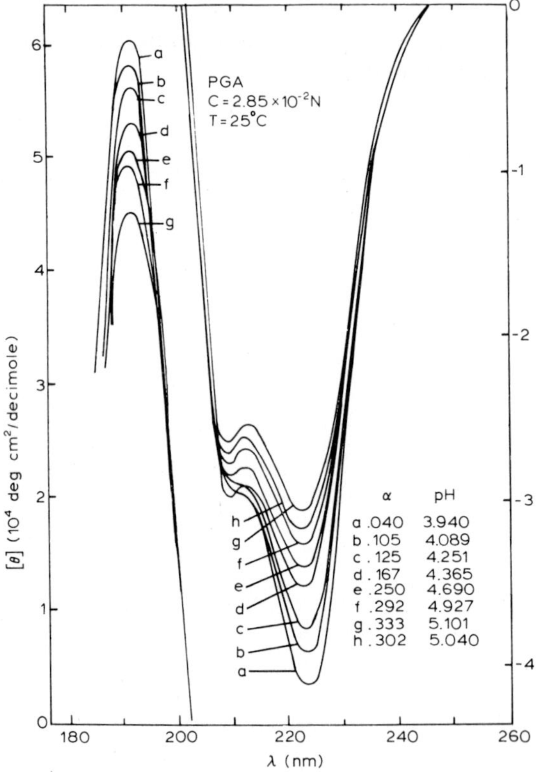

Figure 40. Circular dichroism spectra of α-helical poly-L-glutamic acid (PGA) as a function of pH, i.e., degree of ionization (α) of the carboxyl side chain. As the degree of ionization decreases from 0.3 at pH 5 to 0.04 at pH 3.9, the extrema increase in magnitude. It has been shown that aggregation of helical rods occurs as the pH decreases. The limited aggregation and the charge effect correspond to increases in ellipticity. As will be seen in Figure 41, more extensive aggregation causes dramatic dampening of ellipticity and other effects. Reproduced, with permission, from [37].

ionized side chains from 30 to 5%, such that the side chain effect is essentially complete. On standing at pH 3.9, however, the sample aggregates very slowly. This aggregation process can be speeded up dramatically by sonication, such that careful sonication can be effectively used to control the state of aggregation, i.e., particle size. The effect here is inverse to that of sonication of membranes as sonication of the unstable PGA system at pH 3.9 drives the system toward the stable aggregated state.

As shown in Figure 41, the state of aggregation has a marked effect on the absorption and CD spectra which were obtained simultaneously using the same phototube and the same beam path and, in fact, counting the same photons for both the absorbance and difference absorbance (CD) measurements [110]. In Figure 41A, increasing mean particle size causes the 190 nm absorption band to be increasingly dampened as the light scattering at 240 nm (a wavelength essentially outside of the absorption range for PGA) increases. Similarly, in Figure 41B the 190 nm band of curve a on aggregation progressively decreases in magnitude, as do the 208 and 222 nm bands, and the 222 nm extremum and the 201 nm cross-over of curve a progressively red shift as particle size increases. An immediate concern is whether aggregation is causing a conformational change. This has been shown not to be the

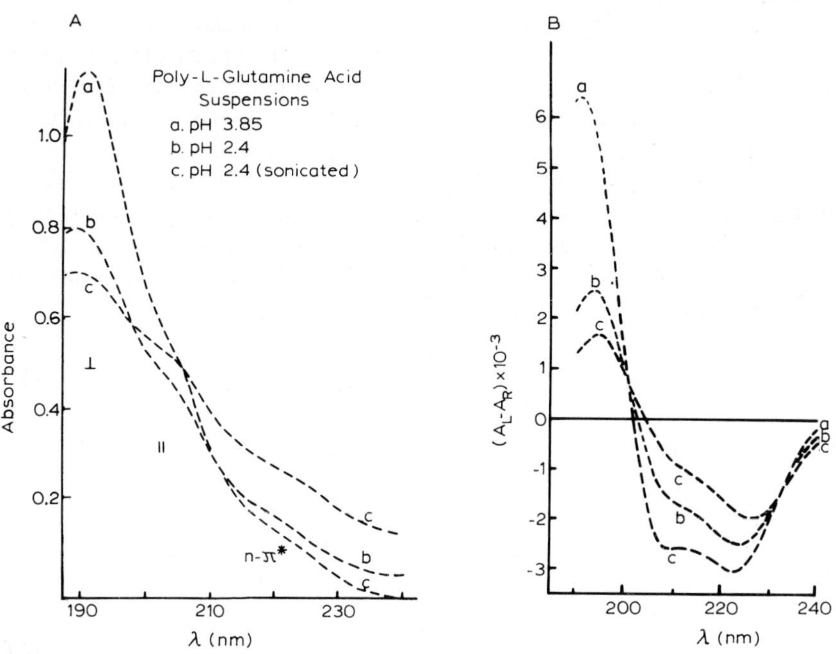

Figure 41. Absorption (A) and circular dichroism (B) spectra of poly-L-glutamic acid, pH 3.85, as a function of particle size, which is achieved by sonication. As particle size increases, the ellipticity and absorption extrema are progressively dampened; there is increased light scattering (curve b and c of A), and there is a marked red shifting of the CD extrema and crossover. These are distortions in the spectra due to the particulate nature of the optically active system. Adapted, with permission, from [110].

case by means of infrared spectroscopy, which demonstrates no changes in the bands characteristic of the α-helix [110]. A second question to be addressed is whether any or part of these effects might be due to helix-helix interactions in the aggregates. This can be discounted, because at pH 4.2, where there are as yet no dampening effects, and in fact where the magnitudes of the bands are still increasing on lowering the pH (see Fig. 40), there is an aggregation to the extent of more than ten helical rods giving an aggregate of the order of 10^6 daltons [111]. Under these conditions (pH 3.9) the effects of helix-helix interactions would already be essentially completely expressed. This leaves particle size as the variable to be considered. Thus, PGA is an ideal model system for studying the effects of optically active particulates on absorption and circular dichroism spectra. Effective description of this system can be expected to provide the essential elements for approaching the corrections for CD spectra of biomembranes. The approach that has been shown to be effective is outlined below [112,113].

(b) Obtaining an equivalent solution absorbance from a suspension absorbance

In the preceding discussions, absorption and circular dichroism data were seen to be valuable in characterizing the conformations and relative geometries of molecules in solution. Where aggregates, such as the heme peptide associations or covalent polymers, were considered it was done under conditions where particulate problems do not arise. If this wealth of information is to be utilized to characterize conformations in biological membranes, it is necessary to correct for the particulate nature of membranes. Accordingly, the objective is to obtain an equivalent solution absorption spectrum from a distorted suspension absorption spectrum. This can be done by considering the absorption process in terms of the probability that a photon entering the sample will be absorbed. In equation 2, the emergent beam intensity is designated as I and the initial beam intensity as I_0. The emergent beam intensity may also be expressed in terms of the intensity loss due to absorption, I_A, i.e.,

$$I = I_0 - I_A \tag{47}$$

Equation 2 is then rewritten as

$$A = -\log [(I_0 - I_A)/I_0] \tag{48}$$

or

$$A = -\log (1 - X_A) \tag{49}$$

where

$$X_A \equiv I_A/I_0 \tag{50}$$

that is, X_A is defined as the fraction of absorbed light or the probability that an entering photon would be absorbed.

When a sample is a suspension of particles, a photon can be scattered in a direction away from the phototube and thus be measured as an absorbance. Loss of emergent beam intensity due to scatter, I_s, can occur. The apparent absorbance due to scatter is designated as A_s and can also be expressed in analogy to equations 48 and 49, i.e.,

$$A_s = -\log [(I_0 - I_s)/I_0] \tag{51}$$

$$A_s = -\log (1 - X_s) \tag{52}$$

where

$$X_s \equiv I_s/I_0 \tag{53}$$

is the fraction of light that is scattered and not measured at the phototube. The correct absorbance of the suspension can be obtained by simply subtracting A_s for a given wavelength from the experimental value of the suspension absorbance at that wavelength. However, the correct absorbance for the suspension cannot be related on a per chromophore basis to the questions of hypo- and hyperchromism and to issues of difference absorbance between left and right circularly polarized light. Each like chromophore must be sampled equally by the beam in order that effects such as arise from dispersion force interactions can be measured accurately. This does not occur in a particulate system. What is required is the absorption spectrum that would occur if the same molecules were molecularly dispersed in solution with the same effective environments for the chromophores.

The problem can be approached in the following way. Consider a solution much like that of curve a in Figure 41A, where there is no light scattering and where the absorbance at the 190 nm peak is about 1.2. The fraction of photons absorbed is obtained by solving for X_A in equation 49,

$$X_A = 1 - 10^{-A} = 1 - 10^{-1.2} = 0.94 \tag{54}$$

The probability that an entering photon will be absorbed is 0.94, i.e., 94 out of every 100 photons entering the sample are absorbed. Now keeping the same total concentration, the molecules are made to aggregate such that the absorbance due to light scattering, A_s, just outside of the absorption range, e.g., at 240 nm, is 0.12 as in curve c of Figure 41A. Under such circumstances the probability that a photon entering the sample will be scattered is

$$X_s = 1 - 10^{-A_s} = 1 - 10^{-0.12} = 0.25 \tag{55}$$

i.e., 25 out of every 100 photons entering the sample will be scattered and not be counted by the phototube. By Mie scattering theory, the quantity X_s is dependent on the difference of the squares of the refractive indices of the particle n_p and the solvent n_s,

$$X_s = K'(\lambda)\left(\frac{n_p^2 - n_s^2}{n_p^2 + 2n_s^2}\right)^2 \tag{56}$$

where $K'(\lambda)$ is a wavelength-dependent factor that in general increases with decreasing wavelength. Since the difference in refractive indices also increases when the necessary transparent solvent is used (see Fig. 12), then X_s can be expected to increase on going from 240 to 190 nm. Thus, on going from solution to suspension, the situation has gone from 96% of photons absorbed to more than 25% of photons scattered at 190 nm. The objective is to do the inverse to correct the suspension absorbance and thereby recover the solution absorbance. Therefore, it is necessary to appreciate that most of the more than 25% of scattered photons will have to be appropriately recovered as absorbed photons. The probability that a photon scattered by the suspension would be absorbed by an equivalent solution is taken as the product of the probabilities of scatter, X_s, and of absorbance, X_A. This defines the lost or suspension obscured absorbance, A_{obsc}, as

$$A_{obsc} = -\log(1 - X_A X_s) \tag{57}$$

The solution absorbance recovered in this manner, A'_{soln}, would be written

$$A'_{soln} = A_{susp} + A_{obsc} - A_s \tag{58}$$

There is yet another effect, which was first described by Duysens [106]. In this effect, the high density of chromophores localized within a particle causes abrupt drops in intensity of the particle-sized pencil of light. This can be viewed, as stated by Duysens, as having the effect of casting a shadow on other subsequent particles in the same pencil of light. These discontinuities in the beam intensity have a flattening effect on the absorption curve. Duysens defined an absorption flattening quotient, Q_A, as the ratio of suspension to solution absorbance at a given wavelength. Adding this effect to the previous arguments gives

$$A'_{soln} = Q_A A_{soln} = A_F \tag{59}$$

Substituting equation 59 into equation 58 and rearranging terms gives the expression for the suspension absorbance, i.e.,

$$A_{susp} = A_F - A_{obsc} + A_s \tag{60}$$

This expression may be written for both the left and right circularly polarized beams to express the molar ellipticity of a suspension (see Eqn. 67 below). Before doing so, however, there are two useful relationships to note.

In equation 57, it is apparent that as an absorption maximum is approached with an absorbance greater than 1, the quantity X_A approaches 1 and in equation 60 A_{obsc} approaches A_s such that at the absorption maxima A_{susp} approaches A_F and Q_A can be estimated, i.e.,

$$Q_A \text{ (absorption maximum)} = A_{susp}/A_{soln} \tag{61}$$

The relationship between A_{obsc} and A_s is given below in Figure 43 for different values of A_F [112]. Knowing Q_A at one wavelength, it is possible with the proper shape of the solution absorption curve to estimate Q_A at all wavelengths in the following way. Considering a spherical particle with an absorbance along the diameter of A_p, Duysens derived the expression for calculating Q_A, i.e.,

$$Q_A = \frac{3}{2A_p}\left(1 - \frac{2[1-(1+A_p)e^{-A_p}]}{A_p^2}\right) \tag{62}$$

Table 3 contains a listing of A_p values and corresponding values of Q_A. If the particle is of uniform density then the absorption spectrum for A_p will at all wavelengths be proportional to the solution absorption spectrum, i.e.,

$$A_p(\lambda_1) = \frac{A_{soln}(\lambda_1)}{A_{soln}(\lambda_2)} A_p(\lambda_2) \tag{63}$$

Because of the approximation that $A_s = A_{obsc}$, this approach will give rise to minimal correction values and an iterative procedure can be used to obtain the desired accuracy [112]. If the particulate system is best described as a vesicle, which is generally the case for membranes, an equivalent A_p for Q_A (vesicles) [114] can also be obtained from Table 3. Of course, the problem at this stage is that in general an equivalent solution absorption spectrum will not be known for the suspension of interest. By also utilizing the circular dichroism data, however, an equivalent solution absorption spectrum can be determined as outlined below. This is called the pseudoreference state approach.

(c) Circular dichroism of suspensions

The experimental mean residue ellipticity for a suspension can be written in analogy to equation 20 as

$$[\theta]_{susp} = 3300 \frac{(A_L - A_R)_{susp}}{Cl} \tag{64}$$

The objective is to obtain the equivalent $[\theta]_{soln}$ which can be referred to as the corrected ellipticities, $[\theta]_{corr}$. The approach is to define an ellipticity distortion quotient, Q_E, by the relation

TABLE 3
Flattening quotients and associated particle absorbances

A_p	Q_A (spheres)[a]	Q_A (vesicles)[a]	A_p	Q_A (spheres)[a]	Q_A (vesicles)[a]
0.02	0.99	0.97	0.86	0.73	0.67
0.04	0.98	0.95	0.88	0.73	0.66
0.06	0.97	0.94	0.90	0.73	0.66
0.08	0.97	0.92	0.92	0.72	0.65
0.10	0.96	0.91	0.94	0.72	0.65
0.12	0.95	0.90	0.96	0.71	0.65
0.14	0.94	0.89	0.98	0.71	0.64
0.16	0.94	0.88	1.00	0.70	0.64
0.18	0.93	0.87	1.02	0.70	0.63
0.20	0.92	0.86	1.04	0.69	0.63
0.22	0.92	0.85	1.06	0.69	0.62
0.24	0.91	0.85	1.08	0.68	0.62
0.26	0.90	0.84	1.10	0.68	0.62
0.28	0.90	0.83	1.12	0.68	0.61
0.30	0.89	0.82	1.14	0.67	0.61
0.32	0.88	0.81	1.16	0.67	0.61
0.34	0.88	0.81	1.18	0.66	0.60
0.36	0.87	0.80	1.20	0.66	0.60
0.38	0.87	0.79	1.22	0.66	0.59
0.40	0.86	0.79	1.24	0.65	0.59
0.42	0.85	0.78	1.26	0.65	0.59
0.44	0.85	0.77	1.28	0.64	0.58
0.46	0.84	0.77	1.30	0.64	0.58
0.48	0.84	0.76	1.32	0.64	0.58
0.50	0.83	0.76	1.34	0.63	0.57
0.52	0.82	0.75	1.36	0.63	0.57
0.54	0.82	0.75	1.38	0.62	0.57
0.56	0.81	0.74	1.40	0.62	0.56
0.58	0.81	0.73	1.42	0.62	0.56
0.60	0.80	0.73	1.44	0.61	0.56
0.62	0.80	0.72	1.46	0.61	0.55
0.64	0.79	0.72	1.48	0.61	0.55
0.66	0.79	0.71	1.50	0.60	0.55
0.68	0.78	0.71	1.52	0.60	0.54
0.70	0.78	0.70	1.54	0.59	0.54
0.72	0.77	0.70	1.56	0.59	0.54
0.74	0.76	0.69	1.58	0.59	0.53
0.76	0.76	0.69	1.60	0.58	0.53
0.78	0.75	0.68	1.62	0.58	0.53
0.80	0.75	0.68	1.64	0.58	0.52
0.82	0.74	0.68	1.66	0.57	0.52
0.84	0.74	0.67	1.68	0.57	0.52
1.70	0.57	0.51	1.86	0.54	0.49
1.72	0.56	0.51	1.88	0.54	0.49
1.74	0.56	0.51	1.90	0.54	0.48
1.76	0.56	0.50	1.92	0.53	0.48
1.78	0.56	0.50	1.94	0.53	0.48
1.80	0.55	0.50	1.96	0.53	0.48
1.82	0.55	0.50	1.98	0.53	0.47
1.84	0.55	0.49	2.00	0.52	0.47

[a]Using the formalism of Gordon and Holzwarth [114] for vesicles to relate Q_A (vesicles) with an associated particle absorbance.

$$[\theta]_{\text{corr}} = [\theta]_{\text{susp}}/Q_{\text{E}} \tag{65}$$

with the definition obviously being

$$Q_{\text{E}} \equiv \frac{(A_{\text{L}} - A_{\text{R}})_{\text{susp}}}{(A_{\text{L}} - A_{\text{R}})} \tag{66}$$

Writing for the obscured absorbance for the left and right circularly polarized beams, A_{OL} and A_{OR}, respectively, and using equation 60, Q_{E} becomes

$$Q_{\text{E}} = \frac{(A_{\text{FL}} - A_{\text{FR}}) - (A_{\text{OL}} - A_{\text{OR}}) + (A_{\text{SL}} - A_{\text{SR}})}{(A_{\text{L}} - A_{\text{R}})} \tag{67}$$

The three terms in equation 67 can be considered individually. The first term is the differential absorption flattening term, the second the differential absorption obscuring term and the third the differential light scattering term.

(i) Differential absorption flattening and differential absorption obscuring
It has been shown [112] that the differential absorption flattening term is well approximated as

$$\frac{(A_{\text{FL}} - A_{\text{FR}})}{(A_{\text{L}} - A_{\text{R}})} \simeq Q_{\text{A}}^2 \tag{68}$$

and that the differential absorption obscuring is approximately

$$\frac{(A_{\text{OL}} - A_{\text{OR}})}{(A_{\text{L}} - A_{\text{R}})} \simeq Q_{\text{A}}^2\,(1 - Q_\sigma) \tag{69}$$

where

$$Q_\sigma = \exp\,(-A_{\text{obsc}}/A_{\text{soln}}) \tag{70}$$

is the absorption obscuring quotient.

Considering the case where differential light scattering is zero, i.e., $A_{\text{SL}} = A_{\text{SR}}$, the ellipticity distortion quotient becomes

$$Q_{\text{E}}^{A_{\text{SL}} = A_{\text{SR}}} = Q_{\text{A}}^2 - Q_{\text{A}}^2\,(1 - Q_\sigma) = Q_{\text{A}}^2 Q_\sigma \tag{71}$$

and the corrected ellipticity for the cases where there is no differential light scattering becomes

$$[\theta]_{\text{corr}}^{A_{\text{SL}} = A_{\text{SR}}} = [\theta]_{\text{susp}}/Q_{\text{A}}^2 Q_\sigma \tag{72}$$

At this stage the relationship between circular dichroism and optical rotatory dispersion becomes of particular interest. As shown in Figure 15 for an isolated band, the ellipticity is at an extremum where the optical rotatory dispersion curve is zero. Referring to equation 22, since α is proportional to $n_L - n_R$, the wavelength at which molar rotation is zero is where $n_L = n_R$. Therefore, even for an optically active particle when $n_L = n_R$ there can be no differential scatter of left and right circularly polarized light and, therefore, equation 72 is correct at this particular wavelength. For a dominantly α-helical protein, this occurs near 224 nm and 190 nm (see curve a of Fig. 42). There is also a wavelength at which the molar ellipticity is zero for an α-helical protein; this occurs near 201 nm. Even for the more complex overlapping bands of an α-helix, the

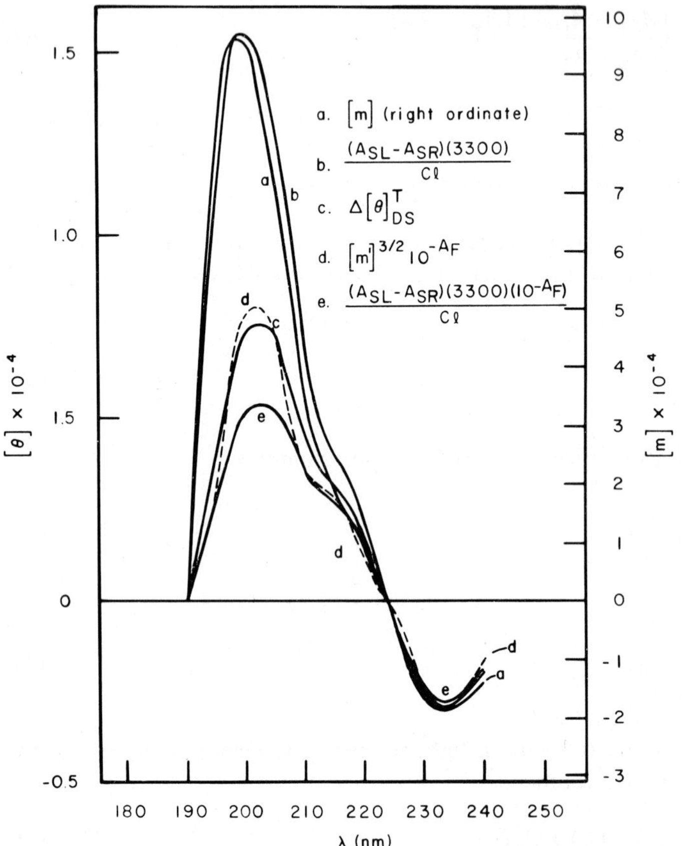

Figure 42. Differential scatter of left and right circularly polarized light by poly-L-glutamic acid particles. Curve a, mean residue rotation of PGA reference state plotted on right-hand ordinate. Curve b, contribution of differential scatter term only, $(A_{SL} - A_{SR})$ 3300/Cl. Curve c, total change in mean residue ellipticity due to differential scatter, includes differential scatter component of the $(A_{OL} - A_{OR})$ term. Curve d, comparison of empirical $k[m']^{3/2}\ 10^{-A_F}$ values with total differential scatter. Curve e, approximation to the total differential scatter which involves 10^{-A_F} flattening of $(A_{SL} - A_{SR})$ term. Note adequacy of the fit of the empirical term $k[m']^{3/2}\ 10^{-A_F}$ (curve d) to the total differential scatter values (curve c). Adapted, with permission, from [130].

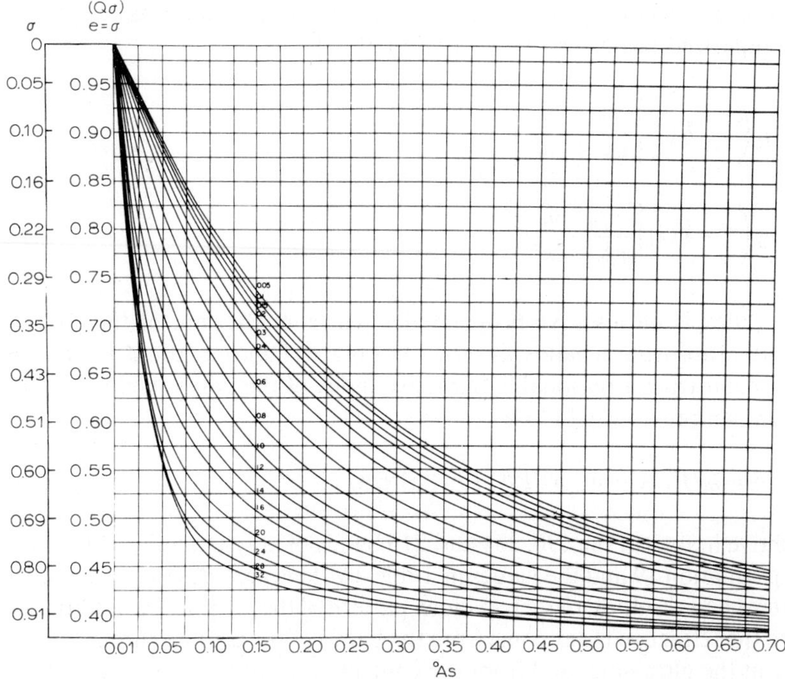

Figure 43. Plot of ^0A_s versus Q_σ. The quantity ^0A_s is defined in equation 83 and the selected curve is for the relevant value of A_F. In general, $A_{susp} - Q_A A_{prs} = {}^0A_s$ where $Q_A A_{prs} = A_F$, and the value of Q_A is arrived at by using equations 61, 62 and 63 and Table 3. See text for discussion, in section 4(d)(i). Adapted, with permission, from [112].

points near maxima on the CD curve (224 and 190 nm) correspond to near zero values on the ORD curve, and vice versa a maximum in the ORD curve near 200 nm corresponds to near zero values in the CD curve. These critical wavelengths can prove useful in achieving the corrections. In Figure 41A, Q_A can be evaluated from the absorption curves at 190 nm and Q_σ can be evaluated from the absorption curves at 224 nm using Figure 43. This presupposes knowledge of the reference solution absorption curve, which of course is known for the PGA model system but in general would not be known. These same critical wavelengths can be used, however, when applied to the CD data, to place close constraints on what values the corrected CD spectrum should have at 224 nm and at 190 nm and these can be used to obtain a pseudoreference state absorption curve.

(ii) Differential light scattering

The contribution to the suspension ellipticity due to differential light scattering, $\Delta[\theta]_{DS}$, can be stated as

$$\Delta[\theta]_{DS} = [\theta]_{susp}^{A_L = A_R} = \frac{3300}{Cl}(A_{SL} - A_{SR})_{susp} \tag{73}$$

From equations 52 and 56 the apparent absorbances due to the scattering of left and right circularly polarized beams can be written [112,115]

$$A_{SL} = -\log\left(1 - K(\lambda)\left[\frac{n_{PL}^2 - n_S^2}{n_{PL}^2 + 2n_S^2}\right]^2\right) \tag{74}$$

$$A_{SR} = -\log\left(1 - K(\lambda)\left[\frac{n_{PR}^2 - n_S^2}{n_{PR}^2 + 2n_S^2}\right]^2\right) \tag{75}$$

Since refractive index is an additive property derived from the sum of the contributions of individual absorption bands, as shown in Figure 11 and equations 18 and 19, it is possible to reconstruct a reasonable refractive index of the poly-L-glutamic acid particle, n_p, as shown previously [115].

$$n_p = 1 + n_b + n_{190}(\text{PLA}) + n_{204}(\text{PLA}) + n_{216}(\text{PLA}) \tag{76}$$

where the refractive indices of the 190, 204 and 216 bands of poly-L-alanine (PLA) are calculated using equation 19 and the values in Table 2A. This provides for the detailed absorption pattern and resulting refractive index pattern of the α-helix. The additional contributions to the refractive index come from the bands of the peptide chromophore that are further in the ultraviolet and from the contributions of the side chain. These are included in the background term, n_b, and can be reconstructed from data on acetic acid and dimethyl formamide [115]. The refractive indices for the PGA particle for left and right circularly polarized light, n_{PL} and n_{PR}, respectively, are derived from the ORD curve, i.e.,

$$(n_L - n_R)_P = \frac{[m]\rho_{PGA}\,\lambda}{18mw} \tag{77}$$

where ρ_{PGA} is the density of the PGA particles, which has been determined to be 1.5 gm/cm^3 [116]; $[m]$ is the mean residue molar rotation, and λ is the wavelength in cm. This gives

$$n_{PL} = n_P + \frac{(n_L - n_R)_P}{2} \tag{78}$$

$$n_{PR} = n_p - \frac{(n_L - n_R)_p}{2} \tag{79}$$

The calculated value for equation 73 is plotted as curve b in Figure 42, where the shape of the ORD curve is seen to parallel it very closely. This component of differential light scattering does not include the effects of the mixed term, i.e., the absorption obscuring term. When this is included the value $\Delta[\theta]_{DS}^{T}$ (curve c of Figure 42) is obtained [112] and an approximation to this [112] includes the factor 10^{-A_F} as seen in curve e of Figure 42. This approximation, however, is not quite as good as the empirical fitting of $[m']^{3/2}\ 10^{-A_F}$ as shown in curve d of Figure 42. Thus, an equation for empirically correcting the differential light scattering effect can be derived from an ORD curve of a solution reference state (again pointing to the interest in finding a way of identifying a pseudoreference state). The approximate equation for the corrected ellipticity, therefore, is written

$$[\theta]_{corr} = \frac{[\theta]_{susp} - (3300/Cl)(A_{SL} - A_{SR})10^{-A_F}}{Q_A^2 Q_\sigma} \tag{80}$$

and an empirical estimate of this can be written

$$[\theta]_{corr} = \frac{[\theta]_{susp} - k[m']_{prs} 10^{-A_F}}{Q_A^2 Q_\sigma} \tag{81}$$

For dominantly α-helical systems k is a constant that can be evaluated at 201 nm because at this wavelength $[\theta]_{corr} = 0$, i.e.,

$$k = \frac{[\theta]_{susp}^{201}}{[m']_{prs}^{201} 10^{-A_F^{201}}} \tag{82}$$

where $[m']_{prs}^{201}$ is the mean residue molar rotation of a suitable pseudoreference state, prs, corrected by the Lorentz factor for solvent refractive index (see Eqn. 24). Since the ellipticity curve is very steep near 201 nm care must be taken that the wavelength is accurate.

(iii) Calculation of $[\theta]_{susp}$ *for poly-L-glutamic acid*

As partly discussed above, there are a number of reasons why poly-L-glutamic acid (PGA) is an especially good model for studying the distortions that occur in the circular dichroism curves of membranes and other particulate systems in biology: (1) PGA is α-helical and the α-helix is the most common regular conformation in globular proteins [85], and is likely to be even more common in intrinsic membrane proteins because a helix capable of providing, where necessary, a hydrophobic exterior is an effective way to interact with or span the lipid layer of a membrane [62]. (2) PGA does not change its conformation on aggregation [110]. (3) PGA at pH 3.9 provides a reference state in which any effects due to helix-helix interaction and changes in side-chain orientation have already been expressed, because at this pH, there is already an association of more than ten helical rods of PGA [111]. (4) At pH 3.9, the degree of ionization is less than 0.05 (see Fig. 40), such that the effects of protonation of the carboxyl are essentially complete: and finally (5) the size of the aggregate can be controlled by sonication to provide sufficiently long-lasting metastable states of aggregates of intermediate size.

With the correct solution absorption spectrum (curve a of Figure 41A), Q_A^{190} is approximated as $A^{190}_{\text{susp}}/A^{190}_{\text{soln}}$. By equation 63 and Table 3, Q_A is calculated for all wavelengths to obtain the flattened absorption curve. The difference is then taken between the flattened absorption curve and the suspension absorption curve to give an ${}^{0}A_s$, which by Figure 43 gives the values for Q_σ for each wavelength [112]. If desired, iterative procedures can be used to refine initial approximate values to obtain more accurate values. Finally, the differential scatter effect is introduced as given in Figure 42 and developed in equations 73–79. The calculations are given in detail in Ref. 112. The results, using both suspension curves b and c of Figure 41A to calculate the experimental curves b and c of Figure 41B, are given in Figure 44A. The effectiveness of the phenomenological approach outlined here can be assessed by

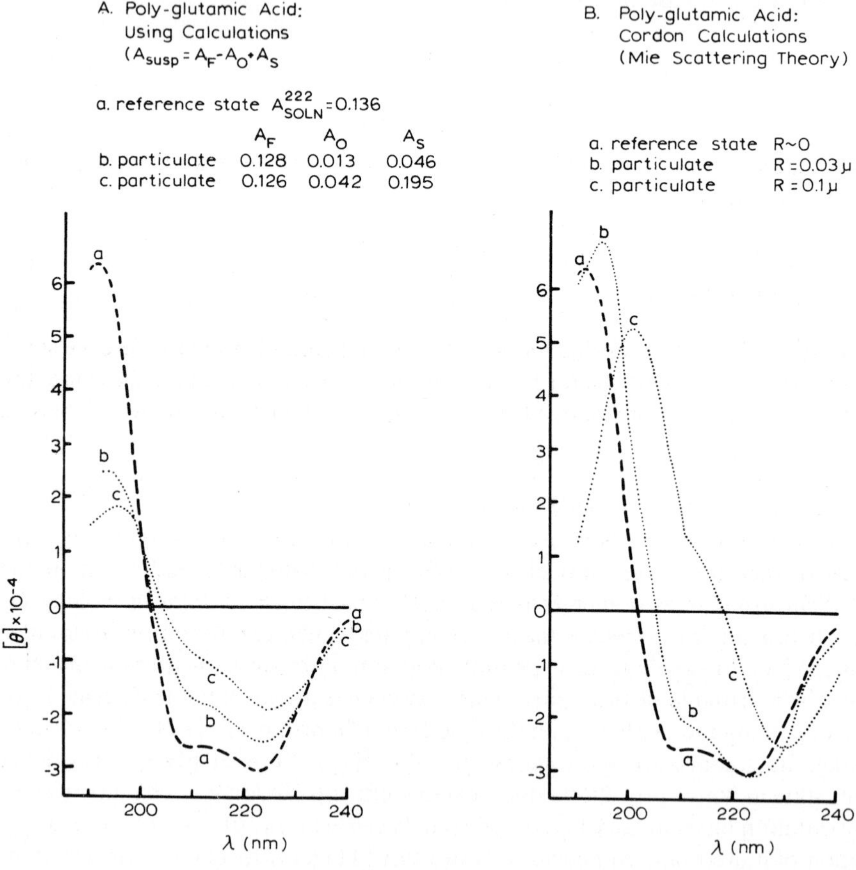

Figure 44. A, Calculation of poly-L-glutamic acid (PGA) suspensions (curves b and c) using the phenomenological approach outlined in the text. Note how closely curves b and c of this figure compare with curves b and c of Figure 41B. B, Calculation of poly-L-glutamic acid suspensions (curves b and c) by Gordon [117] using Mie scattering theory. Comparison of these curves b and c with curves b and c of Figure 41B show the limitations of the Mie scattering theory approach. Reproduced, with permission, from [113].

comparison with calculations on the same system and data using Mie scattering theory [117] (see Fig. 44B). The calculated curves b and c of Figure 44A closely reproduce curves b and c of Figure 41B whereas those of Figure 44B do not. It is apparent that Mie scattering theory cannot give rise to the proper dampening of the 222 nm extremum even when the particle size has become such that the resulting calculated curve (curve c, Fig. 44B) has become dominated by the differential light scatter effect. It is possible that, if the absorption obscuring effect were explicitly considered, the Mie scattering approach would more nearly give rise to the actual distorted shapes. Obviously, dramatic dampening of the 222 nm extremum occurs (see Figs. 39 and 41B) and this dampening is primarily due to what is described above as the absorption obscuring effect. Having developed a phenomenological approach which effectively calculates the distortions in the CD spectra of particulate α-helical poly-L-glutamic acid, the next concern is to see whether this approach can be the basis for correcting spectra on biological membranes.

(d) Application to the purple membrane of Halobacterium halobium: The pseudoreference state approach

The purple membrane is a unique patch of membrane in halobacteria [118–120]; it contains a single protein of 26000 molecular weight and the protein is 75% of the mass of the membrane. The remainder is primarily phospholipid. The protein is called bacteriorhodopsin, in analogy to rhodopsin. Characterization by electron microscopy has shown the protein to be comprised of seven helical rods traversing back and forth across the membrane [121]. This is interpreted to indicate some 70–75% α-helix, the presence of which is also indicated from X-ray diffraction data [122,123]. Accordingly, the purple membrane provides an especially interesting opportunity to apply the preceding analyses for the correction of distortions in the ellipticity patterns of biomembranes.

(i) The pseudoreference state approach

Circular dichroism spectra of purple membrane are shown in Figure 45, where it is seen that sonication of the frozen sample (curve a) results in large increases in ellipticities. Thus, the purple membrane system is subject to the same distortions of the CD pattern as were observed in Figure 39 for mitochondria and Figure 41B for particulate α-helical poly-L-glutamic acid. The first step in making the corrections is to obtain a dissolved solution, i.e., a molecularly dispersed state (mds), of the membrane. From the previous discussions on the intensity of the n–π^* transition and of dispersion force interactions giving rise to hypo- and hyperchromism, it should be appreciated that the n–π^* transition is not subject to large hypo- or hyperchromism effects. This has been demonstrated by dissolving microsomes in a solvent ranging from 0 to 30% trifluoroethanol in water with 0.5% sodium dodecyl sulfate [124]. The change in the 222 nm absorbance was about 25% whereas the change in the absorption peak was 360%. Importantly, estimates of the $[\theta]_{\text{CI}}^{222}$, i.e., an initial estimate of the ellipticity of the negative band obtained as outlined below, changed by less than 15% over the

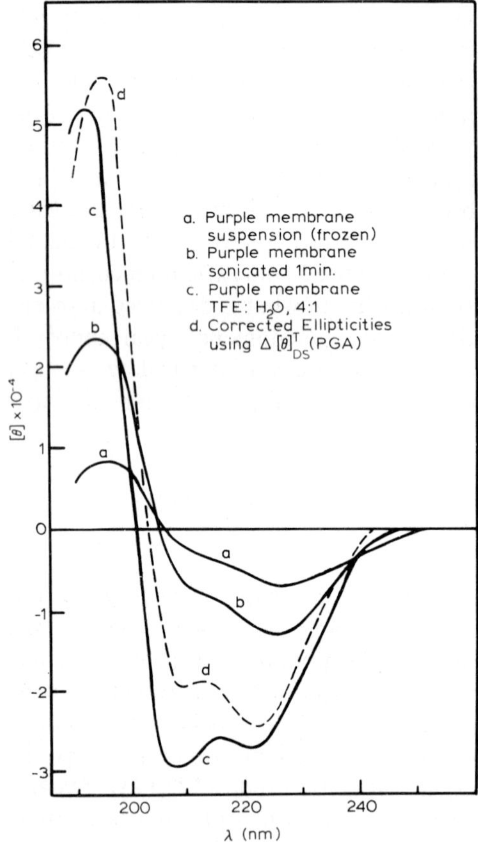

Figure 45. Circular dichroism spectra of purple membrane. Curve a, suspension of previously frozen purple membrane. Curve b, suspension of curve a after sonication for 1 minute. Note the large increases in ellipticity that accompany decrease in particle size due to sonication. Curve c, purple membrane dissolved in 80% trifluoroethanol (TFE)/20% water. Characteristic α-helix CD spectrum is obtained. This is a molecularly dispersed state. Curve d, corrected ellipticities using the pseudoreference state of Figure 46B with differential scatter correction using proportionality to $\Delta[\theta]^{T}_{DS}$ for PGA given as curve c of Figure 42. Adapted, with permission, from [125].

entire range while the actual ellipticity at 222 nm for the dissolved state varied by more than 50%. One can use an appropriate molecularly dispersed state to obtain the absorbance at 224 nm (A^{224}_{mds}) for example. Then the difference is taken with suspension absorbance at the same wavelength, i.e.,

$$A^{224}_{susp} - A^{224}_{mds} = {}^{0}A^{224}_{s} \tag{83}$$

This value, ${}^{0}A_{s}$, is then used in Figure 43 to obtain an estimate of Q^{224}_{σ}, neglecting Q^{224}_{A} and taking A^{224}_{mds} to be A^{224}_{F}. Since experience has shown that Q^{224}_{A} is of the order of 0.9 for the conditions required to obtain spectra down to 190 nm, this can be introduced if desired or it can be appreciated that the initial estimate of ellipticity, $[\theta]_{\mathrm{CI}}^{224}$, is a minimal value, being of the order of 10% low.

The next step is to search for a molecularly dispersed state in which the value of $[\theta]_{\mathrm{mds}}^{224}$ is about 10–20% greater than $[\theta]_{\mathrm{CI}}^{224}$.

$$|[\theta]_{\mathrm{mds}}^{224}| > |[\theta]_{\mathrm{CI}}^{224}| = \left| \frac{[\theta]_{\mathrm{susp}}^{224}}{Q_{\sigma}^{224}} \right| \tag{84}$$

When this dissolved state (mds1) is found, its absorption curve at 190 nm is used to calculate an initial Q_{A}^{190}, which provides an estimate of the ellipticity at 190 nm, i.e.,

$$Q_{\mathrm{A}}^{190} = \frac{A_{\mathrm{susp}}^{190}}{A_{\mathrm{mds1}}^{190}} \tag{85}$$

and

$$[\theta]_{\mathrm{mds1}}^{190} > [\theta]_{\mathrm{CI}}^{190} = \frac{[\theta]_{\mathrm{susp}}^{190}}{(Q_{\mathrm{A}}^{190})^2} \tag{86}$$

This value of $[\theta]_{\mathrm{CI}}^{190}$ is an underestimate of about 10–30%. When a molecularly dispersed state is found which obeys both equations 84 and 86, then it is called the pseudoreference state (prs) and its absorption curve is to be used to calculate the values of Q_{A} and Q_{σ} for the corrected ellipticity curve. This is sufficient to provide reasonable estimates of the mean residue ellipticities at 224 and 190 nm. Since these are the important values for estimating the amount of α-helix, the result is sufficient. The differential light scattering term, however, can be estimated by using the ORD curve of the pseudoreference state as outlined in equations 81 and 82. This is more of a cosmetic effect to see that the entire curve generally conforms to the α-helix pattern implied by the mean residue ellipticities at 224 and 190 nm using equation 44.

The absorption curve for the pseudoreference state of the purple membrane is given in Figure 46A (curve b) along with the suspension absorption curve obtained after sonicating purple membrane for 1 minute [125]. Q_{A}^{190} is then estimated as $A_{\mathrm{susp}}^{190}/A_{\mathrm{prs}}^{190}$. This value is used with equation 63 and Table 3 to obtain Q_{A} for all wavelengths and the flattened absorption curve is calculated. The difference between the suspension and flattened absorption curves gives rise to values of ${}^{0}A_{\mathrm{s}}$ for each wavelength which by Figure 43 give Q_{σ} for each wavelength. The next step is to use curve c of Figure 46B to evaluate k of equation 82 and the second term in the numerator of equation 81 is estimated. The product $Q_{\mathrm{A}}^{2}Q_{\sigma}$ is finally divided into the difference to give the corrected curve, which is plotted in Figure 46B as curve d. The form of the curve is very reasonable and the ellipticity at 224 nm is 2.35×10^{4}. As the reference PGA α-helical state has a value of 3.2×10^{4} at 224 nm and as the ellipticity of non-regular segments can be taken as zero at this wavelength, the estimate of α-helix is simply $2.35/3.2 \simeq 0.73$, or 73%.

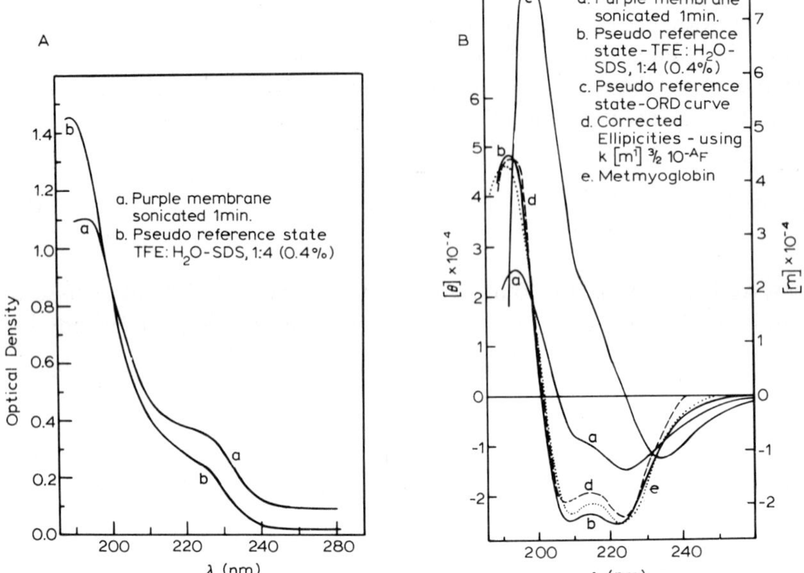

Figure 46. A, Absorption curves for the purple membrane sonicated 1 minute (curve b) and the pseudoreference state as defined in equations 84 and 86, curve a. Reproduced, with permission, from [125]. B, Circular dichroism spectra of purple membrane. Curve a, suspension of fresh purple membrane preparation sonicated for 1 minute. Curve b, pseudoreference state, 20% trifluoroethanol (TFE)/80% water with 0.4% SDS. This state was defined by equations 84 and 86. Curve c, pseudoreference state optical rotatory dispersion curve plotted on the right-hand ordinate. Dashed curve d, corrected ellipticities using $k[m']^{3/2}\ 10^{-A_F}$ as the differential scatter correction with $[\theta]^{201}_{susp} = 1.36 \times 10^4$. The mean residue ellipticity at 224 nm is 2.35×10^4, and using a reference value for the 100% α-helical state of 3.2×10^4, the calculated α-helix is 73%. Adapted, with permission, from [125]. Dotted curve e, mean residue ellipticity curve for metmyoglobin which has been shown in the crystal structure to have an α-helix content of about 77% [126,127]. The correspondence of the corrected purple membrane curve d and the curve for metmyoglobin (curve e) show the two protein structures to have closely similar α-helical contents. Curve e reproduced, with permission, from [70].

It has previously been suggested that when a particulate system is dominantly α-helical that the $\Delta[\theta]^{T}_{DS}$ for poly-L-glutamic acid could be used in proportion to the percentage of α-helix to give the differential light scattering correction [112]. This is included in Figure 45 as curve d for comparison. Curve b of Figure 45 or curve a of Figure 46B uncorrected would lead to estimates of α-helix of the order of 40%. Clearly this is an unsatisfactory and misleading result. The corrected curves give much different values which are closely within the range estimated from the electron micrograph studies [121]. The approach used was developed and applied to other membrane systems years before the purple membrane system was considered [112,113]. This provides substantial support for the usefulness of the corrections in a most favorable case, where the α-helix content is high and where the protein is three-fourths of the mass of the membrane. As the amount of lipid increases and as the amount of α-helix decreases, the approach will be less reliable in a manner not different from the application of absorption and circular dichroism to estimate conformation in a structurally mixed protein or polypeptide. Structurally informative studies of the sort outlined in section 3 can also be applied to biological membranes to obtain similar information when care is taken to account for their particulate nature.

5. Acknowledgments

The author wishes to acknowledge the past and present members of the Laboratory of Molecular Biophysics for their important contributions to the work reviewed here, and to thank particularly the present members of the laboratory; C.M. Venkatachalam for plotting Figures 22, 24 and 27, M.M. Long for her contribution to the purple membrane study and K. Ramer for plotting the curves in Figure 12. This work was supported in part by the National Institutes of Health under Grant Nos. HL-29578 and GM-26898.

References

1 Tinoco, I. (1962) Adv. Chem. Phys. 4, 113–160.
2 Moscowitz, A. (1960) in Optical Rotatory Dispersion (Djerassi, C., ed.), pp. 150–177, McGraw-Hill, New York.
3 Urry, D.W. (1970) in Spectroscopic Approaches to Biomolecular Conformation (Urry, D.W., ed.), pp. 33–121, American Medical Association Press, Chicago, IL.
4 Urry, D.W. (1974) Res. Dev. 25, 20–24.
5 Kasha, M. (1950) Discuss. Faraday Soc. 9, 14–19.
6 Davydov, A.S. (1962) Theory of Molecular Excitons (translated by Kasha, M. and Oppenheimer, Jr., M.), McGraw-Hill, New York.
7 Kasha, M. (1963) Radiat. Res. 20, 55–71.
8 Kasha, M., El-Bayoumi, A. and Rhodes, W. (1961) J. Chem. Phys. 58, 916–925.
9 Rhodes, W. (1961) J. Am. Chem. Soc. 83, 3609–3617.
10 Tinoco, Jr., I. (1960) J. Am. Chem. Soc. 82, 4785–4790; (1961) J. Chem. Phys. 34, 1067.
11 Urry, D.W. (1967) J. Biol. Chem. 242, 4441–4448.
12 Harbury, H.A. and Loach, P.A. (1960) J. Biol. Chem. 235, 3640–3645.
13 Urry, D.W. and Pettegrew, J.W. (1967) J. Am. Chem. Soc. 89, 5276–5283.
14 Urry, D.W. (1967) J. Am. Chem. Soc. 89, 4190–4196.
15 Urry, D.W., Wainio, W.W. and Grebner, D. (1967) Biochem. Biophys. Res. Commun. 27, 625–631.
16 Van Gelder, B.F., Urry, D.W. and Beinert, H. (1968) in Structure and Function of Cytochromes (Okunuki, K., Kamen, M.D. and Sekuzu, I., eds.), pp. 335–339, University Park, Baltimore, MD.
17 Urry, D.W. and Van Gelder, B.F. (1968) in Structure and Function of Cytochromes (Okunuki, K., Kamen, M.D. and Sekuzu, I., eds.), pp. 210–214, University Park Press, Baltimore, MD.
18 Williams, R.J.P. (1974) Nature 249, 425–429.
19 Kuhn, W. (1925) Z. Physik 33, 408–412.
20 Thomas, W. (1925) Naturwissenschaften 13, 627.
21 Pauling, L. (1935) Proc. Natl. Acad. Sci. U.S.A. 21, 186–191.
22 Wyman, J. (1964) Adv. Prot. Chem. 19, 223–286.
23 Krivacic, J.R. and Urry, D.W. (1970) Anal. Chem. 42, 596–599.
24 Velluz, L., Legrand, M. and Grosjean, M. (1965) Optical Circular Dichroism: Principles, Measurements and Applications, Academic Press, New York, NY.

25 Krivacic, J.R. and Urry, D.W. (1970) Anal. Chem. 43, 240–246.
26 Krivacic, J.R. and Urry, D.W. (1971) Anal. Chem. 43, 1508–1510.
27 Urry, D.W. (1969) Proc. Natl. Acad. Sci. U.S.A. 63, 261–265.
28 Condon, E.U. (1937) Rev. Mod. Phys. 9, 432–457.
29 Kirkwood, J.G. (1937) J. Chem. Phys. 5, 479–491.
30 Urry, D.W. (1965) Proc. Natl. Acad. Sci. U.S.A. 54, 640–648.
31 Miles, D.W. and Urry, D.W. (1967) J. Phys. Chem. 71, 4448–4454.
32 Kuhn, W. (1930) Trans. Faraday Soc. 26, 293–309.
33 Condon, E.U., Altar, W. and Eyring, H. (1937) J. Chem. Phys. 5, 753–775.
34 Moffitt, W., Woodward, R.B., Moscowitz, A., Klyne, W. and Djerassi, C. (1961) J. Am. Chem. Soc. 83, 4013–4018.
35 Schellman, J.A. (1966) J. Chem. Phys. 44, 55–63.
36 Kauzmann, W., Walter, J.E. and Eyring, H. (1940) Chem. Rev. 26, 339–407.
37 Urry, D.W. (1968) Annu. Rev. Phys. Chem. 19, 477–530.
38 Litman, B.J. and Schellman, J.A. (1965) J. Phys. Chem. 69, 978–983.
39 Molin-Case, J.A., Fleischer, E. and Urry, D.W. (1970) J. Am. Chem. Soc. 92, 4728–4730.
40 Jones, L.L. and Eyring, H. (1961) Tetrahedron 13, 235–240.
41 Caldwell, D.J. and Eyring, H. (1963) Rev. Mod. Phys. 35, 577–586.
42 Pauling, L. (1949) Proc. Natl. Acad. Sci. U.S.A. 35, 495–499.
43 Linderberg, J. and Michl, J. (1970) J. Am. Chem. Soc. 92, 2619–2625.
44 Boyd, D.B. (1974) Int. J. Quantum Chem.: Quantum Biol. Symp. No. 1, 13–19.
45 Neubert, L.A. and Carmack, M. (1974) J. Am. Chem. Soc. 96, 943–945.
46 Pauling, L., Corey, R.B. and Branson, H.R. (1951) Proc. Natl. Acad. Sci. U.S.A. 37, 205–211.
47 Pauling, L. and Corey, R.B. (1951) Proc. Natl. Acad. Sci. U.S.A. 37, 729–740.
48 Banner, D.W., Bloomer, A.C., Petsko, G.A., Phillips, D.C., Pogson, C.I. and Wilson, I.A. (1975) Nature 255, 609–614.
49 Quicho, F.A. and Lipscomb, W.N. (1971) Adv. Prot. Chem. 25, 1–78.
50 Richardson, J.S. (1981) Adv. Prot. Chem. 34, 167–339.
51 Urry, D.W. (1975) Int. J. Quantum Chem.: Quantum Biol. Symp. No. 2, 221–235.
52 Urry, D.W. (1982) in Proteins in the Nervous System: Structure and Function (Haber, B., Perez-Polo, J.R. and Coulter, J.D., eds.), pp. 87–111, Alan Liss, Inc., New York, NY.
53 Ramachandran, G.N. and Kartha, G. (1954) Nature 174, 269–270.
54 Ramachandran, G.N. and Kartha, G. (1955) Nature 176, 593–595.
55 Rich, A. and Crick, F.H.C. (1955) Nature 176, 915–916.
56 Porter, R.R. and Reid, K.B.M. (1979) Adv. Prot. Chem. 33, 1–71.
57 Sandberg, L.B., Soskel, N.T. and Leslie, J.G. (1981) N. Engl. J. Med. 304, 566–579.
58 Urry, D.W. and Long, M.M. (1976) CRC Crit. Rev. Biochem. 4, 1–45.
59 Cook, W.J., Einspahr, H.M., Trapane, T.L., Urry, D.W. and Bugg, C.E. (1980) J. Am. Chem. Soc. 102, 5502–5505.
60 Urry, D.W., Trapane, T.L., Sugano, H. and Prasad, K.U. (1981) J. Am. Chem. Soc. 103, 2080–2089.
61 Venkatachalam, C.M. and Urry, D.W. (1981) Macromolecules 14, 1225–1229.
62 Urry, D.W. (1971) Proc. Natl. Acad. Sci. U.S.A. 68, 672–676.
63 Urry, D.W., Goodall, M.C., Glickson, J.D. and Mayers, D.F. (1971) Proc. Natl. Acad. Sci. U.S.A. 68, 1907–1911.
64 Ramachandran, G.N. and Chandrasekharan, R. (1972) in Progress in Peptide Research (Lande, S., ed.), Vol. 2, pp. 195–215, Gordon and Breach, New York, NY.
65 Ramachandran, G.N. and Chandrasekharan, R. (1972) Ind. J. Biochem. Biophys. 9, 1–11.
66 Veatch, W.R., Fossel, E.T. and Blout, E.R. (1974) Biochemistry 13, 5249–5256.
67 Lotz, B., Colonna-Cesari, F., Heitz, F. and Spach, G. (1976) J. Mol. Biol. 106, 915–942.
68 Urry, D.W., Long, M.M., Jacobs, M. and Harris, R.D. (1975) Ann. NY Acad. Sci. 264, 203–220.
69 Doty, P. and Gratzer, W.B. (1962) in Polyamino Acids, Polypeptides and Proteins (Stahman, M.A., ed.), pp. 111–118, The University of Wisconsin Press, Madison, WI.

70 Quadrifoglio, F. and Urry, D.W. (1968) J. Am. Chem. Soc. 90, 2755–2760.
71 Moffitt, W. and Moscowitz, A. (1959) J. Chem. Phys. 30, 648–660.
72 Holzwarth, G. and Doty, P. (1965) J. Am. Chem. Soc. 87, 218–228.
73 Quadrifoglio, F. and Urry, D.W. (1968) J. Am. Chem. Soc. 90, 2760–2765.
74 Applequist, J. (1982) Biopolymers 21, 779–795.
75 Madison, V. and Schellman, J. (1972) Biopolymers 11, 1041–1076.
76 Woody, R.W. (1969) Biopolymers 8, 669–683.
77 Pysh, E.W. (1970) J. Chem. Phys. 52, 4723–4733.
78 Balcerski, J.S., Pysh, E.S., Bonora, G.M. and Toniolo, C. (1976) J. Am. Chem. Soc. 98, 3470–3473.
79 Johnson, W.C., Jr. (1978) Annu. Rev. Phys. Chem. 29, 93–114.
80 Ramachandran, G.N. (1981) in Proceedings of the Second SUNYA Conversation in the Discipline Biomolecular Stereodynamics, Vol. II (Sarma, R.H., ed.), pp. 1–36, Adenine Press, New York, NY.
81 Tterlikkis, L., Loxsom, F.M. and Rhodes, W. (1973) Biopolymers 12, 675–684.
82 Mandel, R. and Holzwarth, G. (1973) Biopolymers 12, 655–674.
83 Venkatachalam, C.M. (1968) Biopolymers 6, 1425–1436.
84 Urry, D.W. (1974) in Arterial Mesenchyme and Arteriosclerosis (Wagner, W.D. and Clarkson, T.B. eds.), pp. 211–243, Plenum Publishing Corporation, New York, NY.
85 Chou, Y.P. and Fasman, G.D. (1977) J. Mol. Biol. 115, 135–175.
86 Khaled, M.A. and Urry, D.W. (1976) Biochem. Biophys. Res. Commun. 70, 485–491.
87 Urry, D.W., Long, M.M., Ohnishi, T. and Jacobs, M. (1974) Biochem. Biophys. Res. Commun. 61, 1427–1433.
88 Woody, R.W. (1974) in Peptides, Polypeptides and Proteins (Blout, E.R., Bovey, F.A., Goodman, M. and Lotan, N., eds.), pp. 338–350, John Wiley and Sons, New York, NY.
89 Urry, D.W., Long, M.M., Cox, B.A., Ohnishi, T., Mitchell, L.W. and Jacobs, M. (1974) Biochim. Biophys. Acta 371, 597–602.
90 Urry, D.W. (1982) Methods Enzymol. 82, 673–716.
91 Venkatachalam, C.M., Khaled, M.A., Sugano, H. and Urry, D.W. (1981) J. Am. Chem. Soc. 103, 2372–2379.
92 Khaled, M.A., Venkatachalam, C.M., Sugano, H. and Urry, D.W. (1981) Int. J. Pept. Protein Res. 17, 23–33.
93 Urry, D.W. (1983) Ultrastruct. Pathol. 4, 227–251.
94 Urry, D.W., Venkatachalam, C.M., Long, M.M. and Prasad, K.U. (1982) in Conformation in Biology (Srinivasan, R. and Sarna, R.H., eds.), G.N. Ramachandran Festschrift Volume, pp. 11–27, Adenine Press, New York, NY.
95 Urry, D.W. (1985) in The Enzymes of Biological Membranes (Martonosi, A.N., ed.), pp. 229–257, Plenum Publishing Corporation, New York, NY.
96 Venkatachalam, C.M. and Urry, D.W. (1985) J. Comput. Chem. 5, 64–71.
97 Sarges, R. and Witkop, B. (1965) Biochemistry 4, 2491–2494.
98 Miles, D.W. and Urry, D.W. (1968) J. Biol. Chem. 243, 4181–4188.
99 Miles, D.W. and Urry, D.W. (1968) Biochemistry 7, 2791–2799.
100 Meyer, W.L., Mahler, H.R. and Baker, R.H., Jr. (1962) Biochim. Biophys. Acta 64, 353–358.
101 Jardetzky, O. and Wade-Jardetzky, N.G. (1966) J. Biol. Chem. 241, 85–91.
102 Suzuki, K., Nakano, H. and Suzuki, S. (1967) J. Biol. Chem. 242, 3319–3325.
103 Siegel, J.M., Montgomery, G.A. and Bock, R.M. (1959) Arch. Biochem. Biophys. 82, 288–299.
104 Whitby, L.G. (1953) Biochem. J. 54, 437–442.
105 Urry, D.W., Masotti, L. and Krivacic, J.R. (1971) Biochim. Biophys. Acta 241, 600–612.
106 Duysens, L.N.M. (1956) Biochim. Biophys. Acta 19, 1–12.
107 Urry, D.W. and Ji, T.H. (1968) Arch. Biochem. Biophys. 128, 802–807.
108 Barone, G., Crescenzi, V. and Quadrifoglio, F. (1966) Biopolymers 4, 529–538.
109 Tiffany, M.L. and Krimm, S. (1969) Biopolymers 8, 347–359.
110 Urry, D.W., Hinners, T.A. and Masotti, L. (1970) Arch. Biochem. Biophys. 137, 214–221.
111 Tomimatsu, Y., Vitello, L. and Gaffield, W. (1966) Biopolymers 4, 653–662.

112 Urry, D.W. (1972) Biochim. Biophys. Acta 265, 115–168.
113 Urry, D.W. and Long, M.M. (1974) in Methods in Membrane Biology (Korn, E.D., ed.), Vol. 1, pp. 105–141, Plenum Press, New York, NY.
114 Gordon, D.J. and Holzwarth, G.M. (1971) Arch. Biochem. Biophys. 142, 481–488.
115 Urry, D.W. and Krivacic, J. (1970) Proc. Natl. Acad. Sci. U.S.A. 65, 845–852.
116 Vinograd, J. and Hearst, J.E. (1962) Fortschr. Chem. Org. Naturstoffe 20, 372–422.
117 Gordon, D.J. (1972) Biochemistry 11, 413–420.
118 Stoeckenius, W. (1982) Proceedings of the Robert A. Welch Foundation Conferences on Chemical Research XXIV, The Synthesis, Structure and Function of Biochemical Molecules, pp. 133.
119 Stoeckenius, W., Lozier, R.H. and Bogomolni, R.A. (1979) Biochim. Biophys. Acta 505, 215–278.
120 Stoeckenius, W. (1980) Account Chem. Res. 13, 337–344.
121 Henderson, R. and Unwin, P.N.T. (1975) Nature 257, 28–32.
122 Blaurock, A.E. (1975) J. Mol. Biol. 93, 139–158.
123 Henderson, R. (1975) J. Mol. Biol. 93, 123–138.
124 Long, M.M. and Urry, D.W. (1981) in Membrane Spectroscopy (Grell, E., ed.), Vol. 31, pp. 143–171, Springer-Verlag, Heidelberg.
125 Long, M.M., Urry, D.W. and Stoeckenius, W. (1977) Biochem. Biophys. Res. Commun. 75, 725–731.
126 Kendrew, J.C., Dickerson, R.E., Strandberg, B.E., Hart, R.G., Davies, D.R., Phillips, D.C. and Shore, V.C. (1960) Nature 185, 422–427.
127 Kendrew, J.C., Watson, H.C., Strandberg, B.E., Dickerson, R.E., Phillips, D.C. and Shore, V.C. (1961) Nature 190, 666–670.
128 Urry, D.W. (1973) Res. Dev. 24, 28–36.
129 Miller, M.H. and Scheraga, H.A. (1976) J. Polymer Sci. Symp. No. 54, 171–200.
130 Urry, D.W. and Long, M.M. (1978) in Physiology of Membrane Disorders (Andreoli, T.E., Hoffman, J.F. and Fanestel, D.D., eds.), pp. 107–224, Plenum Publishing Corporation, New York, NY.

Neuberger/Van Deenen (eds.) Modern Physical Methods in Biochemistry, Part A

CHAPTER 5

Protein crystallography

L.N. JOHNSON

Laboratory of Molecular Biophysics, South Parks Road, Oxford OX1 3PS, U.K.

1. Introduction

The first X-ray photographs of a protein crystal were described 50 years ago by Bernal and Crowfoot [1]. These remarkable photographs indicated that a wealth of structural information was available for protein molecules once methods for the solution of the patterns had been developed. At that time the determination of atomic positions even in the crystals of small molecules was a difficult task. In 1954, Perutz and his colleagues [2] showed that the technique of heavy atom isomorphous replacement could be used to solve the phase problem. The method was put on a sound systematic basis by Blow and Crick [3] and extended to include the use of anomalous scattering [4,5]. Until recently, these methods provided the basis for all protein structure determinations. They have been remarkably effective (as illustrated below) and new developments have both increased the size of the problem solvable and provided greater insights.

The first protein crystal structure, myoglobin, was solved in 1960. The second, lysozyme, followed in 1965. In 1967 three structures were solved: ribonuclease, chymotrypsin and carboxypeptidase. Thereafter, the number solved has increased almost exponentially year by year so that by 1979 there were some 161 structures known, at least at the level of tracing the fold of the polypeptide chain [6]. To date, there are well over 200 structures solved, but this number includes several structures of the same protein in a different crystal form or from a different species. Some protein structures are illustrated in Figure 1.

The elucidation of so many structures has allowed a successful classification of protein structures [7,8]; it has laid the basis for certain predictive methods [9,10], and it has given insight into the possible evolutionary origins of proteins [11–13]. Our understanding of biological function in terms of structure has not increased so fast. This has turned out to be a more difficult problem. In the case of enzymes, a description of several different states of the protein complexed with substrate, intermediates and products together with necessary co-factors and activators is required. Often these different states can only be achieved by co-crystallisation, and even then it may be difficult to trap the necessary conformation. To date, it is

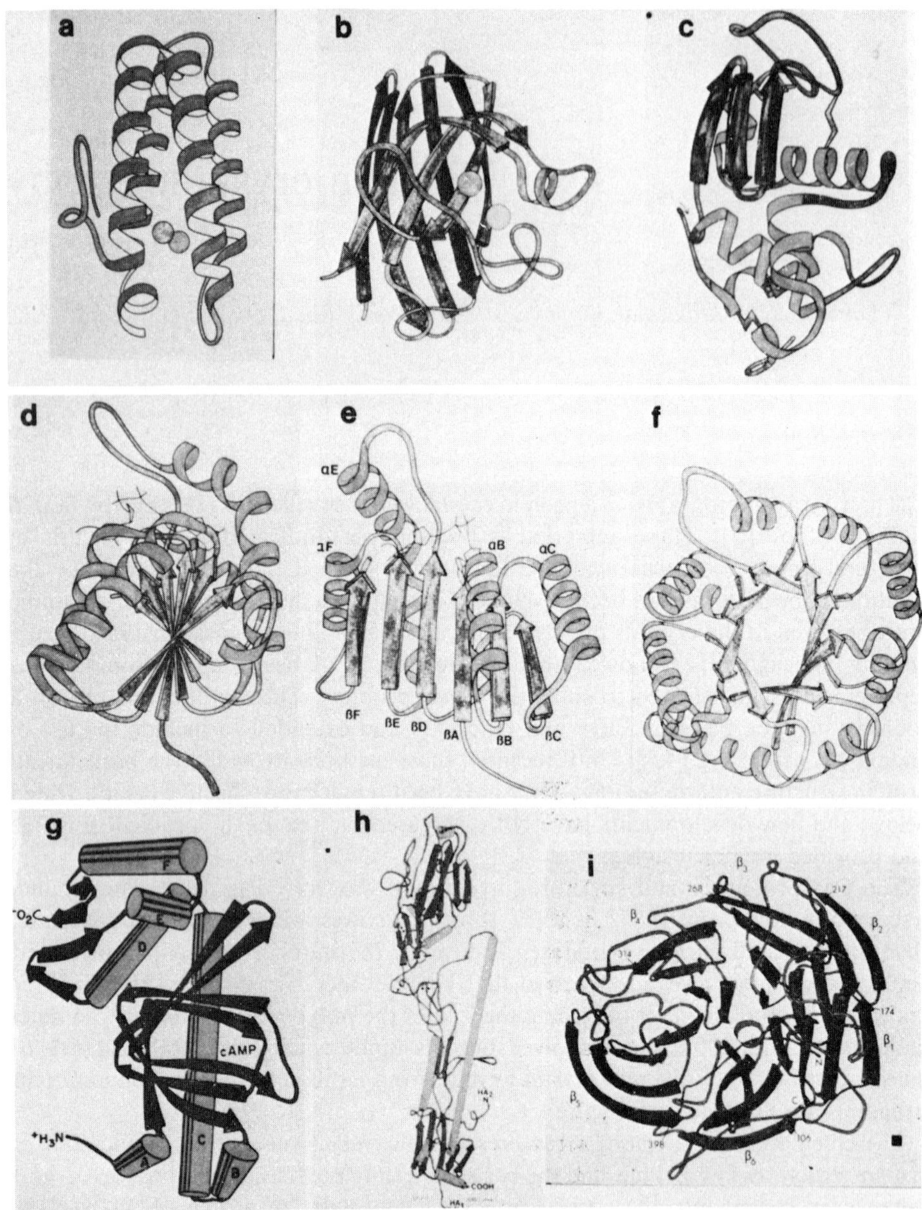

Figure 1. Some protein structures solved by X-ray crystallography. The figure illustrates the different secondary structure assemblies. α-Helices are represented by spirals or cylinders; β-strands by arrows. (a) Hemerythrin, an all α-protein. (b) Superoxide dismutase, an all β-protein. (c) Lysozyme, an $\alpha + \beta$ protein. (d) and (e) Two orthogonal views of the NAD binding domain of lactate dehydrogenase, an α/β protein. (f) Triose phosphate isomerase, an α/β structure. (g) A DNA binding protein, the CAP protein from *E. coli.* (h) Influenza virus haemaglutinin. (i) Influenza virus neuraminidase. From Blake and Johnson [12b], which also contains references to the original sources of these structures.

estimated that the stereochemical basis of biological function is understood for perhaps ten systems, although the essential features of the recognition sites of many other proteins are well established. Allosteric proteins represent an extreme example of the problem. To date, there is only one detailed description of an allosteric protein, that of haemoglobin [14–19]. Progress has been made with one state of several other allosteric enzymes, notably glycogen phosphorylase [20–22], aspartate transcarbamylase [23a,23b] and phosphofructokinase [24], but further advances are held up by lack of suitable crystals of alternative states.

By 1976, the methods of protein crystallography were well established and these were described by Blundell and Johnson [25]. This review is written for the biochemist and the biochemistry undergraduate. In section 2 a summary of the current crystallographic theory and methods, with special emphasis on recent advances, is given. The most notable advance has been in the refinement of protein structures (2(g)). The field most in need of progress is crystallisation (2(b)). In section 3 new developments in the field are described, with reference to the sort of results that can be obtained. Refinement of protein structures, made possible by technological advances in high-speed computers and computer graphics, allows a description of the thermal parameters associated with each atom. This in turn has led to a description of the *spatial* dynamics of the protein structure (3(b)), information which can be correlated with the information on the *temporal* dynamics obtained from nuclear magnetic resonance observations and theoretical molecular dynamics calculations. Low-temperature studies (3(c)) allow a further insight into these processes and also provide the way for a study of true enzyme-substrate or enzyme-intermediate complexes. New sources (synchrotron radiation and neutrons) were just coming into use in 1976. Their greater availability has provided new perspectives and novel results (3(d), (e)). The next breakthrough in protein crystal structure determination may well come from the application of direct methods of phase determination based on the 'maximum entropy' principle, methods which are at present in their infancy (3(f)). Underlying all of these studies is the vexed question of the relationship of the structure in the crystal to the functional state of the molecule. Protein crystallography provides the only method for the solution of the three-dimensional structure of proteins greater than 10 000 molecular weight. Evidence summarised in Section 3(a) suggests that crystallisation does not change the gross structure but that the lattice forces may cause perturbations that restrict the conformational mobility of the protein, and hence lead to alterations of the biological response. Further analysis in this field is needed, especially now that precise descriptions of protein structures can be obtained.

All significant protein crystallographic calculations are performed by computers. The existence of well-developed programmes and programme systems and their general availability has made a major contribution to the advances in protein crystallography. Throughout the history of the subject, the demands made by protein crystallography have been such that they have only been met by developments at the forefront of computer technology.

2. *Protein crystallographic methods*

(a) Basic X-ray diffraction equations

Crystals are three-dimensional arrays distinguished from an amorphous solid by the regular arrangement of molecules. The unit cell of a crystal is defined as the basic parallelipiped shaped block from which the whole volume of the crystal is built by regular assembly of these blocks. The unit cell edges are by convention defined as a, b and c with angles α, β, γ. Crystals obey the laws of symmetry (based on 1, 2, 3, 4, or 6-fold rotations) which govern three-dimensional patterns of any kind. Thus, the unit cell of the crystal may çontain more than one molecule related to the others in the unit cell by symmetry operations (Fig. 2).

There are three fundamental equations in X-ray diffraction: these are described in this section before subsequent discussion of the methods of protein structure determination.

Bragg's Law: In 1913, Bragg provided the explanation for the formation of diffraction spots seen in Von Laue's first X-ray crystal photographs taken in 1912. Bragg visualised the scattering of X-rays by a crystal in terms of reflections from planes of atoms. The condition for diffraction is given by (Fig. 3a)

$$2d \sin \theta = n\lambda \quad (1)$$

where d is the interplanar spacing of the atoms, θ is the Bragg angle defined in Figure 3a, λ is the wavelength of X-rays ($\lambda = 1.5418$ Å for CuK_α radiation), n is an integer ($\pm 1, 2, 3$ ...).

Bragg [26] modestly remarked "I have always felt the association of my name with it to be an easily earned honour because it is merely the familiar optical relation giving colours reflected by thin films, in another guise".

The equation predicts the angular position of a diffracted ray.

In the light microscope (Fig. 3b) resolution is governed by the numerical aperture of the lens: that is, the cone of light accepted by the objective lens. The larger the cone, the better the resolution. According to the Abbé theory, a spacing of distance d can be resolved if the first-order diffraction arising from this spacing can be focussed by the objective lens. The situation with X-rays, which are also part of the electromagnetic spectrum, is similar. A resolution of d Å quoted by a protein crystallographer means

---

Figure 2. The unit cell of a glycogen phosphorylase *b* crystal. The projection is down the crystallographic *c* axis. Some of the symmetry elements of the $P4_32_12$ space group are shown. The 4-fold and 2-fold rotational operations are apparent. The positions of every other Cα atom for the 841 amino acid residues of glycogen phosphorylase have been plotted. It is seen that the protein molecules pack so that there are large solvent channels (diameter approximately 16 Å) which run the length of the crystal. The unit cell size is $a = b = 128.5$ Å, $c = 116$ Å. (I am grateful to P.J. McLaughlin for producing this diagram with the programme FITZ (G.L. Taylor).)

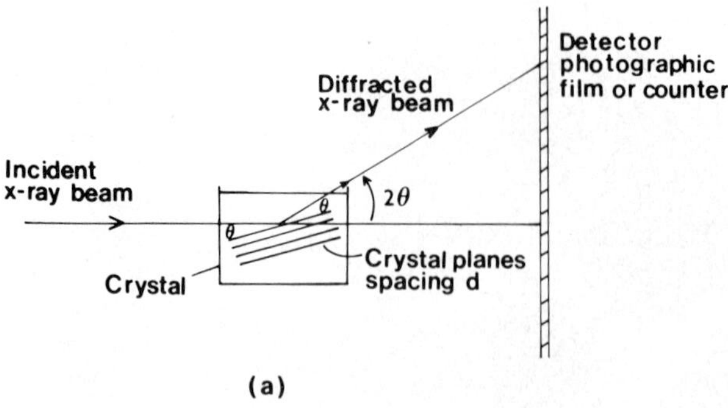

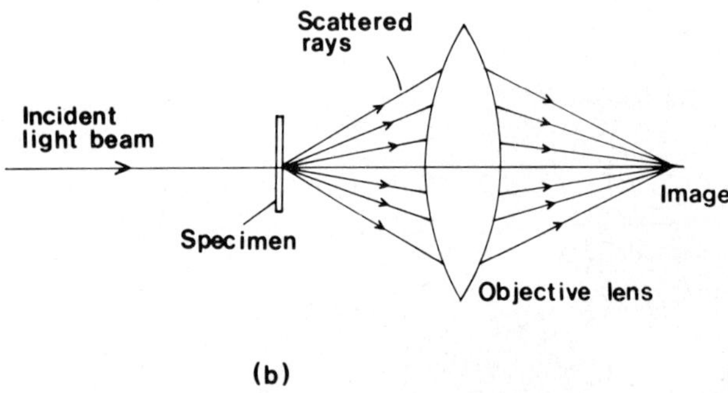

Figure 3a. Bragg's Law. A schematic diagram showing X-ray diffraction by planes of atoms in the crystal. θ is the Bragg angle and d is the interplanar spacing. The crystal size has been exaggerated. b. A schematic diagram showing the formation of an image in a light microscope. The light rays scattered by the specimen are collected by the objective lens and brought to a focus in the image plane. In the X-ray case, the diffracted X-rays cannot be focussed, but instead they are combined mathematically with the aid of a computer to form an image.

that data have been collected to a Bragg angle θ corresponding to $d = \lambda/2 \sin \theta$. The features that can be traced at different resolutions are:
6 Å resolution: outline of the molecule and α-helices can be identified.
3 Å resolution: course of polypeptide chain can be traced and topology of folding established. Amino acid side chains can be identified only with the help of amino acid sequence.
2 Å resolution: details such as identification of carbonyl oxygens allow the conformation of polypeptide to be established with precision. Details of side chains, e.g., forking of isoleucine residues, becomes apparent and in certain cases assignment of amino acid type can be made without sequence data.
1.5 Å resolution: individual atoms (C, N, O, S) are almost resolved. Hydrogen atoms do not become visible until about 1.2 Å resolution (Fig. 4). Series termination errors are significantly reduced, which allows water molecules to be placed with confidence.

Torsion angles for the polypeptide chain and hydrogen bonds can be established with precision.

The electron density equation: very simple structures such as NaCl can be solved by comparison of the relative intensities of the diffraction spots. For more complicated structures, the power of Fourier transform methods was soon appreciated [27]. In order to produce an image of the structure, the diffracted rays must be combined. In the light microscope this is achieved by the focussing power of the objective lens (Fig. 3b). For X-rays the refractive index of almost all substances is close to 1 and it is not possible to construct a lens. The diffracted rays must be combined mathematically. This is achieved with the electron density equation.

The electron density (ρ) in a crystal at a point x,y,z is given by

$$\rho(xyz)=1/V \sum_{h=-\infty}^{\infty} \sum_{k=-\infty}^{\infty} \sum_{l=-\infty}^{\infty} F(hkl) \exp i\, \alpha(hkl) \exp -2\pi i(hx+ky+lz) \quad (2)$$

where V is the volume of the unit cell, hkl are the indices of the diffracted ray arising from planes $d(hkl)$, $F(hkl)$ and $\alpha(hkl)$ are the amplitude and phase of the diffracted ray (for those unfamiliar with complex numbers, note that $\exp i\, \phi = \cos \phi + i \sin \phi$ so the summation can be considered as a sum of cosine and sine functions).

In practice, the summation over hkl will be restricted by the resolution of the data. The electron density will have a high value at atomic positions and will fall to zero between atoms (provided the resolution is sufficient). In the interstitial spaces between protein molecules which is occupied by the bulk solvent the electron density will be close to the corresponding solvent value for a protein crystal.

If the structure factor amplitudes and phases of each of the diffracted rays are known, the electron density of the crystal can be computed and the structure solved (Fig. 4).

The amplitude can be determined from the intensity $I(hkl)$ of the diffracted rays measured from the blackening of a photographic film or with a proportional counter

$$|F(hkl)|^2 \propto I(hkl)$$

However, when the intensities of the X-rays are recorded in this manner all information of the phase is lost. Thus, the fundamental problem in a structure determination is the phase problem. Until recently, the phase problem in protein crystallography has been solved by the heavy atom isomorphous replacement method (sections 2(d) and (e)), but other methods are also available (sections 2(e) and (f)).

The structure factor equation: each atom in the unit cell contributes to each of the diffracted rays according to the following equation for the structure factor.

$$\underline{F}(hkl)= \sum_{n=1}^{N} f_n \exp -(B_n \sin^2\theta/\lambda^2) \exp 2\pi i(hx_n+ky_n+lz_n) \quad (3)$$

where the summation is over the N atoms in the unit cell, x_n, y_n, z_n are the fractional co-ordinates of the nth atom and f_n is the atomic scattering factor for the nth atom.

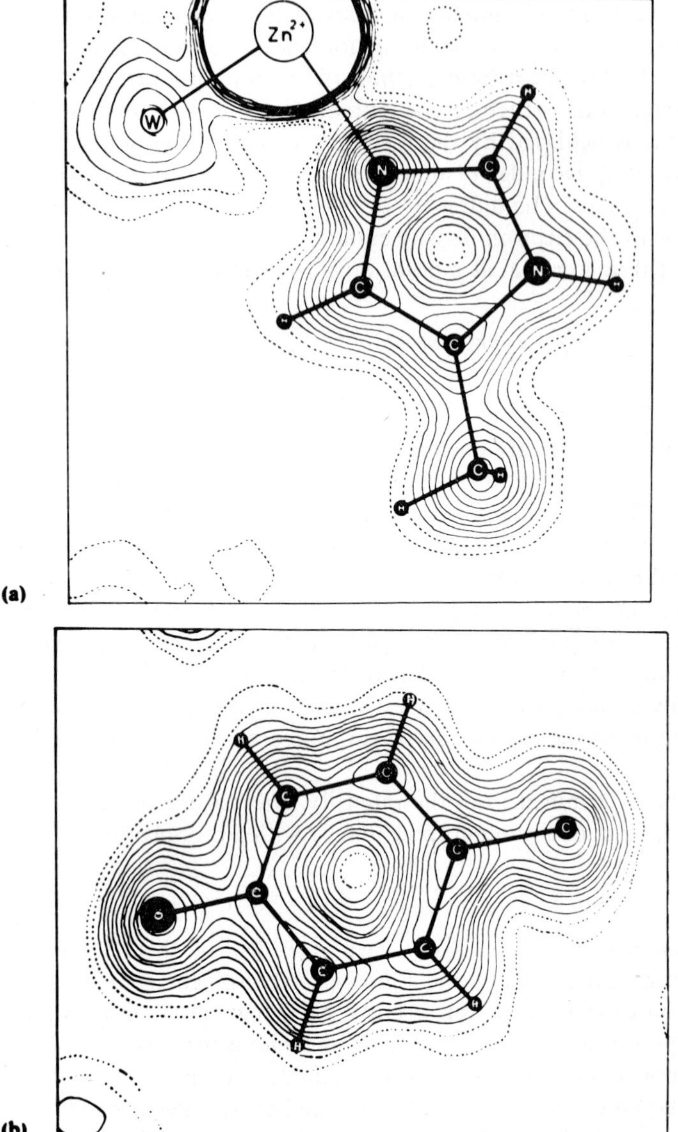

Figure 4. A portion of the electron density map of insulin at 1.2 Å resolution showing (a) histidine B10 and the zinc atom and (b) tyrosine B16. The atomic positions including assumed hydrogen atoms are superimposed. Contours are at 0.4 (dashed), 0.6 (dashed), 0.8, 1.0, 1.2 … e Å^{-3}. From Sakabe et al. [244].

The values of f for each atom have been computed from wave mechanics and tabulated [28]. At $\theta = 0, f = Z$, the atomic number of the atom. At higher Bragg angles

f falls off exponentially as the spatial distribution of the electrons in the atom give rise to interference effects. B_n is the Debye-Waller temperature factor. As early as 1914, Debye had anticipated that the intensity of a reflection would be reduced by temperature and provided experimental proof. $B = 8\pi\bar{u}^2$ where $\bar{u}^2$ is the isotropic mean square amplitude of vibration perpendicular to the reflecting planes and is discussed further in Section 3(b).

Once a crystal structure has been solved, the structure factor $F(hkl) = F(hkl) \exp i\,\alpha(hkl)$ can be computed from the positional and thermal parameters of the atoms.

In this very brief treatment of the fundamental equations of X-ray crystallography, there has been insufficient space to explain the derivations. For a more detailed treatment the reader is referred to reference 25 and to more general crystallographic texts, such as Stout, G.H. and Jenson, L.H. (1968) X-ray Structure Determination, Macmillan, New York; Woolfson, M.M. (1970) Introduction to X-ray Crystallography, Cambridge University Press, Cambridge; Lipson, H. and Cochran, W. (1966) The Determination of Crystal Structures, Bell and Sons, London.

(b) Crystallisation

Crystallisation forms the most important and often the rate-limiting step in any structure determination. Much effort has been put into the crystallisation of small atoms or molecules (e.g., silicon, sugar) because of their technological importance. In comparison, systematic work with proteins is more complex. Proteins differ from small molecules in two respects with reference to crystallisation. (1) Proteins are highly solvated, even in the crystal, and so the relative lattice energies obtained from protein-protein contacts are small (Fig. 2). (2) Proteins possess many potential attachment sites of approximately equivalent energy to the small number of specific sites involved in the crystal and so there may be an equal incentive to form an amorphous precipitate or a crystal. The problem is to obtain a crystal rather than an amorphous precipitate as supersaturation conditions are approached.

Crystallisation was used for many years as a step in the isolation and purification of proteins. Microcrystallinity (the appearance of a silky sheen) was one of the first indications of approaching purity. A good early review of protein crystallisation stemming from this tradition is given by Czok and Bucher [29] for the enzymes of rabbit skeletal muscle. In the past, proteins selected for study were those that were readily available and crystallisable. It is noteworthy that a large number of the enzymes illustrated in Dixon and Webb's [30] crystal atlas have now had their structure solved by X-ray diffraction.

For protein crystal structure analysis, crystals of at least 0.3 mm in dimension are required. With the introduction of the intense synchrotron radiation source (Section 3(d)) this value may be reduced to 0.1 mm. Crystallisation may be divided into three steps: (a) supersaturation of the protein solution; (b) crystal nucleation; (c) post-nucleation growth and cessation of growth. In the following sub-sections these steps are considered briefly. Previous reviews [25,31] contain greater detail. I am indebted to M.S.P. Sansom [32] for the clear summary presented in his thesis.

(i) Supersaturation: Factors affecting the solubility of proteins

The solubility of a protein molecule in water depends on the balance of intermolecular attractions and molecule-solvent interactions. These in turn depend upon protein concentration, ionic strength, pH, temperature, presence of organic solvents and the binding of counter ions or ligands. Each of these parameters may be varied in a crystallisation trial within certain guidelines.

Protein concentration: Concentrations in the range 5 to 50 mg/ml are usually used. Concentrations less than 2 mg/ml usually are not effective [31]. In general, it is best to keep the protein concentration as high as possible and vary other parameters.

Ionic strength: The protein molecule in aqueous solvent is surrounded by a water layer which interacts with exposed polar and charged groups via hydrogen bonds. To a rough approximation the protein can be viewed as a large polyvalent ion and its solubility described in terms of the Debye-Huckel theory which was developed to explain the behaviour of smaller ions. At low ionic strengths an 'atmosphere' of ions of opposite charge tends to form around each exposed ionic group and facilitates the interaction of the protein molecule with water. An increase in solubility at low ionic strength (e.g., 1–10 mM concentration) compared to no electrolyte is a property common to many proteins. The phenomenon is known as 'salting in'. At higher ionic strengths, the ions begin to compete with each other and with the protein for the surrounding water. The resulting removal of water molecules decreases the solubility of the protein. This 'salting out' process is proportional to the ionic strength. Small highly charged ions (e.g., potassium phosphate, sodium sulphate, ammonium sulphate) are more effective than large low-charged ions (e.g., potassium chloride) in salting out [33].

Ammonium sulphate is commonly used as a precipitant because of its great solubility and ready availability. It can liberate ammonia, which reacts with heavy atom derivatives. This problem can in some circumstances be alleviated by changing the mother liquor of the crystals to phosphate. Ammonium sulphate caused further problems in the analysis of the crystal structure of oxyhaemoglobin. In the presence of this salt and under irradiation, crystals of oxyhaemoglobin were oxidised rapidly to aqua-met-haemoglobin. The problem was solved by growing the crystals under usual conditions [34] and then transferring them to 3 M phosphate [16].

Poly(ethylene glycol) (PEG): McPherson [35] has advocated the use of PEG for protein crystallisation. Out of 22 proteins chosen to test PEG as a precipitant, 13 yielded crystals, six for the first time. The general formula of PEG is

$$HOCH_2CH_2\text{-}(OCH_2CH_2)_n\text{-}OCH_2CH_2OH$$

and the average value of n is expressed in terms of the average chain molecular weight. McPherson found that optimal sizes were 1000, 4000 and 6000. The concentration of a given PEG size to yield crystals was in the range 2–20% and did not appear to be as critically sensitive as the concentration of many salts and organic solvents. However, solutions of 10% PEG are quite viscous and so are awkward to handle. Recently, there has been a report that PEG 4000 and 20 000 in the concentration range 10–20%

reduced the sensitivity of crystals to radiation damage [36]. Thus, there is a further incentive to explore the use of this reagent.

Poly(ethylene glycol) produced an unexpected result with haemoglobin. Crystals grown in the presence of air were found to have the protein in the T state, in which the α-subunits were oxygenated and the β-subunits oxygen free [37]. The result has important implications for allostery.

pH: Variation in pH will be limited by the stability of the protein and the wish to study the protein under hydrogen ion concentrations approximately equivalent to those of the protein's natural surroundings. In general, a protein is more soluble the more net charge it contains and therefore is least soluble at its isoelectric point. Crystallisation near the isoelectric point is not often exploited. Lysozyme crystallises well at pH 4.7. Its isoelectric point is 10.5.

Temperature: The temperature coefficient of solubility varies greatly from protein to protein. At high ionic strengths many proteins are less soluble at 25°C than at 4°C. This may be exploited by forming a super-saturated solution at 4°C and then allowing it to warm up slowly. At low ionic strength solubility usually increases with temperature.

Organic solvents: Organic solvents lower solubility by lowering the dielectric constant of the solvent and thus increasing the strength of intermolecular electrostatic interactions and by competition for water [38]. Volatile solvents such as ethanol and acetone are often difficult to handle during subsequent stages of structure analysis. 2-Methyl-2,4-pentanediol (MPD) has proved useful as it is less volatile, miscible with water and does not denature proteins readily. Crystallisation from organic solvents may also be an advantage if low-temperature crystallography is envisaged. Many organic solvents are cryoprotectants [39].

Addition of ligands: In order for crystals to form, the protein solution needs to be homogeneous with respect to both its purity and the conformational state of the protein. Stabilisation of a state can often be achieved by addition of ligands or ions. Allosteric proteins present problems. If the protein is strongly allosteric, so that in the absence of ligand the ratio of T to R state molecules is large (say), then the protein is predominantly in the T state [40]. Conformational homogeneity prevails and crystal growth is favoured. If on addition of ligand the allosteric ratio T/R is altered to 1 (say), then there is a high degree of conformational heterogeneity and crystallisation is likely to prove difficult [32]. Therefore, crystallisation of the less favoured state of a strongly allosteric protein is often problematic.

(ii) Nucleation and seeding

In some circumstances the rate of formation of nuclei is enhanced by the preliminary formation of an amorphous precipitate, which is usually more soluble than the crystal. With enolase, a significant portion of dissolved protein was in equilibrium with the precipitate, and crystals grew from this mixture [29]. However, frequently the outcome is not so happy. There are no general rules which favour crystal growth except that crystals grow best if supersaturation is approached slowly and there are no heterogeneous nucleation sites such as dust particles, impurities, etc.

A theoretical approach has been developed by Kam et al. [41] in terms of competition between growth rates for compact three-dimensional structures and growth rates for long-chain structures leading to amorphous precipitates. The size and shape distribution of small aggregates in a model system (lysozyme) was monitored during the prenucleation stage by quasi-elastic light scattering. Nucleation was considered as addition of monomer to an existing polymer or crystal of j-monomers (j-mer) and an equilibrium constant K_j defined as

$$K_j = c_{j+1}/c_j \cdot c_1$$

where c_j are the equilibrium concentrations of j-mers. For a crystal to form $K_\infty^{\mathrm{xtal}}/K_1^{\mathrm{xtal}} \gg 1$ and a value of 35 was found for lysozyme. For an amorphous precipitate $K_\infty^{\mathrm{Am}}/K_1^{\mathrm{Am}} \simeq 1$. Thus, if c_1 is larger than the solubility of the linear amorphous precipitate, only small size aggregates will be stable and no barrier exists for the growth of a linear aggregate. Kam et al. advocate the determination of K_1 and K_∞ in a systematic crystallisation trial in order to establish favourable crystallisation conditions. (The abbreviations xtal and Am refer to crystal and amorphous precipitate, respectively.)

Controlled nucleation may be achieved once seed crystals are available. Seeding is best carried out using a very few very small crystals. For example, with phosphorylase b, a large ($0.4 \times 0.4 \times 1$ mm) crystal was crushed in a small volume (1 ml) of solution. The solution was left to settle for 10 minutes and the supernatant serially diluted by 10^4. Additions to the crystallisation trial were made so that the final dilution of seed solution was 10^6 to 10^{12}.

With pig aspartate aminotransferase, crystals were obtained by seeding with chicken enzyme crystals [42].

Alber et al. [43], have observed an interesting phenomenon with yeast triosephosphate isomerase when crystallised from poly(ethylene glycol), which they termed 'oiling-out'. At high concentrations of PEG ($>20\%$) the protein formed droplets which coalesced. On dilution most of these dissolved, leaving a few droplets which subsequently converted to crystals. The authors comment that these observations are not limited to triosephosphate isomerase nor to poly(ethylene glycol).

(iii) Crystal growth and cessation of growth

Crystals grow by forming monomolecular layers parallel to crystal planes which have high molecular density. The addition of a further molecule to an incomplete plane is usually favourable, but initiation of a new plane may not be so. Thus, crystal growth is not spatially invariant. The external morphology of a crystal usually gives some indication of the molecular packing. For example, plate-like molecules or arrays of molecules usually stack to form needle-like crystals. Arrays of molecules which have one long direction will assemble to form plate-like crystals. There is often an inverse relationship between the dimensions of the crystal and the dimension of the building block. In fact, as was recognised by the earliest crystallographers, those planes which are most likely to appear as external faces of the crystal are those most densely occupied by atoms. Since the physical process of crystal growth is one in which energy factors are important, the resulting shape should be one of minimum free energy. Crystal planes are more likely to occur parallel to planes of strongest bonding in the molecules or molecular arrays. Despite these rules for simple cases, it is often difficult to give an explanation of the wide diversity of habit exhibited by crystals, including protein crystals.

Kam et al. [41] have noticed a self-limiting phenomenon which caused crystals to stop growing after a certain size. When these terminal size crystals were cleaved, growth occurred at the surface until the original size was restored. Mosaic spread measurements have been used to investigate long-range order in crystals of lysozyme, chymotrypsin and myoglobin [44]. The results indicated almost perfect long-range order in these proteins and the absence of dislocations. This is compatible with the weak binding energies and mechanical softness of protein crystals. Accumulation of strain (for example, arising from protein mobility) which cannot be relieved by dislocations may be a possible mechanism for cessation of growth.

(iv) Practical techniques for crystallisation

Practical techniques for crystallisation have been reviewed [25,31,45,46]. Rapid screening of possible conditions are described by Carter and Carter [47] and Rayment [48], and McPherson [31] has documented nearly 200 crystallisation conditions for different proteins, providing a useful compendium which illustrates the diversity of methods employed.

Variations on the vapour diffusion method have met with considerable success. A solution of the protein containing a salt concentration approximately 10% below that needed for precipitation is equilibrated by vapour diffusion with a larger volume of more concentrated salt solution which is only slightly below the concentration needed for precipitation. With non-volatile precipitants water distils out from the protein solution to the reservoir. With volatile solvents, distillation and equilibration will proceed in the opposite direction. The ‘hanging drop’ version of this method allows numerous trials of different conditions with very little protein material. Plastic tissue culture plates (for example, with 24 cylindrical wells of 2 ml volume) may be used. The precipitant solution (1 ml volume) is placed in the wells. These are then sealed with a coverslip onto which a drop of protein solution (5–20 μl) has been placed and then inverted. A drop of light oil on the rim of the well makes for an air-tight seal. The method allows ready inspection of the drops without disturbing them.

(v) Crystallisation of membrane proteins

There have been several special cases where membrane proteins have been crystallised and high-resolution X-ray structures obtained. Where the protein is anchored in the membrane by a hydrophobic tail the protein may be released, for example, by proteases. The water-soluble component so released can be crystallised like a normal soluble globular protein. Outstanding examples of this approach have been cytochrome b_5 [49], influenza virus haemaglutinin [50] and influenza virus neuraminidase [51]. Alternatively, there have been a few examples of small proteins which can incorporate themselves into membranes and which are sufficiently homogeneous to be crystallised from organic solvents. Notable examples are alamethicin [52] (crystallised from 10-fold dilution with acetonitrile of a solution of the protein in methanol (100 mg/ml)) and crambin [53] (crystallised from 60% aqueous ethanol (v/v)). In addition, the structure of the naturally crystalline purple membrane from *Halobacterium halobium* has been solved to 7 Å resolution by electron microscopy [54,55].

However, despite these advances, there are a large number of membrane proteins, such as transport proteins, receptors and energy-conversion molecules, for which knowledge of the structure is of crucial importance for an understanding of biological function and for which, as yet, no crystals are available. Most membrane proteins have an amphiphilic nature. They are composed of hydrophobic regions that enable them to incorporate into the lipid bilayer and hydrophilic regions that enable them to interact with the cytoplasm or extracellular space. The amphiphilic nature causes difficulties in the purification and crystallisation of membrane proteins. The proteins are not soluble in aqueous buffer solutions and require detergent for solubilisation. Removal of the detergent leads to aggregation and denaturation. Recently, there have been considerable advances in the crystallisation of membrane proteins, and these have been reviewed by Michel [56] and Garavito and Jenkins [57]. These have led to suitable X-ray crystals for bacteriorhodopsin [58–60], photosynthetic reaction centre of *Rhodopseudomonas viridis* [61] and several proteins from the outer membrane of *Escherichia coli* [62–64].

In these studies the protein was solubilised by the use of mild detergents and the protein in the micelle was the starting point for crystallisation using usual precipitants. The realisation that the detergent (its size and chemistry) played a direct role in crystallisation was an important step forward. The size and quality of the crystals were improved by addition of small polar organic molecules, which may act by improving packing of the micelles in the crystal lattice.

Very recently, these methods have resulted in the first crystal structure determination by X-ray diffraction of an integral membrane protein, the photosynthetic reaction centre of *Rps. viridis* [265].

(c) Data collection

Protein molecules crystallise with large unit cells in comparison with small molecules. The number of reflections to be measured at any given resolution is correspondingly large and the intensities are correspondingly weak. The number of reflections to be measured is proportional to the volume of the unit cell and to the inverse cube of the resolution. Thus, for a protein of molecular weight 100 000 there may be some 2500 reflections to be measured at 6 Å resolution, some 20 000 reflections at 3 Å resolution and some 67 000 reflections at 2 Å resolution. Protein crystals also suffer radiation damage so that the maximum exposure for many proteins is restricted to about 20 hours. Strategies for data collection must allow intensity measurements to be recorded to within the required precision (e.g., 5%) and with the minimum of X-ray exposure of the crystal.

The precession camera is still the most useful instrument in the preliminary analysis of a crystal. It enables an undistorted photograph of a plane of the three-dimensional diffraction pattern to be recorded and is invaluable in the estimation of unit cell parameters, the detection of symmetry and for screening heavy atom derivatives (see, for example, Ch. 9 of Ref. 25 for a fuller description).

The diffractometer [65] remains the method of choice for three-dimensional data collection for proteins with unit cell sizes less than 100 Å and crystal sizes better than 0.2 mm. The crystal is oriented by a computer-controlled mechanical system so as to bring each reflection in turn into the reflecting position, and a detector is moved so as to receive this reflection. With computer control of the setting angles and the measurement cycles, data are accumulated automatically. Each reflection is recorded and measured at the same time and the results stored in the computer ready for subsequent analysis. Absorption corrections [66] are readily measured and applied to the data.

Electronic area detectors of the multiwire proportional chamber (MWPC) design have been developed for diffractometers and used to solve structures [67,68]. These offer the combined advantages of the flexibility in setting of the diffractometer and the ability to record all reflections generated at a particular crystal orientation. The use of area detectors is not widespread, probably because of the lack, until recently, of a suitable commercial detector and the not inconsiderable problem of optimising software. Arndt and his colleagues [69,70] have developed a TV detector especially suited for the high flux at synchrotron radiation sources. Both MWPC and TV detectors are now manufactured commercially and development of these detectors is in progress at various synchrotron radiation sources [71,230]. Area detectors are likely to provide the basis for the next leap forward in protein crystal data collection.

The oscillation camera [72,73] has become the most widely used instrument for crystals of large unit cell or for very small crystals. Such an instrument is in operation at all protein crystallography stations on synchrotron radiation sources. The operation of the camera is simple. The crystal is rotated about an axis and all reflections thus generated are recorded on a photographic film. Three-dimensional data are collected by taking a series of small angle oscillation photographs (of the order of 0.5 to 3°) through an appropriate total rotation about the axis (Fig. 5). The oscillation angle is chosen to minimise overlap of reflections and the total rotation range depends on the symmetry of the crystal. Spots which are partially recorded on one film, because they occur too close to the beginning or end of the oscillation range, have their remaining portion recorded on the adjacent film. Provided the camera oscillation mechanism is free of backlash, the individual components may be added to give the total intensity.

A commercial instrument was developed in 1973 and programmes were written for processing the data [73–75]. Further improvements have been the use of profile fitting using the systematic matching of theoretically predicted diffraction patterns against the intensities [76] and in background corrections and the use of learned profiles for weak intensities [77]. The greatest problems have been in the comparison of those reflections fully recorded and those obtained by summation of partially recorded

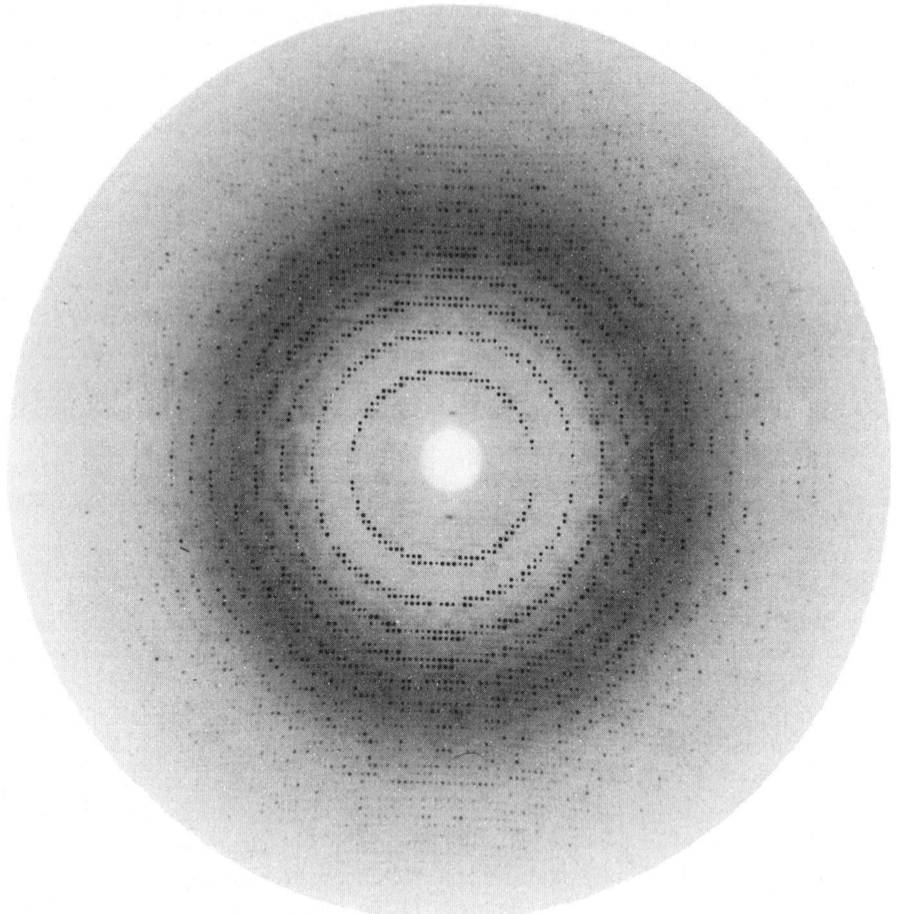

Figure 5. An oscillation photograph of a glycogen phosphorylase *b* crystal recorded at the Wiggler station 9.6, Synchrotron Radiation Source, Daresbury. The tetragonal crystal is mounted about *a*. Synchrotron power: 1.8 GeV, 131 mA; wavelength: 1.0 Å; crystal to film: 100 mm; exposure 12.5 minutes; unit cell $a = b = 128.6$ Å, $c = 116$ Å. The data extend to 1.9 Å resolution.

reflections. There is a trend, especially with $\sin^2\theta/\lambda^2$, for the intensities of fully recorded reflections to appear to be underestimated. This discrepancy is most sensitive to errors in crystal orientation parameters. Two groups [78,79], spurred on by the difficulties in working with virus crystals where only one photograph per crystal may be obtained, have developed 'post-refinement' methods which deal more successfully with the problem of partial reflections. Estimate of setting parameters are obtained by comparing a partial reflection on one film with a full observation of the same reflection on the same or another film. A measure of the observed degree of partiality can be made and related to the calculated value derived from the crystal setting parameters. Least-squares minimisation of the sum of the squares of the differences between observed and calculated values for degree of partiality leads to the refinement of the orientation of each crystal, the unit cell dimensions and the crystals effective mosaic spread. The prediction is sensitive to the X-ray beam divergence and dispersion, a problem which has been treated by Greenhough and Helliwell [80]. With these innovations, the quality of data measured with the oscillation camera is comparable to the quality of data obtained with a diffractometer, and often the oscillation camera provides the only method by which data can be obtained.

(d) Preparation of heavy atom derivatives

Until recently, all protein crystal structures were solved by the method of heavy atom isomorphous replacement. The object is to introduce a heavy atom at one or a few sites on the protein such that the protein and the crystal lattice are perturbed as little as possible. The heavy atom acts as a marker for phase determination (Section 2(f)). Methods for the preparation of heavy atoms were comprehensively reviewed in reference 25. Here a summary is given, with brief notes on new developments.

In metalloproteins, the metal cofactor may be removed and replaced by a heavier atom with similar chemistry. The zinc in insulin was successfully replaced by cadmium or lead, the zinc in carboxypeptidase and carbonic anhydrase by mercury, the zinc in thermolysin by lanthanide ions or strontium or barium and the calcium in staphylococcal nuclease by barium. Success has most usually been achieved by soaking the crystals in a chelating agent and subsequently diffusing in the heavier atom.

Heavy atom-labelled inhibitors have the advantage that the specificity of the active site is exploited to generate a single site derivative. However, such reagents are likely to perturb the region of the enzyme of most interest. They have been useful in providing an approximate phase set that helps interpret a multi-site derivative and have then been discarded when the other derivatives are refined. 5-Iodouridine 2′,3′-phosphate was used in this way with ribonuclease-S [81].

Direct binding of heavy atom salts provides the most universal method of heavy atom preparation. Mercury compounds are widely used. Their high binding constants to sulphydryl groups formed the basis of the first derivative prepared for haemoglobin, *p*-chlormercuribenzoate (PCMB). Most reagents have a covalent mercury-carbon bond and a dissociable chloride or acetate ion. The positively charged mercury reacts with the negatively charged sulphydryl group. Some specificity is achieved by altering the size and shape of the substituent groups to create greater or less activity. Examples include methylmercury chloride, ethylmercury chloride, and dimercury compounds such as dimercuriacetate (DMA) and 1,4-diacetoxy mercuri-2,3-dimethoxybutane (Baker's mercurial) which have their mercury atoms, respectively, 1.7 and 4.9 Å apart. Ethylmercurithiosalycilate reacts with displacement of the thiosalycilate by the sulphydryl, and its reaction is therefore restricted to sites of appropriate geometry and reactivity. It has been found useful in liver alcohol dehydrogenase [82] and glycogen phosphorylase *b* [83], where other mercury reagents tended to be too reactive. Recently, methyl phenylmercury has been shown to react with previously inaccessible sulphydryls in hydrophobic sites [84]. It does not react with the exposed sulphydryls of haemoglobin.

Uranyl acetate which dissociates to UO_2^{2+} in solution was first used successfully with lysozyme, where it bound to a site between the carboxyl-carboxylate pair at the active site [85]. This reagent continues to be useful, but often produces multisites. Lanthanides which also bind to carboxylate side chains tend to be more selective. Samarium and gadolinium have been found to be effective.

Platinum compounds have been widely used. These include the $PtCl_4^{2-}$ ion or the less reactive *cis* or *trans* platinum diaminodichloride compounds. At acid pH they react with methionine, cystine disulphides, N-termini or histidine. In the presence of ammonium sulphate, the chloride ions are rapidly substituted by ammonium ions to form $[Pt(NH_3)_4]^{2+}$, which is unreactive. Square planar negatively charged complexes such as $Pt(CN)_4^{2-}$ have been found to be effective in binding at the nucleotide-binding site in dehydrogenases. The cyanide ligands are firmly bound to the metal and are not displaced by protein atoms.

As larger proteins and protein complexes are studied there is a need for 'heavier' heavy atoms. The problem of a suitable change in intensity produced by the heavy atom has been discussed by Eisenberg [86]. For phosphorylase (molecular weight, 100 000) conventional mercury or platinum derivatives proved sufficient, but for higher molecular weights heavy atom clusters may be necessary. In the recent structure determination of the nucleosome core particle (molecular weight, 206 000) at 7 Å resolution two heavy atom clusters were used [87]. Tetrakis(acetoxy-mercuri) methane (TAMM) bound to a single site on histone H3 close to the dyad axis so that it probably bridges two sulphydryl groups, one from each H3. Di-μ-iodo,bis(diaminoethyl-*N*,*N'*)diplatinum (II) (PIP) bound to three major and several minor sites.

(e) Calculation of phases

(i) Use of heavy atom isomorphous derivatives

For most proteins, the phase information is obtained from isomorphous heavy atom derivatives. A heavy atom (e.g., Hg, U or Pt), introduced into the protein, scatters more than the light atoms (C,N,O) of the protein and is used as a marker atom.

The structure factor of the native protein may be represented by a vector $\underline{F}_p$ and described in terms of its magnitude F_p and phase α_p for a particular reflection. Introduction of the heavy atom leads to a change in intensity. The corresponding structure factor is $\underline{F}_{pH}$ with magnitude F_{pH} and phase α_{pH}. Both the magnitudes, F_p and F_{pH}, can be measured experimentally from the intensities of the native protein and protein plus heavy atom, respectively. $\underline{F}_{pH}$ is related to $\underline{F}_p$ by the vector addition of $\underline{F}_H$ which is the contribution of the heavy atom to the structure factor (Fig. 6a).

$$\underline{F}_{pH} = \underline{F}_p + \underline{F}_H$$

The positions of the heavy atoms can be determined from experimental measurements

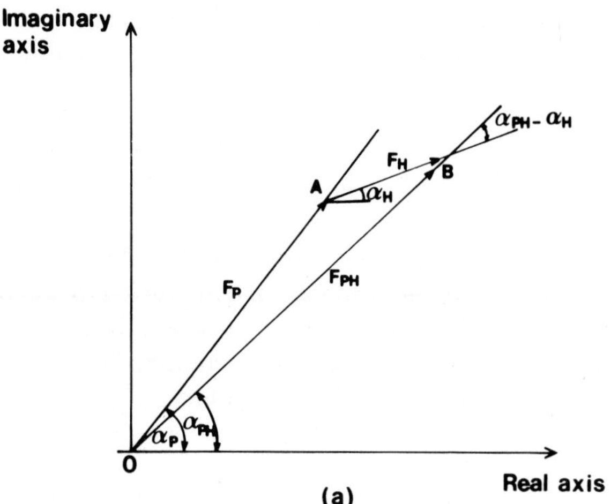

(b)

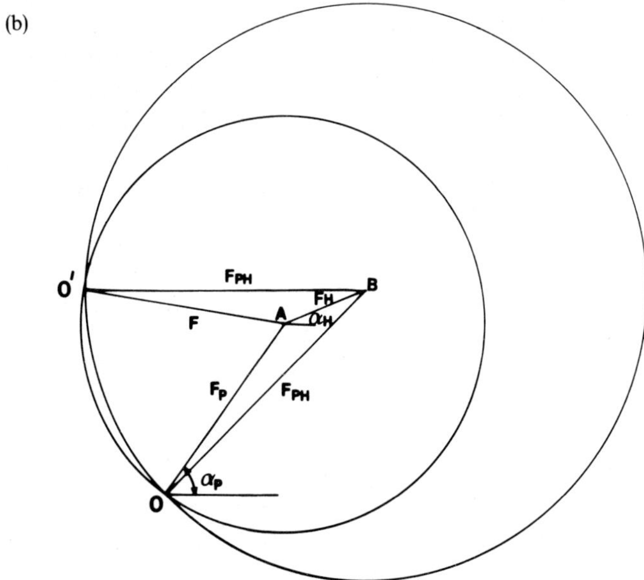

Figure 6a. A vector diagram illustrating the native protein (F_p) and heavy atom (F_H) contributions to the structure factor (F_{pH}) for the heavy atom derivative of the protein. α_p, α_H and α_{pH} are the phases for the native protein, the heavy atom, and the heavy atom derivative of the protein, respectively. b. The Harker construction for the phase calculation by the method of single isomorphous replacement corresponding to the situation shown in Figure 6a. The scale has been reduced slightly. The vector AB represents the amplitude (F_H) and phase (α_H) of the heavy atom. With centre A a circle radius F_p is drawn. Similarly, with centre B a circle radius F_{pH} is drawn. The intersections of the circles at O and O′ represent the two possibilities for α_p. Only one (O) is the correct solution.

(see, for example, Ch. 11 of ref. 25). Thus, the vector $\underline{F}_H$ can be computed from the structure factor equation 3. Thus, both F_H and α_H are known. From Figure 6a and using the cosine law it may be shown

$$\alpha_p = \alpha_H + \cos^{-1}\left(\frac{F_{pH}^2 - F_p^2 - F_H^2}{2F_pF_H}\right) \tag{4}$$

Because of the ambiguity in the $\cos^{-1}$ term, there are two possible values for α_p (Fig. 6b). The ambiguity can be resolved by a second heavy atom derivative. For some reflections, two derivatives may be sufficient to solve the phase problem. In general, more than two derivatives and/or the use of anomalous data are required, because of the effects of errors in the measurements on the phase determination.

(ii) Use of anomalous scattering

When the wavelength of the incident X-ray beam is close to the *K*, *L*, *M* absorption wavelengths of the atoms in the crystal then the electrons of the atom no longer appear to scatter as free atoms and an anomalous component is introduced into the atomic scattering factor. The atomic scattering factor is represented by a real part, $f_0 + \Delta f'$, and an imaginary part, $\Delta f''$, which lags $\Pi/2$ behind the primary wave. Since there is a phase change of Π on scattering, the imaginary part is always $\Pi/2$ in front of the scattered wave. f_0 is the atomic scattering factor in the absence of anomalous effects and $\Delta f'$ and $\Delta f''$ are the anomalous components. Examples of these contributions are given in Table 1.

The structure factor amplitude of the reflection (*hkl*) is equal to the structure factor amplitude of the centro-symmetrically related reflection ($\bar{h}\bar{k}\bar{l}$), i.e.,

$$|F(hkl)| = |F(\bar{h}\bar{k}\bar{l})|$$

(Hereafter, we write these reflections as $F(+)$ and $F(-)$.)

In the presence of an anomalous scatterer this relation no longer holds (Fig. 7). In vector terms we may write

$$\underline{F}_{pH}(+) = \underline{F}_H(+) + i\underline{F}_H''(+) + \underline{F}_p(+)$$

$$\underline{F}_{pH}(-) = \underline{F}_H(-) + i\underline{F}_H''(-) + \underline{F}_p(-)$$

where $\underline{F}_H$ and $\underline{F}_H''$ are the real and imaginary components of heavy atom structure factors.

Again, from application of the cosine rule it may be shown that

$$\alpha_{pH} = \alpha_H + \sin^{-1}\left(\frac{F_{pH}(+) - F_{pH}(-)}{2F_H''}\right)$$

Thus, α_{pH} can be determined and consequently the angle α_p can be found from the

TABLE 1
Atomic scattering factors and absorption edges for selected atoms of interest to protein crystallographers

Atom	Atomic number	$\Delta f'$ ($\sin\theta/\lambda=0$)	$\Delta f''$ ($\sin\theta/\lambda=0$)	Absorption edge (Å)
C	6	0	0	43 (K)
N	7	0	0	30 (K)
O	8	0	0.1	23 (K)
Mg	12	0.1	0.3	9.512 (K)
P	15	0.2	0.5	5.787 (K)
S	16	0.3	0.6	5.018 (K)
Ca	20	0.3	1.4	3.070 (K)
Mn	25	−0.5	3.0	1.896 (K)
Fe	26	−1.1	3.4	1.743 (K)
Co	27	−2.2	3.9	1.608 (K)
Cu	29	−2.1	0.7	1.380 (K)
Zn	30	−1.7	0.8	1.283 (K)
As	33	−1.2	1.2	1.045 (K)
Mo	42	−0.5	3.0	0.6198 (K), 4.912 (L)
Ag	47	−0.5	4.7	0.4858 (K), 3.698 (L)
Cd	48	−0.6	5.0	3.503 (L)
I	53	−0.9	7.2	2.719 (L)
Cs	55	−1.7	8.3	2.474 (L)
Sm	62	−6.6	13.3	1.845 (L)
Gd	64	(−12)	12.0	1.709 (L)
Tb	65	(−11)	8	1.740 (L)
Pt	78	−5	8	1.072 (L)
Au	79	−5	8	1.040 (L)
Hg	80	−5	9	1.009 (L)
Pb	82	−4	10	0.950 (L)
U	92	−4	16	0.722 (L)

$\Delta f'$ and $\Delta f''$ are the real and imaginary components of the anomalous scattering at CuK_α wavelengths (CuK_α radiation $\lambda=1.542$ Å). The L absorption edge is L_{III}. (From International Tables for Crystallography, Vol. III.)

vector triangle defined by $\underline{F}_p$, $\underline{F}_{pH}$ and $\underline{F}_H$. The combination of anomalous and isomorphous information allows the sign ambiguity inherent in either the sine or cosine function to be overcome.

Until recently, the use of anomalous scattering had been restricted to those proteins which contained iron or heavy atom derivatives for which anomalous scattering was appreciable at CuK_α wavelengths (Table 1). Synchrotron radiation provides a tuneable source of X-rays so that the wavelength may be varied to optimise anomalous scattering for the particular atom present in the crystal (section 3(d)).

Anomalous scattering can be used to solve protein structures without the need for other information. There are two methods. In the first the normal scattering contributions of the anomalous scatterer are used as a partial structure to resolve the ambiguity inherent in the phase information from the anomalous scatterer. This

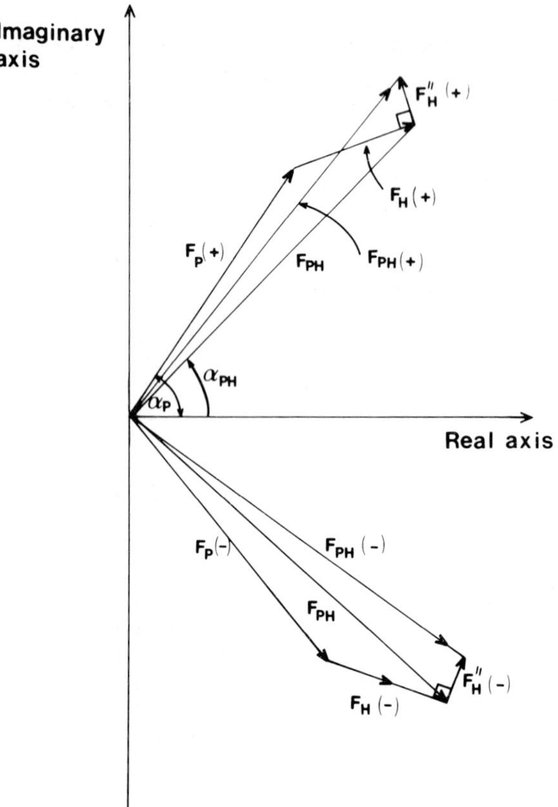

Figure 7. A vector diagram illustrating the effects of heavy atom anomalous scattering on the reflections *hkl* (denoted +) and $\bar{h}\bar{k}\bar{l}$ (denoted −). F_{pH} is the average structure factor amplitude for the heavy atom derivative of the protein. F''_{H} is the imaginary part of the heavy atom structure factor amplitude which arises from anomalous scattering. Because F''_{H} always advances the phase by $\Pi/2$, F_{pH} (+) and F_{pH} (−) are no longer equal. The measured difference between these amplitudes can be used for phase determination.

method was used in the structure determination of crambin [53] (46 amino acids) which utilised anomalous scattering from the six sulphur atoms of the three disulphide bridges at a single wavelength ($\lambda = 1.54$ Å). In another example, the Peking group have solved the structure of despentapeptide insulin again at a single wavelength using the anomalous scattering from a bound cadmium ion [88]. The second method employs multi-wavelengths (at least three) chosen to optimise anomalous effects and leads to an algebraically definitive solution of the phase problem. This method has been made feasible with synchrotron radiation (section 3(d)).

(iii) Molecular replacement

When a protein has more than one molecule or subunit in the asymmetric unit, then redundancies exist in the intensity data that can be used to generate phase information by the solution of algebraic relationships between these intensities. The method, known as molecular replacement, is well described in Rossmann's monograph [89]. Later, it was shown by Bricogne [90] that it is possible to operate on the electron density map by an iterative procedure and obtain the same result more quickly. The procedure is to compute an electron density map with the currently available phases, average the electron density of all crystallographically independent molecules and rebuild the averaged subunit setting the density outside the molecular boundary to its average value. The phase information from the resulting structure may be combined with that obtained from isomorphous derivatives and anomalous scattering to produce a better set of phases. Rossmann's [91] group have also shown that the method of non-crystallographic symmetry, because it essentially flattens the solvent region, can be exploited to extend resolution. Although the method makes the initial assumption that all the subunits are identical, this restraint can be relaxed during subsequent refinement of the structure.

The method has been used to great effect in the analysis of virus structures such as Tobacco mosaic virus [92] (17×2 subunits; molecular weight, 18 000), Southern bean mosaic virus [93] (180 subunits; molecular weight, 28 200) and Tomato bushy stunt virus [94a,94b] (180 subunits; molecular weight, 43 000; +1 RNA, molecular weight, 1.5×10^6; +1 protein, molecular weight, 85 000). It is also widely applied to multimeric protein structures, as for example in the structure determination at 3 Å resolution of the copper-containing haemocyanin [95] (six subunits; molecular weight, 75 000 each) where heavy atom information was used to only 4 Å resolution and in the structure determination at 3 Å resolution of influenza virus haemaglutinin [50] (three subunits; molecular weight, 69 500 each) where data from only one heavy atom derivative were available.

(iv) Treatment of errors

Although the above equations suggest that precise phases may be calculated, in practice this is not the case. Errors arise from inaccuracies in measurements of intensities, lack of isomorphism and incorrect heavy atom positions. Early on, Blow and Crick [3] described the proper treatment of errors and provided criteria for the best choice of weights in the calculation of the electron density map. The method has been developed to include information from anomalous scattering [4,5] and other sources [96].

Blow and Crick define a phase probability function:

$$P(\alpha) = \exp - \sum_j (\varepsilon_j^2(\alpha)/2E_j^2)$$

where $P(\alpha)$ is the probability of the phase angle of the protein structure (α_p) being α, $\varepsilon_j(\alpha)$ is the 'lack of closure' error and represents the difference between the observed and calculated values for F_{pH} for the jth heavy atom. E_j is the root mean square error associated with the distribution ε_j.

The 'best' Fourier is that which has the least mean square error in electron density over the unit cell. This is achieved by choosing the centre of gravity of the probability distribution. The coefficients in the Fourier summation are:

$$F(\text{best}) = F(hkl) \int_{\alpha=0}^{2\pi} \exp i\alpha\, P(\alpha)\, d\alpha \Big/ \int_{\alpha=0}^{2\pi} P(\alpha)\, d\alpha$$

$$= mF(hkl) \exp i\alpha_{\text{best}}$$

where

$$m = \left| \int_{\alpha=0}^{2\pi} P(\alpha) \exp(i\alpha)\, d\alpha \Big/ \int_{\alpha=0}^{2\pi} P(\alpha)\, d\alpha \right|$$

and is known as the figure of merit.

If the probability is sharp, m will have a value close to unity. If the probability is nearly uniform, m will have a value close to zero. m corresponds to the mean value of the cosine of the error in phase angle for each reflection. Thus, a value $m = 0.74$ corresponds to an error of about 42°.

In a protein structure determination the mean figure of merit is often quoted. Usually, for a structure at 3 Å a figure of merit better than 0.6 is required for an interpretable map. Precise values are hard to give because m is a relative value and depends on estimates of the standard errors E. If these are underestimated, m will tend to be overestimated. Even with an $m = 0.82$ there can still be substantial errors in phase determination. Sielecki et al. [97] have shown that some reflections whose experimental phases were found eventually to differ by as much as 170 to 180° from their calculated values had an apparent mean figure of merit of 0.67, which would indicate a mean error of only 48°.

The Blow and Crick formulation does not allow for the easy addition of new information into the phase calculation. This problem has been solved by Hendrickson and Lattman [96] who have cast the probability distribution into a slightly different form.

Hendrickson and Lattman [96] write the probability distribution for the phase angle α in terms of the lack of closure error (δ) in F_{pH}^2 rather than F_{pH}. Thus,

$$P(\alpha) = \exp - \sum_j (\delta_j^2(\alpha)/2D_j^2)$$

where D_j is the standard deviation of the errors associated with δ_j.

This probability function can be cast in the form

$$P(\alpha) = N \exp(A \cos\alpha + B \sin\alpha + C \cos 2\alpha + D \sin 2\alpha)$$

where the coefficients A, B, C, D constitute a complete record of the phase information for a reflection. N is a non-essential normalising factor. Inclusion of new information only requires additions to these coefficients, which greatly simplifies the calculation of the new probability distribution.

(f) Interpretation of electron density maps

The electron density $\rho(xyz)$ (Eqn. 2) is calculated at grid intervals over the unit cell where the choice of interval depends on the resolution of the data. A suitable grid interval is $d/3$, where d is the resolution, i.e., 1 Å grid intervals for a 3 Å map. In the past it was conventional to transfer the contours, generated by the computer programme, section by section to transparent sheets. These were then displayed in a 'Richards Box' [98], which allows simultaneous display of the electron density map and a three-dimensional wire model. Thus, a model (usually on the scale 2 cm/Å) could be constructed. By the early seventies computer graphics were beginning to be used in protein crystallography [99–101]. Computer graphics are now used routinely in the interpretation of maps and have eased some of the labour. The illusion of a three-dimensional object viewed on the two-dimensional screen is achieved by rotating the object about a vertical axis.

In a computer fit one is limited to the display of a relatively small volume of the map. Therefore, a small-scale map (0.25 cm/Å) is usually prepared and the course of the polypeptide chain traced by placing stickers approximately 4 Å apart, corresponding to α-carbon positions. The Fourier synthesis is then displayed as a net corresponding to contours at appropriate intervals on a television screen. The electron density is usually contoured in sections parallel to the faces of the crystallographic unit cell, which gives the resulting image a 'chicken wire' effect (Fig. 8). However, with the computer display the map can be viewed from any direction by use of appropriate computer algorithms. This provides a considerable advantage over the manual displays, which are limited to a view down only one crystallographic axis. The illusion of atoms fitting the density is created by an overlay of the atomic skeleton (Fig. 8). Programmes (e.g., FRODO [102,103], BILDER [104,105] and GRIP [106]) have been developed. These enable the crystallographer to manipulate the polypeptide chain and amino acid side chains so that the best fit of atoms to density can be obtained. First, a particular stretch of polypeptide chain perhaps 5 residues long might be moved so that it roughly fits the density. The (φ,ψ) torsion angles may then be adjusted so that the pentapeptide conformation approximately follows the electron density. Alternatively, individual residues might be broken off and moved separately to their apparent electron density. The separate components are then joined and regularised (for example, by the method of Hermans and McQueen [128]) so that they conform with standard bond lengths and angles within certain specified limits. If the electron density for particular atoms (or groups of atoms) is especially well defined, then those atoms may be fixed in their well determined positions while the remainder of the structure is regularised. One can then proceed to the next stretch of polypeptide, using the regularisation facility to join up the breaks with correct stereochemistry. Inevitably, interpretation is a lengthy procedure. The better the electron density map, the easier is the interpretation. Often there are regions which appear uninterpretable

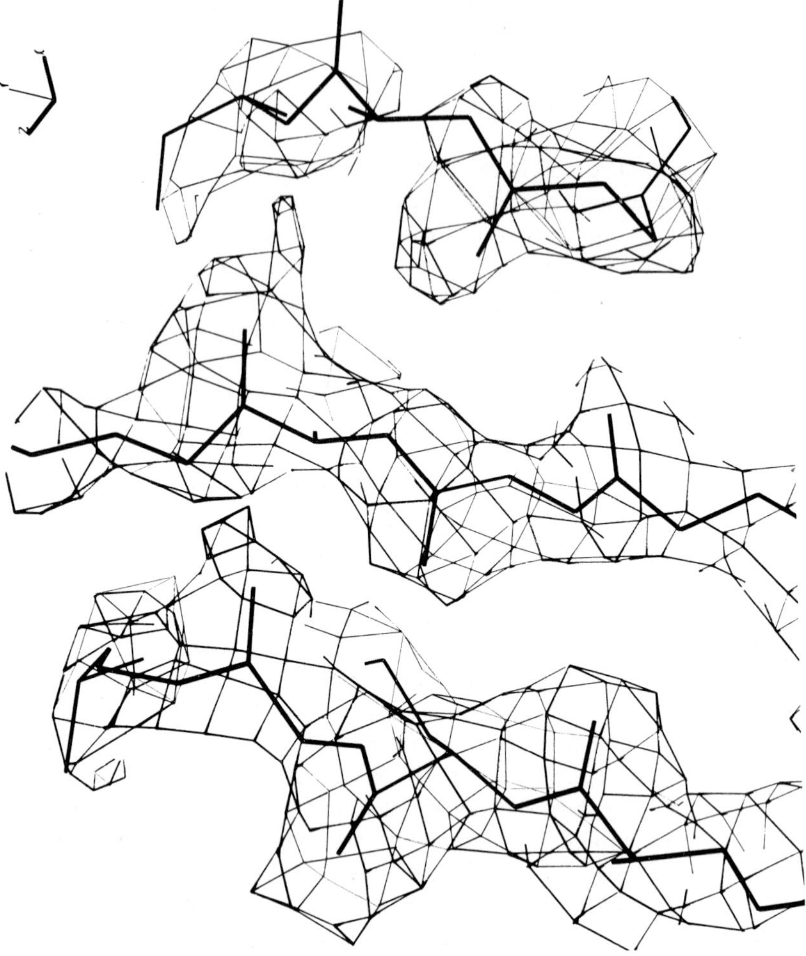

Figure 8. Part of the 2.25 Å electron density map of glycogen phosphorylase *b* with atomic positions superimposed. The picture was taken on an Evans and Sutherland picture system 2 using the programme FRODO. The region displayed shows three strands of a parallel β-pleated sheet. (I am grateful to R. Acharya for producing this drawing.)

either because they are genuinely disordered or because of errors in the map and these are best left until the structure determination is more advanced.

The most popular graphics systems for protein structure research are those manufactured by Evans and Sutherland and by Vector General. These employ a colour vector display system and are linked to a host computer. Manufacturers now provide their own graphic support processor capable of controlling completely the graphics system and of being interfaced to any host computer. The display systems are expensive, but because of their importance for certain industries (e.g., the aircraft industry) and the advances in microprocessors the cost of these systems has dropped over the years. Their use in structural molecular biology finds application not only in the interpretation of electron density maps but also in a whole range of subjects, such as target-drug interactions, evolutionary comparisons of proteins from different species, molecular dynamics simulations, prediction of structures, and any general conformational or macromolecular fitting problems.

(g) Refinement

Once a trial structure has been obtained from an interpretation of the electron density map, a precise description of the atomic positional and thermal parameters and an objective assessment of their correctness is required [107]. This is achieved by refinement of the structure against the observed structure factor amplitudes. By the early seventies, several methods had been developed for the refinement of proteins, notably that of Diamond [108], which utilised real space refinement and was applied to lysozyme [110] (in the development of the programme) and then to trypsin inhibitor [109], and the difference Fourier synthesis least-squares approach which was used for rubredoxin [111]. These studies showed that protein structures could be refined, but the methods were relatively laborious and not easily extended to large proteins. During the late seventies, there was a dramatic development of refinement methods coupled with progress in computer technology so that even proteins with a molecular weight of 100 000 can be refined. These developments have provided the most important advance in the field of protein crystallography during the last decade. They have led to a precise description of protein structures, a study of their dynamics, and a more secure foundation for understanding structure-function relationships.

In a crystallographic problem the parameters to be determined are the positional and thermal parameters for each atom (x,y,z,B). The observables are the structure factor amplitudes $F_{obs}(h)$, where h represents the reflection hkl. Following the method of Lagrange, the best parameters are those for which the sum of the squares of the differences between the observed and calculated structure factors is a minimum, i.e.,

$$\phi = \sum_h \omega_h (F_{obs}(h) - F_{calc}(h))^2$$

is a minimum where ω_h are the weights associated with $F_{obs}(h)$.

The structure factor (Eqn. 3) is not a linear function of the parameters. In order to overcome this problem, trial values of the parameters are assumed and the equation expanded by Taylor's approximation: namely, the value of some function f at $x_0 + \delta x$ is given by

$$f(x_0 + \delta x) \simeq f(x_0) + \delta x \left. \frac{df}{dx} \right|_{x_0}$$

The least-squares equations are solved for small corrections (δx) to the parameters and the process iterated until convergence is reached. The radius of convergence is usually of the order of $d/4$ where d is the resolution.

The progress of the refinement is monitored by the R value.

$$R = \left(\sum_h \left| F_{obs}(h) - F_{calc}(h) \right| \right) \Big/ \sum_h F_{obs}(h)$$

For a random structure, $R = 0.83$ for a centric distribution and $R = 0.59$ for an acentric distribution (which is always the case with proteins in three dimensions) [112,113]. In a small molecule structure R values of <0.10 are routine and many have $R < 0.05$. For proteins an $R \simeq 0.30$ at 2.5 Å resolution usually indicates that most of the structure is correct but several errors may remain. An $R < 0.2$ is usually satisfactory. Luzzati [114] has shown that if the errors in position are normally distributed and that if these errors are the sole cause of differences between observed and calculated structure factors, then at 2 Å resolution a mean error in atomic position of 0.2 Å gives rise to an $R = 0.23$, and an error of 0.1 Å gives rise to an R of 0.12. The 'Luzzati' estimate of errors, which is frequently used in protein crystallography, is usually an overestimate because other sources of error also contribute to the residual R.

The chief obstacles to the routine application of least-squares methods to protein structures are the relatively poor ratio of observables to unknowns and the large demands on computing resources. If it is assumed that there is little correlation between the atomic parameters of different atoms, the computing requirement is proportional to $M \times N$ where M is the number of reflections (of the order of 10 000 (14 000 molecular weight protein) to 60 000 (100 000 molecular weight protein) at 2 Å resolution) and N is the number of parameters (of the order of 4000 (14 000 molecular weight protein) to 28 000 (100 000 molecular weight protein)). Moreover, in a poorly defined structure the shifts to the atoms calculated from the least-squares equations are likely to lead to a gross distortion of stereochemistry. These problems have been overcome in a number of different but related ways. (a) The number of observations has been increased by including known stereochemical features as restraints [115–118]. (b) In addition, the number of parameters has been reduced by constraints which treat certain arrangements of atoms as rigid groups [120–123]. (c) Fast Fourier transform methods have been used to solve the least-squares equations [124–126], and the model has been regularised (subsequently). (d) Least-squares methods have been combined with energy minimisation [127]. There has been no comprehensive comparison of the different methods, but a limited study [37] suggests that methods c and d, although dissimilar mathematically, lead to essentially similar results. The choice of method depends on the problem, the computing resources and the in-house expertise. A good collection of papers on the different methods has been published by the Daresbury Laboratory [119].

(i) Restrained least-squares

The method of Konnert and Hendrickson [116–118] is the most widely used. The known stereochemical features of the amino acids are included as restraints and these provide observations additional to the structure factors. The function minimised is

$$\phi_{\text{tot}} = \phi + \phi' = \sum_h \omega_h (F_{\text{obs}}(h) - F_{\text{calc}}(h))^2 + \sum_l \omega_l (d_l' - d_l)^2$$

where ϕ is defined above, l is the range of the distances to be restrained, d_l' and d_l are the ideal and the calculated distances with appropriate weights ω_l.

The standard amino acid geometries upon which the stereochemical constraints are based are taken from small molecule structures [248]. The different types of restraints, denoted by the parameters d_l' and d_l, include bond distances, bond angles, dihedral angles, planarity (e.g., for peptide, aromatic, carboxylate and guanidinium groups), chirality (for α carbons, threonine and isoleucine), Van der Waals contacts (usually included only as a repulsive potential to prevent atoms being attracted into a false minima) and torsion angles (e.g., for the peptide main chain). Refinement of thermal parameters is accomplished by restraining the variances of the interatomic distances to small values so that the temperature factors of bonded atoms are correlated [118].

For N parameters the least-squares equations lead to an $N \times N$ normal matrix. Because the restraints involve only near neighbour atoms, the matrix is sparse, with the majority of non-diagonal terms being zero and less than 1% of the elements non-zero [117]. For n atoms and m distance restraints the number of elements to be stored is $6n + 9m$. For example, with a small protein of 812 atoms and 2030 restraints (approximately 3 × the number of atoms), the number of elements is 23 142. For phosphorylase *b* with 6640 atoms there are 26 561 parameters and some 229 451 non-zero elements on the normal matrix, which is still only 0.03% of the total matrix elements. In the restrained least-squares refinement (and many of the other refinement methods) the normal equations are solved by the conjugate-gradient algorithm [129].

The relative weights for the X-ray observations and the restraints may be adjusted. In the early stages of refinement, the weights for the restraints might be relatively high in order to achieve a stereochemically sensible model. Gross errors in the structure can be detected by difference Fourier syntheses (section 2(h)). As the refinement progresses, the restraints may be relaxed. The final R value depends upon resolution and the restraints, and it is important that both values are quoted so that the stereochemical reasonableness of the structure can be assessed.

The Konnert-Hendrickson method is relatively expensive in terms of computing power [131], but recent developments that combine the use of fast Fourier transform methods (section iii below) have provided dramatic increases in speed.

As an example of the application of the method we quote values for the refinement of protease A from *Streptomyces griseus* [97]. The R value for some 12 662 reflections in the resolution range 8.0–1.8 Å was 0.139 for some 5912 variable parameters for 1250 protein atoms and 175 water molecules. The final structure differed from ideal bond lengths by an overall root mean square (rms) deviation of 0.02 Å and the probable error in atomic co-ordinates was of the order of 0.15 Å.

(ii) Constrained-restrained refinement

The method of Sussman et al. [121, 122] utilises ideas originally proposed by Scheringer [120]. The model is constructed from rigid groups instead of individual atoms. The groups are joined by flexible links and restrained by distance restraints. Thus, the number of parameters is greatly reduced. For example, a benzene ring normally requires 18 positional and six thermal parameters. Instead it is described in terms of a rigid group specified by three positional, three orientational and one thermal parameters. A possible disadvantage of this approach is that the final errors in the model are not distributed over all the atoms but may accumulate at the flexible regions. The method has considerable power where it is suspected that structural homology exists between the unknown structure and a known structure. Individual subunits, domains or regions of supersecondary structure can be refined as rigid groups. Thus, in its definition of rigid groups the method is most flexible. It is also effective at low resolution, which makes it useful in the preliminary stages of analysis.

The programme (CORELS) was originally devised for nucleic acid structures, but has been extended to proteins [122]. For example, the structure of demetallised concanavalin A has been determined using the structure of native concanavalin A as a starting point [123]. The two crystals belong to different space groups. The resulting crystallographic R factor from some 1800 atoms was 0.26 for 7800 reflections in resolution range 10–3.2 Å. Likewise, the refinement of the intact immunoglobulin Kol was initiated using the previously determined structure of the Kol F_{ab} fragment as a starting point [132]. The determination of this structure has shown that the F_c regions of the antibody are disordered in the crystal, consistent with the idea that the hinge region provides a flexible attachment point for the F_{ab} and F_c segments [133].

(iii) Fast-Fourier least-squares

The method of Isaacs and Agarwal [119,124–126]: in a difference Fourier synthesis based on the difference in amplitudes of the observed and calculated structure factors, the corrections to the atomic positions may be derived from the gradient of the difference density divided by the curvature of the native electron density at the atomic position. Cochran [135] has shown that these shifts obtained in real space from difference Fourier maps are equivalent to those derived in diffraction space by least-squares methods. The method of Isaacs and Agarwal [125] exploits this equivalence. All computations of the structure factors, the vector gradients and the normal matrix are carried out by different applications of fast Fourier transform methods [136,137]. These methods, as their name suggests, are very fast and make relatively small demands on computer storage. The most expensive and most difficult part of the calculations is the modelling of the electron density. Because this is essentially a free atom least-squares process, the structure must be regularised [138]. So cycles of refinement are interspersed with cycles of regularisation. This decreases the rate of convergence. Test calculations [124] have suggested the method is capable of correcting errors of the order of 0.75 Å but the method has been used in the refinement of structure interpreted from a 2.8 Å electron density map. In its present day use, the method is used only to calculate structure factors and their derivatives and these are then used as input into either method a or d.

The method was successfully applied to insulin [125] and actinidin [139] and a number of other proteins [126]. For insulin, the final R for 1.5 Å resolution was 0.11 for some 1077 atoms and 11 890 reflections. For actinidin, the final R for 10–1.7 Å resolution was 0.17 for some 1821 atoms and 23 390 reflections. The standard deviation in bond lengths from ideal values was 0.014 Å.

(iv) Simultaneous energy and least-squares refinement

The method of Jack and Levitt [127] essentially combines the fast Fourier transform least-squares method with minimisation of a potential energy function. This energy function [140] includes terms for bond stretching, bond angle bending, torsion potentials and non-bonded and electrostatic forces. The crystallographic residual, ϕ, and its derivatives are computed by fast Fourier transform methods and used as input to the programme which minimises the energy and the residual by solution of the normal equations by conjugate gradient methods. The method has a large radius of convergence (2–3 Å) and is fast. It was originally devised for the structure refinement of tRNA, but has been extended to proteins.

For example, the crystal structure of human deoxyhaemoglobin [17] has been refined at 1.74 Å resolution by the method of Jack and Levitt using the energy parameters and solvent correction parameters described by Phillips [134]. The resulting crystallographic R value was 0.16 with rms deviation of C–C bond lengths from their ideal values of 0.023 Å. The estimate of error for positional parameters was 0.1 Å for all atoms and 0.05 Å for the iron atoms. The refinement included some 4460 hydrogen atoms, 221 waters and two ions (probably phosphate). The structure of oxyhaemoglobin [16] has been refined to an R value of 0.223 in the resolution range 10.0 to 2.1 Å, with an approximate overall standard deviation of positional parameters of 0.3 Å. The precision achieved by these refinements is essential for an understanding of the stereochemical basis of allostery.

(h) Difference Fourier syntheses

(i) Use in refinement

After several cycles of refinement, a local minimum is usually reached. The model must be examined and large errors, that are outside the radius of convergence of the refinement, corrected. In a difference Fourier synthesis based on coefficients $(F_{obs}(h) - F_{calc}(h)) \exp i\alpha_{calc}$ atoms that are wrongly placed will have corresponding pairs of positive and negative peaks that indicate the direction of the correction (e.g., [25,141]) and atoms that have been omitted from the phase calculation should be represented by peaks that are half their true height [142]. Such difference Fourier syntheses are useful in the final stages of refinement when only a few atoms are misplaced, but in the early stages of refinement they are difficult to interpret without the aid of the protein density. Instead, modified difference Fourier syntheses are used with appropriate weighting of both amplitudes and phases derived from the experimental and calculated values.

The experimental phases, α_{obs}, may be combined with the calculated phases, α_{calc}, by means of a joint probability distribution.

$$P_{com}(\alpha) = P_{obs}(\alpha) \times P_{calc}(\alpha)$$

$P_{obs}(\alpha)$ is the observed probability distribution for phases obtained for heavy atoms, etc., and has been defined in section 2(e). Sim [143a] has shown that

$$P_{calc}(\alpha) = (\exp X \cos(\alpha - \alpha_c))/2\pi I_0(X)$$

where $I_0(X)$ is a modified zero-order Bessel function (this function has a maximum value at the origin ($x=0$) and then falls off like an attenuated cosine wave)

$$X = 2F_{obs}(h)F_{calc}(h)/\beta.$$

In Sim's derivation $\beta = \sum f_j^2$ where the summation is taken over all the atoms not included in the phase determination. For protein structures, a better value for β is the mean square error $\langle |F_{obs}(h) - F_{calc}(h)|^2 \rangle$.

Following Blow and Crick [3], Sim [143b] suggests that the best set of weights are those resulting in the least mean square error in the density map, and hence

$$m_{calc} = \int_{-\pi}^{\pi} \cos\alpha \, P_{calc}(\alpha) \, d\alpha$$

Thus, the information from the calculated phases may be combined with that from the experimental phases to give appropriate values m_{comb} and α_{comb}.

Consideration must now be given to the type of amplitude coefficients to be used in the Fourier synthesis. The object is to reduce spurious peaks resulting from wrongly placed atoms and to enhance peaks for those atoms that have not yet been placed or included in the phase calculation. Main [144] examined several different types of syntheses and concluded that the one based on amplitude coefficients $(2F_{obs}(h) - F_{calc}(h))$ suppressed peaks of wrongly positioned atoms most effectively and most closely approached the true electron density. A qualitative explanation for its success may be given as follows. In a difference Fourier synthesis based on $2(F_{obs}(h) - F_{calc}(h))$ atoms that are wrongly placed will have negative regions around the incorrect positions and positive features around the positions to which the atoms should move. The factor of 2 takes into account that these features are otherwise half their true height [142]. A synthesis based on $F_{calc}(h)$ will have positive contours at approximately their true heights around the atomic positions used in the calculations. Hence, in a synthesis based on $2(F_{obs}(h) - F_{calc}(h)) + F_{calc}(h) = 2F_{obs}(h) - F_{calc}(h)$ the negative features of the $2(F_{obs}(h) - F_{calc}(h))$ map will be cancelled by the positive features of the $F_{calc}(h)$ map and peaks will occur at the correct atomic positions. However, if the number of unknown atoms represents a substantial proportion of the structure, greater weight needs to be given to the difference terms (e.g., $3-2$ should be used in place of the 2 − 1 coefficients [145]. Rice [146] has examined several of these syntheses in the refinement of horse phosphoglycerate kinase and has concluded that the 2 − 1 map with combined phases gave the best signal to noise ratio.

Thus, the most widely used synthesis is that based on coefficients

$$m_{comb}(2F_{obs}(h) - F_{calc}(h)) \exp 2\pi i \alpha_{comb}$$

However, there are two problems which arise from the somewhat arbitrary choice of amplitude coefficients. Firstly, if there is a large discrepancy between $F_{obs}(h)$ and $F_{calc}(h)$, such as occurs for example in low-angle terms when the solvent has not been included in the calculations (i.e., $F_{calc}(h)$ is considerably overestimated), then the map will contain errors. Omission of the low-angle data will destroy the continuity of the electron density and the distinction between the solvent and the protein. Secondly, although the phases α_{comb} have been suitably weighted to minimise bias from the calculated structure, no such weighting has been applied to the amplitude coefficients. These problems have been considered by Stuart and Artymiuk [147], who have shown that a map based on coefficients

$$m_{comb}(F_{obs}(h) + Q_{comb}(F_{obs}(h) - F_{calc}(h))) \exp 2\pi i \alpha_{comb}$$

gives greatly improved results in terms of solvent-protein definition, reduction of spurious density, and indications for improvement in the structure.

Q_{comb} is chosen to minimise the bias and is given by $Q_{comb} = Q_c(\log m_{exp}/(\log m_{exp} + \log m_{calc}))$ where Q_c usually = 1 and m_{exp} and m_{calc} are the figures of merit for the experimental and calculated phases, respectively.

(ii) Use in ligand binding studies

Once a protein structure has been solved, the study of the association of small molecules with the protein may be accomplished relatively easily by means of difference Fourier syntheses. The method has been widely applied in the study of binding of inhibitors and pseudo-substrates to a large number of proteins and has provided the means by which active and allosteric sites may be located. It is assumed that the small ligand does not change the unit cell or perturb the protein substantially, and that the protein phases are approximately equal to those for the protein and ligand. Small changes in conformation can be distinguished as in conventional difference syntheses. The coefficients used are

$$m(F_{PL}(h) - F_p(h)) \exp 2\pi i\, \alpha_p$$

where $F_{PL}(h)$ and $F_p(h)$ are the structure factor amplitudes for the protein plus ligand and the native protein, respectively. m and α_p are the figure of merit and native protein phase and will depend on the current state of refinement of the protein.

Errors in difference syntheses are proportional to the root mean square difference of $|F_{PL} - F_p|$ and hence are much less than those in the native electron density map

[25,148]. Thus, it is possible to distinguish relatively small groups of atoms. For example, in glycogen phosphorylase *b* [149] (molecular weight, 97 000), the binding of a single phosphate ion can be detected at 3 Å resolution.

(i) The solvent structure

The structural stability and the biological activity of protein molecules are dependent upon the interactions of the protein with solvent. Protein crystals typically contain 50% solvent and this component needs to be accounted for in the calculation of structure factors. In turn, the refinement of crystal structures and their solvent content leads to a description of the ordered water molecules and the bulk solvent continuum.

The ordered water molecules are located by difference Fourier syntheses calculated at a stage when the refinement of the protein is reasonably well advanced. Thereafter, the oxygen atoms representing the water molecules may be incorporated into the refinement to obtain precise values of their positions and temperature factors. The number of bound water molecules is approximately 12–15% of the number of protein atoms and will be related to the accessible surface area of the protein. These waters lie in a surface layer of approximate 4.5 Å thickness around the protein surface. They make hydrogen bonds to the polar groups of the protein or are less strongly held by hydrogen bonds to other water molecules. These waters contribute significantly to the high-angle diffraction pattern.

The bulk water has proved more difficult to model satisfactorily. It contributes to the low angle scattering (reflections with spacings between ∞ and 4.5 Å). If it is not accounted for, the calculated structure factor amplitudes are grossly overestimated in this region, leading to very high R values and misleading features in difference maps. In the approach of Phillips [134] and Blake et al. [130], a model electron density map of the protein component is constructed on a grid of approximately 0.5 Å using a Gaussian representation of the atoms. This map is modified by setting those grid points not occupied by protein atoms to the electron density of the solvent. The remaining electron density is set to zero. The modified map is then Fourier transformed to produce a set of structure factors. These are then scaled to the observed structure factors for the low-angle region and also modified by a temperature factor which serves to smooth the discontinuity at the protein-water interface. These modified bulk water structure factors are then combined vectorially with those from the protein and ordered water molecules to give the total structure factors for the crystal. Values for the solvent electron density will depend on the crystallisation conditions. The value for 3 M ammonium sulphate is 0.395 e/Å^3 and for water is 0.33 e/Å^3. The disordered water might comprise some 60–80% of the total water content of the crystal. This method has proved broadly satisfactory, but there is difficulty in modelling the protein solvent interface and concern that some protein cavities may erroneously be assigned solvent density. An illustration of the results for lysozyme is shown in Figure 9.

An alternative representation of the bulk solvent has been developed by D.I. Stuart and P.J. Artymiuk and A. Leslie (unpublished results) following methods used in fibre

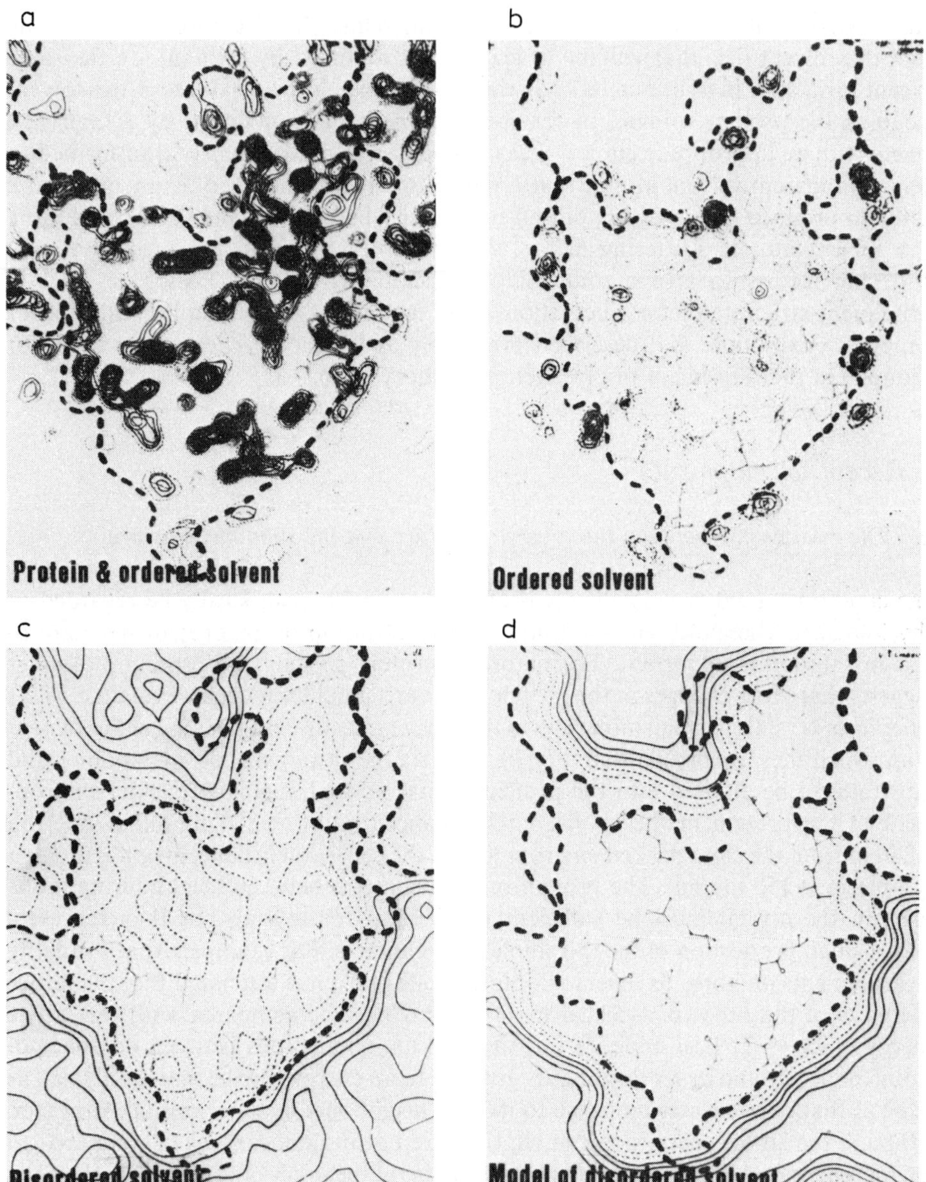

Figure 9. Electron density maps of the same sections of tortoise eggwhite lysozyme showing: (a) the protein component; (b) the ordered water component; (c) the electron density of the disordered water component; and (d) the model of the ordered water component. From [130].

diffraction. The method is based on Babinet's principle. In the scattering of electromagnetic radiation it may be shown that the diffraction pattern of an opaque disc in a transparent mask is the same as that from a transparent disc in an opaque mask (apart from the direct beam contribution). Thus, the structure factors arising from the solvent (i.e., that volume of crystal not occupied by protein) are the same (except for a shift in phase of 180°) as those which would be produced if the solvent occupied the protein volume. In the method an atom is modelled by a Gaussian sphere with an appropriate choice of electron density and the density transformed to yield an apparent solvent atomic scattering factor. This is modified by a temperature factor to produce the necessary fall off in intensity between ∞ and 4.5 Å resolution. The solvent atomic scattering factor is then subtracted from the protein atomic scattering factors to give modified atomic scattering factors. These are used in subsequent structure factor calculations. The method is much simpler and quicker computationally than that described previously and has been successfully used (for example) in the refinement of glycogen phosphorylase *b*.

3. Recent developments

(a) The relationship between the crystal structure and the solution structure

The conditions used to crystallise a protein (section 2(b)) can hardly be regarded as physiological. Therefore, the question of the relationship of the protein structure to that in solution is important. The nature of protein crystals and protein molecules suggests that gross changes in the protein structure should not occur. Protein crystals differ from crystals of small molecules in that they typically contain 50% of the solvent from which they are crystallised (Fig. 2). This enables heavy metals, metabolites and substrates to be diffused into the protein crystal. Indeed, the close packed environment of the protein in the crystal lattice (concentration $\simeq$800 mg/ml) is not too different from the close packed environment in the cell (protein concentration in yeast cytoplasm $\simeq$150 mg/ml). The protein-protein contacts between neighbouring molecules in the crystal must be sufficient to stabilise the crystal, but they represent only a small proportion of the intramolecular contacts. For example, from solubility measurements on horse oxyhaemoglobin crystals [150] it is estimated that the lattice energy is of the order of -5 kcal mole^{-1}. This may be compared with the lattice energy of ice (-11 kcal mole^{-1}), i.e., the packing interactions between one haemoglobin molecule and its six neighbours give rise to an energy change that is only half as large as that of one water molecule to its four neighbours in an ice crystal. However, although the lattice energies are small, they are not insignificant when compared to the free energy for folding of a polypeptide chain (of the order of -10 kcal mole^{-1}) or the free energy for quaternary structure changes (of the order of -13 kcal mole^{-1} for haemoglobin). The function of a protein molecule is sensitive to the environment of the whole molecule, and it may well be that the packing forces, even though they are small, could give rise to local perturbations. These may result in changes in the functional properties of the molecule.

(i) Evidence that the gross structure of the protein is not altered by crystallisation
By 'gross' we mean the protein fold in terms of its topology, its assembly of α-helices and β-sheet and the interior packing of side chains. We exclude conformations of surface loops and exposed side chains.

(1) Comparative studies on proteins from different species show that the structures are essentially the same despite different crystallisation conditions. Examples include sperm whale and seal myoglobin, horse and human haemoglobin, horse, tuna, bonito and rice cytochrome *c*, hen egg white, tortoise egg white and human lysozyme, horse and yeast phosphoglycerate kinase, porcine and hagfish insulin and lobster and *Bacillus stearothermophilus* glyceraldehyde 3-phosphate dehydrogenase. Coordinates for these proteins are held in the Protein Data Bank [151].

(2) Comparative studies on the same protein crystallised under different conditions and in different space groups show that the structures are essentially the same. For example, monoclinic (C2) subtilisin crystallised from 2.1 M ammonium sulphate, pH 5.9, has the same crystal structure as monoclinic ($P2_1$) subtilisin crystallised from a 55% acetone/water mixture at pH 9.1 [152]. Other examples include tetragonal and triclinic lysozyme, orthorhombic and trigonal trypsin, and trigonal and monoclinic ribonuclease.

(3) Comparative studies on homologous proteins show a high conservation of the structural fold and constellation of amino acid side chains at the active site. The best example is that of the serine proteinase family of trypsin, chymotrypsin, elastase and protease A from *Streptomyces griseus* (see, for example, reference 97) (Fig. 10).

(4) Chemical modification studies on the protein in solution can, in general, be rationalised in terms of the structure. For example, alkylation studies with ribonuclease showed that the Nε2 position of His12 was preferentially modified. This position

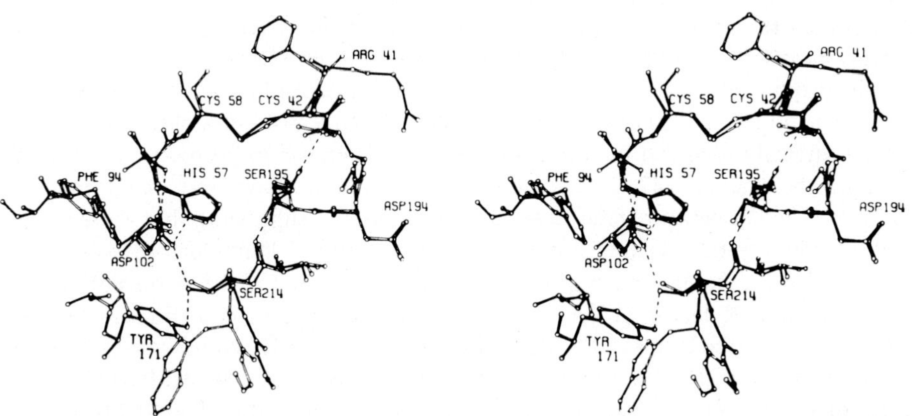

Figure 10. A stereodiagram showing a comparison of the active site geometry of *Streptomyces griseus* protease A (SGPA) with that of bovine β-trypsin. SGPA is presented by solid bonds, trypsin by open bonds. The root mean square deviation after minimisation by application of an appropriate rotation translation matrix is 0.39 Å for some 62 common atom positions. From [97].

is more exposed in the crystal structure [81]. Again, in ribonuclease S oxidation of Met13 reduces the binding of the S peptide to the S protein by some 300-fold. Met13 is buried in a hydrophobic core in the crystal structure: on oxidation it could swing out to accommodate the extra oxygen, but only at the expense of destroying the hydrophobic core [81].

In lysozyme the reactivity of the six tryptophans to various reagents correlates nicely with their positions observed in the crystal structure. This and many other detailed chemical modification studies on lysozyme are summarised in references 153 and 154.

In hydrogen-deuterium exchange studies, surface sites at points of contact of adjacent molecules in the crystal are fully exchanged, indicating that hydrogen exchange is not inhibited by the crystalline state [254].

(5) The radius of gyration of the protein in solution can be measured by low-angle X-ray scattering. The values obtained compare well with those calculated from the crystal co-ordinates. For example [155], the radii of gyration of the hexokinase molecule computed from the crystallographic co-ordinates are 23.66 Å and 22.76 Å for the free and glucose-bound forms of the enzyme, respectively. Since the positions of approximately 300 out of 3600 atoms had not been determined and because these are mostly surface residues, it is expected that the radii are underestimated by about 5%. The corresponding values measured experimentally from low-angle X-ray scattering experiments are 24.73 ± 0.19 Å and 23.78 ± 0.14 Å. It is noted that the change in the radius of gyration from the open native to the closed glucose form of the enzyme is 0.90 Å from the crystallographic co-ordinates and 0.95 ± 0.24 Å from the measured values, showing excellent agreement between X-ray and solution studies for this large conformational change.

(6) Comparison of the molecules, where there is more than one protein molecule in the asymmetric unit, shows the structures to be essentially the same. However, detailed analysis may reveal conformational heterogeneity or the effects of lattice forces. For example, tuna ferricytochrome *c* has been refined at a resolution of 1.8 Å with a final crystallographic *R* factor of 0.208% for the two independent cytochromes in the asymmetric unit and 49 water molecules [156]. The two molecules of the asymmetric unit are virtually identical in conformation (rms difference in positions of all atoms including the haem = 0.47 Å: rms differences for main chain and cβ atoms only = 0.28 Å). The temperature factors are also very similar, suggesting that there is little effect of the crystal packing forces on conformation. Tuna ferrocytochrome *c* crystallises in a different space group and has been refined at 1.5 Å resolution to a crystallographic *R* factor of 0.173%. The rms difference in co-ordinates between reduced and oxidised forms is 0.57 Å. In this instance a detailed comparison of the two structures revealed that the reduced form had a slightly less exposed haem and less hydrophilic environment of the haem crevasse. When reduced cytochrome *c* is oxidised, a buried water molecule and its hydrogen-bonded Asn52 and Tyr67 shift towards the haem [157]. The more polar environment of the haem would tend to favour the loss of an electron by the FeII iron. These results both emphasise the precision which can now be obtained in protein structure determination (positional

standard deviations in co-ordinates of less than 0.1 Å) and demonstrate the order of magnitude of conformational changes that may be of significance in biology. Indeed, the very small difference is surprising in view of the fact that the two redox states have never been crystallised in the same crystal form and that a crystal of one redox state cannot be changed completely into the other with reducing or oxidising reagents without breaking the crystals. On the other hand, crystals of reduced cytochrome *c* can be oxidised to an extent of 67% (as judged by spectrophotometric measurements) and a difference Fourier synthesis reveals the same movements of Asn52 and Tyr67, as observed previously [157]. Examination of plots of differences in atomic positions [157] for the two oxidised molecules shows that about 23 residues have conformational differences greater than 1 Å. Many of these residues are on the surface and the observed heterogeneity may indicate a range of possible structures. These residues also include some which are in regions of the molecule associated with crystal packing. Thus, the detailed comparison does indicate some differences in the structure of the molecules that may arise from crystal packing. This point is considered further in section 3(a)(iv).

In deoxyhaemoglobin, the $P2_1$ unit cell contains one $\alpha_2\beta_2$ tetramer per asymmetric unit. Examination of the intermolecular contacts [17] showed that distortion of the molecule (by less than 1 Å) did occur, as judged by molecular asymmetry, but that the effect was highly localised. Several intermolecular contacts caused no significant asymmetry.

In the case of 2 Zn and 4 Zn porcine insulin, where the protein crystallises with the dimer as the asymmetric unit, the two monomers are similar but not identical in structure, as revealed by refinement at 1.5 Å resolution [158]. One monomer of 2 Zn insulin is almost identical to the corresponding monomer of 4 Zn and to the monomer of hagfish insulin. The differences in structure are best described by movements of packed helices relative to one another, with shifts accommodated by motions of side chains arising from small changes in torsional angles [159]. The root mean square difference in side chain atoms after superposition of main chain atoms is 0.38 Å, and most movements are less than 1.5 Å. This study illustrates both retention of conformation in different crystal forms for one subunit and conformational heterogeneity for the other subunit that may give clues as to how conformational changes in protein molecules can be transmitted.

(ii) Cases where differences have been observed

Apart from the small and rather subtle differences that have been observed when different molecules in the asymmetric unit are compared, most of the differences between the behaviour in the crystal and solution have been detected from changes in functional properties.

The solvent from which the protein was crystallised can alter the properties. In the presence of ammonium sulphate, oxyhaemoglobin is oxidised rapidly to aquamethaemoglobin on irradiation. This problem has been overcome by a change to phosphate salts [16], as discussed previously.

The lattice contacts may promote spurious binding sites. In a diffusion experiment with concanavalin A, iodo-phenyl-β-D-glucopyranoside bound to a site specific for the hydrophobic aglycone moiety. The correct sugar binding site was located subsequently by co-crystallisation studies [160].

In cases where there is more than one subunit per asymmetric unit, the lattice contacts may lead to asymmetric binding. In crystals of chicken triose phosphate isomerase, which contain one dimer per asymmetric unit, only one subunit was found to bind stubstrate but there was no 'half-sites' reactivity observed in solution studies. In crystals of the yeast enzyme, where the lattice contacts are different, both subunits bind substrate and undergo substantial conformational change [161]. In both crystal forms there is a loop of chain, residues 168–177, which, in the native enzyme, exhibits conformational flexibility. On binding substrate the loop moves to close the substrate binding site and becomes ordered [223]. In the crystals of the chicken enzyme, movement of this loop for one subunit is blocked by lattice contacts. In the other subunit, movement of the loop and substrate binding are observed, just as in the yeast triose phosphate isomerase crystals.

In many cases the protein cannot undergo the required conformational change in the crystal lattice. When glucose is diffused into preformed crystals of hexokinase, the sugar binds and substantial conformational changes take place. However, even greater changes that involve movements of the two domains relative to one another are observed in crystals in which hexokinase has been co-crystallised with glucose [162]. Evidently, in the diffusion experiment the lattice forces hindered the 'induced fit' movements.

More puzzling differences have been observed with trypsinogen. Two structures have been solved, one obtained from 1.5 M $MgSO_4$, pH 6.9 (Munich) [163] the other from 30% ethanol, pH 5.8 (Pasadena) [164]. The crystals have the same trigonal space group and unit cell dimensions. The structures are essentially the same, except in the important region of the activation domain (residues 142–153, 184–194, 217–223). In the Munich structure these residues are disordered, but in the Pasadena structure they are loosely organised. The differences do not appear to be a result of misinterpretation or an effect of solvent or lattice forces [133]. A possible source of the discrepancy may be in sample preparation, although this has not been established.

(iii) Activity in the crystal

For many proteins, a direct comparison can be made of functional properties in solution and the crystal. Following the pioneering studies of Doscher and Richards [165], the activity of several enzymes has been assayed in the crystal. From a comparison of many different types of experiments, Rupley [166] concluded that in general the equilibrium properties of the enzyme (e.g., the binding of saccharides to lysozyme) and certain kinetic properties (e.g., hydrogen exchange of lysozyme, diffusion of solutes into β-lactoglobulin crystals) were not altered significantly in the crystal. In most cases, however, enzymic activity was significantly diminished by factors varying between 1- and 1000-fold. While even a thousand-fold reduction in rate must be set against the overall rate enhancement of the enzymic reaction, which is typically of the order of 10^{10}, an explanation for the significant decrease must be sought.

Firstly, not all enzymes exhibit activity in the crystal (e.g., lysozyme) because neighbouring molecules in the crystal lattice block access to the active site. Secondly, in those enzymes where a conformational change is an obligatory part of the reaction, a reduction in rate may be anticipated if these conformational changes cannot be accomplished readily in the crystal. Thirdly, there is a limitation imposed by diffusion of substrate into and products out of the crystal. Quiocho and Richards [167] showed that with carboxypeptidase A_{α}, crystals of 5 μm or less were required before the specific activity of the enzyme in the crystal became independent of crystal size. Rossi and Bernhard [168,169] studied the deacylation rate of co-crystallised acylated α-chymotrypsin using a chromophoric substrate. Under conditions where diffusion away of product was not essential for detection of reaction, they showed that deacylation rates were the same in the crystal as in solution.

The problem of crystal reactivity and diffusion limitations has been considered in detail by Makinen and Fink [170]. They provide a simple treatment for crystals approximated as a plane sheet of material which leads to the definition of a limiting crystal thickness below which kinetic measurements of second-order rate constants are not affected by rate-limiting diffusion processes. For papain [172], ribonuclease A [173] and deoxyhaemoglobin [174], where the crystal thicknesses are comparable to the critical crystal thickness, reactivities are the same in the crystal and solution. In the case of glycogen phosphorylase *b* Kasvinsky and Madsen [175] demonstrated that the K_m values for both substrates, glucose 1-phosphate (37 ± 8 mM) and maltoheptaose (176 ± 20 mM), were the same in the crystal and solution. The 10–100-fold reduction in rate, despite the fact that crystal thickness was only twice the critical thickness, may be attributable partly to the allosteric nature of this enzyme and partly to the fact that the large substrate maltoheptaose (molecular weight, 1152) may not obey the simple diffusion rules in the crystal.

In the case of glyceraldehyde-3-phosphate dehydrogenase, single crystal microspectrophotometric measurements have been used to study the kinetics of acylation of the holoenzyme, binding of NAD^+ to the acylated subunits with activation of the acyl bond and deacylation in the presence of arsenate [171a,b]. The ammonium sulphate present in the crystallisation medium greatly reduces the rates of all the reactions and diffusion is not rate-limiting. It was found that all the reactions proceed in the crystalline state with practically the same rate constants as in solution at high-salt concentration. Thus, protein-protein interactions within the crystal lattice do not affect the catalytic efficiency and do not influence the intersubunit interactions that control the kinetic mechanism leading to half-site reactivity.

(iv) NMR evidence

High-resolution proton nuclear magnetic resonance studies provide a method for assessing the environment of amino acid residues, once assignment of the resonances has been made. In the last few years NMR techniques have become powerful enough to study protein structure at the atomic level, at least with proteins with molecular weight $< 15\,000$. Assignment of a resonance to a particular amino acid in the sequence requires knowledge of the spatial relationships of groups or the ability to relate some special property to the primary sequence, such as chemical modification or substitution of amino acids. In the latter case, there has been the remarkable combination of NMR and crystallographic techniques in the study of the mechanism of the serine proteases, especially with regard to the state of ionisation of the catalytic groups [176]. In this short section, we restrict the discussion to those instances where assignment of residues has led to a three-dimensional interpretation that may be compared with the crystal. Bovine pancreatic trypsin inhibitor (BPTI), cytochrome *c*, lysozyme, calcium-binding protein, neurotoxin and glucagon are some of the systems that have been studied in most detail A recent overview of achievements of NMR in biochemistry has been given by Moore et al. [177], and is discussed in detail in other chapters of this book.

The chemical shift differences that exist between the folded protein spectrum and the random coil spectrum are attributable to specific intramolecular interactions of the globular structure. For the large part, the conformation-dependent shifts of all non-α-carbon protons of diamagnetic proteins can be related to the proton micro-environment by way of ring current effects originating from the aromatic rings. The structural information is restricted to a small volume of space immediately surrounding the aromatic centre (approximately 7 Å in radius) and is limited by the correctness of the ring current model. Application of this method to protein structure determination has been reviewed by Perkins [178]. In general, there is satisfactory agreement between the ring current shift predicted on the basis of the crystal structure and the observed chemical shifts in solution. For example, in lysozyme [179] the observed shifts range in magnitude from −0.39 ppm to 2.13 ppm. The observed root mean square difference between calculated and observed shifts is 0.2 ppm. Differences, where they exist, may be due to inaccuracies in crystal co-ordinates or inadequacies of the models used to account for the chemical shifts. The power of this method has been demonstrated in work with trypsin, where a small correction to an amino acid side chain was deduced correctly [180]. Furthermore, this work produced evidence that trypsinogen activation may involve several subtle differences in conformation in localised regions of the protein previously not thought to be involved on the basis of X-ray evidence.

Estimates of spatial information can be obtained from analysis of nuclear Overhauser enhancement, in which the intensity of a resonance is compared with and without the selective saturation of a second resonance. In lysozyme a good agreement between observed and calculated effects is observed [181]. For 34 resonances, all distances were within 0.8 Å of those determined by X-ray diffraction.

Wagner and Wüthrich [182] have pioneered the sequence specific assignment of resonances for small proteins by two-dimensional mapping using both through-bond J-coupling contributions and through-space dipolar nuclear Overhauser enhancement effects. This has resulted in the sequential resonance assignments for BPTI (58 amino acids) based only on primary structure. Further, the through-space connectivities for the α-helices and β-sheet result in easily identifiable patterns in the two-dimensional maps. Thus, secondary structure can be determined without reference to the X-ray structure. The method has been extended to produce maps of distance constraints between assigned hydrogen atoms. These result in a number of conformers which satisfy the condition that through-space linked H atoms must be within 2–5 Å of one another. The structures may be improved and distinguished by energy refinement. The structure of glucagon (29 amino acids) in solution and when bound to perdeuterated dodecylphosphocholine micelles [183] has been solved in this way. The structure has been compared with the structure of glucagon trimers observed in the crystal in the presence of acetate buffer [184]. Since the hormone is known to exhibit conformational mobility as a function of environment, it is not surprising that significant differences are noted between the NMR and the X-ray structure, especially in the N-terminal 15 residues.

The use of paramagnetic shifts, when available, provides a powerful and sensitive method of structure comparison. This has been applied to cytochrome *c* where ferrocytochrome *c* is diamagnetic and ferricytochrome *c* is paramagnetic [185a]. Assignments have been made for some 212 protons in both redox states, representing 35% of the total from 40 out of 103 amino acids, and a further 28 resonances for six residues assigned for either ferri- or ferrocytochrome *c*. The dipolar pseudo contact shift arising from electron-nucleus interactions is the dominant contribution to the redox state shifts of most of the cross-assigned protons. Correlation of the theoretically predicted shifts based on the X-ray structures with those observed indicates that the main outline of the protein and the arrangement of the haem-packing side chains are remarkably similar in the crystal and solution structures. Some of the residues where pseudo contact shifts differ are close to the haem and within the edge of the shift cone, where small changes in position may cause large changes in pseudo contact shifts. A second group appear to have variable positions either in the crystal (e.g., Ile75, Lys79, Ile85) or in solution (e.g., Ala101). A third group of protons from the side chains of Thr9, Trp59, Leu64, Leu94 and Thr102 have paramagnetic shifts which are not well represented by their calculated pseudo contact shifts. In addition, nuclear Overhauser measurements indicate that the side chains of Ile57 and Thr63 may not be well represented by their crystal structures [185b]. These differences may arise from crystal packing forces, either directly or indirectly. If this inference is substantiated, then it appears that the change in structure that accompanies the change in redox state may be different from that inferred from X-ray evidence.

(v) Summary

It is concluded that the three-dimensional structure observed in the crystal lattice represents a meaningful thermodynamic minimum free energy conformation of the protein and that crystallisation does not change the gross structure. Changes have been observed in functional properties and in some cases these have been resolved by a change in solvent, species, crystallographic space group or by co-crystallisation studies. For many proteins a direct comparison between behaviour in the crystal and solution can be made. Usually the equilibrium binding properties are not altered by crystallisation, but the catalytic rates may be diminished. Crystallisation may trap one conformational state and the rate is diminished because the enzyme cannot respond with the conformational changes along the reaction pathway. Wilkinson and Rose

[186a,b] have exploited this phenomenon in their isotope-trapping methods on rapidly dissolved crystals of hexokinase. They were able to show that the crystalline enzyme-glucose complex contained the glucose in a functional state that appears to be the kinetic equivalent of the soluble enzyme-glucose complex, although the crystalline complex itself was inactive. Isotope trapping of an enzyme-glucose-(ADP) complex, which was active, showed the glucose to be present on an enzyme form that is further advanced to product formation.

The justification for including this topic under new developments is (1) the careful comparison of structures that can now be made following advances in refinement, such as, for example, in the work on insulin and cytochrome *c* and (2) the advances in NMR methods which have permitted detailed comparative studies on certain proteins that allow an assessment of the compatibility of the structural information obtained by the two methods. These have indicated that more work is required by the crystallographers in the description of intermolecular contacts and their possible importance.

(b) Dynamics and flexibility

Proteins are known to exhibit internal motions over a time scale ranging from 10^{-13} second to greater than 1 second [187]. These motions include thermal vibrations (10^{-10}–10^{-13} second), rotations of side chains about single bonds which are relatively unhindered and which occur on a fast time scale (10^{-8}–10^{-11} second); flipping of aromatic rings which require transient concerted movements of adjacent residues (10^{-2}–10^{-5} second); and more general breathing motions or local unfolding which allow penetration of ligands to otherwise inaccessible sites and which permit deuterium exchange with amide hydrogens (10^{-2}–10^{-4} second). In addition, there are large-scale movements, such as disorder to order transitions on binding of ligands or covalent modification, and movements of linked domains relative to one another, which may involve over 15% of the mass of the protein. These movements have excited interest not only for their inherently interesting aspect of a physical structure but also because these fluctuations may play a role in catalysis [188], in allosteric response and in triggering the relay of messages after messenger binding. NMR methods have proved especially effective in elucidating some of these motions and progress in this field has been reviewed [189,190].

Until recently, X-ray crystallography of proteins was regarded as a static technique providing clear results on the three-dimensional positions of the atoms but no information on their fluctuations. The X-ray structure is spatially averaged over the volume of the crystal and temporally averaged over the time taken to collect data (usually several days). However, as a result of advances in refinement of protein structures (section 2(g)), the temperature factors of individual atoms can be determined and these provide information on the atomic displacements. Refinement of protein structures has allowed disordered or flexible regions to be assigned with confidence and shown not to be the result of poor electron density resulting from poor phases. Comparative studies on liganded and unliganded forms of several enzymes have allowed interdomain movements to be described and their functional significance assessed. Protein dynamics and motion and disorder have been reviewed by Petsko and Ringe [191] and Stuart and Phillips [192].

Thermal vibrations of atoms which have a frequency of about 10^{13} second^{-1} are slow compared to the X-ray frequency, which is about 10^{18} second^{-1}. Consequently, the atoms appear to be stationary to the X-rays and the diffraction pattern represents a time average of many instantaneous states. If the motion of the atoms is harmonic so that the restoring force is proportional to the distance of the atom from its rest position and if the motion is isotropic so that the mean square displacements $\bar{U}^2$ of the atom in all directions are the same, then $\bar{U}^2$ is related to the temperature factor, B, by

$$B = 8\pi^2\bar{U}^2$$

Displacements may arise not only from thermal motion but also from static disorder when corresponding atoms in different unit cells take up slightly different mean positions. Certain side chains, especially those exposed, may take up a few radically different conformations in different molecules so that separate images of them can be seen with reduced occupancy in electron density maps. The mean square displacement will also include contributions from lattice disorders but these are usually small in protein crystals that diffract well to high resolution [191]. In principle, the thermal vibrations can be distinguished from static disorder by varying temperature. Simple harmonic vibrations are expected to decrease linearly with temperature.

There are certain practical aspects that need to be taken into account in assessing the significance of temperature factors. Errors in measurement of intensities, arising for example from incomplete correction of radiation damage or absorption, will have more serious effects on temperature factors than on atomic positions. The data must extend to a resolution of better than 2 Å, otherwise temperature factors tend to be underestimated. Restraints in the refinement which assume that positional disorders of bonded atoms are highly correlated may bias the results. However, it is encouraging that in several structures (e.g., rubredoxin [193] and avian pancreatic polypeptide [194]) where the restraints were relaxed these assumptions were found to be valid.

Displacements derived from temperature factors have been compared with those obtained from molecular dynamics. In these calculations, an empirical potential energy function is expressed as a function of the positional co-ordinates of the atoms. This function is then used to obtain the force on each atom (energy is a generalised force × a generalised displacement) and the Newtonian equations of motion are solved for a small time interval, usually a fraction of a picosecond. Good agreement has been obtained for BPTI [195] and cytochrome *c* [196]. There are likely to be significant developments in this field as the sophistication of both refinement and simulation methods is increased.

The thermal motions are probably anisotropic, not isotropic. Much of the intramolecular motion occurs by relatively large movement in soft variables, such as rotations about single bonds rather than fluctuations in bond lengths. If such motions are modelled by isotropic temperature factors, the fluctuations in bond lengths may be overestimated. Six parameters are required to describe an anisotropic thermal ellipsoid and such refinement is only possible with very small proteins at high resolution. For example, with avian pancreatic polypeptide [194] (36 amino acids) the resulting R value at 0.98 Å resolution was 0.156. The anisotropic thermal ellipsoids indicated directions of motion of the polypeptide chain and showed evidence for concerted vibrations of certain groups (Fig. 11). Stuart and Phillips [192] have described a method for refinement of anisotropic temperature factors that requires fewer parameters. The model takes into account rotations about single bonds and the successive librations that will occur along a side chain.

Analysis of isotropic temperature factors derived from the refinement of lysozyme [197] at 2 Å resolution reveal an overall mean square displacement amplitude of 0.23 Å^2 (overall $B = 18$ Å^2). If contributions from experimental errors, crystal disorder or imperfections in the refinement are neglected, then the apparent thermal motion is found to be compatible with pairs of molecules that have strongest interactions in the crystal either moving as a rigid body with vibration about a common axis or vibrating in an intramolecular mode. Side chains exposed to the solvent have greater apparent motion than the remainder of the molecule and there is in general greater mobility in the residues that line the active site cleft. The good correlation between these displacements in crystals of human and hen lysozyme [198] suggests that these intramolecular motions are a property of the lysozyme molecule and that the effects of experimental error and crystal packing are not serious. If each atom is considered as a

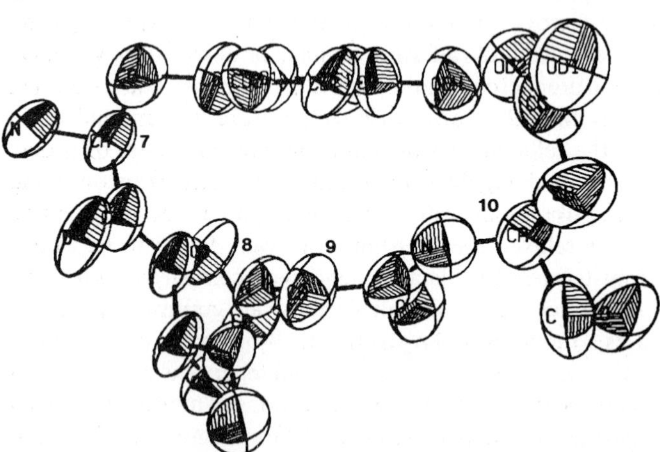

Figure 11. Atomic positions and ellipsoids of anisotropic thermal vibrations for residues 7–10 of avian pancreatic polypeptide (from I. Glover, Ph.D. Thesis, 1984, University of London, see [194]). The side chain of Tyr7 stacks above Gly9. There are indications of concerted thermal motion for these residues, with the largest vibrations in approximately the vertical direction of the page. There is least motion along the bond directions. Atoms at the end of side chains have greater anisotropic motion than main chain atoms.

harmonic oscillator and its classical energy equated with the thermodynamic energy, the crystallographic temperature factors may be transformed to yield a measure of the atomic frequencies. Rogers [206a] has shown that the mode of the distribution for lysozyme is 3×10^{11} second^{-1}. This frequency is in the range where the dielectric behaviour of water in response to a high-frequency field changes significantly [206b].

Observations on the role of mobility at the active site have come from the work of James et al. [199] in the analysis of *Streptomyces griseus* protease A at 1.8 Å resolution ($R = 0.13$; overall $B = 12.4$ Å^2). It was found that certain regions involved in the substrate binding (residues 167–172, 190–194 and 214–227) had some of the largest mean square amplitudes of vibration in the native enzyme ($\bar{U}^2 > 0.2$ Å^2; $B > 16$ Å^2 with a maximum of ≈ 40 Å^2 for Gly223). On formation of a complex with a tetrapeptide substrate, there were small conformational changes (of the order of 0.15 Å) in this region and a concomitant decrease (maximum, 9 Å^2) in the temperature factors. It has been proposed that these molecular vibrations may provide an important factor in enhancing enzyme acticity.

While these displacements in lysozyme and protease A are relatively small, much greater flexibility is observed in certain regions of trypsinogen. The 'activation domain' (residues N-terminus to Gly19; Gly142-Pro152; Gly184-Gly193 and Gly216-Asn223) is disordered with temperature factors greater than 200 Å^2. On formation of the complex between trypsinogen and the pancreatic trypsin inhibitor (a natural transition state analogue), these residues become ordered [163,200]. Analysis of the trypsinogen crystals at 173 and 103 K using synchrotron radiation [201] showed that the overall isotropic temperature factor in methanol-water mixture fell from 16.1 Å^2 at room temperature to 11.6 Å^2 at 173 K, with no further reduction at 103 K. The order of the activation domain of trypsinogen was not increased detectably, except in the N-terminal region. These results suggest in this instance that mobility is associated with static disorder. It is interesting that two serine oxygen atoms in residues outside the activation domain were distributed between two sites at 173 K but 'froze out' into only one of these sites at 103 K, suggesting that their conformation is governed by a potential function with two unequal minima.

The temperature dependence of $\bar{U}^2$ has been investigated in detail with myoglobin [202]. Diffraction studies at four temperatures between 220 and 300 K show the structure to be composed of a condensed core around the haem with displacements of the order of 0.04 Å^2, which are temperature sensitive, and a semiliquid region towards the outside with mean square displacements 0.04–0.25 Å^2, which are essentially temperature independent. The movements of the surface residues point to a possible pathway to the haem group. More detailed analysis at 80 K with crystals cooled by flash-freezing without the use of cryoprotectants showed a decrease in overall B from 14 Å^2 at 300 K to 5 Å^2 at 80 K [203]. Analysis of individual temperature factors showed that 46 out of the 153 residues in myoglobin had average B factors that extrapolated to zero at 0 K (i.e., arose from thermal vibrations alone). The temperature vibration of the remainder of the protein was consistent with the notion that conformational substrates could be frozen out at low temperatures. An additional 51 residues could be modelled with a linear dependence on temperature but with

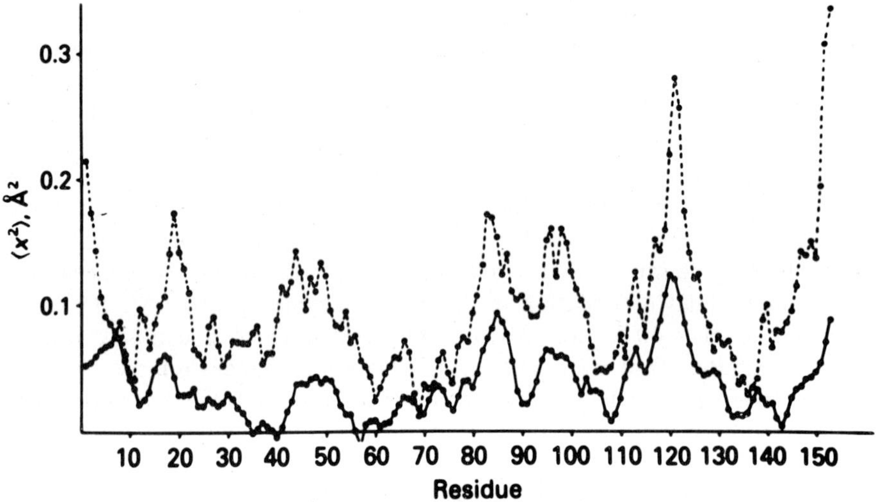

Figure 12. Average backbone mean square displacement values ($\langle x^2 \rangle$) for myoglobin vs. residue number at (●—●) 80 K and (●---●) 300 K. The average is taken over the N, C_α, and the carbonyl C atoms only, since the $\langle x^2 \rangle$ values of the carbonyl O atoms are usually higher. A value of 0.045 Å^2 has been subtracted from the individual observed values to account for lattice disorder. From Hartmann et al. [203].

$\bar{U}^2 \simeq 0.04$ Å at 0 K. A plot of the average mean square displacement of the main chain atoms against residue number at 300 and 80 K is given in Figure 12. If a Gaussian model is assumed for isotropic motion, the observed $\bar{U}^2$ values imply that some regions of the protein have a reasonable probability of undergoing displacements of 1–2 Å in amplitude. It is interesting that the static structure of myoglobin shows no pathway by which oxygen could have access to the haem. A series of transient movements as indicated by the temperature factor analysis could open a channel from the surface to the interior of myoglobin and allow access to the shielded ligand binding site.

The movements discussed above refer to spatial fluctuations that occur on a relatively fast time scale. In addition, proteins exhibit greater conformational flexibility that can be significantly altered by the addition of ligands or covalent modification. These include: the disorder-order phenomena that have been described for the trypsinogen-trypsin [163,200] system or the phosphorylase *b* to *a* transition [20–22]; the movement of two domains relative to one another, as for example with hexokinase on binding glucose [162], alcohol dehydrogenase on forming a ternary enzyme substrate complex [204] or citrate synthase on binding the coenzyme acetylcoenzyme A [205]; and those domain proteins such as the immunoglobulins [133] for which there appear to be a relatively wide range of motions between domains that allows a single molecule to recognise pairs of antigenic determinants with different spatial orientations. These movements of domains and their functional significance have been comprehensively reviewed by Bennet and Huber [133].

(c) Low temperature studies

The study of protein crystal structures at low temperatures (e.g., 223 K (−50°C) to 83 K (−190°C)) allows both greater precision, as the thermal motions of the atoms are damped, and an insight into the dynamics of the structure (section 3(b)). Moreover, since radiation damage is less (maybe reduced 10× by cooling from room temperature to −75°C), data may be collected with greater precision. However, the major incentive for low temperature studies comes from the expectation that enzyme reactions may be sufficiently slow that an X-ray study on an enzyme-substrate or enzyme-intermediate may be carried out. In the past, active sites have been probed using inhibitors or pseudo-substrates, because the time taken to collect a data set (several days) is considerably longer than the life time of the enzyme-substrate complex. The use of the bright synchrotron radiation source (section 3(d)) to reduce data collection times to a few hours together with the use of low temperature makes the study of these complexes feasible.

The temperature dependence of a chemical reaction may be expressed in terms of the Arrhenius equation

$$v = A \exp -(E_a/RT)$$

where v is the rate, A a constant, E_a the activation energy, R the gas constant and T the temperature. Many enzyme-catalysed reactions have rate-limiting steps corresponding to energies of activation between 12 and 20 kcal mol^{-1}. On cooling from 293 K (20°C) to 183 K (−90°C) a reaction with $E_a = 18$ kcal mol^{-1} would be slowed by a factor of 10^8. This may well be sufficient to allow a study of the relevant complex. Of course, the precise nature of the intermediate trapped will depend on the system under study, but it is anticipated that those intermediates that immediately precede the rate-limiting step will accumulate. Experiments on the kinetics of intermediate formation and breakdown in solution and in the crystal under the defined conditions of solvent, pH and temperature are necessary companion studies.

There may be problems in interpretation of results. The mechanisms by which enzymes reduce the energy of activation include conformational responses and dynamic fluctuations of the protein. If the crucial movements of the protein atoms are damped at low temperature then the resulting complex may not necessarily represent the constellation of groups in the productive complex. Critical reviews of cryo-enzymology in the crystalline state, together with much practical information, have been given by Makinen and Fink [170], Fink and Petsko [207] and Douzou and Petsko [218].

Many protein crystals can be cooled to 253 K (−20°C) without formation of ice crystals, but for studies at lower temperatures either flash freezing or the presence of a cryoprotectant is required.

Flash freezing was pioneered by Haas and Rossmann [208] in their early studies on lactate dehydrogenase at 198 K (−75°C) and was also used in the recent studies on myoglobin at 80 K (−193°C) [203]. In this method the crystals are plunged into either liquid N_2 or propane and subsequently stored and examined at liquid nitrogen temperatures. The object is to cool the water in the crystal sufficiently quickly that nucleation of ice crystals is by-passed and a vitreous state of water [209] obtained. Quantitative relationships between the degree of sub-cooling and the rates of ice nucleation and crystallisation have been reviewed [210,211]. Below a certain temperature the mean ice crystal size becomes inversely proportional to the cooling rate. These effects are explicable in terms of heterogeneous nucleation (seeding by impurities) and homogeneous nucleation (organisation of clusters of water molecules in the supercooled liquid). At very fast cooling rate, heterogeneous nucleation is inhibited and the liquid is cooled to a temperature where a high concentration of homogeneous nuclei can develop, leading to a high concentration of very small crystals perhaps 5–10 nm in size. For these processes to occur a cooling rate of the order of 10^6 K second^{-1} is required. This is difficult to achieve in practice. Nevertheless, rates of the order of 10^4 K second^{-1} can be achieved with methods adapted from low-temperature electron microscopy and appear to be sufficient for protein crystallographic purposes. Costello and Corless [212] have described a 'guillotine' arrangement that allows the specimen to be plunged into the cooling liquid under gravitational acceleration with speeds at least twice that achievable by hand. They report that propane (83 K) gave the highest cooling rate, 9.8×10^3 K second^{-1} followed by Freon 13 (88 K), 7.8×10^3 K second^{-1}. Liquid N_2 (77 K) was less effective (cooling rate, 1.6×10^3 K second^{-1}). Once cooled, it is important to prevent the specimen warming because this will result in ice formation. Flash freezing has the advantage that no changes in the crystal mother liquor are required.

The alternative method is slower cooling in the presence of a cryoprotectant. The physical and chemical properties of the most commonly used cryosolvents have been documented by Douzou and co-workers [39,213,218]. Cryosolvent mixtures for protein crystals have been described by Petsko [214]. The choice of cryoprotectant will depend on the crystal, the conditions that it will tolerate and the desired low temperature. For subsequent substrate binding studies it is essential that the solvent is fluid at the low temperature. Mixed aqueous-organic solvents are generally favoured. The organic component is usually methanol, ethanol, ethyleneglycol or dimethylsulphoxide. Methanol-water mixtures have relatively low increase in viscosity with temperature ($\eta = 10$ at -30°C, $\eta = 48$ at -60°C for a 70% methanol-water mixture, compared with $\eta = 18$ at 10°C, $\eta = 125$ at -40°C for a 50% ethylene-glycol mixture, where the units of viscosity (η) are centipoise). A 70% methanol-water mixture freezes at 188 K (-85°C) but may permit even lower temperatures to be reached. Trypsinogen crystals [201] in 70% methanol continued to diffract well even at 103 K (-170°C). For glycogen phosphorylase *b* all the organic solvents mentioned above cracked the crystals. Glycerol, which is not usually chosen because of its high viscosity, was found to be suitable. A 70% glycerol-water mixture enabled measurements to be made at 123 K (-150°C), even though the solvent was almost solid at this temperature. The dielectric constant of a solution increases with decreasing temperature. Since the stabilising forces for a protein and the crystal lattice are dependent upon dielectric constant, it is desirable to compensate changes in dielectric constant with temperature with an increase in the organic component of the solvent. The apparent pH of an organic-aqueous solution tends to be higher than that of the corresponding aqueous solution, and again it is important to compensate for this effect. Douzou and Petsko [218] give an excellent summary of these properties.

In low temperature studies, it is preferable to mount the crystal in a flow cell [215]. This permits the gradual change of cryoprotectant concentration as the temperature is lowered and substrate is introduced into the crystal. A quartz capillary is attached to a brass support by epoxy cement and a tight-fitting polyethylene tube sealed to the bottom of the capillary. The capillary is filled with protein mother liquor solution. The crystal is introduced to the top and allowed to settle either on to a bed of pipe cleaner fibres placed on top of the polyethylene tube as support or on to a support made by previously introducing a constriction into the quartz capillary. The crystal may be secured further by additional fibres. The inlet polyethylene tube is sealed to the top of the capillary. The inlet and outlet tubes are then connected to the reservoir and sink, respectively (Fig. 13).

Several low-temperature devices have been described for protein crystallography. In the device of Marsh and Petsko [216] a jet of cold dry N_2 gas is directed at the crystal. This simple and inexpensive device is good for temperatures around $-4°C$. For diffractometers there are a number of devices manufactured commercially [217], which are suitable for very low temperatures. For the oscillation camera, the device of Bartunik and Schubert [233] is recommended. The flow cell is surrounded by a double-walled cylindrical chamber composed of mylar (8 μm thick). A diffuse stream of N_2 gas passes through the inner chamber and cools the sample. Icing is prevented

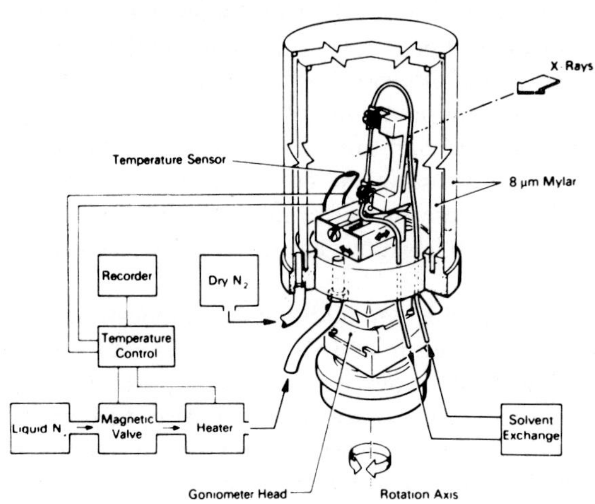

Figure 13. A schematic drawing of the low temperature device of Bartunik and Schubert, with the flow cell arrangement. The flow cell is mounted on a standard goniometer head. Cold N_2 gas passes through the inner chamber, its temperature is controlled by a sensor and heating coil. Dry N_2 circulates through the outer chamber in order to prevent the windows from icing. The walls of the inner and outer parts consist of mylar foils of 8 μm thickness. From Bartunik and Schubert [233].

by circulation of dry N_2 through the outer chamber and a stream of warm air blown on to the outer surface. The device is suitable for temperatures between 323 K (50°C) and 100 K (−173°C) and is finding widespread use on oscillation cameras at synchrotron radiation sources (Fig. 13).

In binding studies with substrate, it is important to know the diffusion times for substrate into the crystal. These times ideally should be several orders of magnitude shorter than the turnover of the reaction. Experimental values of diffusion times taken from the literature and laboratory experience are given in Table 2. It is seen there in a wide range of times, but in general small molecules diffuse into crystals with short times ($\simeq$15 minutes). Diffusion into the crystal depends on solute size and drag through the pores. In a study on the time course of diffusion of a series of bromine-containing solutes into β-lactoglobulin crystals, Bishop and Richards [219] estimated that the effective pore radius for potassium bromide and lithium bromouridylate was between 8 and 13 Å, depending on the hydrodynamic model assumed. They were able to conclude that no special properties need be attributed to the major part of the crystal liquid. The pores in a crystal have a complicated shape and are not uniform over the length of the unit cell. Their precise dimensions can be calculated once the protein structure is known. Bishop and Richards derive equations for the effect of solute size and drag on the diffusion coefficient in which $D_{crystal}/D_{solution}$ = function (a/r) where a is the radius of the diffusing solute and r is the radius of the pores. Fitting of experimental data to the theoretical models showed that the diffusion coefficient was diminished relative to that in free solution, by 0.5 for KBr, 0.16 for lithium bromouridylate (neither of which bound to the protein) and by 0.02 for bromopropanol, which bound fairly strongly ($K_D = 1.5$ mM).

From equations given by Crank [220], Fink and Petsko [207] show that for a one-dimensional sheet of thickness $2l$, substrate will penetrate to the centre of the lattice in a time t where $t = l^2/D'$ and D' is the corrected diffusion coefficient in the crystal. The diffusion coefficient in solution for many compounds is of the order 1×10^{-5} cm^2 second^{-1}. Thus, we see that for a crystal of 0.4 mm thickness, diffusion times should be of the order of 80 seconds to 33 minutes using the correction factors 0.5–0.02 obtained by Bishop and Richards [219]. These times are in reasonable agreement with some of the times given in Table 2. Note the last two entries (7 and 8) also involve a catalytic reaction.

However, in low-temperature work there is an additional complication. The diffusion coefficient (D) is a function of temperature and viscosity (η). $D = RT/6\pi aN\eta$ and $\eta = A \exp(\Delta E_{visc}/RT)$. On cooling from room temperature to −60°C, 70% methanol shows an approximate 5-fold increase in viscosity [213]. Also, Bishop and Richards [219] have shown that diffusion in the crystal is not simply related to viscosity but also to the solvent. An increase of 4-fold in viscosity (arising from increasing sucrose concentration from 0 to 35%) resulted in approximately a 10-fold decrease in diffusion coefficients. This effect can be partly explained by the sucrose molecules blocking the pores and thereby causing a decrease in pore size. These considerations suggest that a rough estimate of diffusion times can be obtained, but because of the uncertainties, especially with regard to models, it is much better to

TABLE 2
Some diffusion times for molecules into protein crystal

Protein	Conditions of experiment	Method of detection	Diffusion time	Reference
1. Ribonuclease S	Change from 75 to 80% ammonium sulphate at room temperature	Monitor of changes in X-ray intensities	$t_{1/2} = 90$ seconds	215
2. Ribonuclease S	Replacement of bound uridine phosphate by 3 mM iodo-uridine phosphate at room temperature	Monitor of changes in X-ray intensities	$t_{1/2} = 11$ hours	215
3. β-Lactoglobulin	Diffusion of 0.02 M KBr	X-ray fluorescence	$t_{1/2} = 72$ seconds	219
	Diffusion of 0.02 M lithium bromouridylate	X-ray fluorescence	$t_{1/2} = 7$ minutes	219
	Diffusion of 5.7 mM bromoproponal	X-ray fluorescence	$t_{1/2} = 2.25$ hours	219
4. Glycogen phosphorylase *b*	Diffusion of 40% glycerol-water into crystals previously in 10 mM magnesium acetate at room temperature	Changes in birefringence of the crystal	$t_{1/2} \cong 4$ minutes	J. Hajdu, unpublished observations
5. Glycogen phosphorylase *b*	Diffusion of 100 mM glucose 1-phosphate or 100 mM maltotriose into crystals approximately 0.4 $\times$ 0.4 $\times$ 1 mm at room temperature	Difference Fourier synthesis, after 10 min soak and subsequent data collection	Apparent saturation reached in less than 10 minutes	J. Hajdu, L. N. Johnson, unpublished observations
6. Elastase	Diffusion of methanol-water mixture at different concentrations and temperatures	Not stated	Equilibration after 10–15 minutes	222
7. Elastase	Diffusion of 3 mM *N*-carbobenzoxy-L-alanyl-*p*-nitrophenol into crystals and formation of the acyl intermediate at $-55°C$	Monitor of changes in X-ray intensities	$t_{1/2} \cong 3.5$ hours (saturation after ≈ 24 hours)	222
8. *Streptomyces griseus* protease A	Diffusion of 5 mM pentapeptide substrate and hydrolysis of the substrate by the enzyme	Change in birefringence in the crystal	Total change in 2–6 hours	199

obtain experimental values. These can be obtained by monitoring the changes in intensities of certain sensitive reflections as substrates are diffused into the crystals.

Applications of low temperature work in structural studies have been described in section 3(b). Application to enzyme action is best exemplified by the pioneering work of Fink and Ahmed [221] and Alber et al. [222] on elastase. *N*-Carbobenzoxy-L-alanyl-*p*-nitrophenol ester was selected for study at −55°C in a 70% methanol-water mixture. Kinetic studies in the presence of cryoprotectant enabled conditions for formation and stabilisation of the acyl-enzyme intermediate to be established. By monitoring changes in intensity of certain reflections as substrate flowed past the crystal at −55°C, it was possible to show that the rate of formation of the acyl-enzyme was comparable to that obtained by monitoring *p*-nitrophenol release spectroscopically. The difference electron density map at 3.5 Å resolution showed a peak consistent with the formation of an acyl-enzyme intermediate, but a detailed mechanistic interpretation requires higher resolution data. When the crystal was warmed to −10°C and the data recollected, the peak in the difference synthesis disappeared, indicating that deacylation had occurred, consistent with the predictions from kinetic studies.

More extensive and detailed studies have been carried out on ribonuclease A by Gilbert and Petsko [266] and these are described in references 218 and 223. Ribonuclease forms a favourable system for such work: it has been widely studied biochemically; the major species on the reaction pathway have been characterised and can be purchased; the protein is crystallised from a cryoprotectant solvent; the crystals diffract well to 1.5 Å resolution and the structure has already been determined in detail by other workers. Solution studies by Fink et al. [267] allowed the determination of conditions of solvent, pH and temperature necessary to stabilise the enzyme-substrate complex for the time required to collect data to high resolution (7–10 days). The structure of the native enzyme was solved at 1.5 Å resolution at −10 and −32°C and a series of binding studies carried out. These included: an enzyme-substrate analogue (a deoxydinucleotide d-C_pA) complex at −10°C; an enzyme-cyclic phosphate intermediate (cytidine 2′,3′-cyclic monophosphate) complex at −70°C; an enzyme transition state complex (uridine vanadate) at −10°C; and an enzyme-product complex at −10°C. The results were consistent with the mechanism for ribonuclease originally suggested by Deavin et al. [268] in 1966. In this mechanism His12 acts as a base and His119 as a general acid to facilitate an in-line attack at the phosphorus atom that leads to the hydrolysis of the phosphate ester bond. Interestingly, Lys41 (whose chemical modification results in inactivation) is found to be disordered in the enzyme-substrate, enzyme-intermediate and enzyme-product complexes. Only in the enzyme-transition state analogue uridine-vanadate complex was the lysine found to be ordered and to make strong interactions with one of the equatorial oxygens of the pentaco-ordinated vanadate ion. (A slightly different contact is observed in the neutron studies [253], section 3(e).) The crucial experiment which required low temperature was that for the cyclic phosphate intermediate. This work provides remarkable structural information on the different species in the reaction and indicates how selection of a single sub-state of a residue, which in the native enzyme has conformational heterogeneity, can lead to stabilisation of a putative transition state on the reaction pathway.

(d) Synchrotron radiation

The encouraging results of the first experimental tests on the potential of synchrotron radiation for protein crystallography (carried out at SPEAR, Stanford) were published in 1976 [224]. Protein crystallography stations are now available at LURE, Orsay [225,226], EMBL Laboratory, Hamburg [227,228], SRS, Daresbury [229,230], and soon will be available at CHESS, Cornell, Photon Factory, Tsukuba, and VEPP-3, Novosibirsk. The synchrotron radiation provides a brilliant source of X-rays with small divergence. Its exploitation results in a reduction of exposure times (40–100-fold), less radiation damage (approximately 5-fold), better resolution and better signal to noise than can be achieved with conventional sources. In addition, the optimum wavelength can be chosen for a particular problem, a feature which is of considerable significance with regard to anomalous scattering.

In a synchrotron storage ring source, electrons (or positrons at LURE) are accelerated in a linear accelerator and injected into a booster synchrotron, where they are constrained to follow a circular path by means of magnetic fields. They are then injected into the storage ring which can be run at energies of the order of 2 GeV and circulating electron currents of several hundreds of milliamps. The electrons emit electromagnetic radiation from each point in the curved orbit in a tangential direction and in a narrow cone whose angle is inversely proportional to the energy. The radiation has wavelengths which are continuous for the range 0.2 Å to 1 mm. The 'critical wavelength' is that wavelength at which half the radiation is radiated above and half below and is given by $\lambda c = 18.6/BE^2$ Å where E is the machine energy in GeV and B is the field of the bending magnet in tesla. In the SRS, Daresbury, $B = 1.2$ T (radius, 5.55 m) and $E = 2$ GeV, hence, $\lambda c = 3.9$ Å. The maximum flux at short wavelengths has been increased at SRS by Wiggler magnets. These consist of three alternating polarity high-field bending magnets in series. The bending radius is reduced locally and λc is altered to give a maximum flux around 1 Å. The electrons circulate around the ring in discrete bunches with a time separation of 2.0 nseconds and bunch length of 0.2 nsecond (at SRS). During a cycle the stored particles are lost, and so re-injection must occur from time to time. Therefore, the protein crystallographer must fit his experiments to approximately 8–12 hour cycles. Most synchrotrons are dedicated partly to high-energy physics and partly to the production of radiation, and hence the crystallographer must wait his turn. At SRS, Daresbury, the synchrotron is dedicated solely to the production of radiation.

The source size varies for the different synchrotrons. At SRS it is 9.9 mm (horizontal) and 0.3 mm (vertical). Since the sample is about 20 m away, careful attention to monochromatisation and focussing optics is required so that the source size matches the crystal size. A curved crystal monochromator which is based on a triangular shaped crystal plate and uses germanium (111) reflection was first described by Lemonier et al. [225] at LURE. Similar focussing monochromators are now in use

at EMBL [231] and SRS [229]. They are usually associated with a single focussing mirror to complement the focussing of the monochromator [229,230]. Germanium has an absorption edge of 1.12 Å, which means heating effects are serious at low wavelengths. Instead, a silicon crystal ((220) reflection) is used [230].

By far the greatest use of synchrotron radiation in protein crystallography has been to obtain data from crystals that either are very small, are radiation sensitive or have large units cells [77,228,230]. The brilliant source allows short data collection times. For example, 2.7 Å data for crystals of phosphorylase *b* (some 120 000 measurements; 23 000 unique reflections) have been obtained in less than 2 hours. Such rapid data collection was advantageous in measurements of activity in the crystal [232].

Reduction in radiation damage by factors of 2–6 have been found with synchrotron radiation when compared with conventional sources. Part of this may be due to the superior optics at synchrotron sources, which results in a small monochromatic beam well matched to the specimen crystal and with very small divergence. Thus, the signal to noise ratio is increased and more data obtained for a given dose. In addition, it may be that a high dose for a short time is generally less damaging than a low dose for a long time. The primary effect of absorption of an X-ray quantum is the production of heat and/or of free radicals. The most abundant sources of the latter are the radiolytic products of water, which can diffuse readily through the aqueous channels in the crystal, inflicting damage on molecules removed from the site of primary absorption. It appears that the deleterious effects of these chain reactions may be reduced during short exposures with intense synchrotron radiation. Further prolongation of the crystal lifetime may be achieved by cooling.

As an example of large unit cell size, data have been recorded from crystals of cowpea mosaic virus to 4.3 Å resolution at LURE [234]. The hexagonal unit cell has dimensions $a = 451$ Å, $c = 1038$ Å. The oscillation range was 0.4°.

Synchrotron radiation also permits time-resolved studies, as for example in the fibre diffraction studies on muscle [235]. The feasibility of this work with protein crystals was demonstrated by the studies of Bartunik [236], who followed the time course of reflection intensities of carbonmonoxy-myoglobin on a time scale of 500 μseconds before and after photodissociation of the ligand by a laser pulse. Thus, synchrotron radiation allows much greater scope in the study of biological molecules.

The ability to select a wavelength for a particular problem has proved advantageous. For example, the anomalous scattering of iron at $\lambda = 1.739$ Å has been used to locate the iron atoms of the four haems in cytochrome c_4 [237]. Crystals of ferritin contain an iron core of 4500 iron atoms, and in this instance absorption and fluorescence of the X-rays at $\lambda = 1.54$ Å led to problems in data collection. A shift of the wavelength to 1.76 Å (the long wavelength side of the absorption edge of iron, $\lambda_K = 1.743$ Å) resulted in approximately 6-fold reduction of absorption and a greatly improved crystal lifetime [238]. The use of the high flux at 1.0 Å with the Wiggler magnets at SRS has proved advantageous with crystals of phosphorylase *b* (Stuart, D.I., McLaughlin, P.J., Hajdu, J. and Acharya, A., unpublished observations). Absorption is decreased 3–4-fold compared with that at $\lambda = 1.54$ Å, radiation damage is less, and high-resolution data (2 Å) can be collected with a flat plate cassette for a crystal with unit cell edges greater than 100 Å (Fig. 5). However, by far the most interesting aspect of the ability to tune the wavelength is the potential application of the multiwavelength method to determine phases based on anomalous scattering methods [230,239–241]. By recording intensities at (at least three) wavelengths close to the absorption edge (Table 1), where the f' and f'' components vary considerably, phase information can be computed. For example, caesium close to its L_{III} absorption edge ($\lambda = 2.474$) exhibits changes in f' from -26.7 to -13.9 and f'' from 4.0 to 16.1 electrons [243a]. The multiwavelength method was used successfully in a neutron diffraction study of cadmium-myoglobin [242]. The technology has now been developed at synchrotron radiation sources [243a,b]. Recently, the structure of parvalbumin, in which the Ca atom was replaced by Tb, has been solved by this method [269].

The broad (0.15 to 4 Å) continuous spread of wavelengths in the radiation from Wiggler beam lines at synchrotron radiation sources allows optimal conditions for the recording of Laue photographs. In this, the oldest method for recording X-ray diffraction patterns, the crystal is kept stationary. The condition for each Bragg reflection is satisfied by the different wavelength (λ) ($\theta = \sin^{-1}(\lambda/2d)$). In the past the method has had considerable success in the examination of crystal defects and in the detection of diffuse scattering which arises from global thermal vibrations of the molecules in the crystal lattice. The Laue method has not been used for data collection involving intensity measurements, mainly because of two limitations.

(1) Many spots recorded on a Laue photograph are composed of multicomponents. This can easily be seen from Bragg's Law. If the Bragg angle θ is satisfied for a particular interplanar spacing d at a wavelength λ, then the same Bragg angle will also be produced by spacings $d/2$ and $\lambda/2$, $d/3$ and $\lambda/3$, etc. Thus, all these spots will superimpose. However, with improved automised methods of spot prediction developed from the routines for the processing of oscillation photographs, it has been shown [270] that in a Laue photograph of a pea lectin crystal approximately 80% of the spots are single component and are resolvable. Thus, a substantial amount of the complete three-dimensional data set can be recorded from one Laue photograph.

(2) The second limitation is rather more serious. The intensity of the incident beam, the scattering power of the crystal and the efficiency of the detector are all wavelength dependent. Thus, there is a very difficult problem in scaling the intensities of the individual spots generated at different wavelengths to a normalised value. If a native data set is already available, then this might serve as a reference. The Laue method is likely to have great power, however, when only differences in intensities are required. This was the case in a recent trial experiment with glycogen phosphorylase *b*. A blocked substrate was diffused into the crystal, a Laue photograph recorded, the substrate unblocked by photolysis and the reaction followed by subsequent Laue photographs (Hajdu, J., Stuart, D.I., McLaughlin, P.J., Acharya, R., Johnson, L.N. and Trentham, D., unpublished results). Each Laue photograph took only 8 seconds to record. Although the results of this experiment are not yet known, they indicate the power of the Synchrotron and the Laue method to record data on a time scale that is relevant to enzymology.

(e) Neutron diffraction

The positions of hydrogen/deuterium atoms are of interest to biochemists both with regard to the state of ionisation of functional groups and identification of exchangeable atoms. The scattering of X-rays is proportional to atomic number. Hence, hydrogen scatters weakly. Only at very high resolution can hydrogen atoms be distinguished in X-ray protein electron density maps, such as the beautiful maps for insulin at 1.2 Å resolution [244] (Fig. 4). Neutrons are scattered by a different process to X-rays. In non-magnetic materials, neutrons are scattered by the nucleus, and the scattering is sensitive to the nuclear radius, the nuclear spin and the existence of more than one isotope. Since the nuclear radius is small compared with the wavelength of neutrons used, there is very little fall off in scattering with Bragg angle. A few atomic nuclei scatter with an additional phase change of 180° to give a negative scattering amplitude. Hydrogen is of this type and appears as negative peaks in neutron Fourier syntheses. Deuterium, on the other hand, scatters with a positive amplitude, and so hydrogen and deuterium can be distinguished. The neutron scattering amplitudes in units of 10^{-12} cm for hydrogen, deuterium and carbon are -0.378, 0.65 and 0.661, respectively. Therefore, neutron diffraction provides a complementary technique to X-ray diffraction, capable of defining hydrogen and deuterium atoms in addition to the heavier atoms of the protein. Neutron diffraction centres for protein crystallography are in operation at Institut Laue-Langevin, Grenoble, Brookhaven National Laboratory, New York, National Bureau of Standards, Maryland, and in the future at Neutron Spallation Source, Rutherford Laboratory, Harwell. Recent advances in methods and applications of neutron protein crystallography have been reviewed well by Kossiakoff [245].

Neutron diffraction studies are not without their drawbacks and limitations. Hydrogen atoms give rise to significant incoherent scattering which produces a high background and limits the precision of measurements. This is especially serious for protein molecules, where about half the atoms are hydrogen. The problem can be overcome by exchange of most of the hydrogens for deuterium by soaking the crystals in D_2O.

The flux generated by neutron sources is generally 4 to 5 orders of magnitude less than a standard X-ray source. This means that large crystals and long exposures are required. As a result of improved instrumentation in detector systems and in data handling, the size of a crystal required for a neutron experiment has been reduced from about 20 to 1.5 mm^3 [246]. Nevertheless, neutron diffraction studies are limited to those proteins for which very large crystals can be obtained. With X-rays, a crystal of this size would lead to severe absorption. Fortunately, the absorption of neutrons by most atoms is very low.

Thermal neutrons have a low energy (0.025 eV) compared to X-rays (10 000 eV) and they are a non-ionising radiation. This means that protein crystals are not damaged by neutrons as they are by X-rays or electrons. Hence, a crystal can remain in the beam for extremely long periods (even years) without suffering radiation damage. This offers a partial compensation for the low flux of neutron sources.

The phases required for a neutron Fourier synthesis can be estimated from the calculation of structure factors using the positional co-ordinates of the atoms obtained from an X-ray electron density map and the corresponding neutron scattering amplitudes. Such phases will be approximate because they do not include hydrogen atoms, but the early work on myoglobin showed that they were sufficiently accurate as a starting point [247]. If an ab initio structure determination is envisaged, a slightly different strategy for preparation of heavy atom derivatives and exploitation of anomalous scattering is required from that used for X-ray diffraction [25]. So far, all neutron studies have exploited existing X-ray structures. The methods used to refine neutron structures are essentially the same as for X-ray structures. The combination of X-ray and neutron information in refinement has been described [248,249]. The range of convergence of a neutron refinement is relatively small ($\approx$0.3 Å at 2.2 Å resolution) because of the proximity of the hydrogen atoms with their negative peaks to their parent atoms that have positive peaks.

Detailed neutron diffraction studies have been carried out with myoglobin, trypsin, ribonuclease, lysozyme and crambin. In the case of trypsin [246,250], the analysis of a deuterated complex of the enzyme with monoisopropyl fluorophosphate showed that in the tetrahedral analogue of a transition state complex, His57 was deuterated while Asp102 was ionised, a crucial piece of evidence for the mechanism of action of serine proteinases (Fig. 14). For lysozyme, Mason et al. [251] have shown that in crystals of triclinic lysozyme Glu35 is protonated, in agreement with the proposals for the mechanism originally made by Phillips [252] in 1967. In ribonuclease, the studies have proved informative but their significance in terms of mechanism is not yet clear. Wlodawer et al. [253] have studied a complex of ribonuclease and uridine vanadate at 2 Å resolution. The hydrogen-deuterium exchange in the crystal was carried out over a period of 3 months. The results showed that both His12 and His119 were protonated, in agreement with the increase in pK of these groups in the presence of

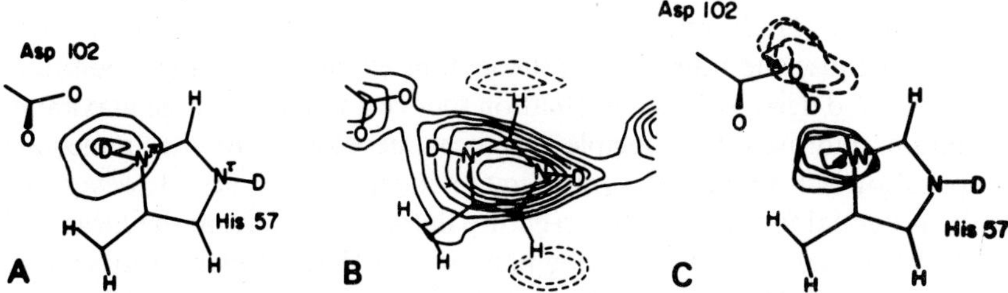

Figure 14. Neutron diffraction studies on trypsin. (A) A difference map $|(F_{obs} - F_{calc}) \exp i\, \alpha_{calc}|$ calculated with only the deuterium between His57 and Asp102 left out of the phase calculation. The difference peak shows the deuterium bound to the imidazole nitrogen. (B) A Fourier synthesis calculated with terms $(2F_{obs} - F_{calc}) \exp i\, \alpha_{calc}$. In this map both deuteriums were omitted from the phase calculation. It is clear that both deuteriums are located on the imidazole. (C) A difference map in which the deuterium was placed by stereochemistry on the Asp102. The difference density peak clearly shows the preferred location of the deuterium on the imidazole of His57. From Kossiakoff and Spencer [246].

inhibitors observed from NMR and that the ε-amino group of Lys41 moved 0.6 Å from its position in the native enzyme and became more ordered (*B* reduced from 20 to 8 Å^2), in agreement with the studies of Petsko and colleagues [218,223]. However, contrary to expectations based on the in-line mechanism of action, His12 hydrogen bonded to an equatorial oxygen of the vanadate ion, and it was Lys41 that hydrogen bonded to the apical oxygen. Since it is unlikely that Lys41 could assume the role of a general-acid base, the authors suggest that proton transfers may not follow the simplest route in the ribonuclease mechanism.

The hydrogen-exchange studies on trypsin crystals, carried out over a period of a year at pH 7, have proved highly informative in terms of structural aspects of the hydrogen-deuterium exchange process [254]. Hydrogen exchange was found to be highly correlated with the extent of hydrogen bonding of the site and adjacent sites and showed little correlation with the hydrophobicity of the side chains or the temperature factors of the site. In β-sheet structures, amide peptide groups were protected from exchange, but in the two short segments of α-helices of trypsin exchange occurred readily. The results have been interpreted in terms of a localised conformational mobility limited mostly to the breaking of a small number of hydrogen bonds.

(f) Maximum entropy and direct methods in protein crystallography

The solution of a protein crystal structure can still be a lengthy process, even when crystals are available, because of the phase problem. In contrast, small molecule (< 100 atoms) structures can be solved routinely by direct methods. In the early fifties it was shown that certain mathematical relationships exist between the phases and the amplitudes of the structure factors if it is assumed that the electron density is positive and atoms are resolved [255]. These mathematical methods have been developed [256,257] so that it is possible to solve a small molecule structure directly from the intensity data [258]. For example, the crystal structure of gramicidin S [259] (a cyclic polypeptide of 10 amino acids, 92 atoms) has been solved using the computer programme MULTAN. Traditional direct methods are not applicable to protein structures, partly because the diffraction data seldom extend to atomic resolution. Recently, a new method derived from information theory and based on the maximum entropy (minimum information) principle has been developed. In the immediate future the application will require an approximate starting phase set. However, the method has the potential for an ab initio structure determination from the measured intensities and a very small sub-set of starting phases, once the formidable problems in providing numerical methods for the solution of the fundamental equations have been solved.

In principle, it is possible to compute a very large number of electron density maps (Eqn. 2, section 2(a)) based on all possible combinations of trial values for the phases. How can the correct solution be selected? Maximum entropy (minimum information) provides a method for introducing constraints which reflect prior knowledge (e.g., the electron density must be positive) and provides a criterion for selection of the best map.

Jaynes [260], in his classic paper, showed how information theory could provide a constructive criterion for setting up a probability distribution on the basis of partial knowledge. The maximum entropy principle leads to a probability distribution which is maximally non-committal with regard to missing information. No possibility is ignored. A positive weight is assigned to every situation that is not absolutely excluded by the given information.

If a quantity x (which we wish to know) is capable of assuming discrete values x_i $(i = 1, 2, \dots n)$ each with a corresponding probability p_i, then, from information theory, it is possible to show that the quantity which is positive, which increases with increasing uncertainty and which is additive for independent sources of uncertainty is

$$S(p_1 \dots p_n) = -K \sum_i p_i \ln p_i$$

where K is a positive constant.

Since this expression is similar to that found for entropy in statistical mechanics, it is called the entropy of the probability distribution p_i. Jaynes [260] shows that the thermodynamic entropy is identical to the information theory entropy except for the presence of the Boltzman constant in the former, made necessary by our arbitrary temperature scale.

Information theory has been developed for X-ray structure determination by Wilkins et al. [261,262]. In the X-ray crystallographic case, the quantity x which we wish to know is ρ, the electron density. This is subject to the constraint that the Fourier transform of the electron density must correspond to the structure factors, the amplitudes of which are directly measureable. In their formulation, Wilkins et al. [262] divide the unit cell into N pixels corresponding to units of volume commensurate with the resolution of the data. n identical discrete units of structure are distributed at random over the N pixels with a prior probability distribution q. These n units correspond to elements of resolution in electron density space, rather than to actual physical entities such as atoms on electrons (although these are not excluded). In the initial stages of analysis, q may be set equal to 1, representing complete ignorance of the structure; later it may be set to reflect the packing of individual molecules in the crystal lattice, the molecular envelope, elements of secondary structure or even greater detail of the structure, depending on the resolution and the state of analysis. These considerations lead to an information function of the form

$$I(p) = \text{constant } (-S(p)) = n \sum_{j=1}^{N} p_j \ln [p_j/q_j]$$

where $p_j = \langle n_j \rangle / n = \langle \rho_i \rangle$ is the average value of the electron density in the jth pixel. The assumptions underlying the derivation ensure that only those structures are considered where the total number of electrons exactly equals the total number of electrons in the unit cell and the electron density is everywhere positive and smoothly varying. The function $I(p)$ may be minimised (i.e., entropy maximised) with respect to p, subject to available further information such as partial knowledge of the phases (constraint 1) or knowledge of the intensities alone (constraint 2). Numerical methods for solving the former (constraint 1) have been developed and coded. Those for constraint 2 are more formidable and work is still in progress.

In a recent paper Bricogne [263] has shown a complete equivalence between the maximum entropy formulation and traditional direct methods. He has developed a theory capable of unifying and simplifying the whole of direct methods while avoiding the limitations and approximations inherent in the traditional approaches. Thus, maximum entropy should become general for both small and large structures.

The maximum entropy method also offers the possibility of 'super-resolution', i.e., better resolution than might be anticipated than the simple analogy with optical systems (section 2(a)). Series termination effects in conventional Fourier syntheses lead to negative regions around the peaks. The maximum entropy principle ensures that the density is everywhere positive and gives much sharper peaks in which series termination effects have been suppressed [261,263].

The maximum entropy principle has been applied already, for example, in the structure determination of the Pf1 filamentous virus at 4 Å resolution [264]. Combination of the data from one heavy atom derivative and the maximum entropy method led to an interpretable map which showed the helical subunits of the virion.

The maximum entropy method achieves a remarkable universality and unification based on 'common sense reduced to calculation'. It has been applied to information theory, statistical mechanics, image processing in radio astronomy, and now to X-ray crystallography. The prospects for a computational solution to the phase problem in protein crystallography appear promising and developments in the field are awaited eagerly.

4. Acknowledgements

I am most grateful to the following for their helpful comments and assistance in the preparation of this review: P.J. Artymiuk, G.R. Moore, D.C. Phillips, N.K. Rogers, D.I. Stuart, S.W. Wilkins and G. Williams. I express special thanks to the phosphorylase team, R. Acharya, J. Hajdu, P.J. McLaughlin and D.I. Stuart, for their help in the preparation of figures and permission to use unpublished material.

References

1 Bernal, J.D. and Crowfoot, D.C. (1934) Nature 133, 794.
2 Green, D.W., Ingram, V.M. and Perutz, M.F. (1954) Proc. Roy. Soc. A225, 287–307.
3 Blow, D.M. and Crick, F.H.C. (1959) Acta Cryst. 12, 794–802.
4 North, A.C.T. (1965) Acta Cryst. 18, 212–216.
5 Matthews, B.W. (1966) Acta Cryst. 20, 82–86.
6 Dickerson, R.E. (1979) Chronological Summary Table of Protein Structure Determinations. Private circulation.
7 Levitt, M. and Chothia, C. (1976) Nature 261, 552–557.

8 Richardson, J.S. (1981) Adv. Prot. Chem. 34, 187–339.
9 Chou, P.Y. and Fasmann, G.D. (1978) Annu. Rev. Biochem. 47, 251–276.
10 Sternberg, M.J.E. and Thornton, J.M. (1978) Nature 271, 15–20.
11 Phillips, D.C., Sternberg, M.J.E. and Sutton, B.J. (1983) in Evolution from Molecules to Men (Bendall, D.C., ed.), pp. 145–173, Cambridge University Press, Cambridge.
12a Blake, C.C.F. (1983) Trends Biochem. Sci. 8, 11–13.
12b Blake, C.C.F. and Johnson, L.N. (1984) Trends Biochem. Sci. 9, 147–151.
13 Rossmann, M.G., Moras, D. and Olsen, K.W. (1974) Nature 250, 194–199.
14 Perutz, M.F. (1970) Nature 228, 726–739.
15 Dickerson, R.E. and Greiss, I. (193) Haemoglobin, Benjamin/Cummings, Menlo Park, CA.
16 Shaanan, B. (1983) J. Mol. Biol. 171, 31–59.
17 Fermi, G., Perutz, M.F. and Shaanan, B. (1984) J. Mol. Biol. 175, 159–174.
18 Baldwin, J. and Chothia, C. (1974) J. Mol. Biol. 129, 175–220.
19 Perutz, M.F. (1979) Annu. Rev. Biochem. 48, 327–386.
20 Sprang, S. and Fleterick, R.J. (1979) J. Mol. Biol. 131, 523–551.
21 Weber, I.T., Johnson, L.N., Wilson, K.S., Yeates, D.G.R., Wild, D.L. and Jenkins, J.A. (1978) Nature 274, 433–437.
22 Fletterick, R.J. and Sprang, S.R. (1982) Acc. Chem. Res. 15, 361–369.
23a Honzatko, R.B., Crawford, J.L., Monaco, H.L., Ladner, J.E., Edwards, B.F.P., Evans, D.R., Warren, S.G., Wiley, D.C., Ladner, R.C. and Lipscomb, W.N. (1982) J. Mol. Biol. 160, 219–263.
23b Lipscomb, W.N. (1983) Annu. Rev. Biochem. 52, 17–34.
24 Evans, P.R., Farrants, G.W. and Hudson, P.J. (1981) Phil. Trans. Roy. Soc. B293, 53–62.
25 Blundell, T.L. and Johnson, L.N. (1976) Protein Crystallography, Academic Press, London.
26 Bragg, W.L. (1975) The Development of X-ray Analysis, G. Bell, London.
27 Bragg, W.L. (1929) Proc. Roy. Soc. A123, 537–559.
28 International Tables for X-ray Crystallography (1962) Vol. III, Physical and Chemical Tables (Macgillavry, C.H., Rieck, G.D. and Lonsdale, K., eds.), Kynoch Press, Birmingham.
29 Czok, R. and Bucher, T.H. (1960) Adv. Prot. Chem. 15, 315–415.
30 Dixon, M. and Webb, E.C. (1967) Enzymes, 2nd edn., Longmans Green, London.
31 McPherson, A. (1976) Methods Enzymol. 23, 249–344.
32 Sansom, M.S.P. (1983) D. Phil. Thesis, University of Oxford.
33 Green, A.A. (1932) J. Biol. Chem. 95, 47–66.
34 Perutz, M.F. (1968) J. Cryst. Growth 2, 54–56.
35 McPherson, A. (1976) J. Biol. Chem. 251, 6300–6303.
36 Cascio, D., Williams, R. and McPherson, A. (1984) J. Appl. Cryst. 17, 209–210.
37 Brzozowski, A., Derewenda, Z., Dodson, E., Dodson, G., Grabowski, M., Liddington, R., Skarzynski, T. and Vallely, D. (1984) Nature 307, 74–76.
38 Pittz, E.P. and Timasheff, S.N. (1978) Biochemistry 17, 615–623.
39 Douzou, P., Hoa, G.H.B., Maurel, P. and Travers, F. (1976) in Handbook of Biochemistry and Molecular Biology (Fasman, G., ed.), 3rd edn. Vol. I. Physical and Chemical Data, pp. 520–540, CRC Press, Cleveland, OH.
40 Monod, J., Wyman, J. and Changeux, J.-P. (1965) J. Mol. Biol. 12, 88–118.
41 Kam, Z., Shore, H.B. and Feher, G. (1978) J. Mol. Biol. 123, 539–555.
42 Eichele, G., Ford, G.C. and Jansonius, J.A. (1979) J. Mol. Biol. 135, 513–516.
43 Alber, T., Hartman, F.C., Johnson, R.M., Petsko, G.A. and Tsernoglou, D. (1981) J. Biol. Chem. 256, 1356–1361.
44 Shaikevitch, A. and Kam, Z. (1981) Acta Cryst. A47, 871–875.
45 Zeppezauer, M. (1971) Methods Enzymol. 22, 253–255.
46 Davies, D.R. and Segal, D.M. (1971) Methods Enzymol. 22, 266–269.
47 Carter, C.W. Jr. and Carter, C.W. (1979) J. Biol. Chem. 254, 12219–12223.
48 Rayment, I. (1981) J. Appl. Cryst. 14, 153–154.
49 Mathews, F.S., Argos, P. and Levine, M. (1972) Cold Spring Harbor Symp. Quant. Biol. 36, 387–397.
50 Wilson, I.A., Skehel, J.J. and Wiley, D.C. (1981) Nature 289, 366–373.

51 Varghese, J.N., Laver, W.G. and Colman, P.M. (1983) Nature 303, 35–44.
52 Fox, R.O. and Richards, F.M. (1982) Nature 300, 325–330.
53 Hendrickson, W.A. and Teeter, M.M. (1981) Nature 290, 107–113.
54 Unwin, P.N.T. and Henderson, R. (1975) J. Mol. Biol. 94, 425–440.
55 Henderson, R. and Unwin, P.N.T. (1975) Nature 257, 28–32.
56 Michel, H. (1983) Trends Biochem. Sci. 8, 56–59.
57 Garavito, R.M. and Jenkins, J.A. (1983) in The Structure and Function of Membrane Proteins (Quagliarello, E. and Palmieri, F., eds.), pp. 205–210, Elsevier/North Holland Biomedical Press, Amsterdam.
58 Michel, H. and Oesterhelt, D. (1980) Proc. Natl. Acad. Sci. U.S.A. 77, 1283–1285.
59 Henderson, R. and Shotton, D. (1980) J. Mol. Biol. 139, 99–109.
60 Michel, M. (1982) EMBO J. 1, 1267–1271.
61 Michel, H. (1982) J. Mol. Biol. 158, 567–572.
62 Garavito, R.M. and Rosenbusch, J.P. (1980) J. Cell. Biol. 86, 327–329.
63 Garavito, R.M., Jenkins, J., Jansonius, J.N., Karlsson, R. and Rosenbusch, J.P. (1983) J. Mol. Biol. 164, 313–327.
64 Rosenbusch, J.P., Dorset, D.C., Garavito, R.M. and Zulauf, M. (1983) in Structure and Function of Membrane Proteins (Quagliarello, E. and Palmieri, F., eds.), pp. 191–197, Elsevier/North Holland Biomedical Press, Amsterdam.
65 Arndt, U.W. and Willis, B.T.M. (1966) Single Crystal Diffractometry, Cambridge University Press, Cambridge.
66 Huber, R. and Kopfman, G. (1969) Acta Cryst. A25, 143–152.
67 Matthews, K.A., Alden, J.T., Bolin, J.T., Freer, S.T., Hamlin, R., Xuong, N., Kraut, J., Poe, M., Williams, M. and Hoogsteen, K.V. (1977) Science 197, 452–455.
68 Xuong, W.H., Freer, S.T., Hamlin, R., Nielsen, C. and Vernon, W. (1978) Acta Cryst. A34, 289–296.
69 Arndt, U.W. and Gilmore, D.J. (1979) J. Appl. Cryst. 12, 1–9.
70 Thomas, D.J. and Arndt, U.W. (1981) Acta Cryst. A37, C-312.
71 Kahn, R., Fourme, R., Caudron, B., Bosshard, R., Benoit, R., Bouclier, R., Charpak, G., Santiard, J.C. and Sauli, F. (1980) Nucl. Instr. Methods 172, 337–344.
72 Arndt, U.W., Champness, J.N., Phizackerley, R.O. and Wonacott, A.J. (1973) J. Appl. Cryst. 6, 457–463.
73 Arndt, U.W. and Wonacott, A.J. (1977) The Rotation Method in Crystallography, North-Holland, Amsterdam.
74 Wilson, K. and Yeates, D. (1979) Acta Cryst. A35, 146–157.
75 Stuart, D.I., Levine, M., Muirhead, H., Stammers, D.K. (1979) J. Mol. Biol. 134, 594–604.
76 Rossmann, M.G. (1979) J. Appl. Cryst. 12, 225–238.
77 Wilson, K.S., Stura, E.A., Wild, D.L., Todd, R.J., Stuart, D.I., Babu, Y.S., Jenkins, J.A., Standing, T.S., Johnson, L.N., Fourme, R., Kahn, R., Gadet, A., Bartels, K.S. and Bartunik, H.D. (1983) J. Appl. Cryst. 16, 28–41.
78 Winkler, F.K., Schutt, C.E. and Harrison, S.C. (1979) Acta Cryst. A35, 901–911.
79 Rossmann, M.G., Leslie, A.G.W., Abdel-Megud, S.S. and Tsukihara, T. (1979) J. Appl. Cryst. 12, 570–581.
80 Greenhough, T.J. and Helliwell, J.R. (1982) J. Appl. Cryst. 15, 338–351.
81 Wyckoff, H.W., Hardman, K.D., Allewell, N.M., Inagami, T., Tsernoglou, D., Johnson, L.N. and Richards, F.M. (1967) J. Biol. Chem. 242, 3984–3988.
82 Branden, C.I., Eklund, H., Nordstrom, B., Boiwe, T., Soderlund, G., Zeppezauer, E., Ohlsson, I. and Aheson, A. (1973) Proc. Natl. Acad. Sci. U.S.A. 70, 2439–2442.
83 Johnson, L.N., Madsen, N.B., Mosley, J.P. and Wilson, K.S. (1974) J. Mol. Biol. 90, 703–717.
84 Abraham, D.J., Phillips, S.E.V. and Kennedy, P.E. (1983) J. Mol. Biol. 170, 249–252.
85 Blake, C.C.F., Koenig, D.F., Mair, G.A.P., North, A.C.T., Phillips, D.C. and Sarma, V.R. (1965) Nature 206, 757.
86 Eisenberg, D. (1970) in The Enzymes, Vol. I, 3rd edn. (Boyer, P.D., ed.), pp. 1–89, Academic Press, New York.

87 Richmond, T.J., Finch, J.T., Rushton, L.B., Rhodes, D. and Klug, A. (1984) Nature 311, 532–537.
88 Stuart, D.I., Jinbi, D., Meizhen, L., Jumming, Y., Todd, R., Jiayao, L., Jiahuai, W. and Dongcai, L. (1984) Scientia Sinica (series B) 27, 572–582.
89 Rossmann, M.G. (1972) Molecular Replacement Method, Gordon and Breach Sciences, New York.
90 Bricogne, G. (1976) Acta Cryst. A32, 832–847.
91 Argos, P., Ford, G.C. and Rossman, M.G. (1975) Acta Cryst. A31, 499–506.
92 Bloomer, A.C., Champness, J.C., Bricogne, G., Staden, R. and Klug, A. (1978) Nature 276, 362–368.
93 Rossmann, M.G., Abad-Zapatero, C., Hermodsen, M.A. and Erickson, J.W. (1983) J. Mol. Biol. 166, 37–83.
94a Harrison, S.C., Olson, A.J., Schutt, C.E., Winkler, F.K. and Bricogne, G. (1978) Nature 296, 368–373.
94b Olson, A.J., Bricogne, G. and Harrison, S.C. (1983) J. Mol. Biol. 171, 61–94.
95 Gaykema, W.P.J., Hol, W.G.J., Vereijken, J.M., Soeter, N.M., Bak, H.J. and Beintema, J.J. (1984) Nature 309, 23–29.
96 Hendrickson, W.A. and Lattman, E.E. (1970) Acta Cryst. B26, 136–143.
97 Sielecki, A.R., Hendrickson, W.A., Broughton, C.G., Delbaere, L.T.J., Brayer, G.D. and James, M.N.G. (1979) J. Mol. Biol. 134, 781–804.
98 Richards, F.M. (1968) J. Mol. Biol. 37, 225–230.
99 Levinthal, C.C., Barry, C.D., Ward, S.A. and Zwick, M. (1968) in Emerging Concepts in Computer Graphics (Nievergelt, J. and Secrest, D., eds.), p. 231, W.A. Benjamin, New York.
100 Barry, C.D. and North, A.C.T. (1971) Cold Spring Harbor Symp. Quant. Biol. 36, 577–584.
101 Jacobi, T.H., Ellis, R.A. and Fritsch, J.M. (1972) J. Mol. Biol. 72, 589–591.
102 Jones, T.A. (1978) J. Appl. Cryst. 11, 268–272.
103 Jones, T.A. (1982) in Crystallographic Computing (Sayre, D., ed.), pp. 303–317, Clarendon Press, Oxford.
104 Diamond, R. (1978) in Symposium on Biomolecular Structure, Conformation, Function and Evolution (Srinivasan, R., ed.), Pergamon Press, Oxford.
105 Diamond, R. (1982) in Crystallographic Computing (Sayre, D., ed.), pp. 318–325, Clarendon Press, Oxford.
106 Wright, H.V. (1982) in Crystallographic Computing (Sayre, D.E., ed.), pp. 294–302, Clarendon Press, Oxford.
107 Lipson, H. and Cochran, W. (1968) The Determination of Crystal Structures, Ch. 12, G. Bell, London.
108 Diamond, R. (1971) Acta Cryst. A27, 436–452.
109 Deisenhofer, J. and Steigemann, W. (1975) Acta Cryst. B31, 238–250.
110 Diamond, R. (1974) J. Mol. Biol. 82, 371–391.
111 Watenpaugh, K.D., Sieker, L.C., Herriott, J.R. and Jensen, L.H. (1973) Acta Cryst. B29, 943–956.
112 Wilson, A.J.C. (1949) Acta Cryst 2, 318.
113 Wilson, A.J.C. (1950) Acta Cryst. 3, 397.
114 Luzzati, V. (1952) Acta Cryst. 5, 802.
115 Waser, J. (1963) Acta Cryst. 16, 1091–1094.
116 Konnert, J.H. (1976) Acta Cryst. A32, 614–617.
117 Hendrickson, W.A. and Konnert, J.H. (1979) in Biomolecular Structure, Conformation, Function and Evolution (Srinivasan, R., ed.), Vol. 1. pp. 43–57, Pergamon Press, New York.
118 Konnert, J.H. and Hendrickson, W.A. (1980) Acta Cryst. A36, 344–350.
119 Refinement of Protein Structures. Proceedings of the Daresbury Study Weekend (1980) (Machin, P.A., Campbell, J.W. and Elder, M., compilers), No. DL/SCI/R16, Daresbury Laboratory, Warrington.
120 Scherringer (1963) Acta Cryst. 16, 546–550.
121 Sussman, J.L., Holbrook, S.R., Church, G.M. and Kim, S.-H. (1977) Acta Cryst. A33, 800–804.
122 Herzberg, O. and Sussman, J.L. (1983) J. Appl. Cryst. 16, 144–150.
123 Shoham, M, Yonath, A., Sussman, J.L., Mout, J., Traub, W. and Kalb, A.J. (1979) J. Mol. Biol. 131, 137–155.
124 Agarwal, R.C. (1978) Acta Cryst. A34, 791–809.

125 Isaacs, N.W. and Agarwal, R.C. (1978) Acta Cryst. A34, 782–791.
126 Isaacs, N.W. (1981) in Structural Studies on Molecules of Biological Interest (Dodson, G., Glusker, J.P. and Sayre, D., eds.), pp. 274–285, Clarendon Press, Oxford.
127 Jack, A. and Levitt, M. (1978) Acta Cryst. A34, 931–935.
128 Hermans, J. and McQueen, J.E. (1974) Acta Cryst. A30, 730–739.
129 Hestenes, M.R. and Stiefel, E. (1952) J. Res. Natl. Bur. Stnd. 49, 409–436.
130 Blake, C.C.F., Pulford, W. and Artymiuk, P.J. (1983) J. Mol. Biol. 167, 693–723.
131 Schmidt, W.C., Girling, R.L. and Amma, E.L. (1977) Acta Cryst. B33, 3618–3620.
132 Marquart, M., Deisenhofer, J., Huber, R. and Palm, W. (1980) J. Mol. Biol. 141, 363–391.
133 Bennet, W.S. and Huber, R. (1984) Crit Rev. Biochem. 15, 291–384.
134 Phillips, S.E.V. (1980) J. Mol. Biol. 142, 531–554.
135 Cochran, W. (1948) Acta Cryst. 1, 138–142.
136 Cooley, J.W. and Tukey, J.W. (1965) Maths. Comput. 19, 297–301.
137 Ten Eyck, L.F. (1973) Acta Cryst. A29, 183–191.
138 Dodson, E.J., Isaacs, N.W. and Rollett, J.S. (1976) Acta Cryst. A32, 311–315.
139 Baker, E.N. and Dodson, E.J. (1980) Acta Cryst. A36, 559–572.
140 Levitt, M. (1974) J. Mol. Biol. 82, 393–420.
141 Lipson, H. and Cochran, W. (1966) The Determination of Crystal Structures, 3rd edn., Bell and Sons, London.
142 Luzzati, V. (1953) Acta Cryst. 6, 142–152.
143a Sim, G.A. (1959) Acta Cryst. 12, 813–815.
143b Sim, G.A. (1960) Acta Cryst. 13, 511–512.
144 Main, P. (1979) Acta Cryst. A35, 779–785.
145 Vijayan, M. (1980) Acta Cryst. A36, 295–298.
146 Rice, D.W. (1981) Acta Cryst. A37, 491–500.
147 Stuart, D.I. and Artymiuk, P. (1985) Acta Cryst. A40, 713–716.
148 Henderson, R.C. and Moffat, J.K. (1971) Acta Cryst. B28, 1414–1420.
149 Lorek, A., Wilson, K., Sansom, M.S.P., Stuart, D.I., Stura, E.A., Jenkins, J.A., Zanotti, G., Hajdu, J. and Johnson, L.N. (1984) Biochem. J. 218, 45–60.
150 Perutz, M.F. (1971) Cold Spring Harbor Symp. Quant. Biol. 37, 295–310.
151 Bernstein, F.C., Koetzle, T.F., Williams, G.J.B., Meyer, E.F., Brice, M.D., Rogers, J.R., Kennard, O., Shimanouchi, T. and Tasumi, M. (1977) J. Mol. Biol. 112, 535–542.
152 Drenth, J., Hol, W.G.J., Jansonius, J.N. and Koekoek, R. (1972) Cold Spring Harbor Symp. 36, 107–116.
153 Imoto, T., Johnson, L.N., North, A.C.T., Phillips, D.C. and Rupley, J.A. (1972) in The Enzymes, 3rd ed. (Boyer, P., ed.), Vol. VII, pp. 665–868, Academic Press, New York.
154 Blake, C.C.F., Grace, D.E.P., Johnson, L.N., Perkins, S.J., Phillips, D.C., Cassels, R., Dobson, C.M., Poulsen, F.M. and Williams, R.J.P. (1978) in CIBA Symposium 60, pp. 137–185, Excerpta Medica, London.
155 McDonald, R.C., Steitz, T. and Engelman, D.M. (1979) Biochemistry 18, 338–342.
156 Takano, T. and Dickerson, R.E. (1981) J. Mol. Biol. 153, 95–115.
157 Takano, T. and Dickerson, R.E. (1982) in Electron Transport and Oxygen Utilisation (Chien, H. ed.), pp. 17–26, Elsevier/North-Holland, Amsterdam.
158 Cutfield, J.F., Cutfield, S.M., Dodson, E.J., Dodson, G.G., Reynolds, C.D. and Vallery, D. (1981) in Structural Studies on Molecules of Biological Interest (Dodson, G., Glusker, J. and Sayre, D., eds.), pp. 525–546, Clarendon Press, Oxford.
159 Chothia, C., Lesk, A.M., Dodson, G.G. and Hodgkin, D.C. (1983) Nature 302, 500–505.
160 Becker, J.W., Reeke, G.N., Cunningham, B.A. and Edelman, G.M. (1976) Nature 259, 406–409.
161 Alber, T., Banner, D.W., Bloomer, A.C., Petsko, G.A., Phillips, D.C., Rivers, P.S. and Wilson, I.A. (1981) Phil. Trans. Roy. Soc. B293, 159–171.
162 Bennet, W.S. and Steitz, T.A. (1980) J. Mol. Biol. 140, 211–230.
163 Fehlhammer, H., Bode, W. and Huber, R. (1977) J. Mol. Biol. 111, 415–438.

164 Kossiakoff, A.A., Chambers, J.L., Kay, L.M. and Stround, R.M. (1977) Biochemistry 16, 654–664.
165 Doscher, M.S. and Richards, F.M. (1963) J. Biol. Chem. 238, 2399–2406.
166 Rupley, J.A. (1969) in Biological Macromolecules (Timasheff, S.N. and Fasman, G.D., eds.), Vol. 2, pp. 291–351, Structure and Stability of Biological Macromolecules, Marcel Dekker, New York.
167 Quiocho, F.A. and Richards, F.M. (1966) Biochemistry 5, 4062–4076.
168 Rossi, G.L. and Bernhard, S.A. (1971) J. Mol. Biol. 55, 215–230.
169 Rossi, G.L. and Bernhard, S.A. (1970) J. Mol. Biol. 49, 85–91.
170 Makinen, M.W. and Fink, A.L. (1977) Annu. Rev. Biophys. Bioeng. 6, 301–342.
171a Vas, M., Berni, R., Mozzarelli, A., Tegoni, M. and Rossi, G.L. (1979) J. Biol. Chem. 254, 8480–8486.
171b Mozzarelli, A., Berni, R., Rossi, G.L., Vas, M., Bartha, F. and Keleti, T. (1982) J. Biol. Chem. 257, 6739–6744.
172 Sluyterman, L.A.A. and De Graaf, M.J.M. (1969) Biochim. Biophys. Acta 171, 277–284.
173 Bello, J. and Nowoswiat, E.F. (1965) Biochim. Biophys. Acta 105, 325–332.
174 Parkhurst, L.J. and Gibson, O.M. (1967) J. Biol. Chem. 242, 5762–5770.
175 Kasvinsky, P.J. and Madsen, N.B. (1976) J. Biol. Chem. 251, 6852–6854.
176 Steitz, T.A. and Shulman, R.G. (1982) Annu. Rev. Biophys. Bioeng. 11, 419–444.
177 Moore, G.R., Ratcliffe, R.G. and Williams, R.J.P. (1983) Essays Biochem. 19, 142–195.
178 Perkins, S.J. (1982) in Biological Magnetic Resonance, Vol. 4 (Berliner, L.J. and Reuben, J., eds.), pp. 193–336, Plenum Press, New York.
179 Perkins, S.J. and Dwek, R.A. (1980) Biochemistry 19, 245–258.
180 Perkins, S.J. and Wüthrich, K. (1980) J. Mol. Biol. 138, 43–64.
181 Dobson, C.M., Hoch, J.C., Olejniczak, E.T. and Poulsen, F.M. (1980) Biophys. J. 32, 625–628.
182 Wagner, G. and Wüthrich, K. (1982) J. Mol. Biol. 155, 347–366.
183 Braun, W., Wider, G., Lee, L.H. and Wüthrich, K. (1983) J. Mol. Biol. 169, 921–948.
184 Sasaki, K., Dockevill, S., Ackmiak, D.A., Tickle, I.J. and Blundell, T.L. (1975) Nature 257, 751–757.
185a Williams, G., Clayden, N.J., Moore, G.R. and Williams, R.J.P. (1985) J. Mol. Biol. 183, in press.
185b Moore, G.R., Robinson, M.N., Williams, G. and Williams, R.J.P. (1985) J. Mol. Biol. 183, in press.
186a Wilkinson, K.D. and Rose, I.A. (1980) J. Biol. Chem. 255, 7569–7574.
186b Wilkinson, K.D. and Rose, I.A. (1981) J. Biol. Chem. 256, 9890–9894.
187 Gurd, F.R.N. and Rothgeb, T.M. (1979) Adv. Prot. Chem. 33, 73–165.
188 Welch, G.R., Somogyi, B. and Damjanovich, S. (1982) Prog. Biophys. Mol. Biol. 39, 109–146.
189 Wüthrich, K. and Wagner, G. (1978) Trends Biochem. Sci. 3, 227–230.
190 Jardetsky, O. (1981) Acc. Chem. Res. 14, 291–298.
191 Petsko, G.A. and Ringe, D. (1984) Annu. Rev. Biophys. Bioeng. 13, 331–371.
192 Stuart, D.I. and Phillips, D.C. (1984) Methods Enzymol., in press.
193 Watenpaugh, K.D., Sieker, L.C. and Jensen, L.H. (1980) J. Mol. Biol. 138, 615–633.
194 Glover, I., Haneef, I., Pitts, J., Wood, S., Moss, D., Tickle, I. and Blundell, T.L. (1983) Biopolymers 22, 293–304.
195 Van Gunsteren, W.F. and Karplus, M. (1982) Biochemistry 21, 2259–2274.
196 Northrup, S.H., Pear, M.R., McCammon, J.A., Karplus, M. and Takano, T. (1980) Nature 287, 659–660.
197 Sternberg, M.J.E., Grace, D.E.P. and Phillips, D.C. (1979) J. Mol. Biol. 130, 231–253.
198 Artymiuk, P.J., Blake, C.C.F., Grace, D.E.P., Oatley, S.J., Phillips, D.C. and Sternberg, M.J.E. (1979) Nature 280, 563–568.
199 James, M.N.G., Sielecki, A.R., Brayer, G.D., Delbaere, L.T.J. and Bauer, C.-A. (1980) J. Mol. Biol. 144, 43–88.
200 Huber, R. and Bode, W. (1978) Acc. Chem. Res. 11, 114–122.
201 Walter, J., Steigemann, W., Singh, T.P., Bartunik, H., Bode, W. and Huber, R. (1982) Acta Cryst. B38, 1462–1472.
202 Freuenfelder, H., Petsko, G.A. and Tsernoglou, D. (1979) Nature 280, 558–563.
203 Hartmann, H., Parak, F., Steigemann, W., Petsko, G.A., Ringe, P.D. and Frauenfelder, M. (1982) Proc. Natl. Acad. Sci. U.S.A. 79, 4967–4971.

204 Branden, C.-I. and Eklund, H. (1980) in Dehydrogenases (Jeffry, J., ed.), pp. 40–84, Birkhauser, Basle.
205 Remington, S., Wiegand, G. and Huber, R. (1982) J. Mol. Biol. 158, 111–152.
206a Rogers, N.K. (1984) D.Phil. Thesis, University of Oxford.
206b Pethig, R. (1979) Dielectric and Electronic Properties of Biological Materials, John Wiley, New York.
207 Fink, A.L. and Petsko, G.A. (1981) Adv. Enzymol. 52, 177–246.
208 Haas, D.J. and Rossmann, M.G. (1970) Acta Cryst. B26, 998–1004.
209 Angell, C.A. (1983) Annu. Rev. Phys. Chem. 34, 593–630.
210 Hobbs, P.J. (1974) Ice Physics, Clarendon Press, Oxford.
211 Franks, F. (1978) J. Microsc. 112, 3116.
212 Costello, M.J. and Corless, J.M. (1978) J. Microsc. 112, 35–55.
213 Douzou, P. (1974) Methods Biochem. Anal. 22, 401–512.
214 Petsko, G.A. (1975) J. Mol. Biol. 96, 381–392.
215 Wychoff, H.W., Doscher, M., Tsernoglou, D., Inagami, T., Johnson, L.N., Hardman, K.D., Allewell, N.M., Kelly, D.M. and Richards, F.M. (1967) J. Mol. Biol. 27, 563–578.
216 Marsh, D.J. and Petsko, G.A. (1973) J. Appl. Cryst. 6, 76–80.
217 Rudman, R. (1976) Low Temperature X-ray Diffraction, Plenum Press, New York.
218 Douzou, P. and Petsko, G.A. (1984) Adv. Prot. Chem. 36, 245–361.
219 Bishop, W.H. and Richards, F.M. (1968) J. Mol. Biol. 38, 315–328.
220 Crank, J. (1956) The Mathematics of Diffusion, Oxford University Press, Oxford.
221 Fink, A.L. and Ahmed, A.I. (1976) Nature 263, 294–297.
222 Alber, T., Petsko, G.A. and Tsernoglou, D. (1976) Nature 263, 294–297.
223 Alber, T., Gilbert, W.A., Ponzi, D.R. and Petsko, G.A. (1983) CIBA Symp. 93, 4–24.
224 Phillips, J.C., Wlodawer, A., Yevitz, M.M. and Hodgson, K.O. (1976) Proc. Natl. Acad. Sci. U.S.A. 73, 128–132.
225 Lemonier, M., Fourme, R., Rousseau, F. and Kahn, R. (1978) Nucl. Inst. Methods 152, 173–177.
226 Kahn, R., Fourme, R., Gadet, A., Janin, J., Dumas, C. and Andre, D. (1982) J. Appl. Cryst. 15, 330–337.
227 Bartunik, H.D., Clout, P.N. and Robrahm, B. (1981) J. Appl. Cryst. 14, 134–136.
228 Bartunik, H.D., Fourme, R. and Phillips, J.C. (1982) in Uses of Synchrotron Radiation in Biology (Stuhrmann, H.B., ed.), Academic Press, New York.
229 Helliwell, J.R., Greenhough, T.J., Carr, P.D., Rule, S.A., Moore, R.R., Thompson, A.W. and Worgan, J.S. (1982) J. Phys. E: Sci. Inst. 15, 1363–1372.
230 Greenhough, T.J. and Helliwell, J.R. (1983) Prog. Biophys. Mol. Biol. 41, 67–124.
231 Hendrix, J., Koch, M. and Bordas, J. (1979) J. Appl. Cryst. 12, 467–472.
232 McLaughlin, P.J., Stuart, D.I., Oihonomakos, N.G., Klein, H. and Johnson, L.N. (1984) Biochemistry 23, 5862–5873.
233 Bartunik, H.D. and Schubert, P. (1982) J. Appl. Cryst. 15, 227–231.
234 Usha, R., Johnson, J.E., Moras, D., Thierry, J.C., Fourme, R. and Kahn, R. (1984) J. Appl. Cryst. 17, 147–153.
235 Huxley, H.E., Simmons, R.M., Faruqi, A.R., Kress, M., Bordas, J. and Koch, M.H.J. (1983) J. Mol. Biol. 169, 469–506.
236 Bartunik, H.D. (1983) Nucl. Inst. Methods 208, 523–533.
237 Sawyer, L. and Helliwell, J. (1984) Unpublished results.
238 Ford, G.C., Harrison, P.M., Rice, D.W., Smith, J.M. and White, J.L. (1984) Daresbury Lab. Info. Q. 13, 15–16.
239 Ramaseshan, S. (1962) in Advanced Methods of Crystallography (Ramachandran, G.N., ed.), Academic Press, New York.
240 Phillips, J.C. and Hodgson, K.O. (1980) Acta Cryst. A36, 856–864.
241 Narayan, R. and Ramaseshan, S. (1981) Acta Cryst. A37, 636–641.
242 Schoenborn, B.P. (1975) in Anomalous Scattering (Ramaseshan, S. and Abrahams, S.C., eds.), p. 407, Monksgaard, Copenhagen.

243a Templeton, D.H., Templeton, L.K., Phillips, J.C. and Hodgson, K.O. (1980) Acta Cryst. A36, 435–442.

243b Arndt, U.W., Greenhough, T.J., Helliwell, J.R., Howard, J.A.K., Rule, S.A. and Thompson, A.W. (1982) Nature 298, 835–838.

244 Sakabe, N., Sakabe, K. and Sasaki, K. (1981) in Structural Studies on Molecules of Biological Interest (Dodson, G., Glusker, J.P. and Sayre, D., eds.), pp. 509–526, Clarendon Press, Oxford.

245 Kossiakoff, A.A. (1983) Annu. Rev. Biophys. Bioeng. 12, 159–182.

246 Kossiakoff, A.A. and Spencer, S.A. (1981) Biochemistry 20, 6462–6484.

247 Norvell, J.C., Nunes, A.C. and Schoenborn, B.P. (1975) Science 190, 568–569.

248 Wlodawer, A. and Hendrickson, W.A. (1982) Acta Cryst. A38, 239–247.

249 Wlodawer, A., (1982) Prog. Biophys. Mol. Biol. 40, 115–159.

250 Kossiakoff, A.A. and Spencer, S.A. (1980) Nature 288, 414–416.

251 Mason, S.A., Bentley, G.A. and McIntyre, G.J. (1982) Brookhaven Symp. Biol. 32.

252 Phillips, D.C. (1967) Proc. Natl. Acad. Sci. U.S.A. 57, 493–495.

253 Wlodawer, A., Miller, M. and Sjolin, L. (1983) Proc. Natl. Acad. Sci. U.S.A. 80, 3628–3631.

254 Kossiakoff, A.A. (1982) Nature 296, 713–721.

255 Hauptman, H. and Karle, J. (1953) The Solution of the Phase Problem, ACA Monograph No. 3, Edwards Brothers, Ann Arbor.

256 Karle, J. and Karle, I.L. (1966) Acta Cryst. 21, 849–859.

257 Germain, G. and Wolfson, M.M. (1968) Acta Cryst. B24, 91–96.

258 Gilmore, C.J. (1982) in Computational Crystallography (Sayre, D., ed.), pp. 126–140, Clarendon Press, Oxford.

259 Hull, S.E., Karlsson, R., Main, P., Woolfson, M.M. and Dodson, E.J. (1978) Nature 257, 206–207.

260 Jaynes, E.T. (1957) Phys. Rev. 106, 620–631.

261 Wilkins, S.W., Varghese, J.N. and Lehmann, M.S. (1983) Acta Cryst. A39, 47–60.

262 Wilkins, S.W., Varghese, J.N. and Steenstrup, S. (1984) in Indirect Imaging (Roberts, J.A., ed.), Cambridge University Press, Cambridge.

263 Bricogne, G. (1984) Acta Cryst. A40, 410–445.

264 Bryan, R.K., Bansal, M., Folkhard, W., Nave, C. and Marvin, D.A. (1983) Proc. Natl. Acad. Sci. U.S.A. 80, 4728–4731.

265 Diesenhoffer, J., Epp, O., Miki, K., Huber, R. and Michel, M. (1984) J. Mol. Biol. 180, 385–398.

266 Gilbert, W.A. and Petsko, G. A. (1985), to be published. Quoted in reference 218.

267 Fink, A.L., Kav, D. and Kolin, R. (1985), to be published. Quoted in reference 218.

268 Deavin, A., Mathias, A.P. and Rabin, B.R. (1966) Biochem. J. 101, 14C.

269 Kahn, R., Fourme, R., Bosshard, R., Wery, J.P., Dideberg, O., Risler, J.L., Brunie, S. and Janin, J. (1984) Symposium on New Methods in X-ray Absorption Scattering and Diffraction for Applications in Structural Biology, Bristol.

270 Hails, J.E., Harding, M.M., Helliwell, J.R., Liddington, R. and Papiz, M.Z. (1984) Daresbury preprint DL/SCI/P428E, Daresbury Laboratory, Warrington.

Subject Index